Biophysics

Biophysics: Tools and Techniques for the Physics of Life covers the experimental, theoretical, and computational tools and techniques of biophysics. It addresses the purpose, science, and application of all physical science instrumentation, theoretical analysis, and biophysical computational methods used in current research labs.

The book first presents the historical background, concepts, and motivation for using a physical science toolbox to understand biology. It then familiarizes undergraduate students from the physical sciences with essential biological knowledge.

The text subsequently focuses on experimental biophysical techniques that primarily detect biological components or measure/control biological forces. The author describes the science and application of key tools used in imaging, detection, general quantitation, and biomolecular interaction studies, which span multiple length and time scales of biological processes both in the test tube and in the living organism.

Moving on to theoretical and computational biophysics tools, the book presents analytical mathematical methods and numerical simulation approaches for tackling challenging biological questions including exam-style questions at the end of each chapter as well as step-by-step solved exercises. It concludes with a discussion of the future of this exciting field.

Future innovators will need to be trained in multidisciplinary science to be successful in industry, academia, and government support agencies. Addressing this challenge, this textbook educates future leaders on the development and application of novel physical science approaches to solve complex problems linked to biological questions.

Features:

- Provides the full, modern physical science toolbox of experimental, theoretical, and computational techniques, such as bulk ensemble methods, single-molecule tools, live-cell and test tube methods, pencil-on-paper theory approaches, and simulations.
- Incorporates worked examples for the most popular physical science tools by providing full diagrams and a summary of the science involved in the application of the tool.
- Reinforces the understanding of key concepts and biological questions.

A solutions manual is available on adoption of qualifying course.

Mark C. Leake holds the Anniversary Chair of Biological Physics and is the Coordinator of the Physics of Life Group at the University of York. He is also Chair of the UK Physics of Life Network (PoLNET). He heads an interdisciplinary research team in the field of single-molecule biophysics using cutting-edge biophotonics, state-of-the-art genetics, and advanced computational and theory tools. His work is highly cited, and he has won many fellowships and prizes.

Biophysics

TOOLS AND TECHNIQUES FOR THE PHYSICS OF LIFE

Second Edition

Mark C. Leake

CRC Press
Taylor & Francis Group
Boca Raton London New York

CRC Press is an imprint of the
Taylor & Francis Group, an **informa** business

Second edition published 2024
by CRC Press
2385 NW Executive Center Drive, Suite 320, Boca Raton FL 33431

and by CRC Press
4 Park Square, Milton Park, Abingdon, Oxon, OX14 4RN

CRC Press is an imprint of Taylor & Francis Group, LLC

© 2024 Mark C. Leake

First edition published by CRC Press 2016

ISBN: 978-1-032-37321-8 (hbk)
ISBN: 978-1-032-37038-5 (pbk)
ISBN: 978-1-003-33643-3 (ebk)

DOI: 10.1201/9781003336433

Typeset in Warnock Pro
by Newgen Publishing UK

Dedicated, with great affection, to A—the yang to my yin, C—the pong to my ping, AR—for the chops, and NTD—for, I think, understanding the jokes before you had to leave

And of course, to Mr. Blue Sky

Contents

Preface

Or, how physical science methods help us understand "life"

And the whole is greater than the part.

Euclid elements, Book I, Common Notion 5, ca. 300 BC

This book concerns one simple question:

What is life?

If only the answer were as simple! In this book, you will find a comprehensive discussion of the experimental and theoretical tools and techniques of biophysics, which can be used to help us address that most innocuous question.

The creation of this second edition was guided mainly by two imperatives: include the feedback from students and teachers made in the classroom, tutorial offices, practical labs, and lecture theaters from the first edition, and to freshen up some key areas and include some new emergent ones in light of recent developments in Physics of Life research. So, a number of sections have been reorganized, condensed, and expanded as appropriate, and the narrative improved in response to reader comments. Several additional worked examples and problem questions are now included. The title has been marginally revised to reflect the intersection of "biophysics" and "biological physics" into a single discipline of the "Physics of Life."

"Interdisciplinary" research between the physical and biological sciences is now emerging as one of the fastest growing fields within both the physical and biosciences—a typical search for "biophysics" in any major Internet search engine now generates several million hits. The trend toward greater investment in this interfacial area is reflected in the establishment of centers of excellence across the world dedicated to interdisciplinary science research and graduate student training that combine the elements of both physical and life sciences. Biophysics/ biological physics is now a standard teaching component option in undergraduate biochemistry and physics courses, and the significant direction of change over the past few years in terms of research direction has been a shift toward a smaller length scale and a far greater physiological realism to highly challenging bioscience experiments—experiments are getting better at imaging molecular processes. On the other hand, there are some theorists who feel that we need biophysical concepts and theories at higher length scales to combat the sometimes suffocating reductionism of modern biology.

Physical science methods historically have been key to providing enormous breakthroughs in our understanding of fundamental biology—stemming from the early development of optical microscopy for understanding the cellular nature of life to complex structural biology techniques to elucidate the shape of vital biomolecules, including essential proteins and DNA, the coding molecule of genes. More recently, physical science developments have involved methods to study single cells in their native context at the single-molecule level, as well as providing ground-breaking developments in areas of artificial tissue bioengineering and synthetic biology, and biosensing and disease diagnosis. But there is also a tantalizing theme emerging for many of the researchers involved to, in effect, reframe their questions across the interface between the physical and life sciences, through a process of "co-creation"; biologists and physicists interacting to generate transformative ways of studying living matter that neither in isolation could achieve. The tools and techniques described in this book resonate with those chords.

This book concisely encompasses the full, modern physical science "toolbox" of experimental and analytical techniques that are in active use. There is an enormous demand for literature to accompany both active research and taught courses, and so there is a compelling, timely argument to produce a concise handbook summarizing the essential details in this broad field, but one which will also focus on core details of the more key areas. This book can be used to complement third- and fourth-year undergraduate courses in physical science departments involving students who are engaged in physical/life sciences modules—there are several of these internationally involving biological physics/biophysics, bioengineering, bionanotechnology, medical/healthcare physics, biomedical sciences, natural sciences, computational biology, and mathematical biology. Similar stage undergraduates in life sciences departments doing physical/life science interfacial courses will also benefit from this text, to accompany lectures covering biophysics, bionanotechnology, systems biology, and synthetic biology. Also, PhD and master's students engaged in physical/life sciences courses and/or research will find significant utility for structuring and framing their study, and expert academics will find this text to be an excellent concise reference source for the array of physical science tools available to tackle biological problems.

In this book I focus specifically on the purpose, the science, and the application of the physical science tools. The chapters have been written to encapsulate all tools active in current research labs, both experimental and analytical/theoretical. Bulk ensemble methods as well as single-molecule tools, and live-cell and test tube methods, are discussed, as all are different components available in the physical scientist's toolbox for probing biology. Importantly, this broad material is comprehensively but concisely mapped into one single volume without neglecting too much detail or any important techniques. Theoretical sections have been written to cover "intermediate" and "advanced" ability, allowing levels of application for students with differing mathematical/computational abilities. The theory sections have been designed to contrast the experimental sections in terms of scale and focus. We need theories not only to interpret experimental techniques but also to understand more general questions around evolution, robustness, nonequilibrium systems, etc., and these are not tied to molecular scales.

Future innovators will need to be trained in multidisciplinary science to be successful, whether in industry, academia, or government support agencies. This book addresses such a need—to facilitate educating the future leaders in the development and application of novel physical science approaches to solve complex challenges linked to biological questions. The importance of industrial application at the physical/life sciences interface should not be underestimated. For example, imaging, tracking, and modeling chemical interactions and transport in industrial biological/biomedical products is an emergent science requiring a broad understanding and application of biophysical sciences. One of the great biopharma challenges is the effective delivery of active compounds to their targets and then monitoring their efficacy. This is pivotal to delivering personalized healthcare, requiring biological processes to be understood in depth. This is also a good example of scales: the necessary molecular scale of target, binding site, etc., and then the much larger organismal scale at which most candidate drugs fail—that of toxicology, side effects, etc. Physically based methods to assess drug delivery and model systems facilitating *in vitro* high-throughput screening with innovative multilength scale modeling are needed to tackle these challenges.

Also, the need for public health monitoring and identifying environmental risks is highly relevant and are key to preventative medicine. An example is exposure to ionizing radiation due to medical imaging, travel, or the workplace. Detecting, monitoring, and understanding radiation damage to DNA, proteins, and cellular systems during routine imaging procedures (such as dental x-rays or cardiothoracic surgery) are key to understanding noncancer effects (e.g., cataracts in the lens of the eye) and the complications of radiation-based cancer therapy. These are core global public health issues.

Understanding the science and application of new instrumentation and analysis methods is vital not only for core scientific understanding of biological process in academia but also for application in many sectors of the industry. Step changes in science and healthcare sectors are always preceded by technological advances in instrumentation. This requires the integration of specialist knowledge with end-users' requirements alongside an appreciation of the potential commercialization process needed to exploit such innovation in both public and private sectors via both multinational companies and small- to medium-sized business enterprises (SMEs). Such real-world challenges require individuals who are skilled scientists and experts in their own field, but who are also equipped to understand how to maximize impact by utilizing the full investigative power of the biophysical sciences. Specific advances are needed in advanced optical microscopy and instrumentation, single-molecule detection and interactions, atomic force microscopy (AFM), image processing, polymer and synthetic chemistry, protein crystallography, microfabrication, and mathematical and finite element modeling to then interface with current biological and healthcare problems.

This book is structured as nine concise chapters, navigating the reader through a historical background to the concepts and motivation for a physical science toolbox in understanding biology, and then dedicating a chapter to orienting readers from a physical science area of expertise to essential biological knowledge. Subsequent chapters are then themed into sections involving experimental biophysical techniques that primarily detect biological components or measure/control biological forces. These include descriptions of the science and application of the key tools used in imaging, detection, general quantitation, and biomolecular interaction studies, which span multiple length and time scales of biological processes, ranging from biological contexts both in the test tube and in the living organism. I then dedicate a chapter to theoretical biophysics tools, involving computational and analytical mathematical methods for tackling challenging biological questions, and end with a discussion of the future of this exciting field.

Each chapter starts with a "General Idea," which is a broad-canvas description of the contents and importance of the chapter, followed by a nontechnical overview of the general area and why it is relevant. Case examples are used throughout for the most popular physical science tools, involving full diagrams and a précis of the science involved in the application of the tool. "Key Points" are short sections used to reinforce key concepts, and "Key Biological Applications" are used to remind the reader of the general utility of different biological questions for the different biophysical tools and techniques discussed. To aid revision, several exam-style questions are also included at the end of each chapter, pitched at a general audience that can consist of readers with a background in either physical or life sciences. Solved exercises are also used and are associated with worked case examples in which study questions are solved in clearly defined steps. Summary bullet points are also used at the end of each chapter to summarize the key concepts.

The writing of this book went through several iterations, not least in part due to the several anonymous expert reviews that came my way, courtesy of Francesca McGowan at Taylor & Francis Group. I am indebted to her and her team, and to the anonymous reviewers who invested their valuable time to improve this textbook. Also, I am indebted to several of my students and academic peers who generated a range of valuable feedback from their own areas of expertise, whether wittingly or otherwise. These include Pietro Cicuta, Jeremy Cravens, Domi Evans, Ramin Golestanian, Sarah Harris, Jamie Hobbs, Tim Newmann, Michelle Peckham, Carlos Penedo, Ehmke Pohl, Steve Quinn, Jack "Emergency Replacement Richard" Shepherd, Christian Soeller, and Peter Weightman.

Also, I extend my gratitude to the past and present committee members of the British Biophysics Society (BBS), the Biological Physics Group (BPG) of the Institute of Physics (UK), and the Light Microscopy section of the Royal Microscopical Society (RMS), who have all been enormously supportive. A special thanks goes to the members of the steering group of the UK Physics of Life network (PoLNET), and to its wider members of over 1,000 researchers in this wonderful field of science - your creativity and passion make the community the envy of nations. And of course, to my dear late colleague and friend Tom McLeish. You left a tough act for me to follow as Chair of PoLNET, and I feel blessed for having known

you... there is, as you know, no such thing as a dumb question.

Special thanks also go to the other members of my own research team who, although not being burdened with generating feedback on the text of this book, were at least generous enough to be patient with me while I finished writing it, including Erik "The Viking" Hedlund, Helen "The Keeper of the Best Lab Books Including a Table of Contents" Miller, Adam "Lens Cap" Wollman, and Zhaokun "Jack" Zhou.

Finally, you should remember that this emergent field of interfacial physical/life science research is not only powerful but genuinely exciting and fun, and it is my hope that this book will help to shine some light on this highly fertile area of science, and in doing so captivate your senses and imagination and perhaps help this field to grow further still.

Perhaps the most important fact that the reader should bear in mind is that the use of physical science methods and concepts to address biological questions is genuinely motivated by *ignorance*; biophysical enquiry is not a box-ticking exercise to confirm what we already know, but rather it facilitates a genuinely new scientific insight. It allows us not only to address the *how*, but more importantly is driven by the most important word in science—*why?*

The most important words in science are "we don't know."

There is no such thing as a stupid question, only fools too proud to ask them.

Additional material is available from the CRC Press website: www.crcpress.com/product/isbn/978149870 2430.

Professor Mark C. Leake
Building Bridges and removing barriers. Not just
'decorating the boundaries'.
University of York, York, United Kingdom

About the Author

Mark Leake is the Chair of Biological Physics, Coordinator of the Physics of Life Group and was the Founder-Director of the Biological Physical Sciences Institute (BPSI) and at the University of York, UK. He heads an interdisciplinary research team in the field of single-molecule biophysics using cutting-edge biophotonics and state-of-the-art genetics, and the UK's Physics of Life network PoLNET. His work is highly cited, and he has won many fellowships and prizes. He holds the strong belief that science progresses best through a combination of tough, open discussion with learned peers which cut across all sciences, using both experimental and theoretical tools, developing science education to make it accessible to all, and fostering a few whacky blue skies ideas that probably won't work, but *might*...

Introduction

Toolbox at the Physical–Life Science Interface

1

physics ≠ biology

> —Prof. Tim Newmann (University of Dundee, UK), in a lecture "Biology Is Simple," *Physics Meets Biology Conference*, September 5, 2014, Martin Wood Lecture Theatre, Clarendon Laboratory, University of Oxford, UK (ca. 2.45 pm…)

General Idea: Here, we describe the historical background, motivation, and scope for research at the interface between the physical and life sciences and introduce the concept of a "toolbox" to address challenging scientific questions at this interface.

1.1 MOTIVATION FOR BIOPHYSICS

There are distinct scientific challenges that lie right at the interface between what we now refer to as the life sciences and the physical sciences. These challenges are as fundamental as any that one can think of in science; they come down to this simple and deceptively innocuous question:

What is life?

The distinction between physics and biology as separate scientific disciplines is a relatively modern invention, though the metaphysical debate as to what is alive and what is not and the existence of vitality and of the concept of the soul go back as far as historical records can attest. The identification of a precise period of divergence between the disciplines of physics and biology is a matter of debate perhaps better suited for learned historians of science, though as a naïve observer I suspect it to be not a very valuable exercise. Developments in scientific paradigms are often only identified several years after they have occurred, through the kaleidoscope of historical hindsight of events of which the protagonists of the day were blissfully unaware. So, defining an absolute line in time inevitably suffers the dangers of having a perspective from the modern age.

That being said, the nineteenth century certainly witnessed key developments in the physical understanding of the nature of thermodynamics through giants such as Carnot, Kelvin, Joule, and then later Boltzmann and Gibbs and in electromagnetism and light largely through the genius efforts of James Clerk Maxwell (Maxwell, 1873), which today still appear as core components of any undergraduate physics degree. Similarly, the publication of *On the Origin of the Species* by Charles Darwin (Darwin, 1859) sparked a scientific and sociological debate that was centered around *living things* and the principle of *selective* environmental pressures resulting in dynamic evolutionary change, which is one of the key principles taught in modern biology university courses. Although these events do not define the explicit invention of modern physics and biology as academic subjects, they are interesting exemplars that at least encourage us to think about the question.

An interesting interdisciplinary historical quirk is that the grandson of Charles Darwin, also called Charles (Galton) Darwin, was a renowned physicist of his time who made important contributions toward our fundamental understanding of the angular momentum of light, which is only now coming full circle, so to speak, back toward biological questions in being utilized as a probe to monitor a variety of molecular machines that operate via

DOI: 10.1201/9781003336433-1

twisting motions (which we will discuss in Chapter 6). It is not really a relevant fact. But it is interesting.

However, the distinctions between the biology and physics disciplines today are often a little contrived. More importantly though, they are often not terribly helpful. It is a tragedy that the majority of high-school-level science students still need to make relatively early choices about whether they wish to specialize primarily in the area of physical/mathematical biased science/engineering or instead in the life sciences. This approach inevitably results in demarcated structures within our higher education factories, faculties of physics, schools of biology, etc., which potentially serve to propagate the distinction further.

Fundamentally, the laws of physics apply to all things, including biological systems, so the need for a core distinction, one might argue, is an intellectual distraction. This distinction between physics and biology also stands out as an historical anomaly. Key developments in the understanding of the natural world by the ancient civilizations of Babylonia and Egypt made no such distinction, neither did Greek or Roman or Arab natural philosophers nor did even the Renaissance thinkers and Restoration experimentalists in the fifteenth to seventeenth centuries, who ultimately gave birth to the concept of "science" being fundamentally concerned with formulating hypotheses and then falsifying them through observation. Isaac Newton, viewed by many as the fountain from which the waters of modern physics flow, made an interesting reference in the final page of his opus *Principia Mathematica:*

> And now we might add something concerning a certain most subtle Spirit which pervades and lies hid in all gross bodies; by force and action of which Spirit the particles of bodies mutually attract one another at near distance, and cohere, if contagious; and electric bodies operate at greater distances, as well repelling as attracting the neighbouring corpuscles; and light is emitted, reflected, refracted, inflected, and heats bodies; and all sensation is excited, and the members of animal bodies move at the command of the will, namely, by the vibrations of this Spirit, mutually propagated along the solid filaments of the nerves, from the outward organs of sense to the brain, and from the brain into the muscles. But these are things that cannot be explained in few words, nor are we furnished with that sufficiency of experiments which is required, to an accurate determination and demonstration of the laws by which this electric and elastic Spirit operates.

This perhaps suggests, with a significant creative interpretation from our modern era, a picture of combined physical forces of mechanics and electric fields, which are responsible for biological properties. Or perhaps not. Again, maybe this is not relevant. Sorry. But it's still very interesting.

We should try to not let disciplinary labels get in the way of progress. The point is that the physical science method has been tremendously successful on a wide range of nonliving problems all the way from nuclei to atoms to solid state to astrophysics. The fundamental theories of quantum mechanics, thermodynamics, electromagnetism, mechanics, optics, etc., are immensely powerful. On the other hand, these theories can sometimes appear to be relatively toothless when it comes to life sciences: for example, a system as small as a cell or as large as an ecosystem. The fact that these objects are emergent from the process of evolution is a fundamental difference. They may be comprehensible, but then again they may not be, depending on whether 4 billion years of life on Earth as we know it is sufficient or not to reach an "evolutionary steady state." So, the very nature of what we are trying to understand, and our likely success in doing so, could be very different in the study of the life sciences. That is the point underpinning the quote at the start of this chapter "physics does not equal biology." We are not trying to make any point about what we call the activities of these disciplines. But it is a point about the nature of the objects under study as described earlier.

"Biophysics" in its current form is a true *multi*disciplinary science. A minimal definition of biophysics is this: it is what biophysicists do. Say this and you are guaranteed a titter of laughs at a conference. But it's as good a definition as any. And the activities of biophysicists use information not only from physics and biology but also from chemistry, mathematics, engineering, and computer science, to address questions that fundamentally relate to how living organisms actually function. The modern reblurring of the lines between biology and physics started around the middle of twentieth century, at a time when several researchers trained originally from a background of the physical sciences made significant advances toward the development of what we now call molecular biology. Biophysics as a new discipline was shaped significantly from the combined successes of physiology and structural biology at around the same time. The former involved, for example, the pioneering work of Alan Hodgkin and Andrew Huxley, which revealed how the fundamental mechanism of conduction of sensory signals in nerves is achieved (Hodgkin and Huxley, 1952). The latter applied emerging physics tools to study the scattering of x-rays from crystals made of biological molecules, first exemplified in the work, from the 1930s onward, of one of the great women of modern science, Dorothy Hodgkin (née Crowfoot), in her determination of the structures of a number of biologically important small molecules, including cholesterol, penicillin, and vitamin B12, but then later on much larger molecules called proteins, first shown on one found in muscles called myoglobin (Kendrew et al., 1958).

Hodgkin and Huxley's seminal paper, at the time of my writing this sentence, has been cited over 16,000 times and deserves its place as one of the pioneering publications of *modern biophysics*. To some biomathematicians and physiologists, this might seem controversial, since they may claim this paper as one from their own fields, especially since neither Hodgkin nor Huxley necessarily identified as being a "physicist" (their respective areas of primary expertise were biochemistry and physiology, respectively). But with the wisdom of historical hindsight, it is clear that their work sits very much at the cutting-edge interdisciplinary interface between biology and physics.

The beauty of this exemplar study, in particular, is that it used multiple biophysical tools to solve a challenging biological question. It investigated the fundamental properties of electrical nerve conduction by reducing the problem to being one of the ion channels in cell membranes (Figure 1.1a) that could be characterized by experimental measurements using biophysical technology of time-resolved voltage signals by electrodes placed inside and outside the nerve fiber during a stimulated nerve conduction (Figure 1.1b). But these experimental signals could then be modeled using the physics of electrical circuitry (Figure 1.1c), for example, by modeling the electric current due to ions flowing through an ion channel in the cell membrane as being equivalent to a resistor in a standard electrical circuit, with a voltage applied across it, which is equivalent to the voltage across the cell membrane, that is, the difference in electrical potential per unit charge between the inside of the cell and the outside of the cell, denoted by V_m, and the cell membrane acting as a dielectric, thus functioning as a capacitor (discussed in Chapter 2), here of capacitance per unit area C_m. In its simplest form, the electric current flow I across the cell membrane can be modeled mathematically as simply

(1.1)
$$I = C_m \frac{\mathrm{d}V_m}{\mathrm{d}t} + I_i$$

where I_i is the electric current flow through the actual ion channel. This model can be easily expanded using multiple sets of differential equations to account for multiple different ion channels (Figure 1.1d), and this also fits the experimental observations very well.

However, the point here of these two approaches is that they illustrate not just the achievements of new biological insight made with physical science *experimental* tools and that they were both coupled closely to advances in methods of physical science *analysis* techniques, in the case of nerve conduction to a coupled series of analytical differential equations called the "Hodgkin–Huxley model," which describes the physics of the propagation of information in the nerves via electrical conduction of sodium and potassium ions across the nerves' outer electrically insulating membranes, an electric phenomenon known by biologists as the "action potential." X-ray crystallography, the analysis concerned with building mathematical tools that could in essence generate the inverse of the x-ray scatter pattern produced by protein crystals to reveal the underlying spatial coordinates of the constituent atoms in the protein molecule, is a process involving Fourier transformation coupled with additional mathematical techniques for resolving the phase relationship between the

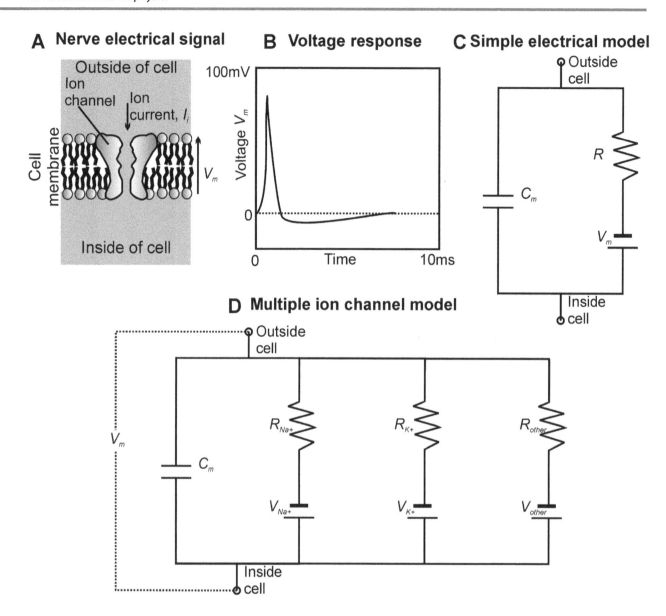

FIGURE 1.1   Hodgkin–Huxley model of membrane current in the nerve. (a) Typical ion channel in a nerve cell membrane. (b) The action of ion channels in the membrane results in a spikelike response in the measured membrane voltage. A single-ion channel in a cell membrane can be modeled (c) as a resistor–capacitor electrical circuit, which (d) can be generalized to account for different types of ion flow across the membrane including sodium (Na+) and potassium ions (K+), plus the effects of any other types.

individual scattering components. This tight coupling in the life sciences between analytical and experimental physical science tools is far from unique; rather, it is a general rule.

But, paradoxically, a key challenge for modern biophysics has been, in the same manner in which it is difficult for the child to emerge from the shadows of an achieving parent, to break free from the enormous successes made through the application of structural biology methods, in particular, over half a century ago. Structural biology, in its ability to render atomic level detail of the molecular machinery of living organisms, has led to enormous insight in our understanding of the way that biological molecules operate and interact with each other. However, what structural biology in its existing form cannot offer is any direct information about the *dynamic* process in living cells. This is a core weakness since biological processes are anything but static. *Everything* in biology is dynamic; it is just a matter of over what time scale. However, dynamic processes are something that single-molecule tools can now address exquisitely well. The wind's direction is already beginning to change with a greater investment in *functional imaging* tools of novel techniques of light microscopy in particular, some of which may indeed develop into new methods of dynamic structural biology themselves.

The benefits of providing new insight into life through the bridging of physics and biology are substantial. *Robust* quantitation through physical science offers a precise route into *reductionist inference* of life, namely, being able to address questions concerning the real underlying mechanisms of natural processes, how tissues are made up of cells and the cells from molecules, and how these all work together to bring about something we call a living organism. The absence of quantifying the components of life precisely, both in space and in time, makes the process of reducing biological phenomena to core processes that we can understand quite challenging, especially in light of the complexity of even the simplest organism. And quantifying physical parameters precisely is one thing in particular that physics does very well.

One of the most cited achievements of biophysics is that of the determination of the structure of the biological molecule called "deoxyribonucleic acid" (DNA), which in the 1950s was resolved through an exquisite combination of biology and physics expertise, in that it required challenging biochemical techniques to purify DNA and then form near-perfect crystalline fibers, and then applying innovative physical science x-ray crystallography tools on these DNA fibers followed by a bespoke physical science analysis to infer the double-helical structure of DNA. But there are also lesser known but equally important examples in which physical science tools that initially had no intended application in the life sciences were eventually utilized for such and which today are very tightly coupled with biology research but decoupled from their original invention.

For example, there is small-angle x-ray scattering (SAXS). SAXS was a physical science tool developed in the 1980s for the investigation of material properties of, originally, non-biological composites at the nanometer length scale, in which a sample's elastic scattering of x-rays is recorded at very low angles (typically $\lesssim 10°$). However, this technique found later application in the investigation of some natural polymers that are made by living cells, for example, the large sugar molecule, starch, that is made by plants. These days, SAXS has grown into a very powerful tool for investigating the mesoscopic periodic length scale features over a range of typically ~5–150 nm (a "nm" or nanometer is 1000 million times smaller than a meter) of several different biological filamentous/polymeric structures and in fact is largely viewed now as being primarily a biophysical technique.

Worked Case Example 1.1—Biomolecular Springs

As you will discover in Chapter 2, DNA molecules are an essential and special type of bio-polymer which carry the genetic code, and you would already be aware that they often adopt an intriguing double-helical structure. What you may be less familiar with is that they also act as a tiny spring when you stretch them…

Stretch experiments were performed on a type of DNA called lambda DNA (a type of DNA produced by a virus that infects bacteria… but it is commonly used in biophysics experiments on DNA) at room temperature by tethering opposite ends of one or more DNA molecule in parallel between two tiny plastic beads around three thousandths of a millimeter in diameter using a tool called "optical tweezers," also known as an "optical trap" (see Chapter 6). Assume that DNA acts as a Hookean spring, and that the optical tweezers work by producing an attractive force on a bead toward the center of a focused near infrared laser beam of wavelength one thousandth of a millimeter, which is proportional to the displacement of the bead from the center of the laser focus whose constant of proportionality is known as the trap stiffness, k.

a   *If the maximum extent of this optical trap is "diffraction-limited," which you can assume here is symmetrical, so roughly the same diameter as the wavelength of the laser, write down an expression for the maximum force that can be exerted on a*

DNA molecule if just one molecule of DNA is trapped between the two beads. What happens if there is more than one DNA molecule tethered between the two beads?

b   *It was found that when the center of one of the trapped beads moved from the center of the focused laser by 10% of its diameter, the measured force on the bead increased from zero to 3×10^{-11} N, while the distance between the two beads increased from zero to 1.3×10^{-6} m. How much energy do you think is required to fully extend a molecule whose fully extended length is 1.5×10^{-5} m? Roughly, how many molecules of water would be needed to transfer all their kinetic energy to extend the DNA fully? Assume that the average thermal energy of a molecule of water is of the order of $k_B T$ where k_B is the Boltzmann constant (1.38×10^{-23} J/K) and T is the absolute temperature.*

c   *In practice, when the DNA was continuously stretched, one of the beads was pulled out of one of the optical traps before the DNA could be fully extended. Any ideas why?*

Answers

This question is a little unfair, since it involves some concepts that you have not properly encountered yet in this book—a bit of a spoiler, sorry! But it's here more to get you in the mood of thinking about biological systems as *physical problems…*

a   We're told that the "extent" of the optical trap is about the same as the wavelength of the laser light. You might hazard a guess and assume that, given no more information, the optical trap was a spherically symmetrical sphere whose diameter was the same as the laser's wavelength. If we assume this, then the maximum distance from the center to the edge of the trap is half of 1/1,000th of a millimeter. This 1/1,000th of a millimeter is encountered a lot at the length scale of cells and is called a micron (symbol μm), equivalent to 1×10^{-6} m. Half of this is 0.5 μm, or 500 nanometers (unit nm—this again you will encounter as the length scale relevant to single molecules and is equal to one billionth of a meter, 1×10^{-9}m). We're told that the optical trap itself is a Hookean spring of stiffness k (i.e., stress is linearly proportional to strain, or trapping force is proportional to displacement of bead from trap center, call it x), so the trapping force is equal to kx. Maximum force is when x is maximum, so equal to k (0.5×10^{-6}) assuming k is in SI units. With one DNA molecule, this then is the maximum force applied, with n tethered molecules between the two beads, which we assume are simply acting in parallel, the maximum force on each is then just $k(0.5 \times 10^{-6})/n$. Note, in reality, when a laser is focused, the shape of this diffraction-limited intensity volume of light in 3D, often referred to as the "confocal volume," has more of a pigskin or rugby ball type shape (depending on which side of the Atlantic you are…), which you will learn about in Chapter 3.

b   The information given here allows you to determine what k is: 10% of the bead diameter is 300 nm, called this Δx, resulting in a force change, ΔF of 3×10^{-11} N, or 30×10^{-12} N, or 30 pN, where 1 pN or piconewton is 1×10^{-12} N, and is the typical force scale for single biomolecules in cells. So, k is just dF/dx, or $\Delta F/\Delta x$ for this Hookean spring case where k is a constant, or 1×10^{-3} N/m—this is also equivalent to 0.1 pN/nm, where the pN/nm is often a unit of stiffness used in the content of mechanical experiments on single biomolecules. The energy required to stretch a Hookean spring a distance L is the work done in moving through a force kx from x=0 to x=L. This is the integral $\int kx.dx$ from x=0 to L, or $1/2kL^2$ since $L = 1.5 \times 10^{-5}$ m (or 15 μm) this is equal to about 1×10^{-13} J. The thermal energy of a water molecule is around $k_B T$ (for more details, see Chapter 2), which is roughly the energy scale of single biomolecules is cells. At room temperature, T is around 300 K, so $k_B T$ is roughly 4.1×10^{-21} J, or 4.1 pN.nm if you like (this can be a useful number to remember….). This is around 27 million times smaller than the estimate above for the energy required to stretch a single DNA molecule, and so since kinetic energy is the same as thermal energy here then 27 million water molecules would be needed… the point here is that biology at the molecular scale often works on small nanoscale level fluctuations as opposed to making dramatic changes orders of magnitude higher…

c   I wonder what speculations you came up with! The real reason is that DNA is not actually a Hookean spring; it only appears to be over relatively short stretches. For longer stretches, DNA behaves non-linearly, so that the force is no longer directly proportional to the end-to-end extension of the molecule and it instead behaves, like many such filamentous polymers, as something called an "entropic spring" (see Chapter 8) such that when the end-to-end extension approaches the fully extended length of the molecule (also known as the contour length), then the spring force of molecular retraction becomes, in principle, infinitely high (at least until the chemical bonds that link the polymer together themselves are broken) since it only has one conformation to adopt (i.e., very ordered) as opposed to at lower extensions (i.e., very disordered), which is a driving entropic force working to retract an extended molecule like DNA back to its relaxed state. This means that you can never completely extend the DNA molecule fully (it would take an infinite amount of energy…) and so the DNA spring force in practice will keep on increasing as the molecule is extended by the two optically trapped beads until the maximum trapping force is reached at which point either bead is at the very end of their respective optical trap and thermal fluctuations are sufficient to allow one of the beads to simply escape out of the trap.

1.2 WHAT DO WE MEAN BY A "TOOLBOX?"

There are some notable differences between the core strategy of scientific research between physics and biology. Research in the life sciences, in its modern form, is very much optimized toward *hypothesis-driven research*. This means, in essence, setting out theories as models of scientific phenomena and constructing empirical investigations that can *falsify* these hypotheses (for interested readers in this basic area of the *scientific method*, see Popper, 1963), though can never actually prove them. To many scientists, this approach is simply what science is all about and is simply what separates science from pseudoscience. But a potential issue lies in investigating the phenomena that are currently undertheorized, to the extent that one can ask key scientific questions that are testable; but this is not necessarily the same as a "hypothesis" that a biologist might construct, which often involves an elaborate existing theory from that particular area of the life sciences. Many of these undertheorized areas are relevant to physics research, involving fundamental physical phenomena as opposed to the "fine-tuning" details of different aspects of testable existing model, and might better be described as *exploratory-driven research*. I mention this here simply to forewarn the reader of the differences in perception of what constitutes "good science" by some biologists and some physicists, since this, hopefully, will assist in a better genuine communication of scientific concepts and research between the two sciences.

A key challenge to finding answers in the biological sciences is the trick of knowing the right questions to ask. Physical science techniques applied to the life sciences, whether experimental or analytical, have often arisen from the need to address highly focused and specific questions. However, these techniques in general have capabilities to address a range of different questions in biology, not necessarily related to the biological systems of the original questions. As such, new emerging questions can be addressed by applying a *repertoire* of dynamically emerging physical tools, and this repertoire of physical science tools is essentially a "toolbox." But as any master craftsman knows, of key importance with any toolbox is knowing which are the best tools for the job in hand.

Not only knowing what are the right questions to ask but also knowing what are the best modern physical science tools to address these questions can, in all but exceptional cases, be difficult to address simultaneously. The former is the realm of the expert biologist in an often highly focused area of the life sciences, whereas the latter lies within the expertise of the physical scientist, engineer, or mathematician who has a professionally critical knowledge of the techniques that could be applied. A more efficient approach, which has been demonstrated in numerous exemplar case studies across the world, is to engage in a *well-balanced* collaborative research between biologists and physical scientists. This strategy, however, is far from a smooth path to navigate. There are not insignificant cultural, language, and skill differences between these two broad areas of science, as well as potential political challenges in determining who is actually "driving" the research. The strategy is decoupled from the approach of a biologist who generates a list of questions they wants to address and asks the physical scientist to find the best tool, as it is from the physical scientist who makes less than robust efforts to explain the tools to the biologist in ways that do not demand a grasp of advanced mathematics.

In an ideal world, the interaction between biologists and physicists would be a genuinely *dialectical* process that could involve, as an example, the biologist explaining the background to their questions and the physical scientist engaging the biologist in the science behind their techniques that ultimately may not only alter the choice of tools to use from the toolbox but also may change the biological questions that are actually asked in light of what specific tasks the tools can, and cannot, perform. It is not rocket science. It is a simple manifestation of respect for the expertise of others.

However, a key challenge for the physical scientist is to encourage the expert biologist to be bold and think "big"—what are the really difficult questions that often are the elephants in the room that do not get addressed during a typical grant proposal? This freedom of scientific thought can be genuinely infectious to tackling difficult biological questions once started. One cannot simply cut and paste physics onto biology, that's not how it works. But it is possible to inspire people to think beyond their comfort zones, and in doing so achieve something much higher.

For example, what drives the development of tools to add to the physical sciences toolbox? Thinking on the theme of hypothesis-driven *versus* exploratory research, the development of new tools is often *driven by exploration*. This, in turn, then leads to a greater space over which to *develop hypotheses*. An interesting contemporary example of this is the *terahertz spectroscopy*, which we will discuss in Chapter 5. The original conception of the terahertz spectroscopy had no biological hypothesis behind it but stemmed from the scientific exploration into the states of condensed matter; however, it is now emerging as a method that can inspire biological hypotheses, that is, do certain biomolecules inside cell membranes convey biological information using quantum mechanical (QM) processes?

Another important word of warning, for which the reader should be duly cautious, is the differences between these physical science tools and the tools that someone embarking on a DIY project may utilize. Physical science tools for investigating challenging biological questions are often expensive, challenging to use, and require sometimes significant levels of expertise to operate. They are thus decoupled from a relatively simple hammer for fixing fence posts or a cheap screwdriver used for assembling flat-pack furniture.

However, provided one is appropriately mindful of these differences, the concept of a suite of physical science tools that can be applied to investigate the complexities of life sciences is useful in steering the reader into learning a sufficient background of the science and applications of these techniques to inform their choices of using them further. This is not to say that this book will allow you to instantly become an expert in their use, but rather might offer sufficient details to know which tools might, or might not, be relevant for particular problems, thus permitting you to explore these tools further directly with the experts in these appropriate areas.

1.3 MAKEUP OF THE SUBSEQUENT CHAPTERS IN THIS BOOK

Although much of the book describes *experimental biophysical tools*, a substantial portion is dedicated to *theoretical biophysics tools*. Naturally, this is an enormous area of research that crosses into the field of applied mathematics. Since students of the physical sciences have often a broad spectrum of abilities and interests in math, these theoretical tools may be of interest to somewhat differing extents. However, what I have included here are at least the *key* theoretical techniques and methods of relevance to biophysics, as I see them, so that the more theoretically inclined student can develop these further with more advanced texts as appropriate, and the less theoretically inclined student at least has a good grounding in the core methods to model biophysical systems and equations of relevance that *actually work*.

In Chapter 2, an orientation is provided for physical scientists by explaining the key basic concepts in biology. For more advanced treatments, the reader is referred to appropriate texts in the reference list. But, again, at least this biological orientation gives physical sciences readers the bare bones knowledge to properly understand the biological context of the techniques described in the rest of this book, without having to juggle textbooks.

Chapters 3 through 6 are themed into different experimental biophysics techniques. These are categorized on the basis of the following:

1.3.1 DETECTION, SENSING, AND IMAGING TECHNIQUES

Chapter 3: Basic, *foundational* techniques that use optical/near-optical spectroscopy and/or light microscopy. Many of these basic optical tools are relatively straightforward; however, they are enormously popular and generate much insight into a range of biological processes.

Chapter 4: More advanced *frontier* techniques of optical/near-optical spectroscopy and microscopy; although there are a range of biophysical tools utilizing various physical phenomena, the tools currently making use of optical methods are significant, especially the more modern techniques, and this is reflected with this additional advanced optics chapter here.

Chapter 5: Biophysical detection methods that are primarily *not* optical or near optical. These encompass the robust, traditional methods of *molecular biophysics*, also known as "structural biology," for example, x-ray crystallography, nuclear magnetic resonance, and electron microscopy, as well as other x-ray diffraction and neutron diffraction methods. But there are also emerging spectroscopic methods included here, such as the terahertz radiation spectroscopy.

1.3.2 EXPERIMENTAL BIOPHYSICAL METHODS PRIMARILY RELATING ESPECIALLY TO *FORCE*

Chapter 6: Methods that mainly measure and/or manipulate biological forces. These cover a range of tools including many modern single-molecule force manipulation techniques such as optical and magnetic tweezers and atomic force microscopy. But there are also force-based techniques included that cover a range of much higher length scales, from cells up to tissues and beyond.

1.3.3 COMPLEMENTARY EXPERIMENTAL TECHNOLOGIES

Chapter 7: Lab-based methods that are not *explicitly* biophysics, but which are invaluable to it. This is a more challenging chapter for the physical science student since it inevitably includes details from other areas of science such as molecular and cell biology, chemistry, engineering, and computer science. However, to really *understand* the machinations of the methods described in the other chapters, it requires some knowledge of the peripheral nonbiophysical methods that complement and support biophysics itself. This includes various genetics techniques, chemical conjugation tools, high-throughput methods such as microfluidics, how to make crystals of biomolecules, the use of model organisms, and physical tools associated with biomedicine, in particular.

1.3.4 THEORETICAL BIOPHYSICS TOOLS

Chapter 8: Methods of computational biophysics and theoretical approaches requiring a pencil and paper. This is a large chapter, since it not only involves both the discrete methods used in advanced computational approaches, such as those of molecular simulations, but also discusses a wide range of continuum approaches covering biopolymer mechanics, fluid dynamics, and reaction–diffusion analysis.

And finally, Chapter 9 discusses the future outlook of the physical science toolbox in the life sciences. This encompasses emerging methods that are not yet fully established in mainstream biophysics but show enormous future promise. They include systems biophysics approaches, synthetic biology and bionanotechnology, biophysics methods that enable personalized healthcare, and tools that extend the length scales of biophysics into the quantum and ecosystem regimes.

It is a challenge to know how to best structure "biophysics techniques" in one's own mind. But my advice is to avoid the pitfall of solely theming these techniques along the lines of very specific physical phenomena for their modes of operation (e.g., techniques that use fluorescence and those that utilize electromagnetism) that underlie a technique's primary mode of action, and also, avoid the pitfall of structuring techniques along their modes of action solely in a particular biological context (e.g., a cellular technique and an *in vitro* technique). The truth, I believe, is that biophysical techniques can operate using a complex armory of many different physical processes in a highly combinatorial fashion, and many of these techniques can be applied in several different biological contexts.

All these make it a challenge to theme into chapters for a textbook such as this one, to say nothing of the challenges for the reader new to the subject.

But my *suggestion* to you, dear reader, when trying to make sense of the vast array of biophysical techniques available, is simply this:

> Always *try* to consider the actual biological *questions* being addressed, as opposed to solely thinking of the techniques as being simply interesting *applications* of physics to biology, which if you're not careful can end up reading like a shopping list of boring acronyms. If you strive to do this, you may find it an invaluable catalyst for the process of integrating physics with biology, *if* that is your wish.

To aid revision, each chapter has a brief "General Idea" at the start, several "Key Points" of importance highlighted specifically throughout, plus concise bullet-point style summaries at the end of each chapter, along with associated questions relating to each chapter that can be used by students or tutors alike. Full references are given at the end of each chapter, and of these, one particularly important "Key Reference" is indicated, which is the reference that all students should make an effort to read. In addition, there are a number of "Key Biological Applications" sections highlighted to summarize the core general biological applications for the different types of biophysical tools and techniques discussed. Also, and most importantly, there are detailed model-worked examples of questions throughout. These, I hope, will really help the student, and the lecturer, to understand many of the challenging physical science concepts that form the basis of the biophysical techniques.

1.4 ONCE MORE, UNTO THE BREACH

I will not bore you with the political challenges of the collaborative process between physics and biology, such as who *drives* and *leads* the research, *where* you do the research, the orders of *authors' names* on a research paper, and where the next *grant* is coming from, those kinds of things. I have theories, yes, but to expound on them requires the structure of a nearby public house, and in the absence of such I will move on.

> But I will say this.
> Let your scientific nose do the leading, *constantly* and *fervently*.

Question dogma. *Always* question dogma.
Be curious about the natural world. *Endlessly* curious. *Keep on asking questions.*
And when you think you have found the answer, well... *think again....*
Enjoy this book. Please let me know if you have or haven't. I'll aim to make it better the next time.

1.5 SUMMARY POINTS

- We can treat physical science techniques applied to biology problems as a "toolbox," but we must be mindful that these tools are often hard to use and expensive.
- Physical scientists might wish to help biologists to think of the big questions and can consider building established balanced, long-term, dialectical collaborations as an alternative to providing transient technical support services.
- Relentless progress on producing new physical science tools emerges, which then can offer different biological insights, but many *old* tools can still provide powerful details.

QUESTIONS

1.1 One criticism by some physical scientists in liaising with biologists, at the risk of *propagating a stereotype*, is that biologists just "don't have the math!" Do you think this is true? Is it actually possible to convey complex mathematical ideas without, for example, writing down an equation?

1.2 Some biologists complain, at the risk of *propagating a stereotype*, that physical scientists simply do not have a good appreciation of the complexities of biology and lack the required knowledge of basic biological and chemical details to help them to understand even the simplest processes inside living cells. Do you think this is correct? Is there anything that physical scientists could do to improve this hypothetical situation?

1.3 Ernest Rutherford, described by many as the *father of nuclear of physics*, suggested famously that "all science is either physics or stamp collecting." Discuss how modern biology is stamp collecting. Then think again, and discuss how modern biology is more than just stamp collecting.

1.4 I've suggested that there is no such thing as a dumb question. But do you agree? In terms of biophysics and the physics of life, are there some things which are so obvious that any questions concerning them really are stupid?

REFERENCES

KEY REFERENCE

Hodgkin, A.L. and Huxley, A.F. (1952). A quantitative description of membrane current and its application to conduction and excitation in nerve. *J. Physiol. 177*:500–544.

MORE NICHE REFERENCES

Darwin, C. (1859). *On the Origin of Species by Means of Natural Selection, or the Preservation of Favoured Races in the Struggle for Life*, Project Gutenberg Literary Archive Foundation, Salt Lake City, UT.

Kendrew, J.C. et al. (1958). A three-dimensional model of the myoglobin molecule obtained by x-ray analysis. *Nature 18*:662–666.

Maxwell, J.C. (1873). *A Treatise on Electricity and Magnetism*, 3rd ed. (June 1, 1954). Dover Publications, New York.

Popper, K.R. (1963). *Conjectures and Refutations: The Growth of Scientific Knowledge.* Routledge & Kegan Paul, London.

Orientation for the Bio-Curious

2

The Basics of Biology for the Physical Scientist

If you want to understand function, study structure. [I was supposed to have said in my molecular biology days.]

—Francis Crick, *What Mad Pursuit: A Personal View of Scientific Discovery* (1988, p. 150)

General Idea: This chapter outlines the essential details of the life sciences that physical scientists need to get to grips with, including the architecture of organisms, tissues, cells, and biomolecules as well as the core concepts of processes such as the central dogma of molecular biology, and discusses the key differences in the scientific terminology of physical parameters.

2.1 INTRODUCTION: THE MATERIAL STUFF OF LIFE

The material properties of *living things* for many physical scientists can be summarized as those of *soft condensed matter*. This phrase describes a range of physical states that in essence are relatively easily transformed or deformed by thermal energy fluctuations at or around room temperature. This means that the free energy scale of transitions between different physical states of the soft condensed matter is similar to those of the thermal reservoir of the system, namely, that of $\sim k_B T$, where k_B is the *Boltzmann constant* of 1.38×10^{-23} m^2 kg s^{-2} K^{-1} at absolute temperature T. In the case of this *living* soft condensed matter, the thermal reservoir can be treated as the surrounding water solvent environment. However, a key feature of living soft matter is that it is not in thermal equilibrium with this water solvent reservoir. Biological matter, rather, is composed of structures that require an external energy input to be sustained. Without knowing anything about the fine details of the structures or the types of energy inputs, this means that the system can be treated as an example of nonequilibrium statistical thermodynamics. The only example of biological soft condensed matter, which is in a state of thermal equilibrium, is something that is *dead*.

KEY POINT 2.1

Thermal equilibrium = death

Much insight can be gained by modeling biological material as a subset of nonequilibrium soft condensed matter, but the key weakness of this approach lies in the coarse graining and statistical nature of such approximations. For example, to apply the techniques of statistical

DOI: 10.1201/9781003336433-2

thermodynamics, one would normally assume a bulk ensemble population in regard to thermal physics properties of the material. In addition, one would use the assumptions that the material properties of each soft matter component in a given material mix are reasonably homogenous. That is not to say that one cannot have multiple different components in such a soft matter mix and thus hope to model complex heterogeneous biological material at some appropriate level, but rather that the minimum length scale over which each component extends assumes that each separate component in a region of space is at least homogenous over several thousand constituent molecules. In fact, a characteristic feature of many soft matter systems is that they exhibit a wide range of different phase behaviors, ordered over relatively long length scales that certainly extend beyond those of single molecules.

The trouble is that there are many important examples in biology where these assumptions are belied, especially so in the case of discrete molecular scale process, exhibited clearly by the so-called molecular machines. These are machines in the normal physicist definition that operate by transforming the external energy inputs of some form into some type of useful *work*, but with important differences to everyday machines in that they are composed of sometimes only a few molecular components and operate over a length scale of typically ~1–100 nm. Molecular machines are the most important drivers of biological processes in cells: they transport cargoes; generate cellular fuel; bring about replication of the genetic code; allow cells to move, grow, and divide; etc.

There are intermediate states of *free energy* in these molecular machines. Free energy is the thermodynamic quantity that is equivalent to the capacity of a system to do mechanical work. If we were to plot the free energy level of a given molecular machine as a function of some "reaction coordinate," such as time or displacement of a component of that machine, for example, it typically would have several peaks and troughs. We can say therefore that molecular machines have a *bumpy free energy landscape*. Local minima in this free energy landscape represent states of *transient stability*. But the point is that the molecular machines are dynamic and can switch between different transiently stable states with a certain probability that depends upon a variety of environmental factors. This implies, in effect, that molecular machines are *intrinsically unstable*.

Molecular free energy landscapes have many local minima. "Stability," in the explicit thermodynamic sense, refers to the *curvature* of the free energy function, in that the greater the local curvature, the more unstable is the system. Microscopic systems differ from macroscopic systems in that relative fluctuations in the former are large. Both can have landscapes with similar features (and thus may be similar in terms of stability); however, the former diffuses across energy landscapes due to intrinsic thermal fluctuations (embodied in the *fluctuation–dissipation theorem*). It is the fluctuations that are intrinsic and introduce the transient nature. Molecular machines that operate as a *thermal ratchet* (see Chapter 8) illustrate these points.

This is often manifested as a molecular machine undergoing a series of molecular conformational changes to bring about its *biological function*. What this really means is that a given population of several thousands of these molecular machines could have significant numbers that are in each of different states at any given time. In other words, there is *molecular heterogeneity*.

KEY POINT 2.2

Molecular machines function through instability, resulting in significant molecular heterogeneity.

This molecular heterogeneity is, in general, in all but very exceptional cases of molecular synchronicity of these different states of time, very difficult to capture using bulk ensemble biophysical tools, either experimental or analytical, for example, via soft matter modeling approaches. Good counterexamples of this rare synchronized behavior have been utilized by a variety of biophysical tools, since they are exceptional cases in which single-molecule detection precision is not required to infer molecular-level behavior of a biological system. One such can be found *unnaturally* in x-ray crystallography of biological molecules and another more *naturally* in muscles.

In x-ray crystallography, the process of crystallization forces all of the molecules, barring crystal defects, to adopt a single favored state; otherwise, the unit cells of the crystals would not tessellate to form macroscopic length scale crystals. Since they are all in the same state, the effective signal-to-noise detection ratio for the scattered x-ray signal from these molecules can be relatively high. A similar argument applies to other structural biology techniques, such as *nuclear magnetic resonance,* (see Chapter 5) though here single energetic states in a large population of many molecules are imposed via a resonance effect due to the interaction of a large external magnetic field with electron molecular orbitals.

In muscle, there are molecular machines that act, in effect, as motors, made from a protein called "myosin." These *motor proteins* operate by undergoing a *power stroke–type* molecular conformational change, allowing them to impose force against a filamentous track composed of another protein called "actin," and in doing so cause the muscle to contract, which allows one to lift a cup of tea from a table to our lips, and so forth. However, in a normal muscle tissue, the activity of many such myosin motors is synchronized in time by a chemical trigger consisting of a pulse of calcium ions. This means that many such myosin molecular motors are in effect in phase with each other in terms of whether they are at the start, middle, or end of their respective molecular power stroke cycles. This again can be manifested in a relatively high signal-to-noise detection ratio for some bulk ensemble biophysical tools that can probe the power stroke mechanism, and so again this permits molecular-level biological inference without having to resort to molecular-level sensitivity of detection. This goes a long way to explaining why, historically, so many of the initial pioneering advances in biophysics were made through either structural biology or muscle biology research or both.

KEY POINT 2.3

Exceptional examples of biological systems exhibiting molecular synchronicity, for example, in muscle tissue, can allow single-molecule interferences from ensemble average data.

To understand the nature of a biological material, we must ideally not only explore the soft condensed matter properties but also focus on the fine structural details of living things, through their makeup of constituent cells and extracellular material and the architecture of subcellular features down to the length scale of single constituent molecules.

But life, as well as being highly complex, is also short. So, the remainder of this chapter is an ashamedly whistle-stop tour of everything the physicist wanted to know about biology but was afraid to ask. For readers seeking further insight into molecular- and cell-level biology, an ideal starting point is the textbook by Alberts et al. (2008). One word of warning, however, but the teachings of biology can be rife with classification and categorization, much essential, some less so. Either way, the categorization can often lead to confusion and demotivation in the uninitiated physics scholar since one system of classification can sometimes contradict another for scientific and/or historical reasons. This can make it challenging for the physicist trying to get to grip with the language of biological research; however, this exercise is genuinely more than one in semantics, since once one has grasped the core features of the language at least, then intellectual ideas can start to be exchanged between the physicist and the biologist.

2.2 ARCHITECTURE OF ORGANISMS, TISSUES, AND CELLS AND THE BITS BETWEEN

Most biologists subdivide living organisms into three broad categories called "domains" of life, which are denoted as *Bacteria, Eukaryotes,* and *Archaea. Archaea* are similar in many ways to bacteria though typically live in more extreme environmental conditions for combinations of external acidity, salinity, and/or temperature than most bacteria, but they also have some biochemical and genetic features that are actually closer to eukaryotes than to bacteria. Complex *higher organisms* come into the eukaryote category, including plants and animals, all of which are composed of collections of organized living matter that is made from multiple unitary structures called "cells," as well as the stuff that is between cells or collections of cells, called the "extracellular matrix" (*ECM*).

2.2.1 CELLS AND THEIR EXTRACELLULAR SURROUNDINGS

The ECM of higher organisms is composed of molecules that provide mechanical support to the cells as well as permit perfusion of small molecules required for cells to survive or molecules that are produced by living cells, such as various nutrients, gases such as oxygen and carbon dioxide, chemicals that allow the cells to communicate with each other, and the molecule most important to all forms of life, which is the universal biological solvent of water. The ECM is produced by the surrounding cells comprising different protein and sugar molecules. Single-celled organisms also produce a form of extracellular material; even the simplest cells called *prokaryotes* are covered in a form of slime capsule called a "glycocalyx," which consists of large sugar molecules modified with proteins—a little like the coating of *M&M's* candy.

The traditional view is that the cell is the *basic unit* for all forms of life. Some *lower organisms* (e.g., the archaea and bacteria and, confusingly, some eukaryotes) are classified as being *unicellular*, meaning that they appear to function as single-celled life forms. The classical perspective is typically hierarchical in terms of length scale for more complex *multicellular* life forms, cells, of length scale $\sim$10–100 µm (1 µm or *micron* is one millionth of a meter), though there are exceptions to this such as certain nerve cells that can be over a meter in length.

Cells may be grouped in the same region of space in an organism to perform specialist functions as *tissues* (length scale $\sim$0.1 mm to several centimeters or more in some cases), for example, muscle tissue or nerve tissue, but then a greater level of specialization can then occur within *organs* (length scales >0.1 m), which are composed of different cells/tissues with what appear to be a highly specific set of roles in the organisms, such as the brain, liver, and kidneys.

This traditional stratified depiction of biological matter has been challenged recently by a more complicated model of living matter; what seems to be more the case is that in many multicellular organisms, there may be multiple layers of feedback between different levels of this apparent structural hierarchy, making the concept of independent levels dubious and a little arbitrary.

Even the concept of unicellular organisms is now far from clear. For example, the model experimental unicellular organisms used in biological research, such as *Escherichia coli* bacteria found ubiquitously in the guts of mammals, and *budding yeast* (also known as "baker's yeast") formally called *Saccharomyces cerevisiae* used for baking bread and making beer, spend by far the majority of their natural lives residing in complex 3D communities consisting of hundreds to sometimes several thousands of individual cells, called "biofilms," glued together through the cells' glycocalyx slime capsules.

(An aside note is about how biologists normally name organisms, but these generally consist of a *binomial nomenclature* of the organism's *species* name in the context of its *genus*, which is the collection of closely related organisms including that particular species, which are all still distinctly different species, such that the name will take the form "*Genus species*." Biologists will further truncate these names so that the genus is often denoted simply by its first letter; for example, *E. coli* and *S. cerevisiae*.)

Biofilms are intriguing examples of what a physicist might describe as an *emergent structure*, that is, something that has different collective properties to those of the isolated building blocks (here, individual cells) that are often difficult, if not impossible, to predict from the single-cell parameters alone—cells communicate with each other through both chemical and mechanical stimuli and also respond to changes in the environment with collective behavior. For example, the evolution of antibiotic resistance in bacteria may be driven by *selective pressures* not at the level of the single bacterial cell as such, but rather targeting a population of cells found in the biofilm, which ultimately has to feedback down to the level of replicating bacteria cells.

It is an intriguing and deceptively simple notion that putatively *selfish genes* (Dawkins, 1978), at a length scale of $\sim 10^{-9}$ m, propagate information for their own future replication into subsequent generations through a vehicle of higher-order, complex emergent structures at much higher length scales, not just those of the cell that are three orders of magnitude greater but also those that are one to three orders of magnitude greater still. In other words, even bacteria seem to function along similar lines to a more complex multicellular organism, and in many ways, one can view a multicellular organism as such an example of a complex, emergent structure. This begs a question of whether we can truly treat an *isolated* cell as the basic unit of life, if its natural life cycle demands principally the proximity of other cells. Either way, there is no harm in the reader training themselves to question dogma in academic textbooks (the one you are reading now is not excluded), especially those of *classical* biology.

KEY POINT 2.4

"Unicellular" organisms exist in the context of many cells, of their own species, and of others, and so our understanding of their biology should take this into account.

Cells can be highly dynamic structures, growing, dividing, changing shape, and restructuring themselves during their lifetime in which biologists describe as their *cell cycle*. Many cells are also *motile*, that is, they move. This can be especially obvious during the *development* stages of organisms, for example, in the formation of tissues and organs that involve programmed movements of cells to correct positions in space relative to other cells, as well in the immune response that requires certain types of cell to physically move to sites of infection in the organism.

2.2.2 CELLS SHOULD BE TREATED ONLY AS A "TEST TUBE OF LIFE" WITH CAUTION

A common misconception is that one can treat a cell as being, in essence, a handy "test tube of life." It follows an understandable reductionist argument from bottom-up *in vitro* experiments (*in vitro* means literally "in glass," suggesting test tubes, but is now taken to mean any experiment using biological components taken outside of their native context in the organism). Namely, that if one has the key components for a biological process in place *in vitro*, then surely why can we not use this to study that process in a very controlled assay that is decoupled from the native living cell. The primary issues with this argument, however, concern space and time.

In the real living cell, the biological processes that occur do so with an often highly intricate and complex spatial dependence. That is, it matters *where* you are in the cell. But similarly, it also matters *when* you are in the cell. Most biological processes have a history dependence. This is not to say that there is some magical memory effect, but rather that even the most simple biological process depends on components that are part of other processes, which operate in a time-dependent manner in, for example, certain key events being triggered by different stages in the cell cycle, or the history of what molecules in a cell were detected outside its cell membrane in the previous 100 ms.

So, although *in vitro* experiments offer a highly controlled environment to understand biology, they do not give us the complete picture. And similarly, the same argument applies to a single cell. Even unicellular organisms do not really operate in their native context solely on their own. The real biological context of any given cell is in the physical vicinity presence of other cells, which has implications for the physics of a cell. So, although a cell is indeed a useful self-enclosed vessel for us to use biophysical technique to monitor biological process, we must be duly cautious in how we interpret the results of these techniques in the absence of a truly native biological context.

2.2.3 CELLS CATEGORIZED BY THE PRESENCE OF NUCLEI (OR NOT)

A cell itself is physically enclosed from its surrounding *cell membrane*, which is largely impervious to water. In different cell types, the cell membrane may also be associated with other membrane/wall structures, all of which encapsulate the internal chemistry in each cell. However, cells are far more complex than being just a boring bag of chemicals. Even the simplest cells are comprised of intricate subcellular architectures, in which the biological process can be compartmentalized, both in space and time, and it is clear that the greater the number of compartments in a cell, the greater its *complexity*.

The next most significant tier of biological classification of cell types concerns one of these subcellular structures, called the "nucleus." Cells that do not contain a nucleus are called "prokaryotes" and include both *bacteria* and *archaea*. Although such cells have no nucleus, there is some ordered structure to the deoxyribonucleic acid (DNA) material, not only due mainly to the presence of proteins that can condense and package the DNA but also due to a natural *entropic spring* effect from the DNA, implying that highly elongated structures in the absence of large external forces on the ends of the DNA are unlikely. This semistructured region in prokaryotes is referred to as the *nucleoid* and represents an *excluded volume* for many other biological molecules due to its tight mesh-like arrangement of DNA, which in many bacteria, for example, can take up approximately one-third of the total volume of the cell.

Cells that do contain a nucleus are called "eukaryotes" and include those of relatively simple "unicellular" organisms such as yeast and *trypanosomes* (these are *pathogen* cells that ultimately cause disease and which result in the disease *sleeping sickness*) as well as an array of different cells that are part of complex multicellular organisms, such as you and I.

KEY POINT 2.5

Basic definition of the "Cell": (1) functionally autonomous, (2) physically enclosed, (3) structurally independent unit of "life," either from unicellular or multicellular organisms, (4) contains biomolecules that have the capacity to self-replicate independently to form a new cell.

2.2.4 CELLULAR STRUCTURES

In addition to the cell membrane, there are several intricate architectural features to a cell. The nucleus of eukaryotes is a vesicle structure bounded by a lipid bilayer (see Section 2.2.5) of diameter 1–10 μm depending on the cell type and species, which contains the bulk of the genetic material of the cell encapsulated in DNA, as well as proteins that bind to the DNA, called "histones," to package it efficiently. The watery material inside the cell is called the "cytoplasm" (though inside some cellular structures, this may be referred to differently, for example, inside the nucleus this material is called the "nucleoplasm"). Within the cytoplasm of all cell types are cellular structures called "ribosomes" used in making proteins. These are especially numerous in a cell, for example, *E. coli* bacteria contain ~20,000 ribosomes per cell, and an actively growing mammalian cell may contain ~10^7 ribosomes.

Ribosomes are essential across all forms of life, and as such their structures are relatively well *conserved*. By this, we mean that across multiple generations of organisms of the same species, very little change occurs to their structure (and, as we will discuss later in this chapter, the DNA sequence that encodes for this structure). The general structure of a ribosome consists of a *large subunit* and a *small subunit*, which are similar between prokaryotes and eukaryotes. In fact, the DNA sequence that encodes part of the small subunit, which consists of a type of nucleic acid (which we will discuss later called the "ribosomal RNA" (rRNA)—in prokaryotes, referred to as the *16S rRNA subunit*, and in eukaryotes as the slightly larger *18S rRNA subunit*), is often used by evolutionary biologists as a *molecular chronometer* (or *molecular clock*) since changes to its sequences relate to abrupt evolutionary changes of a species, and so these differences between different species can be used to generate an evolutionary lineage between them (this general field is called "phylogenetics"), which can be related to absolute time by using estimates of spontaneous mutation rates in the DNA sequence.

The region of the nuclear material in the cell is far from a static environment and also includes protein molecules that bind to specific regions of DNA, resulting in *genes* being switched on or off. There are also protein-based molecular machines that bind to the DNA to replicate it, which is required prior to cell dividing, as well as molecular machines that read out or *transcribe* the DNA genetic code into another type of molecular similar to DNA called "ribonucleic acid" (*RNA*), plus a host of other proteins that bind to DNA to *repair* and *recombine* faulty sections.

Other subcellular features in eukaryotes include the *endoplasmic reticulum* and *Golgi body* that play important roles in the assembly or proteins and, if appropriate, how they are packaged to facilitate their being exported from cells. There are also other smaller *organelles* within eukaryotic cells, which appear to cater for a subset of specific biological functions, including *lysosomes* (responsible for degrading old and/or foreign material in cells), *vacuoles* (present in plant cells, plus some fungi and unicellular organisms, which not only appear to have a regulatory role in terms of cellular acidity/pH but also may be involved in waste removal of molecules), *starch grains* (present in plant cells of sugar-based energy storage molecules), storage capsules, and *mitochondria* (responsible for generating the bulk of a molecule called "adenosine triphosphate" [*ATP*], which is the universal cellular energy currency).

There are also invaginated cellular structures called "chloroplasts" in plants where light energy is coupled into the chemical manufacturing of sugar molecules, a process known as *photosynthesis*. Some less common prokaryotes do also have structured features inside their cells. For example, *cyanobacteria* perform photosynthesis in organelle-type structures composed of protein walls called "carboxysomes" that are used in photosynthesis. There is also a group of aquatic bacteria called "planctomycetes" that contain semicompartmentalized cellular features that at least partially enclose the genetic DNA material into a nuclear membrane–type vesicle.

Almost all cells from the different domains of life contain a complex scaffold of protein fibers called the "cytoskeleton," consisting of *microfilaments* made from actin, *microtubules* made from the protein *tubulin*, and *intermediate filaments* composed of several tens of different types of protein. These perform a mechanical function of stabilizing the cell's dynamic 3D structure in addition to being involved in the transport of molecular material inside cells, cell growth, and division as well as movement both on a whole cell motility level and on a more local level involving specialized protuberances such as *podosomes* and *lamellipodia*.

2.2.5 CELL MEMBRANES AND WALLS

As we have seen, all cells are ultimately encapsulated in a thin film of a width of a few nanometers of the cell membrane. This comprises a specialized structure called a "lipid bilayer," or more accurately a *phospholipid* bilayer, which functions as a sheet with a hydrophobic core enclosing the cell contents from the external environment, but in a more complex fashion serves as a locus for diverse biological activity including attachments for molecular detection, transport of molecules into and out of cells, the cytoskeleton, as well as performing a vital role in unicellular organisms as a dielectric capacitor across which an electrical and charge gradient can be established, which is ultimately utilized in generating the cellular fuel of ATP. Even in relatively simple bacteria, the cell membrane can have significant complexity in terms of localized structural features caused by the heterogeneous makeup of lipids in the cell membrane, resulting in dynamic phase transition behavior that can be utilized by cells in forming nanoscopic *molecular confinement* zones (i.e., yet another biological mechanism to achieve compartmentalization of biological matter).

The cell membrane is a highly dynamic and heterogeneous structure. Although structured from a phospholipid bilayer, native membranes include multiple proteins between the phospholipid groups, resulting in a typical *crowding density* of 30%–40% of the total membrane surface area. Most biomolecules within the membrane can diffuse laterally and rotationally, as well as phospholipid molecules undergoing significant vibration and transient flipping conformational changes (unfavorable transitions in which the polar head group rotates toward the hydrophobic center of the membrane). In addition, in eukaryotic cells, microscale patches of the cell membrane can dynamically invaginate either to export chemicals to the outside world, a process known as *exocytosis*, which creates phospholipid vesicle buds containing the chemicals for export, or to import materials from the outside by forming similar vesicles from the cell membrane but inside the cell, a process known as *endocytosis*, which encapsulates the extracellular material. The cell membrane is thus better regarded as a complex and dynamic fluid.

The most basic model for accounting for most of the structural features of the cell membrane is called the "Singer–Nicholson model" or "fluid mosaic model," which proposes that the cell membrane is a fluid environment allowing phospholipid molecules to diffuse laterally in the bilayer, but with stability imparted to the structure through the presence of transmembrane proteins, some of which may themselves be mobile in the membrane.

Improvements to this model include the *Saffman–Delbrück model*, also known as the *2D continuum fluid model*, which describes the membrane as a thick layer of viscous fluid surrounded by a bulk liquid of much lower viscosity and can account for microscopic dynamic properties of membranes. More recent models incorporate components of a protein skeleton (parts of the cytoskeleton) to the membrane itself that potentially generates semistructured compartments with the membrane, referred to as the *membrane fence model*, with modifications to the fences manifested as "protein pickets" (called the "transmembrane protein picket model"). Essentially though, these separately named models all come down to the same basic phenomenon of a self-assembled phospholipid bilayer that also incorporates interactions with proteins resulting in a 2D partitioned fluid structure.

Beyond the cell membrane, heading in the direction from the center of the cell toward the outside world, additional boundary structures can exist, depending on the type of cell. For example, some types of bacteria described as *Gram-negative* (an historical description relating to their inability to bind to a particular type of chemical dye called "crystal violet" followed by a counterstrain called "safranin" used in early microscopy studies in the nineteenth century by the Danish bacteriologist Hans Christian Gram, which differentiated them from cells that did bind to the dye combination, called "Gram-positive" bacteria) possess a second outer cell membrane.

Also, these and many other unicellular organisms, and plant cells in general, possess an outer structure called the "cell wall" consisting of tightly bound proteins and sugars, which functions primarily to withstand high *osmotic pressures* present inside the cells. Cells contain a high density of molecules dissolved in water that can, depending on the extracellular environment, result in *nonequilibrium* concentrations on either side of the cell boundary that is manifested as a higher internal water pressure inside the cell due to pores at various points in the cell membrane permitting the diffusion of water but not of many of the larger solute molecules inside the cell (it is an example of *osmosis* through a *semipermeable* membrane).

Cells from animals are generally in an *isotonic* environment, meaning that the extracellular osmotic pressure is regulated to match that of the inside of the cells, and small fluctuations around this can be compensated for by small changes to the volume of each cell, which the cell can in general survive due to the stabilizing scaffold effect of its cytoskeleton. However, many types of nonanimal cells do not experience an isotonic environment but rather are bathed in a much lower *hypotonic* environment and so require a strong structure on the outside of each cell to avoid bursting. For example, *Staphylococcus aureus* bacteria, a modified form of which results in the well-known *MRSA superbug* found in hospitals, need to withstand an internal osmotic pressure equivalent to ~25 atmospheres.

2.2.6 LIQUID–LIQUID PHASE-SEPARATED (LLPS) BIOMOLECULAR CONDENSATES

A feature of life is information flow across multiple scales, yet the physical rules that govern how this occurs in a coordinated way from molecules through to cells are unclear; there is not, currently, a Grand Unified Information Theory of Physical Biology. However, observations from recent studies implicate liquid–liquid phase separation (LLPS) in cell information processing (Banani, 2017). Phase transitions are everywhere across multiples scales, from cosmological features in the early universe to water boiling in a kettle. In biomolecular LLPS, a mixture of biomolecules (typically proteins and RNA, which you will find out about later in this chapter) coalesce inside a cell to form liquid droplets inside the cytoplasm. The transition of forming this concentrated liquid state comprising several molecules from previously isolated molecules that are surrounded by solvent molecules of water and ions involves an increase in overall molecular order, so the reduction in entropy since the number of accessible free energy microstates is lower.

In essence, the biomolecules are transitioning from being *well-mixed* to *demixed*. Such a process would normally be thermodynamically unfavorable, however, in this case it is driven by a net increase in the free energy due to attractive enthalpic interactions between the molecules in the liquid droplet on bringing them closer together. When considering the net enthalpic increase, we need to sum up all the possible attractive interactions (often interactions between different types of molecules) and subtract all of the total repulsive interactions (often interactions between the same type of molecule)—see *Worked Case Example 2.1*.

These liquid droplets are broadly spherical but have a relatively unstable structure; their shape can fluctuate due to thermal fluctuations of the surrounding molecules in the cytoplasm, they can also grow further by accumulating of "nucleating" more biomolecules, and also shrink reversibly, depending upon factors such as the local bimolecular concentrations and the mixture of biomolecules and the physicochemical environment inside the cell. They comprise components held by weak noncovalent interactions, imparting partial organization via emergent liquid crystallinity, microrheology, and viscoelasticity, qualities that enable cooperative interaction over transient timescales. Weak forces permit dynamics of diffusion and molecular turnover in response to triggered changes of fluidity to facilitate the release of molecular components. A traditional paradigm asserts that compartmentalization, which underpins efficient information processing, is confined to eukaryotic cells' use of membrane-bound organelles to spatially segregate molecular reagents. However, an alternative picture has recently emerged of membraneless LLPS droplets as a feature of all cells that enable far more dynamic spatiotemporal compartmentalization. Their formation is often associated with the cell being under stress, and the big mystery in this area of research is what regulates their size (anything from tiny droplets of a few nanometers of diameter up to several hundred nanometers), since classical nucleation physics theory would normally predict that under the right conditions a liquid–liquid phase transition goes to completion, that is, a droplet will continue to grow in size until all the biomolecular reagents are used up, but this is not what occurs (see Chapter 8 for more details on this).

If we consider the pressure difference between the inside and outside of a droplet or radius r as ΔP, then the force due to this exerted parallel to any circular cross-section is simply the total area of that cross-section multiplied by ΔP, or $F_p = \Delta P.\pi r^2$. This is balanced by an opposing force due the surface tension T per unit length (a material property relating to the biomolecular droplet and the surrounding water solvent) that acts around the circumference of this cross-section, of $F_T = T.2\pi r$. In steady state, $F_p = F_T$ so $\Delta P = 2T/r$. What this simple analysis shows is that smaller droplets have a higher pressure difference between the inside and the outside, so more work must be done for droplet molecules to escape. However, the total work for a finite volume of all droplets is small for larger droplets due to a lower overall surface area to volume ratio, so overall surface tension favors droplet growth, and this growth becomes more likely the larger droplets become.

So, there is some interesting size regulation occurring, which links droplet biophysics and their biological functions. LLPS droplets in effect are a very energy efficient and a rapid way to generate spatial compartmentalization in the cell since they do not require a bounding lipid membrane, which is often slow to form and requires energy input. Instead, LLPS droplets can form rapidly and reversibly in response to environmental triggers in the cell and can package several biomolecules into one droplet to act as a very efficient nano-reactor biochemical vessel since the concentration of the reactants in such a small volume can be very high. LLPS droplets research is very active currently, with droplets now being found in many biological systems and being associated with both normal and disease processes. As you will see from Chapter 4, research is being done using super-resolution microscopy to investigate these droplets experimentally, but as you will also see from Chapter 8 much modeling computational simulation research tools are being developed to understand this interesting phenomenon.

2.2.7 VIRUSES

Worked Case Example 2.1: Biomolecular Liquid Condensates

One type of biomolecule B dissolved in solvent S has an exothermic interaction enthalpy of $3k_BT$ between 2 B molecules, $2k_BT$ between 2 S molecules, and $1.5\ k_BT$ between 1 B and 1 S molecule. If there are no differences between the number of accessible microstates between well-mixed and demixed, what is the probability that a well-mixed solution of B will phase separate? Assume each B or S molecule must either bind to another B molecule or an S molecule.

Answer:

With no difference in the number of accessible microstates between well-mixed and demixed, this implies that here there is no entropy difference upon phase transition, and so the likelihood of phase transition occurring is determined solely by the net enthalpic differences. To determine the probability of a transition for any thermodynamic process, we use the Boltzmann factor since the probability of any transition occurring, which has a total free energy activity barrier of E, is proportional to the Boltzmann factor of $\exp(-E/k_BT)$ where k_B is the Boltzmann constant and T the absolute temperature. So, the total probability for phase separation occurring is given by the sum of all relevant Boltzmann factors for phase separation to occur, divided by the total sum of all possible Boltzmann factors (i.e. for all transitions for demixing of the biomolecules and the solvent molecules in phase separation, but also for those for well-mixed solvent with biomolecule)—for those acquainted with statistical physics, this sum is often referred to as the parameter Z, known as the canonical *partition function* (often omitting the word "canonical") and here in effect serves as a normalization constant to generate the probability. We are told that each B or S molecule must either bind to another B or S molecule. Thus, the possible combinations of molecule pairs are SS, BB, SB or BS. SS and BB are demixed, SB and BS are well-mixed. Interaction enthalpies here are all exothermic (attractive), so the associated energy barriers are all negative: −2.0, −1.5. −1.5. −3.0 k_BT. The phase transition probability therefore goes as:

$$\frac{\exp(-E_{SS}/k_BT) + \exp(E_{BB}/k_BT)}{\exp(-E_{SS}/k_BT) + \exp(-E_{BB}/k_BT) + \exp(-E_{SB}/k_BT) + \expt(E_{BS}/k_BT)}$$

After substituting in the values given, this probability comes out as ~0.75, or 75%. This sort of analysis doesn't of course give you any information about the spatial dependence of droplet formation. But it does illustrate that relatively small energy differences equivalent to the typical thermal energies of a few water molecules can make the difference between phase transition occurring or not. In other words, LLPS droplets are relatively unstable.

A minority of scientists consider viruses to be a minimally sized unit of life; the smallest known viruses having an effective diameter of ~20 nm (see Figure 2.1 for a typical virus image, as well as various cellular features). Viruses are indeed self-contained structures physically enclosing biomolecules. They consist of a protein coat called a "capsid" that encloses a simple viral genetic code of a nucleic acid (either of DNA or RNA depending on the virus type). However, viruses can only replicate by utilizing some of the extra genetic machinery of a *host*

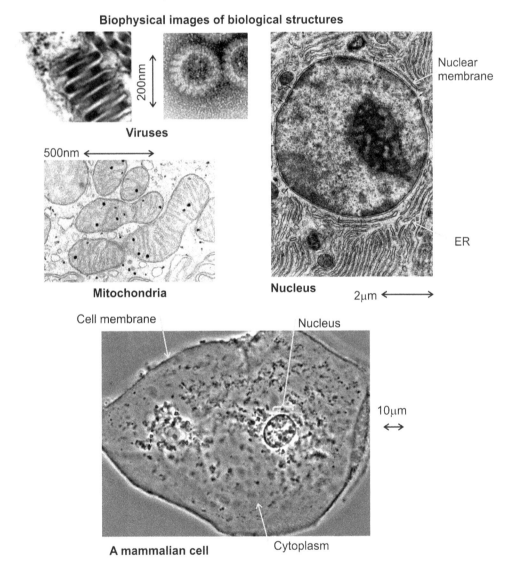

FIGURE 2.1 The architecture of biological structures. A range of typical cellular structures, in addition to viruses. (a) Rodlike maize mosaic viruses, (b) obtained using negative staining followed by transmission electron microscopy (TEM) (see Chapter 5); (c) mitochondria from guinea pig pancreas cells, (d) TEM of nucleus with endoplasmic reticulum (ER), (e) phase contrast image of a human cheek cell. (a: Adapted from Cell Image Library, University of California at San Diego, CIL:12417 c: Courtesy of G.E. Palade, CIL:37198; d: Courtesy of D. Fawcett, CIL:11045; e: Adapted from CIL:12594.)

cell that the virus infects. So, in other words, they do not fulfil the criterion of *independent* self-replication and cannot thus be considered a basic unit of life, by this semiarbitrary definition. However, as we have discussed in light of the selfish gene hypothesis, this is still very much an area of debate.

2.3 CHEMICALS THAT MAKE CELLS WORK

Several different types of molecules characterize living matter. The most important of these is undeniably water, but, beyond this, carbon compounds are essential. In this section, we discuss what these chemicals are.

2.3.1 IMPORTANCE OF CARBON

Several different atomic elements have important physical and chemical characteristics of biological molecules, but the most ubiquitous is carbon (*C*). Carbon atoms, belonging to Group IV of the *periodic table*, have a normal typical maximum *valency* of 4 (Figure 2.2a) but have the lowest *atomic number* of any Group IV element, which imparts not only a relative stability to carbon–carbon *covalent bonds* (i.e., bonds that involve the formation of dedicated molecular bonding electron orbitals) compared to other elements in that group such as silicon, which contain a greater number of protons in their nuclei with electrons occupying outer molecular orbitals more distant from the nucleus, but also an ability to form relatively long chained molecules, or to *catenate* (Figure 2.2b). This property confers a unique versatility in being able to form ultimately an enormous range of different molecular structures, which is therefore correlated to potential *biological functions*, since the structural properties of these carbon-based molecules affect their ability to stably interact, or not, with other carbon-based molecules, which ultimately is the primary basis of all biological *complexity* and *determinism* (i.e., whether or not some specific event, or set of events, is triggered in a living cell).

KEY POINT 2.6

Carbon chemistry permits complex catenated biological molecules to be made, which have intermediate chemical stability, that is, they are stable enough to perform biological functions but can also be relatively easily chemically modified to change their functional roles.

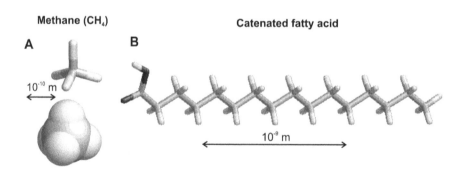

FIGURE 2.2   Carbon chemistry. (a) Rod and space-filling tetrahedral models for carbon atom bound to four hydrogen atoms in methane. (b) Chain of carbon atoms, here as palmitic acid, an essential fatty acid.

The general field of study of carbon compounds is known as "organic chemistry," to differentiate it from *inorganic chemistry* that involves noncarbon compounds, but also confusingly can include the study of the chemistry of pure carbon itself such as found in *graphite, graphene,* and *diamond. Biochemistry* is largely a subset or organic chemistry concerned primarily with carbon compounds occurring in biological matter (barring some inorganic exceptions of certain metal ions). An important characteristic of biochemical compounds is that although the catenated carbon chemistry confers stability, the bonds are still sufficiently labile to be modified in the living organism to generate different chemical compounds during the general process of *metabolism* (defined as the collection of all biochemical transformations in living organisms). This dynamic flexibility of chemistry is just as important as the relative chemical stability of catenated carbon for biology; in other words, this stability occupies an optimum regime for life.

The chemicals of life, which not only permit efficient functioning of living matter during the normal course of an organism's life but also facilitate its own ultimate replication into future generations of organisms through processes such as cellular growth, replication, and division can be subdivided usefully into types mainly along the lines of their chemical properties.

2.3.2 LIPIDS AND FATTY ACIDS

By chemically linking a small alcohol-type molecule called "glycerol" with a type of carbon-based acid that contain typically 20 carbon atoms, called "fatty acids," *fats*, also known as *lipids*, are formed, with each glycerol molecule in principle having up to three sites for available fatty acids to bind. In the cell, however, one or sometimes two of these three available binding sites are often occupied by an electrically polar molecule such as *choline* or similar and/or to charged *phosphate* groups, to form *phospholipids* (Figure 2.3a). These impart a key physical feature of being *amphiphilic*, which means possessing both *hydrophobic*, or water-repelling properties (through the fatty acid "tail"), and hydrophilic, or *water-attracting* properties (through the polar "head" groups of the choline and/or charged phosphate).

This property confers an ability for stable structures to form via *self-assembly* in which the head groups orientate to form electrostatic links to surrounding electrically polar water molecules, while the corresponding tail groups form a buried hydrophobic core. Such stable structures include at their simplest globular *micelles*, but more important biological structures can be formed if the phospholipids orient to form a *bilayer*, that is, where two layers of phospholipids form in effect as a mirror image sandwich in which the tails are at the sandwich center and the polar head groups on the outside above and below (Figure 2.3b). Phospholipid bilayers constitute the primary boundary structure to cells in that they confer an ability to stably compartmentalize biological matter within a liquid water phase, for example, to form spherical *vesicles* or *liposomes* (Figure 2.4a) inside cells. Importantly, they form smaller organelles inside the cells such as the cell nucleus, for exporting molecular components generated inside the cell to the outside world, and, most importantly, for forming the primary boundary structure around the outside of all known cells, of the *cell membrane*, which arguably is a larger length scale version of a liposome but including several additional nonphospholipid components (Figure 2.4b).

A phospholipid bilayer constitutes a large free energy barrier to the passage of a single molecule of water. Modeling the bilayer as a dielectric indicates that the electrical permittivity of the hydrophobic core is 5–10 times that of air, indicating that the free energy change, ΔG, per water molecule required to spontaneously translocate across the bilayer is equivalent to ~65 $k_B T$, one to two orders of magnitude above the characteristic thermal energy scale of the surrounding water solvent reservoir. This suggests a likelihood for the process to occur given by the *Boltzmann factor* of $\exp(-\Delta G/k_B T)$, or $\sim 10^{-28}$. Although gases such as oxygen, carbon dioxide, and nitrogen can diffuse in the phospholipid bilayer, it can be thought of as being practically *impermeable* to water. Water, and molecules solvated in

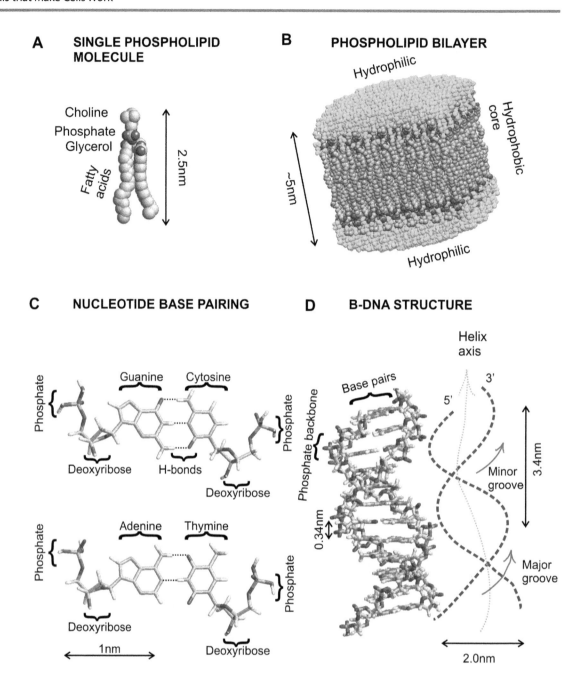

A SINGLE PHOSPHOLIPID MOLECULE

Choline
Phosphate
Glycerol
Fatty acids
2.5nm

B PHOSPHOLIPID BILAYER

Hydrophilic
Hydrophobic core
~5nm
Hydrophilic

C NUCLEOTIDE BASE PAIRING

Phosphate
Guanine Cytosine
Phosphate
Deoxyribose H-bonds
Deoxyribose

Phosphate
Adenine Thymine
Phosphate
Deoxyribose
Deoxyribose
1nm

D B-DNA STRUCTURE

Helix axis
Base pairs
3′
5′
Phosphate backbone
0.34nm
Minor groove
3.4nm
Major groove
2.0nm

FIGURE 2.3 Fats and nucleic acids. (a) Single phospholipid molecule. (b) Bilayer of phospholipids in water. (c) Hydrogen-bonded nucleotide base pairs. (d) B-DNA double-helical structure.

water, requires assistance to cross this barrier, through protein molecules integrated into the membrane.

Cells often have a heterogeneous mixture of different phospholipids in their membrane. Certain combinations of phospholipids can result in a phase transition behavior in which one type of phospholipid appears to pool together in small *microdomains* surrounded by a sea of another phospholipid type. These microdomains are often dynamic with a temperature-sensitive structure and have been referred to popularly as *lipid rafts*, with a range of effective diameters from tens to several hundred nanometers, and may have a biological relevance as transient zones of molecular confinement in the cell membrane.

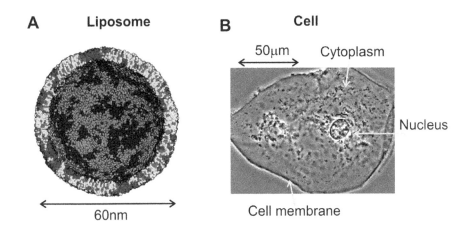

FIGURE 2.4 Structures formed from lipid bilayers. (a) Liposome, light and dark showing different phases of phospholipids from molecular dynamics simulation (see Chapter 8). (b) The cell membrane and nuclear membranes, from a human cheek cell taken using phase contrast microscopy (Chapter 3).

(a: Courtesy of M. Sansom; b: Courtesy of G. Wright, CIL:12594.)

KEY POINT 2.7

Phospholipid bilayers are ubiquitous self-assembled structures in cells that represent an enormous barrier for water molecules.

2.3.3 AMINO ACIDS, PEPTIDES, AND PROTEINS

Amino acids are the building blocks of larger important biological polymers called "peptides" or, if more than 50 amino acids are linked together, they are called "polypeptides" or, more commonly, "proteins." Amino acids consist of a central carbon atom from which is linked an amino (chemical base) group, $-NH_2$, a carboxyl (chemical acid) group, $-COOH$, a hydrogen atom $-H$, and one of 23 different side groups, denoted usually as $-R$ in diagrams of their structures (Figure 2.5a), which defines the specific type of amino acid. These 23 constitute the *natural* or *proteinogenic* amino acids, though it is possible to engineer artificial side groups to form unnatural amino acids, with a variety of different chemical groups, which have been utilized, for example, in bioengineering (see Chapter 9). Three of the natural amino acids are usually classed as nonstandard, on the basis of either being made only in bacteria and archaea, or appearing only in mitochondria and chloroplasts, or not directly being coded by the DNA, and so many biologists often refer to just 20 natural amino acids, and from these the mean number of atoms per amino acid is 19.

It should be noted that the α-carbon atom is described as *chiral*, indicating that the amino acid is *optically active* (this is an historical definition referring to the phenomenon that a solution of that substance will result in the rotation of the plane of polarization of incident light). The α-carbon atom is linked in general to four different chemical groups (barring the simplest amino acid *glycine* for which R is a hydrogen atom), which means that it is possible for the amino acid to exist in two different optical isomers, as mirror images of each other—a left-handed (*L*) and right-handed (*D*) isomers—with chemists often referring to optical isomers with the phrase *enantiomers*. This isomerism is important since the ability for other molecules to interact with any particular amino acid depends on its 3D structure and thus is specific to the optical isomer in question. By far, the majority of natural amino acids exist as L-isomers for reasons not currently resolved.

The natural amino acids can be subdivided into different categories depending upon a variety of physical and chemical properties. For example, a common categorization is *basic*

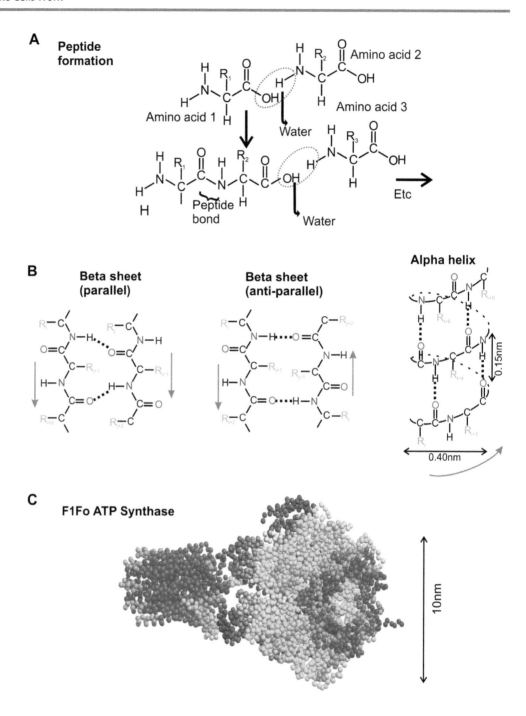

FIGURE 2.5   Peptide and proteins. (a) Formation of peptide bond between amino acids to form the primary structure. (b) Secondary structure formation via hydrogen bonding to form beta sheets and alpha helices. (c) Example of a complex 3D tertiary structure, here of an enzyme that makes ATP.

or *acidic* depending on the concentration of hydrogen H^+ ions when in water-based solution. The chemistry term *pH* refers to $-\log_{10}$ of the H^+ ion concentration, which is a measure of the acidity of a solution such that solutions having low values (0) are strong *acids*, those having high values (14) are strong *bases* (i.e., with a low acidity), and *neutral* solutions have a pH of exactly 7 (the average pH inside the cells of many living organism is around 7.2–7.4, though there can be significant localized deviations from this range).

Other broad categorizations can be done on the basis of overall electrical *charge* (positive, negative, neutral) at a neutral pH 7, or whether the side groups itself is electrically polar or not, and whether or not the amino acid is hydrophobic. There are also other structural features such as whether or not the side groups contain benzene-type ring structures (termed *aromatic* amino acids), or the side groups consist of chains of carbon atoms (*aliphatic* amino acids), or they are cyclic (the amino acid *proline*).

Of the 23 natural amino acids, all but two of them are encoded in the cell's DNA genetic code, with the remaining rarer two amino acids called "selenocysteine" and "pyrrolysine" being synthesized by other means. Clinicians and food scientists often make a distinction between *essential* and *nonessential* amino acids, such that the former group cannot be synthesized from scratch by a particular organism and so must be ingested in the diet.

Individual amino acids can link through a chemical reaction involving the loss of one molecule of water via their amino and carboxyl group to form a covalent *peptide bond*. The resulting *peptide* molecule obviously consists of two individual amino acid subunits, but still has a free $-NH_2$ and $-COOH$ at either end and is therefore able to link at each with other amino acids to form longer and longer peptides. When the number of amino acid subunits in the peptide reaches a semiarbitrary 50, then the resultant polymer is termed a "polypeptide or protein." Natural proteins have as few as 50 amino acids (e.g., the protein hormone *insulin* has 53), whereas the largest protein is found in muscle tissue and is called "titin," possessing 30,000 amino acids depending upon its specific type or *isomer*. The median number of amino acids per protein molecule, estimated from the known natural proteins, is around 350 for human cells. The specific sequence of amino acids for a given protein is termed as "primary structure."

Since free rotation is permissible around each individual peptide bond, a variety of potential *random coil* 3D protein conformations are possible, even for the smallest proteins. However, *hydrogen bonding* (or *H-bonding*) often results in the primary structure adopting specific favored generic conformations. Each peptide has two independent bond angles called "phi" and "psi," and each of these bond angles can be in one of approximately three stable conformations based on empirical data from known peptide sequences and stable phi and psi angle combinations, depicted in clusters of stability on a *Ramachandran plot*. Hydrogen bonding results from an electron of a relatively electronegative atom, typically either nitrogen $-N$ or oxygen $-O$, being shared with a nearby hydrogen atom whose single electron is already utilized in a bonding molecular orbital elsewhere. Thus, a bond can be formed whose length is only roughly twice as large as the effective diameter of a hydrogen atom (~0.2 nm), which, although not as strong a covalent bond, is still relatively stable over the 20°C–40°C temperatures of most living organisms.

As Figure 2.5b illustrates, two generic 3D *motif* conformations can result from the periodic hydrogen bonding between different sections of the same protein primary structure, one in which the primary structure of the two bound sections run in opposite directions, which is called a "β-strand," and the other in which the primary structure of the two bound sections run in the same direction, which results in a spiral-type conformation called an "α-helix." Each protein molecule can, in principle, be composed of a number of intermixed random coil regions, α-helices and β-strands, and the latter motif, since it results in a relatively planar conformation, can be manifest as several parallel strands bound together to form a *β-sheet*, though it is also possible for several β-strands to bond together in a curved conformation to form an enclosed *β-barrel* that is found in several proteins including, for example, *fluorescent proteins*, which will be discussed later (see Chapter 3). This collection of random coil regions, α-helices and β-strands, is called the protein's "secondary structure."

A further level of bonding can then occur between different regions of a protein's secondary structure, primarily through longer-range interactions of electronic orbitals between exposed surface features of the protein, known as *van der Waals* interactions. In addition, there may be other important forces that feature at this level of structural determination. These include hydrophobic/hydrophilic forces, resulting in the more hydrophobic amino acids being typically buried in the core of a protein's ultimate shape; *salt bridges*, which are a type of *ionic bond* that can form between nearby electrostatically polar groups in a protein of opposite charge (in proteins, these often occur between negatively charged, or *anionic*, amino acids of *aspartate* or *glutamate* and positively charged, or *cationic*, amino acids of *lysine* and *arginine*); and the so-called disulfide bonds (–S–S–) that can occur between two nearby *cysteine* amino acids, resulting in a covalent bond between them via two sulfur (–S) atoms. Cysteines are often found in the core of proteins stabilizing the structure. Living cells often contain *reducing agents* in aqueous solution, which are chemicals that can *reduce* (bind hydrogen to or remove oxygen from) chemical groups, including a disulfide bond that would be broken by being reduced back into two cysteine residues (this effect can be replicated in the test tube by adding artificial reducing agents such as dithiothreitol [DTT]). However, the *hydrophobic core* of proteins is often inaccessible to such chemicals. Additional nonsecondary structure hydrogen bonding effects also occur between sections of the same amino acids, which are separated by more than 10 amino acids.

These molecular forces all result in a 3D fine-tuning of the structure to form complex features that, importantly, define the shape and extent of a protein's structure that is actually exposed to external water-solvent molecules, that is, its *surface*. This is an important feature since it is the interface at which physical interactions with other biological molecules can occur. This 3D structure formed is known as the protein *tertiary structure* (Figure 2.5c). At this level, some biologists will also refer to a protein being *fibrous* (i.e., a bit like a rod), or *globular* (i.e., a bit like a sphere), but in general most proteins adopt real 3D conformations that are somewhere between these two extremes.

Different protein tertiary structures often bind together at their surface interfaces to form larger multimolecular *complexes* as part of their biological role. These either can be separate tertiary structures all formed from the same identical amino acid sequence (i.e., in effect identical *subunit* copies of each other) or can be formed from different amino acid sequences. There are several examples of both types in all domains of life, illustrating an important feature in regard to biological complexity. It is in general not the case that one simple protein from a single amino acid sequence takes part in a biological process, but more typically that several such polypeptides may interact together to facilitate a specific process in the cell. Good examples of this are the *modular architectures* of *molecular tracks* upon which *molecular motors* will translocate (e.g., the *actin* subunits forming *F-actin* filaments over which *myosin* molecular motors translocate in muscle) and also the protein *hemoglobin* found in the *red blood cells* that consists of four polypeptide chains with two different primary structures, resulting in two α-*chains* and two β-*chains*. This level of multimolecular binding of tertiary structures is called the "quaternary structure."

Proteins in general have a net electrical charge under physiological conditions, which is dependent on the pH of the surrounding solution. The pH at which the net charge of the protein is zero is known as its *isoelectric point*. Similarly, each separate amino acid residue has its own isoelectric point.

Proteins account for 20% of a cell by mass and are critically important. Two broad types of proteins stand out as being far more important biologically than the rest. The first belongs to a class referred to as *enzymes*. An enzyme is essentially a *biological catalyst*. Any catalyst functions to lower the effective *free energy barrier* (or *activation barrier*) of a chemical reaction and in doing so can dramatically increase the rate at which that reaction proceeds. That is the simple description, but this hides the highly complex detail of how this is actually achieved in practice, which is often through a very complicated series of intermediate reactions, resulting in the case of biological catalysts from the underlying molecular heterogeneity of enzymes, and may also involve quantum tunneling effects (see Chapter 9). Enzymes, like all catalysts, are not consumed *per se* as a result of their activities and so can function efficiently at very low cellular concentrations. However, without the action of enzymes, most chemical reactions in a living cell would not occur spontaneously to any degree of efficiency over the time scale of a cell's lifetime. Therefore, enzymes are essential to life. (Note that although by far the majority of biological catalysts are proteins, another class of *catalytic RNA* called "ribozymes" does exist.) Enzymes in general are named broadly after the biological process they primarily catalyze, with the addition of "ase" on the end of the word.

The second key class of proteins is known as *molecular machines*. The key physical characteristic of any general machine is that of transforming energy from one form into some type of *mechanical work*, which logically must come about by changing the force vector (either in size or direction) in some way. Molecular machines in the context of living organisms usually take an input energy source from the controlled breaking of high-energy chemical bonds, which in turn is coupled to an increase in the local thermal energy of surrounding water molecules in the vicinity of that chemical reaction, and it is these thermal energy fluctuations of water molecules that ultimately power the molecular machines.

Many enzymes act in this way and so are also molecular machines; however, at the level of the energy input being most typically due to thermal fluctuations from the water solvent, one might argue that all enzymes are types of molecular machines. Other less common forms of energy input are also exhibited in some molecular machines, for example, the absorption of photons of light can induce mechanical changes in some molecules, such as the protein complex called "rhodopsin," which is found in the retina of eyes.

There are several online resources available to investigate protein structures. One of these includes the Protein Data Bank (www.pdb.org); this is a data repository for the spatial coordinates of atoms of measured structures of proteins (and also some biomolecule types such as nucleic acids) acquired using a range of structural biology tools (see Chapter 5). There are also various biomolecule structure software visualization and analysis packages available. In addition, there are several bioinformatics tools that can be used to investigate protein structures (see Chapter 8), for example, to probe for the appearance of the same sequence repeated in different sets of proteins or to predict secondary structures from the primary sequences.

2.3.4 SUGARS

Sugars are more technically called "carbohydrates" (for historical reasons, since they have a general chemical formula that appears to consist of water molecules combined with carbon atoms), with the simplest natural sugar subunits being called "monosaccharides" (including sugars such as *glucose* and *fructose*) that mostly have between three and seven carbon atoms per molecule (though there are some exceptions that can have up to nine carbon atoms) and can in principle exist either as chains or in a conformation in which the ends of the chain link to each other to form a *cyclic molecule*. In the water environment of living cells, by far the majority of such monosaccharide molecules are in the cyclic form.

Two monosaccharide molecules can link to each other through a chemical reaction, similar to the way in which a peptide bond is formed between amino acids by involving the loss of a molecule of water, but here it is termed as *glycosidic bond*, to form a *disaccharide* (Figure 2.6a). This includes sugars such as *maltose* (two molecules of glucose linked together) and *sucrose* (also known as *table sugar*, the type you might put in your tea, formed from linking one molecule of glucose and one of fructose).

All sugars contain at least one carbon atom which is chiral, and therefore can exist as two optical isomers; however, the majority of natural sugars exist (confusingly, when compared with amino acids) as the –*D* form. Larger chains (Figure 2.6b) can form from more linkages to multiple monosaccharides to form polymers such as *cellulose* (a key structural component of plant cell walls), *glycogen* (an *energy storage molecule* found mainly in muscle and the liver), and starch.

KEY POINT 2.8

Most sugar molecules are composed of *D*-optical isomers, compared to most natural amino acids that are composed of *L*-optical isomers.

These three examples of *polysaccharides* happen all to be comprised of glucose monosaccharide subunits; however, they are all structurally different from each other, again illustrating how subtle differences in small features of individual subunits can be manifest as big differences as *emergent properties* of larger length scale structures. When glucose molecules

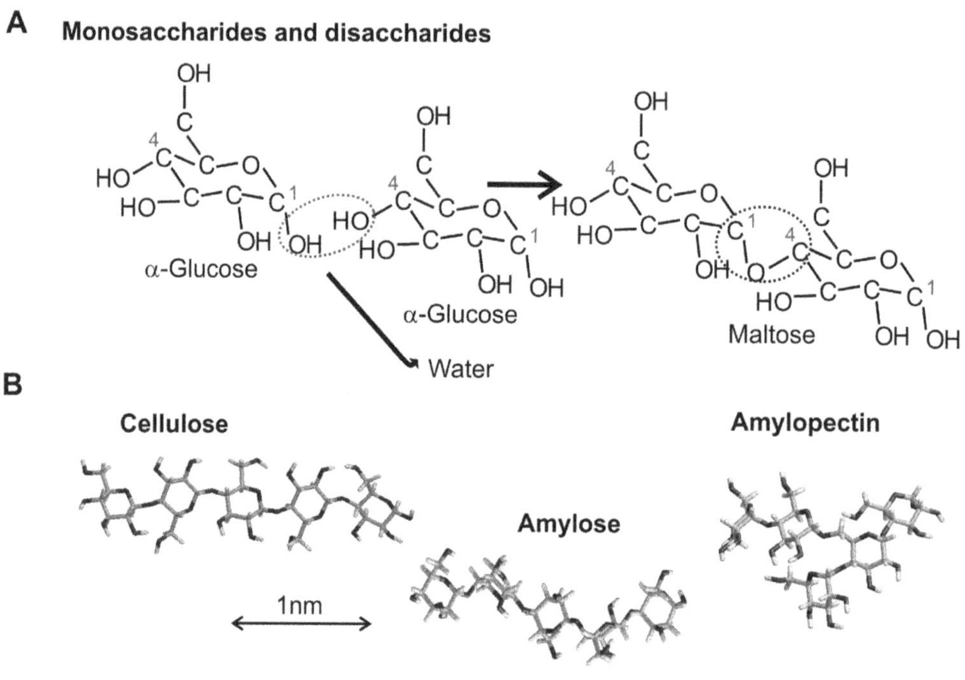

A   **Monosaccharides and disaccharides**

B

FIGURE 2.6   Sugars. (a) Formation of larger sugars from monomer units of monosaccharide molecules via loss of water molecule to form a disaccharide molecule. (b) Examples of polysaccharide molecules.

bind together, they can do so through one of two possible places in the molecule. These are described as either $1 \rightarrow 4$ or $1 \rightarrow 6$, referring to the numbering of the six carbon atoms in the glucose molecule.

In addition, the chemical groups that link to the glycosidic bond itself are in general different, and so it is again possible to have two possible *stereoisomers* (a chemistry term simply describing something that has the same chemical formula but different potential spatial arrangements of the constituent atoms), which are described as either α or β; cellulose is a linear chain structure linked through β($1 \rightarrow 4$) glycosidic bonds containing as few as 100 and as high as a few thousand glucose subunits; starch is actually a mixture of two types of polysaccharide called "amylose" linked through mainly α($1 \rightarrow 4$) glycosidic bonds, and *amylopectin* that contains α($1 \rightarrow 4$) and as well as several α($1 \rightarrow 6$) links resulting in branching of the structure; glycogen molecules are primarily linked through α($1 \rightarrow 4$), but roughly for every 10 glucose subunits, there is an additional link of α($1 \rightarrow 6$), which results in significant branching structure.

2.3.5   NUCLEIC ACIDS

Nucleic acids include molecules such as *DNA* and various forms of *RNA*. These are large polymers composed of repeating subunits called "nucleotide bases" (Figure 2.3c) characterized by having a *nucleoside* component, which is a cyclic molecule containing nitrogen as well as carbon in a ringed structure, bound to a five-carbon-atom monosaccharide called either "ribose," in the case of RNA, or a modified form of ribose lacking a specific oxygen atom called "deoxyribose," in the case of DNA, in addition to bound phosphate groups. For DNA, the nucleotide subunits consist of either *adenine* (A) or *guanine* (G), which are based on a chemical structure known as *purines*, and *cytosine* (C) or *thymine* (T), which are based on a smaller chemical structure known as *pyrimidines*, whereas for RNA, the thymine is replaced by *uracil* (U).

The nucleotide subunits can link to each other in two places, defined by the numbered positions of the carbon atoms in the structure, in either the 3′ or the 5′ position (Figure 2.3d), via a *nucleosidic bond*, again involving the loss of a molecule of water, which still permit further linking of additional nucleotides from both the end 3′ and 5′ positions that were not utilized in internucleotide binding, which can thus be subsequently repeated for adding more subunits. In this way, a chain consisting of a potentially very long sequence of nucleotides can be generated; natural DNA molecules in live cells can have a contour length of several microns.

DNA strands have an ability to stably bind via *base pair interactions* (also known as *Watson–Crick base pairing*) to another *complementary strand* of DNA. Here, the individual nucleotides can form stable multiple hydrogen bonds to nucleotides in the complementary strand due to the tessellating nature of either the C–G (three internucleotide H-bonds) or A–T (two internucleotide H-bonds) structures, generating a double-helical structure such that the H-bonds of the base pairs span the axial core of the double helix, while the negatively charged phosphate groups protrude away from the axis on the outside of the double helix, thus providing additional stability through minimization of electrostatic repulsion.

This base pairing is utilized in DNA replication and in reading out of the genetic code stored in the DNA molecule to make proteins. In DNA replication, errors can occur spontaneously from base pairing mismatch for which noncomplementary nucleotide bases are paired, but there are *error-checking machines* that can detect a substantial proposal of these errors during replication and correct them. Single-stranded DNA can exist, but in the living cell, this is normally a transient state that is either stabilized by the binding of specific proteins or will rapidly base pair with a strand having a complementary nucleotide sequence.

Other interactions can occur above and below the planes of the nucleotide bases due to the overlap of delocalized electron orbitals from the nucleotide rings, called "stacking interactions," which may result in heterogeneity in the DNA helical structures that are dependent upon both the nucleotide sequence and the local physical chemistry environment, which may result in different likelihood values for specific DNA structures than the base pairing interactions along might suggest. For the majority of time under normal conditions inside the cell, DNA will adopt a *right-handed* helical conformation (if the thumb of your right hand was aligned with the helix axis and your relaxed, index finger of that hand would follow the grooves of the helix as they rotate around the axis) called "B-DNA" (Figure 2.3d), whose *helical width* is 2.0 nm and *helical pitch* is 3.4 nm consisting of a mean of 10.5 base pair turns. Other stable helical conformations exist including *A-DNA*, which has a smaller helical *pitch* and wider *width* than B-DNA, as well as *Z-DNA*, which is a stable *left-handed* double helix. In addition, more complex structures can form through base pairing of multiple strands, including *triple-helix* structures and *Holliday junctions* in which four individual strands may be involved.

The importance of the *phosphate backbone* of DNA, that is, the helical lines of phosphate groups that protrude away from the central DNA helix to the outside, should not be underestimated, however. A close inspection of native DNA phosphate backbones indicate that this repeating negative charge is not only used by certain enzymes to recognize specific parts of DNA to bind to but perhaps more importantly is essential for the structural stability of the double helix. For example, replacing the phosphate groups chemically using noncharged groups results in significant structural instability for any DNA segment longer than 100 nucleotide base pairs. Therefore, although the Watson–Crick base pair model includes no role for the phosphate background in DNA, it is just as essential.

KEY POINT 2.9

Although DNA can adopt stable double-helical structures by virtue of base pairing, several different double-helical structures exist, and DNA may also adopt more complex nonhelical structures.

The genetic code is composed of DNA that is packaged into functional units called "genes." Each gene in essence has a DNA sequence that can be read out to manufacture a specific type of peptide or protein. The total collection of all genes in a given cell in an organism is in general the same across different tissues in the organism (though note that some genes may have altered functions due to local environmental nongenetic factors called "epigenetic modifications") and referred to as the *genome*. Genes are marked out by start (*promoter*) and end points (*stop codon*) in the DNA sequence, though some DNA sequences that appear to have such start and end points do not actually code for a protein under normal circumstances. Often, there will be a cluster of genes between a promoter and stop codon, which all get read out during the same gene *expression burst*, and this gene cluster is called an "operon."

This presence of large amounts of noncoding DNA has accounted for a gradual decrease in the experimental estimates for the number of genes in the *human genome*, for example, which initially suggested 25,000 genes has now, at the time of writing, been revised to more like 19,000. These genes in the human genome consist of 3×10^9 individual base pairs from each parent. Note, the *proteome*, which is the collection of a number of different proteins in an organism, for humans is estimated as being in the range $(0.25-1) \times 10^6$, much higher than the number of genes in the genome due to *posttranscriptional modification*.

KEY POINT 2.10

Genes are made from DNA, which code for proteins. The genome is the collection of all individual genes in a given organism.

DNA also exhibits higher-order structural features, in that the double helix can stably form coils on itself, or the so-called supercoils, in much the same way as the cord of a telephone handset can coil up. In nonsupercoiled, or relaxed B-DNA, the two strands twist around the helical axis about once every 10.5 base pairs. Adding or subtracting twists imposes strain, for example, a circular segment of DNA as found in bacteria especially might adopt a figure-of-eight conformation instead of being a relaxed circle. The two lobes of the figure-of-eight conformation are either clockwise or counterclockwise rotated with respect to each other depending on whether the DNA is positively (overwound) or negatively (underwound) supercoiled, respectively. For each additional helical twist being accommodated, the lobes will show one more rotation about their axis.

In living cells, DNA is normally negatively supercoiled. However, during DNA replication and transcription (which is when the DNA code is read out to make proteins, discussed later in this chapter), positive supercoils may build up, which, if unresolved, would prevent these essential processes from proceeding. These positive supercoils can be relaxed by special enzymes called "topoisomerases."

Supercoils have been shown to propagate along up to several thousand nucleotide base pairs of the DNA and can affect whether a gene is switched on or off. Thus, it may be the case that mechanical signals can affect whether or not proteins are manufactured from specific genes at any point in time. DNA is ultimately compacted by a variety of proteins; in eukaryotes these are called "histones," to generate higher-order structures called "chromosomes." For example, humans normally have 23 pairs of different chromosomes in each nucleus, with each member of the pair coming from a maternal and paternal source. The paired collection of chromosomes is called the "diploid" set, whereas the set coming from either parent on its own is the *haploid* set.

Note that bacteria, in addition to some archaea and eukaryotes, can also contain several copies of small enclosed circles of DNA known as *plasmids*. These are separated from the main chromosomal DNA. They are important biologically since they often carry genes that benefit the survival of the cell, for example, genes that confer resistance against certain *antibiotics*. Plasmids are also technologically invaluable in molecular cloning techniques (discussed in Chapter 7).

It is also worth noting here that there are nonbiological applications of DNA. For example, in Chapter 9, we will discuss the use of *DNA origami*. This is an engineering nanotechnology that uses the stiff properties of DNA over short (ca. nanometer distances) (see Section 8.3) combined with the smart design principles offered by Watson–Crick base pairing to generate *artificial* DNA-based nanostructures that have several potential applications.

RNA consists of several different forms. Unlike DNA, it is not constrained solely as a double-helical structure but can adopt more complex and varied structural forms. *Messenger RNA (mRNA)* is normally present as a single-stranded polymer chain of typical length of a few thousand nucleotides but potentially may be as high as 100,000. Base pairing can also occur in RNA, rarely involving two complementary strands in the so-called RNA duplex double helices, but more commonly involving base pairing between different regions of the same RNA strand, resulting in complex structures. These are often manifested as a short motif section of an *RNA hairpin*, also known as a *stem loop*, consisting of base pair interactions between regions of the same RNA strand, resulting in a short double-stranded stem terminated by a single-stranded RNA loop of typically 4–8 nucleotides. This motif is found in several RNA secondary structures, for example, in *transfer RNA (tRNA)*, there are three such stem loops and a central double-stranded stem that result in a complex characteristic *clover leaf* 3D conformation. Similarly, another complex and essential 3D structure includes *rRNA*. Both tRNA and rRNA are used in the process of reading and converting the DNA genetic code into protein molecules.

One of the subunits of rRNA (the *light subunit*) has catalytic properties and is an example of an RNA-based enzyme or *ribozyme*. This particular ribozyme is called "peptidyl transferase" that is utilized in linking together amino acids during protein synthesis. Some ribozymes have also been demonstrated to have self-replicating capability, supporting the *RNA world hypothesis*, which proposes that RNA molecules that could self-replicate were in fact the precursors to life forms known today, which ultimately rely on nucleic acid-based replication.

2.3.6 WATER AND IONS

The most important chemical to life as we know is, undeniably, water. Water is essential in acting as the *universal biological solvent*, but, as discussed at the start of this chapter, it is also required for its thermal properties, since the thermal fluctuations of the water molecules surrounding molecular machines fundamentally drive essential molecular conformational changes required as part of their biological role. A variety of electrically charged *inorganic ions* are also essential for living cells in relative abundance, for purposes of electrical and pH regulation, and also as being utilized as in cellular signaling and for structural stability, including sodium (Na^+), potassium (K^+), hydrogen carbonate (HCO_3^-), calcium (Ca^{2+}), magnesium (Mg^{2+}), chloride (Cl^-), and water-solvated protons (H^+) present as *hydronium* (or *hydroxonium*) ions (H_3O^+).

KEY POINT 2.11

To maintain the pH of a solution requires a chemical called a "pH buffer." The cell contains many natural pH buffers, but in test tube experiments, we can use a range of artificial buffers at low millimolar concentrations that can constrain the pH of a solution to within a specific narrow range. For example, a simple pH buffer used is "phosphate-buffered saline," which contains phosphate ions, plus additional sodium ions, which can maintain the pH over a range of 6–8.

Other transition metals are also utilized in a variety of protein structures, such as zinc (Zn, e.g., present in a common structural motif involving in protein binding called a "zinc finger motif") as well as iron (Fe, e.g., located at the center of hemoglobin protein molecules used to bind oxygen in the blood). There are also several essential enzymes that utilize higher atomic number transition metal atoms in their structure, required in comparatively small quantities in the human diet but still vital.

2.3.7 SMALL ORGANIC MOLECULES OF MISCELLANEOUS FUNCTION

Several other comparatively small chemical structures also perform important biological functions. These include a variety of *vitamins*; they are essential small organic molecules that for humans are often required to be ingested in the diet as they cannot be synthesized by the body; however, some such vitamins can actually be synthesized by bacteria that reside in the guts of mammals. A good example is *E. coli* bacteria that excrete *vitamin K* that is absorbed by our guts; the living world has many such examples of two organisms benefiting from a mutual *symbiosis*, *E. coli* in this case benefiting from a relatively stable and efficacious external environment that includes a constant supply of nutrients.

There are also *hormones*; these are molecules used in signaling between different tissues in a complex organism and are often produced by specialized tissues to trigger emergent behavior elsewhere in the body. There are *steroids* and *sterols* (which are steroids with alcohol chemical groups), the most important perhaps being *cholesterol*, which gets a bad press in that its excess in the body lead to a well-reported dangerous narrowing of blood vessels, but which is actually an essential stabilizing component of the eukaryote cell membrane.

There are also the so-called neurotransmitters such as *acetylcholine* that are used to convey signals between the junctions of nerve cells known as *synapses*. Nucleoside molecules are also very important in cells, since they contain highly energetic phosphate bonds that release energy upon being chemically split by water (a process known as *hydrolysis*); the most important of these molecules is *adenosine triphosphate* that acts as the *universal cellular fuel*.

Worked Case Example 2.2: DNA "Information" Storage

The human haploid genome contains ca. 3000 million DNA nucleotide base pairs.

a   *What are the possible stable DNA base pairs? How many raw bits of information are there in a single DNA base pair?*
b   *Is one Blue-ray disk sufficient to store the human genome information from a typical family? (A typical Blue-ray dual-layer disk, of the type you might watch a movie on at home, has a storage capacity of 50 GB. A recent U.S. census figure suggests the average household contains 2.58 people.)*
c   *If a hypothetical storage device of similar volume to a USB hard drive could be made using DNA is in its B form to store information assuming that storage units of B-DNA are tightly packed cylinders whose diameter and height are equal to the double-helical width and pitch, respectively, what % more data could it save compared to a typical high capacity 2 TB USB hard drive? (A typical 2 TB USB hard drive has, at the time of writing, dimensions ca. 12 cm × 8 cm × 2 cm.)*
d   *What is the maximum number of copies of a single complete human genome that could fit into a cell nucleus of diameter 10 μm if it were structured in similar storage units? Comment on the result.*

Answers:

a   Each base pair (AT or TA, CG or GC) can have a total of 2 bits of information since they result in 2^2, that is, 4 possible combinations.
b   1 byte (B) contains 8 bits; therefore, each base pair contains 0.25 bytes of information. 1 kB = 2^{10} B, 1 MB = 2^{10} kB, 1 GB = 2^{10} MB = 2^{30} B ≈ 1.1×10^9 B.

Therefore, the whole human haploid genome contains $(0.25 \times 3 \times 10^9)/(1.1 \times 10^9)$ GB = 0.68 GB. However, a normal human genome is diploid; thus, the information stored is 2×0.68 GB ≈ 1.4 GB.

One Blue-ray disk thus can hold 50/1.4 = 3.6 genomes or 3 complete genomes. The "average" family consists of 2.58 people, and so the storage capacity of 1 Blue-ray disk is sufficient.

c   One complete "storage unit" is a turn of a single double-helical pitch of B-DNA involving 10.5 base pairs enclosed within a cylinder of diameter 2.0 nm and length 3.4 nm.

The number of bytes of information in 1 turn $\approx 10.5 \times 0.25 = 2.62$ B.

The volume of the cylinder enclosing one double-helical turn $= \pi \times (2.0 \times 10^{-9} \times 0.5)^2 \times (3.4 \times 10^{-9}) = 1.1 \times 10^{-26}$ m^3.

Thus, the storage capacity density of B-DNA $s_D = 2.62/(1.1 \times 10^{-26}) = 2.5 \times 10^{26}$ B m^{-3}.

For a 2 TB USB hard drive, 2 TB $= 2 \times 2^{10}$ GB $= 2^{41}$ B $= 2.2 \times 10^{12}$ B.

The volume of the hard drive $= 12 \times 8 \times 2$ cm^3 = 192 cm^3 = 1.92×10^{-4} m^3.

Thus, storage capacity density of the hard drive $s_H = (2.2 \times 10^{12})/(1.92 \times 10^{-4}) = 1.1 \times 10^{16}$ B m^{-3}.

The percentage value of more data on the DNA device $= s_D/s_H = 100 \times (2.5 \times 10^{26})/(1.1 \times 10^{16}) = 2.2 \times 10^{12}\%$

d   The volume of the nucleus $V_n = 4 \times \pi \times (10 \times 10^{-6} \times 0.5)^3/3 = 1.3 \times 10^{-16}$ m^3.

The volume of the DNA "cylinder" per base pair $= (1.1 \times 10^{-26})/10.5 = 1.0 \times 10^{-25}$ m^3.

Thus, the volume of the whole diploid genome $V_g = 2 \times (3 \times 10^9) \times (1.0 \times 10^{-25}) = 6.0 \times 10^{-16}$ m^3.

Therefore, the number of complete diploid genomes that could be stored in the nucleus $= V_n/V_g = (1.3 \times 10^{-16})/(6.0 \times 10^{-16}) = 0.22$. Since this value is <1, this suggests that either DNA is present as a more compact form than B-DNA or (more likely) DNA inside the nucleus is compacted to a greater extent than the simple cylinder-based storage unit suggests (e.g., using histone proteins).

2.4 CELL PROCESSES

Cells can regulate their behavior, or *phenotype*, by ultimately controlling the number of protein molecules of different types that are present inside the cell at any one time. This is important since cells inside an organism may all have the same ultimate set of genes made from the DNA inside each of their cell nuclei but may need to perform very diverse roles in the organism. For example, in the human body there are roughly 200 different types of cells, as classified by biological experts, cells that will have different sizes and shapes and have catered biochemical and mechanical properties to be specialized in specific parts of the body, such as in the nerves, bones, muscles, skin, and blood.

KEY POINT 2.12

Cells in an organism in general have the same set of genes but may be many different cell types that are specialized for particular biological roles.

Most of the very smallest cells belong to the archaea domain in a *genus* subdivision also called *Mycoplasma*, found commonly in soil, which are roughly 200 nm in diameter, very close to the theoretical minimum size predicted on the basis of estimating the length of DNA genetic code in principle required to generate the very barest essential components necessary for a cell to replicate itself and thus be "alive" and using the polymer physics properties of DNA to predict its typical end-to-end distance. Mycoplasma "ghost" cell membranes (cells minus their native DNA genetic material) were also used in generating the first self-replicating *artificial cell* (Gibson et al., 2010). The longest cells known are *nerve cells*, which in some animals can be several meters in length.

The way that the number, or ultimately the concentration, of each type of protein molecule in a cell is controlled is through dynamic fine-tuning of the rate of production of proteins and the rate at which they are removed, or degraded, from the cell. There are mechanisms to controllably degrade proteins in cells, for example, eukaryotes have a mechanism of tagging proteins (with another protein called "ubiquitin"), leading to their being ultimately captured inside subcellular organelles and subsequently degraded by the action of the so-called proteolytic enzymes, with other similar mechanisms existing for prokaryotes but with the absence of dedicated subcellular organelles. However, the most control that is imparted by cells for regulating the equilibrium concentration of cellular proteins is through the direct regulation of the rate at which they are manufactured by the cell from the genes. The fine-tuning of the rate of production of proteins in a cell is done through a process called gene regulation, and to understand how this is achieved, we must explore the concept of the *central dogma of molecular biology*.

2.4.1 CENTRAL DOGMA OF MOLECULAR BIOLOGY

For reasons that arguably are more metaphysical than scientific, the process, which is considered by many expert biologists to be the most important of all biology, which governs *how the DNA genetic code is ultimately read out and transformed into different proteins*, is referred to as a *central dogma* as opposed to a *law*. Either way, the process itself is ubiquitous across all domains of life, and essential, summarized in its simplest form in Figure 2.7. In essence, the following applies:

1 The *genetic code* of each cell is encapsulated in its DNA, into a series of *genes*.
2 *Genes* can be *transcribed* by molecular machinery to generate molecules of mRNA.
3 *mRNA* molecules can be *translated* by other molecular machinery involving the binding of molecules of *tRNA* to the mRNA to generate peptides and proteins.

This is an enormous simplification of what is a very complex process requiring the efficient coordination of multiple different molecular machine components. The principal *flow of information* from the genes incorporated into DNA molecules to the rest of the organism is through the route DNA → mRNA → protein. The proteins that are then generated can

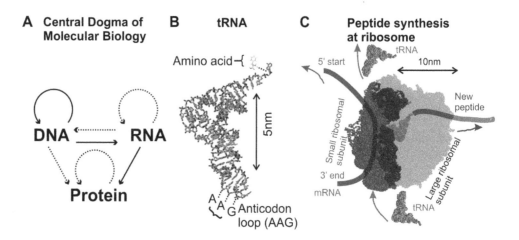

FIGURE 2.7 Central dogma of molecular biology, (a) Schematic of the flow of information between nucleic acids and proteins. (b) Structure of tRNA. (c) Interaction of tRNA with ribosome during peptide manufacture.

feature in, most importantly, various different enzymes that potentially catalyze thousands of different biochemical reactions in thousands of biological processes in an organism, as well as a vast range of molecular machines that drive a variety of energy-dependent systems inside cells, not to mention an enormous range of essential structural cellular components as well as those involved in the detection of chemical signals both inside and outside the cell.

As Figure 2.7 suggests, there are other mechanisms for information to flow from, and to, nucleic acids, as well as directly from protein to protein. For example, *DNA replication*, an essential process, which ultimately allows daughter cells from newly divided cells to receive a copy of the parental cell's genetic code, involves DNA → DNA information flow. Protein → protein information flow can occur through the generation of *prions;* peptide-based self-replicating structures requiring no direct transfer of information from nucleic acids, which when incorrectly folded, are implicated in various pathologies of the brain including *Creutzfeldt–Jakob disease*, more commonly referred to by its equivalent disorder in cattle of *mad cow's disease*. Note also that there is evidence that correctly folded prions may also have a functional role in information flow. For example, certain damaged nerve cells appear to cleave correctly folded prion molecules whose fragments then act as a signal to neighboring cells called "Schwann cells," which stimulates them to repair the damaged nerve cell by manufacturing an increased amount of a substance called the "myelin sheath," which is a fatty-based dielectric that acts as an electrical insulator around nerve cells.

RNA → DNA information flow can occur through an enzyme called "reverse transcriptase," which is utilized by some types of *viruses* called "retroviruses" that store their genetic material in the form of RNA but then use reverse transcriptase to convert it to DNA prior to integrating this into the DNA of a host-infected cell (a well-known example is the *human immunodeficiency virus* [*HIV*]). RNA → RNA information flow can also occur through a direct replication of RNA from an RNA template using another viral enzyme called "RNA replicase" (studied most extensively in the *polio virus*).

The key stages of the principal information flow route of DNA → mRNA → protein for the central dogma are as follows:

1   A molecular machine enzyme called "RNA polymerase" (*RNAP*) binds to a specific region of the DNA at the start of a particular gene, called the promoter, whose binding core contains a common nucleotide sequence that is present in all domains of life of 5′-TATAAA-3′ and is also known as the *TATA box*. A series of proteins called "transcription factors" (*TFs*) can also compete for binding of the RNAP through specific binding to the particular sequence of a given gene's promoter region and in doing so can specifically inhibit the binding of the RNAP in the promoter region of that gene. This is thought to be the primary way in which the *expression* of proteins and peptides from genes, that is, whether or not a gene is switched on, is regulated, in that if a TF is bound to the promoter region, then the gene will not *express* any protein, and so is *switched off*, whereas in the absence of any bound TF, the gene is *switched on*. Expression from a single gene is thus *stochastic* and occurs in bursts of activity.

2   The RNAP is a good example of a multicomponent enzyme. One component is responsible for first unwinding the double helix in the vicinity of the RNAP.

3   The RNAP then moves along one of the single strands of DNA specifically in the 3′–5′ direction; this process in itself is highly complex and far from completely understood but is known from a variety of single-molecule experiments performed in a test tube environment (i.e., *in vitro* techniques) to require a chemical energy input from the hydrolysis of ATP, resulting in molecular conformational changes to the RNAP that fuel its movement along the DNA. The *transcription speed* along the DNA varies typically from 20 to 90 nucleotides per second (though note that some viruses can adapt the cell's RNAP to increase its effective speed of transcription by a factor of 20).

4   As the RNAP moves along the single strand of DNA, each nucleotide base of the DNA is copied by generating a complementary strand of mRNA.

5   Once the RNAP reaches a special stop signal in the DNA code, the copying is stopped and the completed mRNA is released from the RNAP.

6   In the case of eukaryotic cells, the mRNA molecule first diffuses out of the nucleus through specialized nanoscale holes in the nuclear membrane called "nucleopores" and can then be modified by enzyme-medicated *splicing* reactions called posttranscriptional modifications that can result in significant variability from the original mRNA molecule manifested as sequence differences in the proteins or peptides that are ultimately generated. Each eukaryotic gene, in general, consists of *coding* DNA regions called "exons" interspersed with noncoding regions called "introns," and splicing of the mRNA involves in effect differential shuffling and relinking of the equivalent exon regions in the mRNA by a complex molecular machine called the "spliceosome." In prokaryotes, there are no introns and no established mechanisms for posttranscriptional modifications.

7   The mRNA molecule, whether modified or not, ultimately then binds to a ribosome in the cytoplasm. Ribosomes are structures roughly 20 nm in average diameter, composed of a mixture of RNA called "rRNA" and proteins. Once bound to a ribosome, the mRNA molecule is *translated* into a peptide, or protein, sequence, at a typical rate of 8 amino acids per second in eukaryotes, and more likely twice this in prokaryotes.

The mRNA is actually read off in chunks of three consecutive bases, called a "codon." In principle, this equates to 4^3, or 64, possible combinations; however, since there are only 20 natural amino acids, there is *degeneracy*, in that multiple codons may code for the same amino acids (typically between two and six codons exist per amino acid mainly by variation in the third base pair, but two amino acids of methionine in eukaryotes, or formylmethionine in bacteria, and tryptophan are specified by just a single codon). The mRNA sequence for methionine/formylmethionine is denoted AUG, since it consists of adenine, uracil, and guanine, and is a special case since it acts as the start codon in most organisms. Similarly, there are also stop codons (UAA, UAG, and UGA), which consist of other combinations of these three nucleotide bases, also called "nonsense codons or termination codons," which do not code for an amino acid but terminate mRNA translation. The region between the start and the nearest upstream stop codon is called the "open reading frame," which generally, but not always, codes for a protein, and in which case is called a "gene."

Each tRNA molecule acts as an *adapter*, in that there is a specific tRNA that is attached to each a specific amino acid (Figure 2.7b). Each tRNA molecule then binds via an anticodon binding site to the appropriate codon on the mRNA bound to a ribosome (Figure 2.7c). The general structure of the ribosome consists of a large and small subunit stabilized by base pairing between rRNA nucleotides, which assembles onto the start sequence of each mRNA molecule, sandwiching it together.

The sandwich acts as the site of translation, such that tRNA molecules bind transiently to each active codon and fuse their attached amino acid residues with the nearest upstream amino acid coded by the previous mRNA codon, with the site of active translation on the mRNA shunted forward by moving the mRNA molecule through the ribosome by another codon unit in a process that is energized by the hydrolysis of the molecule GTP (similar to ATP). Multiple ribosomes may bind to the same mRNA molecule to form a *polysome* (also known as a *polyribosome*) cluster that can manufacture copies of the same protein from just a single mRNA template each individual ribosome outputting proteins in parallel.

2.4.2 DETECTION OF SIGNALS

Cells can detect the presence of external chemicals with remarkable specificity and efficiency. The typical way this is achieved is through a highly specific *receptor* that is integrated into the cell membrane composed mainly of protein subunits. The unique 3D spatial conformation adopted by the receptor can allow specific binding of a *ligand* molecule if it has a conformation that can efficiently fit into the 3D binding site of the receptor; biologists sometimes describe this as a *lock-and-key* mechanism and is also the way that enzymes are believed to operate on binding to intermediate structures in a biochemical reaction. This correct binding can then trigger subsequent chemical events inside the cell.

The exact mechanism for achieving this is not fully understood but is likely to involve some conformational change to the receptor upon ligand binding. This conversion of the original external chemical signal to inner cellular chemical events is an example of *signal transduction*. These inner chemical events can then trigger other biological processes and so in effect represents a means of *flowing information* from the extracellular environment to the inside of the cell. There is scope for similar-shaped molecules outside the cell to compete for binding with the true ligand, and in fact, this is the basis for the action of many pharmaceutical drugs, which are explicitly designed to "block" receptor binding sites in this way.

There is an increasing evidence now for several different cell types possessing an ability to also detect nonchemical signals of mechanical origin. In tightly packed populations of cells, such as in certain tissues and microbial biofilms, the magnitude and direction of mechanical forces are dependent on spatial localization in the matrix of cells. In other words, mechanical signals could potentially be utilized as a cellular metric for determining where it is in relation to other cells. This has relevance to how higher-order multicellular structures emerge from smaller discrete cell components, for example, in microbial biofilms and many different types of animals and plant tissues. As to how such mechanical signals are detected, and ultimately transduced, is not clear. There is evidence of *mechanoreceptors* whose conformation appears to be dependent on local stresses in the vicinity of its localization in the cell membrane. There is also evidence that mechanical forces on DNA can affect its supercoiling topology in a controlled way.

2.4.3 TRAPPING "NEGATIVE" ENTROPY

A useful thermal physics view of living matter is that this is characterized by pockets of locally trapped "negative" entropy. The theoretical physicist Erwin Schrödinger wrote a useful treatment on this (Schrödinger, 1944) discussing how *life feeds off negative entropy*. By this, he was really referring to the concept of minimizing free energy to form a stable state, as opposed to some mysterious quantity of negative entropy *per se*. Life in essence results in pockets of locally ordered matter. This appears to be decoupled from the spirit of the *second law of thermodynamics*, though note that we cannot consider biological systems to be thermally closed, and instead when we consider the entropy of the whole universe, this will never decrease due to any biological process. But life can be thought of as being local reductions of entropy.

How is this achieved? What does "life" actually do to create local order? Ultimately, living organisms chemically combine carbon with other chemicals to form the various molecular forms of carbon-based living matter alluded to previously, all of which have greater order than their respective reactants. But where does this carbon come from? Organisms can eat other organisms of course and assimilate their biochemical contents, but somewhere at the very bottom of the food chain, the carbon has to come from a nonbiological source. This involves extracting carbon dioxide from the atmosphere by chemically combining it with water, fueled by energy from the sun, in a process called "photosynthesis," which occurs in plants and some microbial organisms.

The first key stage in photosynthesis involves an enzyme called "ribulose-1,5-bisphosphate carboxylase oxygenase" (*RuBisCO*), which is the most abundant known protein on Earth. It catalyzes the reaction of carbon dioxide into a precursor of sugars in a process called the "Calvin cycle," fueled through ATP hydrolysis. RuBisCO in prokaryotes is often found in specialized cellular organelles of carboxysomes. The initial absorption of light occurs either in the cell membrane directly (in photosynthetic cyanobacteria) or invaginated membrane *thylakoids* of chloroplasts (in plant eukaryotes) in *light-harvesting complexes*, which are multiprotein machines that operate as antennae to absorb visible light photons in combination with pigments (e.g., *carotenoids* and *chlorophylls*). This results in an effective spatial *funneling* of the incident photons through transfer of their energy to surrounding molecules via a nonradiative electronic molecular orbital resonance effect that generates high-energy electrons from the *photosynthetic reaction center*.

Quantum tunneling of these excited electrons (see Chapter 9) occurs in a series of electron transfer reactions with a drop in electron energy coupled at each stage to a series of chemical reactions, which results in the of pumping protons across a membrane in which consequent electrochemical energy is used to fuel the reaction of carbon dioxide with water, to oxygen as a by-product as well as produce small sugar molecules, which lock up the energy of the originally excited electrons into high-energy chemical bonds.

On the basis of simple thermodynamics, for any process to occur spontaneously requires a negative change in free energy, which is a (thermal) nonequilibrium condition. This is true for all processes, including those biological. Living matter in effect delays the dispersion of their free energy toward more available microstates (which moves toward a condition of thermal equilibrium at which the change in free energy is precisely zero) by placing limits on the number of available microstates toward which the free energy can be dispersed. This is achieved by providing some form of continuous energy input into the system.

Ultimately, this energy input comes principally from the sun, though in some archaea, this can be extracted from heat energy from thermal vents deep in the ocean. Another way to view this is that energy inputted into the local thermal system of a living organism is utilized to perform mechanical work in some form to force the system away from its natural tendency of a state of maximum disorder as predicted from the second law, which is done by fueling a variety of subcellular, cellular, and multicellular processes to regulate the organism's stable internal environment, a process that biologists describe as *homeostasis* (from the Greek, meaning literally "standing still").

KEY POINT 2.13

To overcome a tendency toward maximum entropy, a local thermal system requires energy input to perform work against this entropic force. In most living organisms, this energy ultimately comes from the sun, but may then also be provided in the form of chemical energy from food that has been synthesized from other organisms using the energy from the sun. This work against the sum of entropic forces in a living organism, through various processes that constitute homeostasis, results in a more stable and ordered internal environment.

Either way, this energy is trapped in some chemical form, typically in relatively simple sugar molecules. However, releasing all of the energy trapped in a single molecule of glucose in one go (equivalent to >500 $k_B T$) is excessive in comparison to the free energy changes encountered during most biological processes. Instead, cells first convert the chemical potential energy of each sugar molecule into smaller bite-sized chunks by ultimately manufacturing several molecules of ATP. The hydrolysis of a single molecule of ATP, which occurs normally under catalytic control in the presence of enzymes called "kinases," will release energy locally equivalent to ~18 $k_B T$, which is then converted into increased thermal energy of surrounding water solvent molecules whose bombardment on biological structures ultimately fuels mechanical changes to biological structures.

This input of free energy into biological systems can be thought of as a delaying tactic, which ultimately only slows down the inevitable process of the system reaching a state of thermal equilibrium, equivalent to a state of maximum entropy, and of death to the biological organism. (Note that many biological processes do exist in a state of *chemical equilibrium*, meaning that the rate of forward and reverse reactions are equal, as well as several cellular structures existing in a state of *mechanical equilibrium*, meaning that the sum of the kinetic and potential energy for that structure is a constant.)

But how are ATP molecules actually manufactured? Most sugars can be relatively easily converted in the cell into glucose, which is then broken down into several chemical steps releasing energy that is ultimately coupled to the manufacture of ATP. Minor cellular

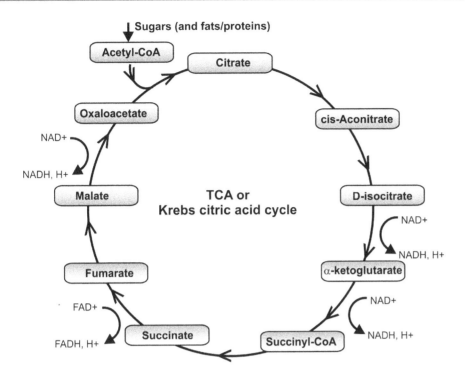

FIGURE 2.8   Schematic of the tricarboxylic acid or Krebs citric acid cycle.

processes that achieve this include *glycolysis* as well as *fermentation* (in plant cells and some prokaryotes), but the principle ATP manufacturing route, generating over 80% of cellular ATP, is via the *tricarboxylic acid (TCA) cycle* (biologists also refer to this variously as the *citric acid, Krebs*, or the *Szent–Györgyi–Krebs cycle*), which is a complex series of chemical reactions in which an intermediate breakdown product of glucose (and also ultimately of fats and proteins) called "acetyl-CoA" is combined with the chemical *acetate* and then converted in a cyclic series of steps into different organic acids (all characterized as having three – COOH groups, hence the preferred name of the process).

Three of these steps are coupled to a process, which involves the transfer of an electron (in the form of atomic hydrogen H as a bound H^+ proton and an electron) to the nucleoside *nicotinamide adenine dinucleotide (NAD$^+$)*, or which ultimately forms the hydrogenated compound *NADH*, with one of the steps using a similar *electron-carrier protein* or *flavin adenine dinucleotide (FAD$^+$)*, which is hydrogenated to make the compound *FADH* (Figure 2.8). The TCA cycle is composed of reversible reactions, but is driven in the direction shown in Figure 2.8 by a relatively high concentration of acetyl-CoA maintained by reactions that breakdown glucose.

Prokaryotes and eukaryotes differ in how they ultimately perform the biochemical processes of manufacturing ATP, known generally as oxidative phosphorylation (OXPHOS), but all use proteins integrated into a phospholipid membrane, either of the cell membrane (prokaryotes) or in the inner membrane of mitochondria (eukaryotes). The electron-carrier proteins in effect contain one or more electrons with a high electrostatic potential energy. They then enter the *electron transport chain (ETC)* and transfer the high-energy electrons to/from a series of different electron-carrier proteins via quantum tunneling (biologists also refer to these electron-carrier proteins as *dehydrogenases*, since they are enzymes that catalyze the removal of hydrogen). Lower-energy electrons, at the end of the series of ETCs, are ultimately transferred to molecular oxygen in most organisms, which then react with protons to produce water; some bacteria are *anaerobic* and so do not utilize oxygen, and in these instances an terminal electron acceptor of either sulfur or nitrogen is typically used.

Chemists treat electron gain and electron loss as *reduction* and *oxidation* reactions, respectively, and so such a series of sequential electron transfer reactions are also called electrochemical "redox reactions." Each full redox reaction is the sum of two separate *half reactions* involving reduction and oxidation, each of which has an associated *reduction potential* (E_0), which is the measure of the equivalent electrode voltage potential if that specific chemical half reaction was electrically coupled to a *standard hydrogen electrode* (the term *standard* means that all components are at concentrations of 1 M, but confusingly the *biochemical standard state* electrode potential is the same as the standard state electrode potential apart from the pH being 7; the pH is defined as $-\log_{10}[H^+ \text{ concentration}]$ and thus indicates a concentration of H^+ of 10^{-7} M for the biochemical standard state).

The reduction half reaction for the electron acceptor NAD^+ is

$$(2.1) \qquad NAD^+ + H^+ + 2e^- \rightleftharpoons NADH \quad E_0 = -0.315\,V$$

An example of a reduction half reaction at one point in the TCA cycle (see Figure 2.8) involves an electron acceptor called of "oxaloacetate⁻," which is reduced to *malate⁻*:

$$(2.2) \qquad Oxaloacetate^- + 2H^+ + 2e \rightleftharpoons Malate^- \quad E_0 = -0.166\,V$$

These two reversible half reactions can be combined by taking one away from the other, so malate⁻ then acts as an electron donor and in the process is oxidized back to oxaloacetate⁻, which is exactly what occurs at one point in the TCA cycle (two other similar steps occur coupled to the reduction of NAD+, and another coupled to FAD^+ reduction, Figure 2.8). The concentrations of oxaloacetate and malate are kept relatively low in the cell at 50 nM and 0.2 mM, respectively, and these low concentrations compared to the high concentration of acetyl-CoA result in a large excess of NAD^+.

A general reduction half reaction can be written as a chemical state O being reduced to a chemical state R:

$$(2.3) \qquad O + nH^+ + ne^- \rightleftharpoons R$$

where the free energy change per mole associated with this process can be calculated from the electrical and chemical potential components:

$$(2.4) \qquad \Delta G = \Delta G_0 + RT\ln\frac{[R]}{[O][H^+]^n} = -nFE$$

where F is *Faraday's constant*, 9.6×10^4 C mol⁻¹, equivalent to the magnitude of the electron charge q multiplied by Avogadro's number N_A, n electrons in total being transferred in the process. This also allows the molar equilibrium constant K to be calculated:

$$(2.5) \qquad K = \exp\left(\frac{nFE_0}{RT}\right) = \exp\left(\frac{nqE_0}{k_BT}\right)$$

where R is the molar gas constant, equal to k_BN_A, with absolute temperature T. Equation 2.4 can be rewritten by dividing through by $-nF$:

$$(2.6) \qquad E = E_0 - \frac{k_BT}{nq}\ln\frac{[R]}{[O][H^+]^n}$$

Equation 2.6 is called the "Nernst equation."

The free energy of oxidation of NADH and FADH is coupled to molecular machines, which pump protons across either the mitochondrial inner membrane (eukaryotes) or cytoplasmic membrane (prokaryotes) from the inside to the outside, to generate a *proton motive force (pmf)*, V_{pmf}, of typical value −200 mV relative to the inside. The free energy required to pump a single proton against this pmf can be calculated from Equation 2.4 equating V_{pmf} to

E. This proton motive force is then coupled to the rotation of the FoF1–ATP synthase in the membrane to generate ATP. For the TCA cycle, each molecule of glucose is ultimately broken down into a theoretical maximum of 38 molecules of ATP based on standard relative chemical stoichiometry values of the electron-carrier proteins and how many electrons can be transferred at each step, though in practice the maximum number is less in a living cell and more likely to be 30–32 ATP molecules per glucose molecule.

The pmf is an example of a *chemiosmotic* proton gradient (for a historical insight, see Mitchell, 1961). It constitutes a capacitance electrostatic potential energy. This potential energy can be siphoned off by allowing the controlled translocation of protons down the gradient through highly specific proton channels in the membrane. In a mechanism that is still not fully understood, these translocating protons can push around a paddle-wheel-type structure in a molecular machine called the "FoF1ATP synthase." The FoF1ATP synthase is a ubiquitous molecular machine in cells composed of several different protein subunits, found inside bacteria, chloroplasts in plants, and most importantly to us humans in mitochondria. The machine itself consists of two coupled rotary motors (see Okuno et al., 2011). It consists of an inner *water-soluble* F1 motor exposed to the cellular cytoplasm with a rotor shaft protein called γ surrounded by six *stator units* composed of alternating α and β proteins (Figure 2.5c). There is also an outer *hydrophobic* Fo motor linked to the rotor shaft. Under more normal conditions, the Fo motor couples the chemiosmotic energy stored in the proton gradient across the cell membrane lipid bilayer to the rotation of the F1 motor that results in ATP being *synthesized* from ADP and inorganic phosphate (but note that under conditions of oxygen starvation the motors can hydrolyze ATP and rotate in the opposite direction, causing the protons to be pumped *up* the proton gradient).

KEY POINT 2.14

ATP is the universal cellular fuel, made by transforming the chemical and electrostatic potential energy across specific dielectric phospholipid bilayers into mechanical rotation of the FoF1 ATP synthase molecular machine (really two counter-rotating motors of Fo and F1), which is coupled to chemically synthesizing ATP from ADP and inorganic phosphate.

2.4.4 NATURAL SELECTION, NEO-DARWINISM, AND EVOLUTION

Neo-Darwinism, which evokes classical *natural selection* concepts of *Darwinism* in the context of modern genetics theory, has been described by some life scientists as the *central paradigm of biology*. The essence of the paradigm is that living organisms experience a variety of *selective pressures*, and that the organisms best adapted to overcome these pressures will survive to propagate their genetic code to subsequent generations. By a "selective pressure," biologists mean some sort of local environmental parameter that affects the stochastic chances of an organism surviving, for example, the abundance or scarcity of food, temperature, pressure, the presence of oxygen and water, and the presence of toxic chemicals. In any population of organisms, there is a distribution of many different biological characteristics, which impart different abilities to thrive in the milieu of these various selective pressures, meaning that some will survive for longer than others and thus have a greater chance of propagating their genetic code to subsequent generations either through asexual cell division processes or through sexual reproduction.

This in essence is the nuts and bolts of natural selection theory, but the devil is very much more in the detail! Neo-Darwinism accounts for the distribution in biological characteristics of organisms through genetics, namely, in the underlying variation of the DNA nucleotide sequence of genes. Although the cellular machinery that causes the genetic code in DNA to be replicated includes error-checking mechanisms, there is still a small probability of, for example, a base pairing mismatch error (see Question 2.7), somewhere between 1 in 10^5 (for certain viruses) and 10^9 (for many bacteria and eukaryotic cells) per replicated nucleotide pair depending on the cell type and organism. If these errors occur within a gene, then they can be manifested as a *mutation* in the *phenotype* due to a change resulting from the physical, chemical, or structural properties of the resulting peptide or protein that is expressed from that particular gene.

Such a change could affect one or more biological processes in the cell, which utilize this particular protein, resulting in an ultimate distribution of related biological properties, depending on the particular nature of the mutated protein. If this mutated DNA nucleotide sequence is propagated into another cellular generation, then this biological variation will also be propagated, and if this cell happens to be a so-called germ cell of a multicellular organism, then this mutation may subsequently be propagated into offspring through sexual reproduction. Hence, selective pressures can bias the distribution of the genetic makeup in a population of cells and organisms of subsequent generations resulting, over many, many generations, in the *evolution* of that species of organism.

However, there is increasing evidence for some traits, which can be propagated to subsequent cellular generations not through alteration of the DNA sequence of the genetic code itself but manifested as functional changes to the genome. For example, modification of histone proteins that help to package DNA in eukaryotes can result in changes to the expression of the associated gene in the region of the DNA packaged by these histones. Similarly, the addition of *methyl* chemical groups to the DNA itself are known to affect gene expression, but without changing the underlying nucleotide sequence. The study of such mechanisms is called "epigenetics." An important factor with many such epigenetic changes is that they can be influenced by external environmental factors.

This concept, on the surface, appears to be an intriguing reversion back to redundant theories exemplified by the so-called Lamarckism, which essentially suggested erroneously that, for example, if a giraffe stretched its neck to reach leaves in a very tall tree, then the offspring from that giraffe in subsequent generations would have slightly longer necks. Although epigenetics does not make such claims, it does open the door to the idea that what an organism experiences in its environment may affect the level of expression of genes in subsequent generations of cells, which can affect the behavior of those cells in sometimes very dramatic ways.

This is most prominently seen in *cellular differentiation*. The term "differentiation" used by biologists means "changing into something different" and is not to be confused with the term used in calculus. This is the process by which *nongerm cells* (i.e., cells not directly involved in sexual reproduction, also known as *somatic cells*) turn into different cell types; these cells all have the same DNA sequence, but there are significant differences in the timing and levels of gene expressions between different cell types, now known to be largely due to epigenetics modifications. This process is first initiated from the so-called stem cells, which are cells that have not yet *differentiated* into different cell types. The reason why stem cells have such current interest in biomedical applications is that if environmental external physical and chemical triggers can be designed to cause stem cells to controllably and predictably change into specific cell types, then these can be used to replace cells in damaged areas of the body to repair specific physiological functions in humans, for example.

The exact mechanisms of natural selection, and ultimately species evolution, are not clear. Although at one level, natural selection appears to occur at the level of the whole organism, on closer inspection, a similar argument could be made at both larger and smaller length scales. For example, at larger length scales, there is natural selection at the level of populations of organisms, as exhibited in the *selfless* behavior of certain insects in appearing to sacrifice their own individual lives to improve the survival of the colony as a whole. At a smaller length scale, there are good arguments to individual cells in the same tissue competing with each other for nutrients and oxygen, and at a smaller length scale, still an argument for completion occurring at the level of single genes (for a good background to the debate, see Sterelny, 2007).

An interesting general mechanism is one involving the so-called emergent structures, a phenomenon familiar to physicists. Although the rules of small length and time scale interaction, for example, at the level of gene expression and the interactions between proteins, can be reduced to relatively simple forces, these interactions can lead to higher-order structures of enormous complexity called emergent structures, which often have properties that are difficult to predict from the fundamental simple sets of rules of individual interacting units. There is good evidence that although evolution is driven at the level of DNA molecules, natural selection occurs at the level of higher-order emergent structures, for example, cells, organisms, colonies, and biofilms. This is different from the notion that higher-order structures are simply "vehicles" for their genes, though some biologists refer to these emergent structures as *extended phenotypes* of the gene. This area of evolutionary biology is still hotly contested with some protagonists in the field formulating arguments, which, to the lay observer, extend beyond the purely scientific, but what is difficult to deny is that natural selection exists, and that it can occur over multiple length scales in the same organism at the same time.

The danger for the physicist new to biology is that in this particular area of the life sciences, there is a lot of "detail." Details are important of course, but these sometimes will not help you get to the pulsing heart of the complex process, that is, *adaptation* from generation to generation in a species in response the external environment. Consider, instead, an argument more aligned with thermal physics:

KEY POINT 2.15

Biological complexity is defined by local order (e.g., more complex cellular structures, or even more complexity of cells inside tissues, or indeed more complexity of individual organisms within an ecology) implying a local decrease in entropy, which thus requires an energy input to maintain. This is a selective disadvantage since competing organisms without this local increase in order benefit from not incurring a local energy loss and thus are more likely not to die (and thus reproduce and propagate their genetic code). Therefore, there has to be a good reason (i.e., some ultimately energy gain) to sustain greater complexity.

This, of course, does not "explain" evolution, but it is not a bad basis from which the physicist has to start at least. Evolutionary change is clearly far from simple, however.

One very common feature, which many recent research studies have suggested, which spans multiple length scales from single molecules up through to cells, tissues, whole organisms, and even ecologies of whole organisms, seems to be that of a phenomenon known as *bet hedging*. Here, there is often greater variability in a population than one might normally expect on the grounds of simple efficiency considerations—for example, slightly different forms of a structure of a given molecule, when energetically it may appear to be more efficient to just manufacture one type. This variability is in one sense a form of "noise"; however, in many cases, it confers robustness to environmental change, for example, by having multiple different molecular forms that respond with different binding kinetics to a particular ligand under different conditions, and in doing so that organism may stand a greater chance of survival even though there was a greater upfront "cost" of energy to create that variability. Unsurprisingly, this increase in noise is often seen in systems of particularly harsh/competitive environmental conditions.

Note that although natural selection when combined with variation in biological properties between a competing population, at whatever disputed length scale, can account for aspects of incremental differences between subsequent generations of cell cycles and organism life spans, and ultimately evolutionary change in a population, this should not be confused with *teleological/teleonomic* arguments. In essence, these arguments focus on the *function* of a particular biological feature. One key difference in language between biology and physics is the use of the term "function"—in physics we use it to mean a mapping between different sets of parameters, whereas in biology the meaning is more synonymous with *purpose* or *role*. It is not so much that new features evolve to perform a specific role, though some biologists may describe it as such, rather that selective pressure results in better *adaptation* to a specific set of environmental conditions.

Also, natural selection *per se* does not explain how "life" began in the first place. It is possible to construct speculative arguments on the basis, for example, of *RNA replicators* being the "primordial seed" of life, which forms the basis of the RNA world hypothesis. RNA is a single-stranded nucleic acid unlike the double-stranded DNA and so can adopt more complex 3D structures, as seen, for example, in the clover leaf shape of tRNA molecules and in the large complex RNAP, both used in transcribing the genetic code. Also, RNA can form a version of the genetic code, seen in RNA viruses and in mRNA molecules that are the translated versions of coding DNA. Thus, RNA potentially is an *autocatalyst* for its own replication, with a by-product resulting in the generation of peptides, which in turn might ultimately evolve over many generations into complex enzymes, which can catalyze the formation of other types of biological molecules. There is a further question though of how cell membranes came into being since these are essential components of the basic cellular unit of life. However, there is emerging evidence that micelles, small, primordial lipid bilayer vesicles, may also be autocatalytic, that is, the formation of a micelle makes it more likely for micelles to form further. But a full discussion of this theory and others of creation myths of even greater speculation are beyond the scope of this book but are discussed by Dawkins and elsewhere.

2.4.5 "OMICS" REVOLUTION

Modern genetics technology has permitted the efficient sequencing of the full genome of several organisms. This has enabled the investigation of the structure, function of, and interactions between whole genomes. This study is *genomics*. The equivalent investigation between the functional interactions of all the proteins in an organism is called "proteomics." Many modern biophysical techniques are devoted to genomics and proteomics investigations, which are discussed in the subsequent chapters of this book. There are now also several other *omics* investigations. *Epigenomics* is devoted to investigating the *epigenome*, which is the collection of epigenetics factors in a given organism. *Metabolomics* studies the set of metabolites within a given organism. Other such fields are lipidomics (the characterization of all lipids in an organism), similarly *transcriptomics* (the study of the collected set of all TFs in an organism), *connectomics* (study of the neural connections), and several others. An interesting new omics discipline is *mechanomics* (Wang et al., 2014); this embodies the investigation of all mechanical properties in an organism (especially so at the level of cellular mechanical signal transduction), which crosses into gene regulation effects, more conventionally thought to be in the regime of transcriptomics, since there is now emerging evidence of mechanical changes to the cell being propagated at the level of the local structure of DNA and affecting whether genes are switched on or off. Arguably, the most general of the omics fields of study is that called simply "interactomics"; this investigates the *interactome*, which is the collection of all interactions within the organism and so can span multiple length and time scales and the properties of multiple physical parameters and, one could argue, embodies the collection of all other omics fields.

2.5 PHYSICAL QUANTITIES IN BIOLOGY

Many of the physical quantities in biological systems have characteristic origins and scales. Also, part of the difference in language between physical scientists and biologists involves the scientific units in common use for these physical quantities.

2.5.1 FORCE

The forces relevant to biology extend from the high end of tissue supporting the weight of large organisms; adult blue whales weigh ~200 tons, and so if the whale is diving at terminal velocity, the frictional force on the surface will match its weight, equivalent to 2×10^6 N. For plants, again the forces at the base of a giant sequoia tree are in excess of 1×10^7 N. For humans, the typical body weight is several hundred newtons, so this total force can clearly be exerted by *muscles* in legs. However, such macrolength scale forces are obviously distributed throughout the cells of a tissue. For example, in muscle, the tissue may be composed of 10–20 *muscle fibers* running in parallel, each of which in turn might be composed of ~10 *myofibrils*, which are the equivalent cellular level units in muscle. With the additional presence of *connective tissue* in between the fibers, the actual cellular forces are on the order of sub-newtons. This level of force can be compared with that required to break a covalent chemical bond such as a carbon–carbon bond of ~10^{-9} N, or a weaker *noncovalent* bond such as those of an *antibody* binding of ~10^{-10} N.

At the lower end of the scale are forces exerted by individual molecular machines, typically around the level of a few multiples of 10^{-12} N. This unit is the *piconewton* (pN), which is often used by biologists. The weakest biologically relevant forces in biology are due to random thermal fluctuations of surrounding water-solvent molecules, which is an example of the *Langevin force* (or *fluctuation force*), depending upon the length and time scale of observation. For example, a nucleus of diameter ~10^{-6} m observed for a single second will experience a net Langevin force of ~10^{-14} N. Note that in biology, gravitational forces of biological molecules are normally irrelevant (e.g., ~10^{-17} N).

One of the key *attractive* forces, which are essential to all organisms, is that of *covalent bonding*, which allows strong chemical bonds to form between atoms, of carbon in particular. Covalent bonds involve the sharing of pairs of electrons from individual *atomic orbitals* to form stable *molecular orbitals*. The strength of these bonds is often quantified by the energy required to break them (as the bond force integrated over the distance of the bond). The bond energy involving carbon atoms typically varies in a range of 50–150 $k_B T$ energy units.

Cells and biomolecules are also affected by several weaker *noncovalent forces*, which can be both attractive and repulsive. An ideal way to characterize a vector force F is through the grad function of the respective potential energy landscape of that force, U:

$$(2.7) \qquad \vec{F} = -\nabla U$$

Electrostatic forces in the water-solvated cellular environment involve layers of polar water molecules and ions, in addition to an *electrical double layer* (EDL) (the *Gouy–Chapman layer*) composed of ions adsorbed onto molecular surfaces with a second more diffuse weakly bound to counter charges of the first layer. The governing equation for the electrostatic potential energy is governed by *Coulomb's law*, which in its simplest form describes the electric potential V_e due to a single point of charge q at a distance r:

$$(2.8) \qquad V_e = \frac{q}{4\pi\varepsilon_0 \varepsilon_m r}$$

where
  ε_0 is the electrical permittivity in a vacuum
  ε_m is the relative electrical permittivity in the given medium

Therefore, for example, for two charges of equal magnitude q but opposite sign, separated by a distance d, each will experience an attractive electrostatic force F_e toward the other parallel to the line that joins the two charges of

$$(2.9) \qquad F_e = \frac{q^2}{4\pi\varepsilon_0 \varepsilon_m r^2}$$

For dealing with multiple electrical charges in a real biological system, a method called "Ewald summation" is employed, which treats the total electrical potential energy as the sum of short-range and long-range components. The usual way to solve this often complex equation is to use the *particle mesh Ewald method* (*PME method*), which treats the *short-range* and *long-range* components separately in *real space* and *Fourier space* (see Chapter 8).

Van der Waals forces (*dispersion-steric repulsion*), as discussed, are short-range steric repulsive potential energy with $1/r^{12}$ (distance r) dependence fundamentally due to the *Pauli exclusion principle* in *quantum mechanics.* The exclusion principle disallows the overlap of electron orbitals. When we combine this short-range repulsive component with a longer-range attractive component from interactions with nonbonding electrons inducing electrical dipoles, a $1/r^6$ dependence, we get the so-called "Lennard–Jones potential," which is also referred to as the *L-J, 6-12,* and *12-6 potential*) $U_{\text{L-J}}$:

(2.10)
$$U_{\text{L-J}} = \frac{A}{r^{12}} - \frac{B}{r^6}$$

Here, A and B are the constants of the particular biological system.

Hydrogen (or H–) *bonding*, already referred to, is a short-range force operating over ~0.2–0.3 nm. These are absolutely essential to forming the higher-order structures of many different biological molecules. The typical energy required to break an H-bond is ~$5k_{\text{B}}T$.

Hydrophobic forces are largely entropic based resulting from the tendency of nonpolar molecules to pool together to exclude polar water molecules. There is no simple law to describe hydrophobic forces, but they are the strongest at 10–20 nm distances, and so are generally perceived as long range. Hydrophobic bonds are very important in stabilizing the structural core of globular protein molecules.

Finally, there are *Helfrich forces*. These result from the thermal fluctuations of cell membranes due to random collisions of solvent water molecules. They are a source entropic force, manifest as short-range repulsion.

2.5.2 LENGTH, AREA, AND VOLUME

At the high end of the biological length scale, for single organisms at least, is a few tens of meters (e.g., the length of the largest animal is the *blue whale* at ~30 m, the largest plant is the *giant sequoia tree* at almost 90 m in height). Colonies of multiple organisms, and whole *ecosystems*, can clearly be much larger still. At the low end of the scale are single biological molecules, which are typically characterized by a few *nanometers* (unit nm or 10^{-9} m; i.e., 1 m/1000 million) barring exceptions such as filamentous biopolymers, like DNA, which can be much longer. Crystallographers also use a unit called the "Angstrom" (Å), equal to 10^{-10} m, since this is the length scale of the hydrogen atom diameter, and so typical covalent and hydrogen bond lengths will be a few Angstroms.

Surface area features investigated in biology often involve cell membranes, and since the length scale of cell diameter is typically a few microns (μm), the μm^2 area unit is not uncommon. For volume, biochemistry in general refers to liter units (L) of 10^{-3} m^3, but typical quantities for biochemical assays often involve volumes in the range 1–1000 μL (microliters), though more bulk assays potentially use several milliliters (mL).

2.5.3 ENERGY AND TEMPERATURE

Molecular scale (pN) forces integrated over with nanometer spatial displacements result in an energy scale of a few *piconewton nanometers*. The piconewton nanometer unit (pN nm) equals 10^{-21} J.

Organisms have a *high* temperature, from a physics perspective, since quantum energy transitions are small relative to classical levels, and so the *equipartition theorem*, that each independent quadratic term, or *degree of freedom*, in the energy equation for a molecule in an ensemble at absolute temperature T has an average energy $k_{\text{B}}T/2$. A water molecule has three translational, three rotational, and three intrinsic vibrational energy modes (note each vibrational mode has two degrees of freedom of potential and kinetic energy) plus up to three additional extrinsic vibrational modes since each atom can, in principle, independently form a hydrogen bond with another nearby water molecule, indicating a mean energy of ~9 $k_B T$ per molecule.

Following collision with a biological molecule, some of this kinetic energy is transferred, resulting in $k_B T$ scale energy fluctuations; $k_B T$ itself is often used by biologists as a standard unit of energy, equivalent to 4.1 pN nm at room temperature. This is roughly the same energy scale as molecular machines undergoing typical displacement transitions of a few nanometers through forces of a few piconewtons. This is because molecular machines have evolved to siphon-off energy from the thermal fluctuations of surrounding water to fuel their activity. The hydrolysis of one molecule of ATP in effect releases 18 $k_B T$ of chemical potential energy from high-energy phosphate bonds to generate thermal energy.

Note that some senior biologists still refer to an energy unit of the *calorie* (*cal*). This is defined as the energy needed to raise the temperature of 1 g of water through 1°C at a pressure of one atmosphere. Intuitively, from the discussion earlier, this amount of energy, at room temperature, is equivalent to 4.1 J. Many bond energies, particularly in older, but still useful, biochemistry textbooks are often quoted in units of *kcal*.

The temperature units used by most biologists are degrees Celsius (°C), that is, 273.15 K higher than absolute temperature. *Warm-blooded* animals have stable body temperatures around 35°C–40°C due to complex *thermoregulation* mechanisms, but some may enter *hibernation* states of more like 20°C–30°C. Many cold-blooded animals thrive at a *room temperature* of 20°C, and some plants can accommodate close to the full temperature range of liquid water. Microbes have a broad range of optimal temperatures; many lie in the range of 20°C–40°C, often optimized for living in the presence of, or symbiotically with, other multicellular organisms. However, some, including the so-called extremophiles, many from the archaea domain of organisms, thrive at glacial water temperatures in the range 0°C–5°C, and at the high end some can thrive at temperatures of 80°C–110°C either in underwater thermal vents or in atmospheric volcanic extrusions. The full temperature range when considered across all organisms broadly reflects the essential requirement of liquid water for all known forms of life. Many proteins begin to *denature* above 50°C, which means that their tertiary and quaternary structures are disrupted through the breaking of van der Waals interactions, with the result of the irreversible change of their 3D structure and thus, in general, destruction of their biological function.

There are some well-known protein exceptions that occur in types of extremophile cells that experience exceptionally high temperatures, known as *thermophiles*. One such is a thermophilic bacterium called *Thermus aquaticus* that can survive in hot water pools, for example, in the vicinity of lava flow, to mean temperatures of 80°C. An enzyme called "Taq polymerase," which is naturally used by these bacteria in processing of DNA replication, is now routinely used in *polymerase chain reactions* (*PCR*) to amplify a small sample of DNA, utilized in biomedical screening and forensic sciences, as well as being routinely used in abundance for biological research (see Chapter 7). A key step in PCR involves cycles of heating up replicated (i.e., amplified) DNA to 90°C to denature the two helical strands from each DNA molecule, which each then acts as a template for the subsequent round of amplification, and Taq polymerase facilitates this replication at a rate >100 nucleotide base pairs per second. The advantage of the Taq polymerase is that, unlike DNA polymerases from non-thermophilic organisms, it can withstand such high heating without significant impairment, and in fact even at near boiling water temperatures of 98°C, it has a stability half-life of 10 min.

2.5.4 TIME

Time scales in biology are broad. Ultimately, the fastest events are quantum mechanical concerning electronic molecular orbitals, for example, a covalent bond vibrates with a time scale of 10^{-15} s, but arguably the most rapid events, which make detectable differences to biomolecular components, are collisions from surrounding water molecules, whose typical separation time scale is 10^{-9} s. Electron transfer processes between molecules are slower at 10^{-6} s.

Molecular conformational changes occur over more like 10^{-3} s. Molecular components also turnover typically over a time scale from a few seconds to several minutes. The lifetime of molecules in cells varies considerably with cell type, but several minutes to hours is not atypical. Cellular lifetimes can vary from minutes through years, as therefore do organism lifetimes, though certain microbial spores can survive several hundred million years. Potentially, the range of time scale for biological activity, at a conservative estimate, is ~20 orders of magnitude. In exceptional cases, some biological molecules go into quiescent states and remain dormant for potentially up to several hundred million years. In principle, the complete time scale could be argued to extend from 10^{-15} s up to the duration over which life is thought to have existed on Earth—4 billion years, or 10^{18} s.

2.5.5 CONCENTRATION AND MASS

Molecules can number anything from just 1 to 10 per cell, up to over 10^4, depending on the type of molecule and cell. The highest concentration, obviously involving the largest number of molecules in the smallest cells, is found in some proteins in bacteria that contain several tens of thousands of copies of a particular protein (many bacteria have a typically small diameter of 1 μm). Biologists often refer to concentration as *molarity* (M), which is the number of *moles* (mol) of a substance in 1 L of water solvent, such that 1 *mol* equals *Avogadro's number* of particles (6.022×10^{23}, the number of atoms of the C-12 isotope present in 12 g of pure carbon-12). Typical cellular molarity values for biological molecules are 10^{-9} M or *nanomolar* (nM). Some biologists also cite *molality*, which is the moles of dissolved substance divided by the solvent mass used in kg (units mol kg^{-1}). Dissolved salts in cells have higher concentrations, for example, the concentration of sodium chloride in a cell is about 200 mM (pronounced *millimolar*) that is also equal in this case to a "200 millimolal" molality.

Mass, for biochemical assays, is often referred to in milligram units (mg). The molecular mass, also called the "molecular weight" (M_w), is the mass of the substance in grams, which contains Avogadro's number of molecules, but is cited in units of the *Dalton* (Da) or more commonly for proteins the *kilodalton* (kDa). For example, the mean molecular weight taken from all the natural amino acids is 137 Da. The largest single protein is an isomer of titin that has a molecular weight of 3.8 MDa. The "molecular weight" of a single ribosome is 4.2 MDa, though note that a ribosome is really not a single molecule but is a complex composed of several subunits. *Mass concentration* is also used by biologists, typical units of being how many milligrams of that substance is dissolved in an milliliters of water, or milligrams per milliliter (mg mL^{-1}; often pronounced "miggs per mill").

An alternative unit, which relates to mass but also to length scale, is the *svedberg* (S, sometimes referred to as Sv). This is used for relatively large molecular complexes and refers to the time it takes to sediment the molecule during centrifugation (see Chapter 6), and so is dependent on its mass and frictional drag. A common example of this includes ribosomes, and their component subunit; the prokaryote ribosome is 70 S. The svedberg is an example of a sedimentation coefficient (see Chapter 6), which is the ratio of a particle's acceleration to its speed and which therefore has the dimensions of time; 1 S is equivalent to exactly 100 femtoseconds (i.e., 100 fs or 10^{-13} s). Svedberg units are not directly additive, since they depend both on the mass of the components and to their fractional drag with the surrounding fluid environment that scales with the exposed surface area, which obviously depends on how separate components are bound together in a complex. When two or more particles bind together, there is inevitably a loss of surface area. This can be seen again in the case of the ribosome; the 70 S prokaryotic ribosome has a sedimentation coefficient of 70 S, but is composed of a large subunit of 50 S and a small subunit of 30 S (which in turn includes the 16 S rRNA subunit).

The concentration of protons [H$^+$] is one of the most important measures in biology, though in practice, as discussed, a proton in a liquid aqueous environment, such as inside a cell, is generally coupled through hydrogen bonding to a water molecule as the hydronium/hydroxonium ion H$_3$O$^+$. As discussed, the normal biological representation of proton concentration is as $-\log_{10}$[H$^+$], referred to as the pH, with *neutral* pH = 7, *acids* <7, and *bases* (or *alkalis*) >7 assuming [H$^+$] is measured in M units.

2.5.6 MOBILITY

The wide speed range of different biological features obviously reflects the broad time scale of the process in life. At the molecular end of the biology length scale, a key speed measure is that of the translocation of molecular machines, which is typically in the range of a few microns per second, μm s^{-1}. In the cellular regime, there are motile cells, such as self-propelling bacteria that swim an order of magnitude faster. And then at the whole organism scale, there are speeds of more like meters per second, m s^{-1}.

To characterize the net flow of matter due to largely random motions, we talk about *diffusion*, which has a dimensionality not the same as that of speed, which is $[L]/[T]$, but rather of $[L]^2/[T]$, and so conceptually more equivalent to rate at which an "area" is explored in a given time. For purely *random-walk* behavior of particles, we say that they are exhibiting *Brownian diffusion* (or *normal diffusion*). The effective Brownian diffusion coefficient of a biomolecule, D, assuming free and unrestricted diffusion, relates to the variation of effective frictional drag coefficient, γ, through the *Stokes–Einstein relation* of

(2.11)
$$D = \frac{k_B T}{\gamma}$$

For the simple case of a sphere of radius r diffusing in a fluid of viscosity η (sometimes referred to more fully as the "dynamic" viscosity, to distinguish it from the "kinematic" viscosity, which is the dynamic viscosity divided by the fluid density), γ is given by $6\pi\eta r$. This is often a good approximation for a globular-like protein diffusing in the cytoplasm, however different biomolecules in different environments need to be approximated with different shape factors (for example, integrated membrane proteins in a lipid bilayer will typically rotate rapidly over a microsecond timescale or faster perpendicular to the plane of the membrane, and so the effective shape when averaged over a timescale of millisecond or more, which is appropriate for typical light microscopy sampling, is closer to a cylinder perpendicular to the membrane).

In the watery part of the cell, such as the cytoplasm, values of D of a few μ^2m s^{-1} are typical, whereas for a molecule integrated into phospholipid bilayer membranes (30% of all proteins come into this category), the local viscosity is higher by a factor of 100–1000, with resultant values of D smaller by this factor. The theoretical *mean squared displacement* $\langle R^2 \rangle$, after a time t of a freely diffusing particle in n-dimensional space (e.g., in the cytoplasm $n = 3$, in the cell membrane $n = 2$, for a molecular motor diffusing on track, for example, a kinesin molecule on a stiff *microtubule* filament track, $n = 1$) is given by

(2.12)
$$\langle R^2 \rangle = 2nDt$$

But note, in reality, blurring within experimental time sample windows as well as detection precision error leads to a correction for experimental measurements, which involve single particle tracking, and also note that in general there can be several other more complex modes of diffusion inside living cells due to the structural heterogeneity of the intracellular environment.

Worked Case Example 2.3: Intracellular Concentrations

In one experiment, the numbers of 40 different protein molecules per cell were estimated from a population of many E. coli bacteria and found to have mean values in the range 9–70,000 molecules per cell. (An E. coli cell is rod-shaped approximated for much of its cell cycle as a cylinder of length 2 μm capped at either end with a hemisphere of diameter 1 μm.)

a   What is the volume of a single E. coli cell in L?

b   What does the mean data for protein numbers suggest in the range of molarity values for different proteins in units of nM?

c   What is the smallest nonzero protein molarity you might measure if you had the sensitivity to perform the experiment on just a single cell with arbitrarily high time resolution?

d   How does this compare with the molarity of pure water (hint: what is the formal definition of molarity)?

Answers:

a   The volume of the cell $V_c = \{(1.0 \times 10^{-6} \times 0.5)^2 \times (2.0 \times 10^{-6})\} + \{2 \times 2 \times \pi \times (1.0 \times 10^{-6} \times 0.5)^3/3\} = 1.0 \times 10^{-18}$ m$^3 = 1.0 \times 10^{-15}$ L.

b   The number of moles for total number of n molecules is $n/(6.022 \times 10^{23})$.

   Thus, the range in number of moles for proteins studied is $\{9–70{,}000\}/(6.022 \times 10^{23}) = 1.5 \times 10^{-23}$ to 1.2×10^{-19} mol.

   Thus, the range in molarity is $\{1.5 \times 10^{-23}$ to $1.2 \times 10^{-19}\}/(1.0 \times 10^{-15}) = 1.5 \times 10^{-8}$ to 1.2×10^{-4} M = 18–120,000 nM.

c   For a single-cell measurement with high time resolution, one might be able to detect the presence of just a single protein molecule in a cell since expression from a single gene is stochastic. This minimal nonzero molarity $= \{1/(6.022 \times 10^{23})\}/(1.0 \times 10^{-15}) = 1.7$ nM.

d   Water has molecular weight equivalent to 18 Da (i.e., 18 g, each molecule consisting of 1 atom of oxygen, atomic mass 16 g, and 2 atoms of hydrogen, atomic mass 1 g). The mass of 1 L of water is 1 kg; thus, the number of moles in 1 L of water is $1/(0.018) \approx 56$ M.

2.6 SUMMARY POINTS

- Carbon chemistry permits catenated compounds that form the chemicals of life.
- Biology operates at multiple length and time scales that may overlap and feedback in complex ways.
- The cell is a fundamental unit of life, but in general it needs to be understood in the context of several other cells, either of the same or different types.
- Even simple cells have highly localized architecture, which facilitates specialist biological functions.
- The most important class biomolecules are biological catalysts called "enzymes," without which most chemical reactions in biology would not happen with any degree of efficiency.
- The shape of molecules is formed from several different forces, which leads to differences in their functions.
- The key process in life is the central dogma of molecular biology, which states that proteins are coded from a genetic code written in the base pair sequence of DNA.

QUESTIONS

2.1   In a reversible isomerization reaction between two isomers of the same molecule, explain what proportion of the two isomers might be expected to exist at chemical equilibrium, and why. In an autocatalytic reaction, one of the reacting molecules itself acts as a catalyst to the reaction. Explain what might happen with a small excess of one isomer to the relative amounts of each isomer as a function of time. How is this relevant to the D/L isomers amino acids and sugars? Discuss how might evolution affect the relative distribution of two isomers? (For a relevant, interesting read, see Pross, 2014.)

2.2   *Staphylococcus aureus* is a spherical bacterium of 1 μm diameter, which possesses just one circular chromosome. A common form of this bacterium was estimated as having 1.2% of its cellular mass taken up by DNA.

a   What is the mass of the cell's chromosome?

b   The bases adenine, cytosine, guanine, and thymine have molecular weights of 0.27, 0.24, 0.28, and 0.26 kDa, respectively, excluding any phosphate groups. The molecular weight of a phosphate group is 0.1 kDa. Estimate the contour length of the *S. aureus* genome, explaining any assumptions.

2.3   The primary structure of a human protein compared to that of budding yeast, *S. cerevisiae*, which appears to carry out the same specific biological function, was found to have 63% identical sections of amino acids based on short sequence sections of at least five consecutive amino acids in length. However, the respective DNA sequences were found to be only 42% identical. What can account for the difference?

2.4   The cell doubling time, a measure of the time for the number of cells in a growing population to double, for *E. coli* cells, which is a rich nutrient environment, is 20 min. What rate of translation of mRNA into amino acids per second can account for such a doubling time? How does this compare to measured rates of mRNA translation? Comment on the result.

2.5   What is the relation between force and its potential energy landscape? Why is it more sensible to consider the potential energy landscape of a particular force first and then deduce the force from this, as opposed to considering just a formulation for the force directly?

2.6   What are the van der Waals interactions, and how do these relate to the Lennard–Jones potential? Rewrite the Lennard–Jones potential in terms of the equilibrium distance r_m in which the net force is zero and the depth of potential parameter is in V_m, which is the potential energy at a distance r_m.

2.7   A DNA molecule was found to have a roughly equal mix of adenine, cytosine, guanine, and thymine bases.

a   Estimate the probability for generating a mismatched base pair in the DNA double helix, stating any assumptions you make. (*Hint:* use the Boltzmann factor.) When measured in a test tube, the actual mismatch error was found to be 1 in 10^5. Comment on the result.

b   In a living cell, there is typically one error per genome per generation (i.e., per cell division). What error does this equate to for a human cell? How, and why, does this compare with the value obtained earlier?

2.8   Calculate, with reasoning, the free energy difference in units of $k_B T$ required to translocate a single sodium ion Na^+ across a typical cell membrane. Show, by considering the sodium ion to be a unitary charge q spread over a spherical shell of radius r, that the activation energy barrier required to spontaneously translocate across the lipid bilayer of electrical relative permittivity ε_r is given by $q/8\pi\, r\varepsilon_r\varepsilon_0$. (Note, this is known as the "electrical self energy"). There is an initial concentration of 150 mM of sodium chloride both inside and outside a roughly spherical cell of diameter 10 μm, with a sodium ion diameter of 0.2 nm. The cell is immersed in a specific kinase inhibitor that prevents ATP hydrolysis, which is normally required to energize the pumping of sodium ions across the cell membrane through sodium-specific ions channels (see Chapter 4), and the cell is then suddenly immersed into pure water. Calculate the number of Na^+ ions that the cell loses due solely to spontaneous translocation across the phospholipid bilayer. State any assumptions you make. (The electrical permittivity in a vacuum is $8.9 \times 10^{-12}\ C^2\ m^{-2}\ N^{-2}$, the relative electrical permittivity of a phospholipid bilayer is 5, and the charge on a single electron is $-1.6 \times 10^{-19}\ C$.)

2.9 The complete oxidation of a single molecule of glucose to carbon dioxide and water in principle involves a total free energy change equivalent of $-2870\ kJ\ mol^{-1}$ (the minus sign indicates that energy is released, as opposed to absorbed), whereas that for ATP hydrolysis to ADP and inorganic phosphate (the opposite of ATP synthesis) is equivalent to $18\ k_B T$ per molecule.

   a If the TCA cycle synthesizes a total of 38 ATP molecules from every molecule of glucose, what is the energetic efficiency?

   b If ATP synthesis occurs inside a mitochondrion whose volume is comparable to a bacterial cell and the concentration of glucose it utilizes is 5 mM, calculate the theoretical rise in temperature if all glucose were instantly oxidized in a single step. Comment on the result in light of the actual cellular mechanisms for extracting chemical energy from glucose.

2.10 At one point in the TCA inside a mitochondrion in a eukaryotic cell cycle, NAD^+ is reduced to NADH oxaloacetate$^-$, and the free energy of NADH is coupled to the pumping of a proton across the inner membrane against the proton motive force.

   a How much free energy in units of $k_B T$ per molecule is required to pump a single proton across the inner membrane?

   b Write down this full electrochemical for the reduction of NAD^+ by oxaloacetate$^-$ and calculate the standard free energy change for this process in units of $k_B T$ per molecule. Stating any assumptions, calculate how many protons NADH pump across the inner membrane in a single TCA cycle.

   c Inside a mitochondrion, the pH is regulated at 7.5. If all nicotinamide is present either as NAD^+ or NADH, calculate the relative % abundance of NAD^+ or NADH in the mitochondrion. Comment on the result.

2.11 The definition of a cell as outlined in this chapter is as being the minimal structural and functional unit of life that can self-replicate and can exist "independently." But in practice, a cell is not isolated, for example, there are pores in the membrane that convey molecules in/out of the cell. Does this alter our notion of independence? Are there more robust alternative definitions of a cell?

2.12 There are $\sim 10^{15}$ cells in a typical human body, but only $\sim 10^{14}$ of them are human. What are the others? Does this alter our view of the definition of a "human organism"?

2.13 If you go to the Protein Data Bank (www.pdb.org), you can download "pdb" coordinates for many, many molecular structures, the following PDB IDs being good examples: 1QO1 and 1AOI. You can install a variety of free software tools (e.g., RasMol, but there are several others available) and open/display these pdb files.

   a Where appropriate, use the software to display separate strands of the DNA double helix red and blue, with nonpolar amino acids residues yellow and polar amino acid residues magenta.

   b Using the software, find out what is the maximum separation of any two atoms in either structure.

2.14

   a What are the attractive and repulsive forces relevant to single biomolecule interactions, and how do they differ in terms of the relative distance dependence and magnitude?

   b What forces are most relevant to protein folding of a protein, and why?

2.15 The average and standard deviation of heights measured from a population of adult women of the same age and ethnic background were 1.64 and 0.07 m, respectively. Comment on how this compares to the expected variation between the sets of genes from the same population.

2.16 A 10 mL of culture containing a virus that infects bacteria was prepared from a culture of growing *E. coli* bacteria at the peak of viral infection activity, and 1 mL of the culture was divided into 10 volumes of 100 μL each. Nine of these were grown with nine separate fresh uninfected bacterial cultures; all of these subsequently developed viral infections. The 10th volume was then added to fresh culture medium to make up to the same 100 μL volume as the original virus culture and mixed. This diluted virus culture was divided into 10 equal volumes as before, and the previous procedure repeated up to a total of 12 such dilutions. In the first nine dilutions, all nine fresh bacterial cultures subsequently developed virus infections. In the 10th dilution, only six of the nine fresh bacterial cultures developed viral infections; in the 11th dilution, only two of the nine bacterial cultures developed viral infections; and in the 12th dilution, none of the nine bacterial cultures developed viral infection.

a   Estimate the molarity of the original virus culture.

b   If this culture consisted of virus particles tightly packed, such that the outer coat of each virus was in contact with that of its nearest neighbors, estimate the diameter of the virus. (*Hint:* a virus culture is likely to cause a subsequent infection of a bacteria culture if there is at least one virus in the culture.)

2.17 A key feature of biology is that components of living matter appear to have specific functions. However, the laws of physics are traditionally viewed as being objective and devoid of "purpose." Discuss this apparent contradiction.

2.18 Liquid–liquid phase-separated biomolecular condensates appear to have a preferred length scale in cells, whereas "abiotic" classical nucleation theory for the phase transition process predicts that a transition would, given sufficient time, go to completion until all the relevant molecules are demixed. What reasons can you think of that could account for this difference?

2.19 Video-rate fluorescence microscopy with a sampling time of 40 ms per image frame could track a membrane-integrated protein reasonably well, however, images of a similar sized fluorescently labeled protein in the cytoplasm using similar microscopy looked blurry and couldn't be tracked. Why is this?

REFERENCES

KEY REFERENCE

Alberts, B. et al. (2008). *Molecular Biology of the Cell*, 5th ed. Garland Science, New York.

MORE NICHE REFERENCES

Banani, S.F. et al. (2017). Biomolecular condensates: organizers of cellular biochemistry. *Nat Rev. Mol. Cell Biol.* 18:285–298.

Dawkins, R. (1978). *The Selfish Gene*, 30th Anniversary ed. (May 16, 2006). Oxford University Press, Oxford, U.K.

Gibson, D.G. et al. (2010). Creation of a bacterial cell controlled by a chemically synthesized genome. *Science* 329:52–56.

Mitchell, P. (1961). Coupling of phosphorylation to electron and hydrogen transfer by a chemi-osmotic type of mechanism. *Nature 191*:144–148.

Okuno, D., Iino, R., and Noji, H. (2011). Rotation and structure of FoF1-ATP synthase. *J. Biochem. 149*:655–664.

Pross, A. (2014). *What Is Life?: How Chemistry Becomes Biology*. Oxford University Press, Oxford, U.K.

Schrödinger, E. (1944). *What Is Life—The Physical Aspect of the Living Cell*. Cambridge University Press, Cambridge, U.K. Available at http://whatislife.stanford.edu/LoCo_files/What-is-Life.pdf. Accessed on 1967.

Sterelny, K. (2007). *Dawkins vs. Gould: Survival of the Fittest (Revolutions in Science)*, 2nd revised ed. Icon Books Ltd., Cambridge, U.K.

Wang, J. et al. (2014) Mechanomics: An emerging field between biology and biomechanics. *Protein Cell 5*:518–531.

Making Light Work in Biology

3

Basic, Foundational Detection and Imaging Techniques Involving Ultraviolet, Visible, and Infrared Electromagnetic Radiation Interactions with Biological Matter

I don't suppose you happen to know
Why the sky is blue?...
Look for yourself. You can see it's true.

—John Ciardi (American poet 1916–1986)

General Idea: Here, we discuss the broad range of experimental biophysical techniques that primarily act through the detection of biological components using relatively routine techniques, which utilize basic optical photon absorption and emission processes and light microscopy, and similar methods that extend into the ultraviolet (UV) and infrared (IR) parts of the electromagnetic spectrum. These include methods to detect and image cells and populations of several cells, as well as subcellular structures, down the single-molecule level both for *in vitro* and *in vivo* samples. Although the techniques are ubiquitous in modern biophysics labs, they are still robust, have great utility in addressing biological questions, and have core physics concepts at their heart that need to be understood.

3.1 INTRODUCTION

Light microscopy, invented over 300 years ago, has revolutionized our understanding of biological processes. In its modern form, it involves much more than just the magnification of images in biological samples. There are invaluable techniques that have been developed to increase the image contrast. Fluorescence microscopy, in particular, is a very useful tool for probing biological processes. It results in high signal-to-noise ratios (SNRs) for determining the localization of biological molecules tagged with a fluorescent dye but does so in a way that is relatively noninvasive. This minimal perturbation to the native biology makes it a tool of choice in many biophysical investigations.

There has been enormous development of visible (VIS) light microscopy tools, which address biological questions at the level of single cells in particular, due in part to a bidirectional development in the operating range of sample length scales over recent years. *Top-down* improvements in *in vivo* light microscopy technologies have reduced the scale of spatial resolution down to the level of single cells, while *bottom-up* optimization of many emerging

DOI: 10.1201/9781003336433-3

single-molecule light microscopy methods originally developed for *in vitro* contexts has been applied now to living cells.

3.2 BASIC UV-VIS-IR ABSORPTION, EMISSION, AND ELASTIC LIGHT SCATTERING METHODS

Before we discuss the single-molecule light microscopy approaches, there are a number of basic spectroscopy techniques that are applied to bulk *in vitro* samples, which not only primarily utilize VIS light but also extend into UV and IR. Some of these may appear mundane at first sight, but in fact they hold the key to generating many preliminary attempts at robust physical quantification in the biosciences.

3.2.1 SPECTROPHOTOMETRY

In essence, a *spectrophotometer* (or *spectrometer*) is a device containing a photodetector to monitor the *transmittance* (or conversely the *reflectance*) of light through a sample as a function of wavelength. Instruments can have a typical wavelength range from the long UV (~200–400 nm) through to the VIS (~400–700 nm) up into the mid and far IR (~700–20,000 nm) generated from one or more broadband sources in combination with wavelength filters and/or *monochromators*. A monochromator uses either *optical dispersion* or *diffraction* in combination with mechanical rotation to select different wavelengths of incident light. Light is then directed through a solvated sample that is either held in a sample cuvette or sandwiched between transparent mounting plates. They are generally made from glass or plastic for VIS, sodium chloride for IR, or quartz for UV to minimize plate/cuvette absorption at these respective wavelengths. Incident light can be scanned over a range of wavelengths through the sample to generate a characteristic light absorption spectral response.

Scanning IR spectrophotometers exclusively scan IR wavelengths. A common version of this is the *Fourier transform infrared (FTIR) spectrometer*, which, instead of selecting one probe wavelength at any one time as with the scanning spectrophotometer, utilizes several in one go to generate a *polychromatic interference pattern* from the sample, which has some advantage in terms of SNR and spectral resolution.

The absorption signal can then be inverse Fourier transformed to yield the IR absorption spectrum. Such spectra can be especially useful for identifying different organic chemical motifs in samples, since the vibrational stretch energy of the different covalent bonds found in biomolecules corresponds to IR wavelengths and will be indicated by measurable absorption peaks in the spectrum. The equivalent angular frequency for IR absorption, ω, can be used to estimate the mean stiffness of a covalent bond (a useful parameter in molecular dynamics simulations, see Chapter 8), by modeling it as a simple harmonic oscillator of two masses m_1 and m_2 (representing the masses of the atoms either end of the bond) joined by a spring of stiffness k_r:

$$(3.1) \qquad\qquad k_r = \mu\omega^2$$

where μ is the *reduced mass* given by

$$(3.2) \qquad\qquad \mu = \frac{m_1 m_2}{m_1 + m_2}$$

Technical experts of IR spectrometers generally do not cite absorption wavelengths but refer instead to wavenumbers in units of cm^{-1}, with the range ~700–4000 cm^{-1} being relevant to most of the different covalent bonds found in biomolecules, which corresponds to a wavelength range of ~2.5–15 μm. Although broadband IR sources are still sometimes used in older machines, it is more common now to use IR laser sources.

The pattern of IR absorption peaks in the spectrum, their relative position in terms of wavelength and amplitude, can generate a unique *signature* for a given biochemical component and so can be invaluable for sample characterization and purity analysis. The principal drawback of *IR spectroscopy* is that water exhibits an intense IR absorption peak and samples need to be in a dehydrated state. IR absorption ultimately excites a transition between different bond vibrational states in a molecule (Figure 3.1a), which implies a change in electrical dipole moment, due to either electrical polar asymmetry of the atoms that form the bond or the presence of delocalized electronic molecular orbitals (Table 3.1).

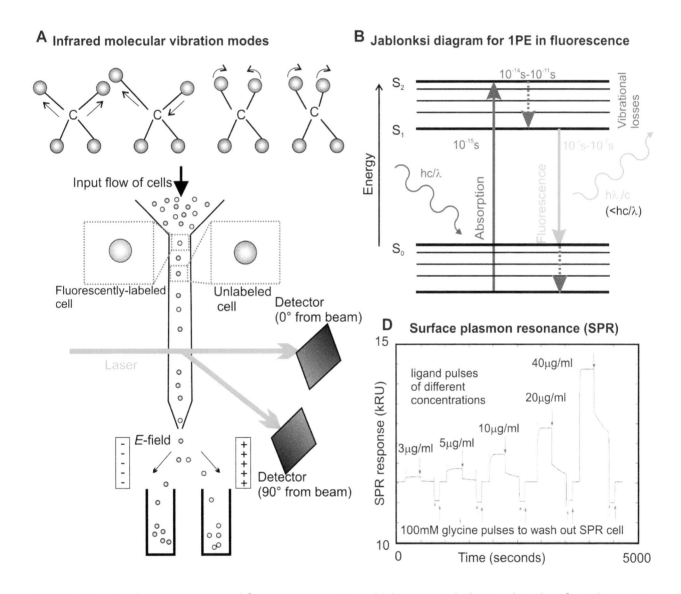

FIGURE 3.1   Simple spectroscopy and fluorescence excitation. (a) Some typical vibrational modes of a carbon atom–centered molecular motif that are excited by infrared absorption. (b) Jablonski energy level transition diagram for single-photon excitation resulting in fluorescence photon emission, characteristic time scales indicated. (c) Schematic of a typical fluorescence-assisted cell sorting device. (d) Typical output response from an SPR device showing injection events of a particular ligand, followed by washes and subsequent ligand injections into the SPR chamber of increasing concentration.

TABLE 3.1 Common Covalent Bonds in Biological Matter, Which Absorb Infrared Electromagnetic Radiation, with Associated Typical Absorption Ranges Indicated

Peak IR Absorption Range (cm⁻¹)	Bond in Biomolecule
730–770	C—H
1180–1200	C—O—C
1250–1340	C—N
1500–1600	C=C
1700–1750	C=O
2500–2700 (and other peaks)	O—H
3300–3500 (and other peaks)	N—H

Conventional FTIR is limited by a combination of factors including an inevitable trade-off between acquisition time and SNR and the fact that there is no spatial localization information. However, recent developments have utilized multiple intense, collimated IR beams from *synchrotron radiation* (see Chapter 5). This approach permits spatially extended detection of IR absorption signals across a sample allowing diffraction-limited time-resolved high-resolution *chemical imaging*, which has been applied to tissue and cellular samples (Nasse et al., 2011).

Many biomolecules that contain chemical bonds that absorb in the IR will also have a strong *Raman signal*. The *Raman effect* is one of the inelastic scattering of an excitation photon by a molecule, resulting in either a small increase or decrease in the wavelength of the scattered light. There is a rough *mutual exclusion principle*, in that strong absorption bands in the IR correspond to relatively weak bands in a Raman spectrum, and vice versa. Raman spectroscopy is a powerful biophysical tool for generating molecular signatures, discussed fully in Chapter 4.

Long UV light (~200–400 nm) is also a useful spectrophotometric probe especially for determining proteins and nucleic acid content in a sample. Peptide bonds absorb most strongly at ~280 nm wavelength, whereas nucleic acids such as RNA and DNA have a peak absorption wavelength of more like ~260 nm. It is common therefore to use the rate of absorption at these two wavelengths as a metric for protein and/or nucleic acid concentration. For example, a ratio of 260/280 absorbance of ~1.8 is often deemed as "pure" by biochemists for DNA, whereas a ratio of ~2.0 is deemed "pure" for RNA. If this ratio is significantly lower, it often indicates the presence of protein (or potentially other contaminants such as phenol that absorb strongly at or near 280 nm). With suitable calibration, however, the 260 and 280 nm absorption values can be used to determine the concentrations of nucleic acids and proteins in the absence of sample contaminants.

In basic spectrophotometers, the transmitted light intensity from the sample is amplified and measured by a photodetector, typically a photodiode. More expensive machines will include a second reference beam using an identical reference cuvette with the same solvent (generally water, with some chemicals to stabilize the pH) but no sample, which can be used as a baseline against which to reference the sample readings. This method finds utility in measuring sample density containing relatively large biological particulates (e.g., cells in suspension, to determine the so-called growth stage) to much smaller ones, such as molecules in solution.

To characterize attenuation, if we assume that the rate absorption of light parallel to the direction of propagation, say z, in an incrementally small slice through the sample is proportional to the total amount of material in that thin slice multiplied by the incident light intensity $I(z)$, then it is trivial to show for a homogeneous tissue:

$$(3.3) \qquad I(z) = I(0)\exp(-\sigma(\lambda)Cz)$$

This is called the "Beer–Lambert law," a very simple model that follows from the assumption that the drop in light intensity upon propagating through a narrow slice of sample is proportional to the incident intensity and the slice's width. It is empirically obeyed up to high scatterer concentrations beyond which electrostatic interactions between scatterers become significant. Here, σ is the mean absorption cross-sectional area of the tissue, which depends on the wavelength λ, and C is the concentration of the absorbing molecules. The absorbance A is often a useful quantity, defined as

$$(3.4) \qquad A = -\log\left(\frac{I}{I_0}\right)$$

For a real tissue over a wide range of z, there may be heterogeneity in terms of the types of molecules, their absorption cross-sectional areas, and their concentrations. The Beer–Lambert law can be utilized to measure the concentration of a population of cells. This is often cited as the *optical density* (*OD*) measurement, such that

$$(3.5) \qquad OD = \frac{A}{L}$$

where L is the total path length over which the absorbance measurement was made. Many basic spectrophotometers contain a cuvette, which is standardized at $L = 1$ cm, and so it is normal to standardize OD measurements on the assumption of a 1 cm path length.

Note that the *absorbance* measured from a spectrophotometer is not exclusively due to photon absorption processes as such, though photon absorption events may contribute to the reduction in transmitted light intensity, but rather scattering. In simple terms, general light scattering involves an incident photon inducing an oscillating dipole in the electron molecular orbital cloud, which then reradiates isotropically. The measured reduction in light intensity in passing through a biological sample in a standard VIS light spectrophotometer is primarily due to *elastic scattering* of the incident light. For scattering particles, the size of single cells is at least an order of magnitude greater than the wavelength of the incident light; this phenomenon is due primarily to *Mie scattering*. More specifically, this is often referred to as "Tyndall scattering": Mie scattering in a colloidal environment in which the scattering particles may not necessarily be spherical objects. A good example is the rod-shaped bacteria cells. Differences in scatterer shape results in apparent differences in OD; therefore, caution needs to be applied in ensuring that like is compared to like in terms of scatterer shape when comparing OD measurements, and if not, then a shape correction factor should be applied. Note that some absorbance spectrometers are capable of correcting for scattering effects.

These absorbance measurements are particularly useful for estimating the density of growing microbial cultures, for example, with many bacteria an OD unit of 1.0 taken at a conventional wavelength of 600 nm corresponds to ~10^8 cells mL^{-1}, equivalent to a typical cloudy looking culture when grown to "saturation." Spectrophotometry can be extended into *colorimetry* in which an indicator dye is present in the sample, which changes color upon binding of a given chemical. This can then be used to report whether a given chemical reaction has occurred or not, and so monitoring the color change with time will indicate details of the kinetics of that chemical reaction.

3.2.2 FLUORIMETRY

A modified spectrophotometer called a fluorimeter (or fluorometer) can excite a sample with incident light over a narrow band of wavelengths and capture fluorescence emissions. For bulk ensemble average *in vitro* fluorimetry investigations, several independent physical parameters are often consolidated for simplicity into just a few parameters to characterize the sample. For example, the absorption cross-section for a fluorescent sample is related to its extinction coefficient ε (often cited in non-SI units of M^{-1} cm^{-1}) and the molar concentration of the fluorophore c_m by

$$(3.6) \qquad \sigma(\lambda) = \frac{\varepsilon(\lambda)c_m}{C}$$

Therefore, the Beer–Lambert law for a fluorescent sample can be rewritten as

$$(3.7) \qquad I(z) = I(0)\exp\left(-c_m \varepsilon(\lambda)z\right)$$

The key physical process in bulk fluorimetry is *single-photon excitation*, that is, the processes by which energy from one photon of light is absorbed and ultimately emitted as a photon of lower energy. The process is easier to understand when depicted as a *Jablonski diagram* for the various energy level transitions involved (Figure 3.1b). First, the photon absorbed by the electron shells of an atom of a *fluorophore* (a fluorescent dye) causes an electronic transition to a higher energy state, a process that takes typically ~10^{-15} s. Vibrational relaxation due to *internal conversion* (in essence, excitation of an electron to a higher energy state results in a redistribution of charge in the molecular orbitals resulting in electrostatically driven oscillatory motion of the positively charged nucleus) relative movements can then occur over typically 10^{-12} to 10^{-10} s, resulting in an electronic energy loss. Fluorescence emission then can occur following an electronic energy transition back to the ground state over ca. 10^{-9} to 10^{-6} s, resulting in the emission of a photon of light of lower energy (and hence longer wavelength) than the excitation light due to the vibrational losses.

In principle, an alternative electronic transition involves the first excited triplet state energy level reached from the excited state via *intersystem crossing* in a classically forbidden transition from a net spin zero to a spin one state. This occurs over longer time scales than the fluorescence transition, ca. 10^{-3}–100 s, and results in emission of a lower-energy *phosphorescence* photon. This process can cause discrete photon bunching over these time scales, which is not generally observed for typical data acquisition time scales greater than a millisecond as it is averaged out. However, there are other fluorescence techniques using advanced microscopy in which detection is performed over much faster time scales, such as *fluorescence lifetime imaging microscopy* (FLIM) discussed later in this chapter, for which this effect is relevant.

The electronic energy level transitions of ground-state electron excitation to excited state, and from excited state back to ground state, are *vertical transitions* on the Jablonski diagram. This is due to the quantum mechanical *Franck–Condon principle* that implies that the atomic nucleus does not move during these two opposite electronic transitions and so the vibration energy levels of the excited state resemble those of the ground state. This has implications for the symmetry between the excitation and emission spectra of a fluorophore.

But for *in vitro* fluorimetry, a cuvette of a sample is excited into fluorescence often using a broadband light source such as a mercury or xenon arc lamp with fluorescence emission measured through a suitable wavelength bandwidth filter at 90° to the light source to minimize detection of incident excitation light. Fluorescence may either be emitted from a fluorescent dye that is attached to a biological molecule in the sample, which therefore acts as a "reporter." However, there are also naturally fluorescent components in biological material, which have a relatively small signal but which can be measurable for *in vitro* experiments, which often include purified components at greater concentrations that occur in their native cellular environment.

For example, *tryptophan fluorescence* involves measuring the native fluorescence of the aromatic amino acid tryptophan (see Chapter 2). Tryptophan is very *hydrophobic* and thus is often buried at the center of folded proteins far away from surrounding water molecules. On exposure to water, its fluorescence properties change, which can be used as a metric for whether the protein is in the folded conformational state. Also, chlorophyll, which is a key molecule in plants as well as many bacteria essential to the process of photosynthesis (see Chapter 2), has significant fluorescent properties.

Note also that there is sometimes a problematic issue with *in vitro* fluorimetry known as the "inner filter effect." The primary inner filter effect (*PIFE*) occurs when the absorption of a fluorophore toward the front of the cuvette nearest the excitation beam entry point reduces the intensity of the beam experienced by fluorophores toward the back of the cuvette and so can result in apparent nonlinear dependence of measured fluorescence with sample concentration. There is also a secondary inner filter effect (*SIFE*) that occurs when the fluorescence intensity decreases due to fluorophore absorption in the emission region. PIFE in general is a more serious problem than SIFE because of the shorter wavelengths for excitation compared to emission. To properly correct these effects requires a controlled titration at different fluorophore concentrations to fully characterize the fluorescence response. Alternatively, a mathematical model can be approximated to characterize the effect. In practice, many researchers ensure that they operate in a concentration regime that is sufficiently low to ignore the effect.

3.2.3 FLOW CYTOMETRY AND FLUORESCENCE-ASSISTED CELL SORTING

The detection of scattered light and fluorescence emissions from cell cultures are utilized in powerful *high-throughput* techniques of *flow cytometry* and *fluorescence-assisted cell sorting* (*FACS*). In flow cytometry, a culture of cells is flowed past a detector using controlled *microfluidics*. The diameter of the flow cell close to the detector is $\sim 10^{-5}$ m, which ensures that only single cells flow past the detector at any one time. In principle, a detector can be designed to measure a variety of different physical parameters of the cells as they flow past, for example, electrical impedance and optical absorption. However, by far the most common detection method is based on focused laser excitation of cells in the vicinity of a sensitive photodetector, which measures the fluorescence emissions of individual cells as they flow past.

Modern commercial instruments have several different wavelength laser sources and associated fluorescence detectors. Typically, cells under investigation will be labeled with a specific fluorescent dye. The fluorescence readout from flow cytometry can therefore be used as a metric for purity of subsequent cell populations, that is, what proportion of a subsequent cell culture contains the original labeled cell. A common adaptation of flow cytometry is to incorporate the capability to sort cells on the basis of their being fluorescently labeled or not, using FACS. A typical FACS design involves detection of the fluorescence signature with a photodetector that is positioned at 90° relative to the incident laser beam, while another photodetector measures the direct transmission of the light, which is a metric for size of the particle flow past the detector that is thus often used to determine if just a single cell is flowing past as opposed to, more rarely, two or more in the line with the incident laser beam (Figure 3.1c).

Cells are usually sorted into two populations of those that have a fluorescence intensity above a certain threshold, and those that do not. The sorting typically uses rapid electrical feedback of the fluorescence signal to electrostatics plates; the flow stream is first interrupted using piezoelectric transducers to generate nanodroplets, which can be deflected by the electrostatic plates so as to shunt cells into one of two output reservoirs. Other commercial FACS devices use direct mechanical sorting of the flow, and some bespoke devices have implemented methods based on optical tweezers (OTs) (Chapter 6).

FACS results in a very rapid sorting of cells. It is especially useful for generating purity in a heterogeneous cell population. For example, cells may have been genetically modified to investigate some aspect of their biology; however, the genetic modifications might not have been efficiently transferred to 100% of the cells in a culture. By placing a suitable fluorescent marker on only the cells that have been genetically modified, FACS can then sort these efficiently to generate a pure culture output that contains only these cells.

3.2.4 POLARIZATION SPECTROSCOPY

Many biological materials are birefringent, or *optically active*, often due to the presence of repeating molecular structures of a given shape, which is manifested as an ability to rotate the plane of polarization of incident light in an *in vitro* sample. In the *linear dichroism* (*LD*) and *circular dichroism* (*CD*) techniques, spectrophotometry is applied using polarized incident light with a resultant rotation of the plane of polarization of the *E*-field vector as it propagates through the sample. LD uses a linearly polarized light as an input beam, whereas CD uses circularly polarized light that in general results in an *elliptically polarized* output for propagation through an optically active sample. The *ellipticity* changes are indicative of certain specific structural motifs in the sample, which although not permitting fine structural detail to be explored at the level of, for example, atomistic detail, can at least indicate the relative proportions of different generic levels of secondary structure, such as the relative proportions of β-sheet, α-helix, or random coil conformations (see Chapter 2) in a protein sample.

CD spectroscopic techniques display an important difference from LD experiments in that biomolecules in the sample being probed are usually free to diffuse in solution and so have a random orientation, whereas those in LD have a fixed or preferred molecular orientation. A measured CD spectrum is therefore dependent on the intrinsic asymmetric (i.e., *chiral*) properties of the biomolecules in the solution, and this is useful for determining the secondary structure of relatively large biomolecules in particular, such as biopolymers of proteins or nucleic acids. LD spectroscopy instead requires the probed biomolecules to have a fixed or preferred orientation; otherwise if random molecular orientation is permitted, the net LD effect to rotate the plane of input light polarization is zero.

To achieve this, the preferred molecular orientation flow can be used to comb out large molecules (see Chapter 6) in addition to various other methods including magnetic field alignment, conjugation to surfaces, and capturing molecules into gels, which can be extruded to generate preferential molecular orientations. LD is particularly useful for generating information of molecular alignment on surfaces since this is where many biochemical reactions occur in cells as opposed to free in solution, and this can be used to generate time-resolved information for biochemical reactions on such surfaces.

LD and CD are complementary biophysical techniques; it is not simply that linearly polarized light is an extreme example of circularly polarized light. Rather, the combination of both techniques can reveal valuable details of both molecular structure and kinetics. For example, CD can generate information concerning the secondary structure of a folded protein that is integrated in a cell membrane, whereas LD might generate insight into how that protein inserts into the membrane in the first place.

Fluorescence excitation also has a dependence on the relative orientation between the *E*-field polarization vector and the *transition dipole moment* of the fluorescent dye molecule, embodied in the *photoselection rule* (see Corry, 2006). The intensity *I* of fluorescence emission from a fluorophore whose transition dipole moment is oriented at an angle θ relative to the incident *E*-field polarization vector is as follows:

$$(3.8) \qquad\qquad I(\theta) = I(0)\cos^2\theta$$

In general, fluorophores have some degree of freedom to rotate, and many dyes in cellular samples exhibit in effect *isotropic emissions*. This means that over the timescale of a single data sampling window acquisition, a dye molecule will have rotated its orientation randomly many times, such that there appears to be no preferential orientation of emission in any given sampling time window. However, as the time scale for sampling is reduced, the likelihood for observing *anisotropy, r,* that is, preferential orientations for absorption and emission, is greater. The threshold time scale for this is set by the *rotational correlation time* τ_R of the fluorophore in its local cellular environment attached to a specific biomolecule. The anisotropy can be calculated from the measured fluorescence intensity, either from a population of fluorophores such as in *in vitro* bulk fluorescence polarization measurements or from a single fluorophore, from the measured emission intensity parallel or perpendicular (I_2) to the incident linear E-field polarization vector after a time t:

$$(3.9) \qquad r(t) = \frac{I_1(t) - I_2(t)}{I_2(t) + 2I_2(t)}$$

In fluorescence anisotropy measurements, the detection system will often respond differently to the polarization of the emitted light. To correct for this, a *G-factor* is normally used, which is the ratio of the vertical polarization detector sensitivity to the horizontal polarization detector sensitivity. Thus, in Equation 3.9, the parameter $I_2(t)$ is replaced by $GI_2(t)$.

Note that another measure of anisotropy is sometimes still cited in the literature as a parameter confusingly called the polarization, P, such that $P = 3r/(2 + r) = (I_1 - I_2)/(I_1 + I_2)$, to be compared with Equation 3.9. The anisotropy decays with time as the fluorophore orientation rotates, such that for freely rotating fluorophores

$$(3.10) \qquad r(t) = r_0 \exp\left[\frac{-t}{\tau_R}\right]$$

where r_0 is called the "initial anisotropy" (also known as the "fundamental anisotropy"), which in turn is related to the relative angle θ between the incident E-field polarization and the transition dipole moment by

$$(3.11) \qquad r_0 = \frac{(3\cos^2\theta - 1)}{5}$$

This indicates a range for r_0 of –0.2 (perpendicular dipole interaction) to +0.4 (parallel dipole interaction). Anisotropy can be calculated from the measured fluorescence intensity, for example, from a population of fluorophores such as in *in vitro* bulk fluorescence polarization measurements. The rotational correlation time is inversely proportion to the rotational diffusion coefficient D_R such that

$$(3.12) \qquad \tau_R = \frac{1}{6D_R}$$

The rotational diffusion coefficient is given by the Stokes–Einstein relation (see Chapter 2), replacing the drag term for the equivalent rotational drag coefficient. Similarly, the mean squared angular displacement $\langle\theta^2\rangle$ observed after a time t relates to D_R in an analogous way as for lateral diffusion:

$$(3.13) \qquad \langle\theta^2\rangle = 2D_R t$$

For a perfect sphere of radius r rotating in a medium of viscosity η at absolute temperature T, the rotational correlation time can be calculated exactly as

$$(3.14) \qquad \tau_R = \frac{4\pi r^3 \eta}{3k_B T}$$

Molecules that integrate into phospholipid bilayers, such as *integrated membrane proteins* and membrane-targeting organic dyes, often orientate stably parallel to the hydrophobic tail groups of the phospholipids such that their rotation is confined to that axis with the frictional drag approximated as that of a rotating cylinder about its long axis using the

Saffman–Delbrück equations (see Saffman, 1975; Hughes, 1981). Here, the frictional drag γ of a rotating cylinder is approximated as

$$(3.15) \qquad \Upsilon = 4\pi\left(\mu_1 + \mu_2\right)rC(\varepsilon) = 8\pi\eta_c rC(\varepsilon)$$

We assume the viscosities of the watery environment just outside the cell membrane and just inside the cell cytoplasm (μ_1 and μ_2, respectively) are approximately the same, η_c (typically ~0.001–0.003 Pa·s). The dimensionless parameter ε is given by

$$(3.16) \qquad \varepsilon = \frac{(r/h)\left(\mu_1 + \mu_2\right)}{\eta_m} = \frac{2r\eta_c}{h\eta_m}$$

The viscosity of the phospholipid bilayer is given by the parameter η_m (~100–1000 times greater than η_c depending on both the specific phospholipids present and the local molecular architecture of nonlipids in the membrane). The parameter C can be approximated as

$$(3.17) \qquad C(\varepsilon)\left\{\varepsilon\left(\frac{\ln(2/\varepsilon) - c + 4\varepsilon/\pi - \varepsilon^2\ln(2/\varepsilon)}{2} + O(\varepsilon^2)\right)\right\}^{-1}$$

Here c is *Euler–Mascheroni constant* (approximately 0.5772). The effective rotational diffusion coefficient can then be calculated in the usual way using the Stokes–Einstein relation and then the rotational correlation time is estimated.

Typical nanometer length scale fluorophores in the watery cytoplasm of cells have rotational correlation times of a few nanoseconds (ns), compared to a few microseconds (μs) in a typical phospholipid bilayer. These parameters can be measured directly using *time-resolved anisotropy*, with a suitable fluorescence polarization spectrometer that can typically perform sub-nanosecond sampling. The application of fluorescence anisotropy to cellular samples, typically in a culture medium containing many thousands of cells, offers a powerful method to probe the dynamics of protein complexes that, importantly, can be related back to the actual structure of the complexes (see Piston, 2010), which has an advantage over standard fluorescence microscopy methods.

3.2.5 OPTICAL INTERFEROMETRY

There are two principal bulk *in vitro* sample optical interferometry techniques: *dual polarization interferometry (DPI)* and *surface plasmon resonance (SPR)*. In DPI, a reference laser beam is guided through an optically transparent sample support, while a sensing beam is directed through the support at an oblique angle to the surface. This steep angle of incidence causes the beam to be totally internally reflected from the surface, with a by-product of generating an *evanescent field* into the sample, generally solvated by water for the case of biophysical investigations, with a characteristic depth of penetration of ~100 nm. This is an identical process to the generation of an evanescent field for *total internal reflection fluorescence (TIRF)* microscopy, which is discussed later in this chapter. Small quantities of material from the sample that bind to the surface have subtle but measureable effects upon polarization in this evanescent field. These can be detected with high sensitivity by measuring the interference pattern of the light that results between sensing and reference beams. DPI gives information concerning the thickness of the surface-adsorbed material and its refractive index.

SPR operates similarly in that an evanescent field is generated, but here a thin layer of metal, ~10 nm thick, is first deposited on the outside surface (usually embodied is a commercial SPR chip that can be removed and replaced as required). At a certain angle of incidence to the surface, the sensing beam reflects slightly less back into the sample due to a resonance effect via the generation of oscillations in the electrons at the metal surface interface, *surface plasmons*. This measured drop in reflected intensity is a function of the absolute amount of the adsorbed material on the metal surface from the sample, and so DPI and SPR are essentially complementary. Both yield information on the stoichiometry and binding kinetics of biological samples. These can be used in investigating, for example, how cell membrane receptors function; if the surface is first coated with purified receptor proteins and the sample chamber contains a ligand thought to bind to the receptor, then both DPI and SPR can be used to measure the strength of this binding, and subsequent unbinding, and to estimate the relative numbers of ligand molecules that bind for every receptor protein (Figure 3.1d).

3.2.6 PHOTOTHERMAL SPECTROSCOPY

There are a group of related *photothermal spectroscopy* techniques that, although perhaps less popular now than toward the end of the last century due to improvements in sensitivity of other optical spectroscopy methods, are still very sensitive methods that operate by measuring the optical absorption of a sample as a function of its thermal properties. Photothermal spectroscopy is still in use to quantify the kinetics of biochemical reactions, which are initiated by light, for example, by direct photochemical reactions or by environmental changes induced by light such as changes in the cellular pH. The time scales of these processes typically span a broad time range from 10^{-12} to 10^{-3} s that are hard to obtain by using other spectroscopy methods.

Incident light that is not scattered, absorbed, or converted into fluorescence emission in optical spectroscopy is largely converted to heat in the sample. Therefore, the amount of temperature rise in an optical absorption measurement is a characteristic of the sample and a useful parameter in comparing different biological materials. *Photothermal deflection spectroscopy* can quantify the changes in a sample's refractive index upon heating. It uses a laser beam probe on an optically thin sample and is useful in instances of highly absorbing biomolecule samples in solution, which have too low transmission signal to be measured accurately. *Photothermal diffraction* can also be used to characterize a biological material, which utilizes the interference pattern produced by multiple laser sources in the sample to generate a diffraction grating whose aperture spacing varies with the thermal properties of the sample.

In *laser-induced optoacoustic spectroscopy*, a $\sim 10^{-9}$ s VIS light laser pulse incident on a sample in solution results in the generation of an acoustic pressure wave. The time evolution of the pressure pulse can be followed by high-bandwidth piezoelectric transducers, typically over a time scale of $\sim 10^{-5}$ s, which can be related back to time-resolved binding and conformational changes in the biomolecules of sample. The technique is a tool of choice for monitoring time-resolved charge transfer interactions between different amino acids in a protein since there is no existing alternative spectroscopic technique with the sensitivity to do so.

KEY BIOLOGICAL APPLICATIONS: BASIC OPTICAL SPECTROSCOPY

Quantifying molecular and cell concentrations *in vitro*; Determining biomolecular secondary structures *in vitro*; Measuring molecular conformational changes from bulk solutions; Characterizing chemical bond types from bulk solutions.

3.3 LIGHT MICROSCOPY: THE BASICS

Light microscopy in some ways has gone full circle since its modern development in the late seventeenth and early eighteenth centuries by pioneers such as Robert Hooke (Hooke, 1665; but see Fara, 2009 for a modern discussion) and Antonj van Leeuwenhoek (see van Leeuwenhoek, 1702). In these early days of modern microscopy, different whole organisms were viewed under the microscope. With technical advances in light microscopy, and in the methods used for sample preparation, the trend over the subsequent three centuries was to focus on smaller and smaller length scale features.

3.3.1 MAGNIFICATION

The prime function of a light microscope is to magnify features in a biological sample, which are illuminated by VIS light, while maintaining acceptable levels of image clarity, contrast, and exhibiting low optical aberration effects. Magnification can be performed most efficiently using a serial combination of lenses. In a very crude form, a single lens is in effect a very simple light microscope but offering limited magnification. In its most simple practical form, a light microscope consists of a high *numerical aperture* (*NA*) objective lens placed very close to the sample, with a downstream imaging lens focusing the sample image onto a highly sensitive light detector such as a high-efficiency charge-coupled device (CCD) camera, or sometimes a photomultiplier tube (PMT) in the case of a scanning system such as in confocal microscopy (Figure 3.2a). Most microscopes are either *upright* (objective lens positioned above the sample stage) or *inverted* (objective lens positioned below the sample stage).

The two-lens microscope operates as a simple telescope system, with magnification *M* given by the ratio of the imaging lens focal length to that of the objective lens (the latter typically being a few millimeters):

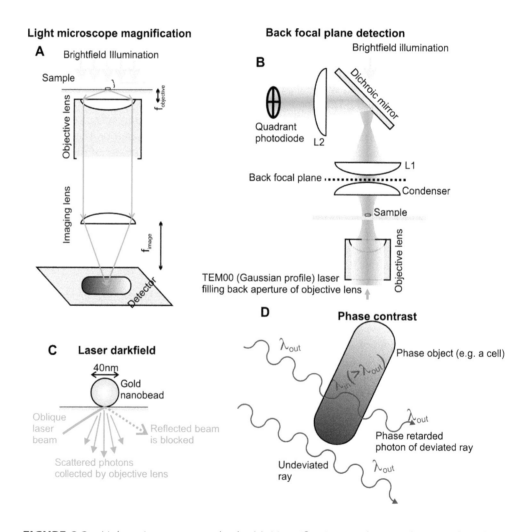

FIGURE 3.2 *Light microscopy methods. (a) Magnification in the simplest two-lens light microscope. (b) Back focal plane detection (magnification onto quadrant photodiode is f_2/f_1 where f_n is the corresponding focal length of lens **Ln**). (c) Laser dark field. (d) Phase retardation of light through a cell sample in phase contrast microscopy.*

(3.18)
$$M = \frac{f_{image}}{f_{objective}}$$

In practice, there are likely to be a series of several lens pairs placed between the imaging lens shown and the detector arranged in effect as telescopes, for example, with focal lengths $f_1, f_2, f_3, f_4, ..., f_n$ corresponding to lenses placed between the objective lens and the camera detector. The magnification of such an arrangement is simply the multiplicative combination of the separate magnifications from each lens pair:

(3.19)
$$M = \frac{f_2 \cdot f_4, ..., f_n}{f_1 \cdot f_3, ..., f_{n-1}}$$

Such additional lens pairs allow higher magnifications to be obtained without requiring a single imaging lens with an exceptionally high or low focal length, which would necessitate either an impractically large microscope or would result in severe optical aberration effects. A typical standard light microscope can generate effective total magnifications in the range 100–1000.

3.3.2 DEPTH OF FIELD

The *depth of field* (d_f, also known as the "depth of focus") is a measure of the thickness parallel to the optical axis of the microscope over which a sample appears to be in focus. It is conventionally defined as one quarter of the distance between the intensity minima parallel to the optic axis above and below the exact focal plane (i.e., where the sample in principle should be most sharply in focus) of the diffraction image that is produced by a single-*point source* light emitting object in the focal plane. This 3D diffraction image is a convolution of the *point spread function (PSF)* of the imaging system with a delta function. On this basis, d_f can be approximated as

(3.20)
$$d_f = \frac{\lambda n_m}{NA^2} + \frac{d_R n_m}{M_L NA}$$

where
  λ is the wavelength of light being detected
  n_m is the refractive index of the medium between the microscope objective lens or numerical aperture NA and the glass microscope coverslip/slide (either air, $n_m = 1$, or for high-magnification objective lenses, immersion oil, $n_m = 1.515$
  d_R is the smallest length scale feature that can be resolved by the image detector (e.g., the pixel size of a camera detector) such that the image is projected onto the detector with a total lateral magnification of M_L between it and the sample.

3.3.3 LIGHT CAPTURE FROM THE SAMPLE

The NA of an objective lens is defined $n_m \sin \theta$, where n_m is the refractive index of the imaging media. The angle θ is the maximum half-angle subtended ray of light scattered from the sample, which can be captured by the objective lens. In other words, higher NA lenses can capture more light from the sample. In air, $n_m = 1$ so to increase the NA, further high-power objective lenses use immersion oil; a small blob of imaging oil is placed in optical contact between the glass microscope slide or coverslip and the objective lens, which has the same high value of refractive index as the glass.

The solid angle Ω subtended by this maximum half angle can be shown using simple integration over a sphere to be

(3.21) $$\Omega = 2\pi(1 - \cos\theta)$$

Most *in vivo* studies, that is, those done on living organisms or cells, are likely to be low magnification $M_L \sim 100$ using a low numerical aperture objective lens of $NA \sim 0.3$ such as to encapsulate a large section of tissue on acquired images, giving a d_f of ~10 μm. Cellular studies often have a magnification an order of magnitude greater than this with NA values of up to ~1.5, giving a d_f of 0.2–0.4 μm.

Note that the human eye has a maximum numerical aperture of ~0.23 and can accommodate typical distances between ~25 cm and infinity. This means that a sample viewed directly via the eye through a microscope eyepiece unit, as opposed to imaged onto a planar camera detector, can be observed with a far greater depth of field than Equation 3.20 suggests. This can be useful in terms of visual inspection of a sample prior to data acquisition from a camera device.

KEY POINT 3.1

Higher-magnification objective lenses have higher NA values resulting in a superior optical resolution, but with a much smaller depth of field than lower magnification lenses.

3.3.4 PHOTON DETECTION AT THE IMAGE PLANE

The technology of photon detection in light microscopes has improved dramatically over the past few decades. Light microscopes use either an array of pixel detectors in a high-sensitivity camera, or a single detector in the form of a *PMT* or *avalanche photodiode* (*APD*). A PMT utilizes the *photoelectric effect* on a primary photocathode metal-based scintillator detector to generate a primary electron following absorption of an incident photon of light. This electrical signal is then amplified through secondary emission of electrons in the device. The electron multiplier consists of a series of up to 12 anodes (or dynodes) held at incrementally higher voltages, terminated by a final anode. At each anode/dynode, ~5 new secondary electrons are generated for each incident electron, indicating a total amplification of $\sim 10^8$. This is sufficient to generate a sharp current pulse, typically 1 ns, after the arrival of the incident photon, with a sensitivity of single-photon detection.

An APD is an alternative technology to a PMT. This uses the photoelectric effect but with semiconductor photon detection coupled to electron–hole avalanche multiplication of the signal. A high reverse voltage is applied to accelerate a primary electron produced following initial photon absorption in the semiconductor with sufficient energy to generate secondary electrons following impact with other regions of the semiconductor (similarly, with a highly energetic electron hole traveling in the opposite direction), ultimately generating an enormous amplification of free electron–hole pairs. This is analogous to the amplification stage in a PMT, but here the amplification occurs in the same semiconductor chip. The total multiplication of signal is $>10^3$, which is less sensitive than a PMT, however still capable of single-photon detection with an advantage of a much smaller footprint, permitting in some cases a 2D array of APDs to be made, similar to pixel-based camera detectors.

Many light microscopes utilize camera-based detection over PMT/APD detection primarily for advantages in sampling speed in not requiring slow mechanically scanning over the sample. Several standard light microcopy investigations that are not photon limited (e.g., bright-field investigations) use *CCD* image sensors, with the most sensitive light microscopes using *electron-multiplying CCD* (*EMCCD*) detection or complementary MOS (CMOS) technology. A CCD image sensor contains a 2D array composed of individual p-doped metal-oxide semiconductor (MOS) pixels. MOS pixels act as micron length scale capacitors with a voltage bias set just above the threshold for inversion, which thus generates electrons and holes on absorption of incoming photons.

Each MOS capacitor accumulates an electric charge proportional to the light absorbed, and control circuitry transfers this charge to its neighbor along each 1D line of the pixel, such that the last MOS capacitor in the line dumps its charge into the MOS pixel in the next line up or down, ultimately with all the charge transferred into an amplifier. This serial voltage data stream can then be subsequently reconstructed as a 2D image. A variant on the CCD includes the *intensified CCD* (*ICCD*) that comprises an initial detection step on a phosphor screen, with this phosphor light image then detect by a CCD behind it, which improves the ultimate photon detection efficiency to >90%.

Many cheaper cameras utilize a CMOS chip (these are now found ubiquitously in webcams, mobile phone cameras, and also in microprocessors in nonimaging applications). The core feature of a CMOS chip is a symmetrical back-to-back combination of n- and p-type *MOS field effect transistors*, requiring less additional circuitry with greater power efficiency compared to CCD pixels, manifest ultimately substantially faster imaging speeds. A scientific CMOS camera has an inferior photon collection efficiency of ~50% compared to ICCDs or EMCCDs, but can acquire data faster by an order of magnitude or more, equivalent to several thousand image frames per second.

An EMCCD utilizes a solid-state electron-multiplying step at the end of each line of CCD pixels. This amplifies relatively weak electrical signal above any *readout noise* that is added from the final output amplification step. This electron multiplication has normally a few hundred stages during which electrons are transferred by impact ionization, which generates multiple secondary electrons to amplify the signal. The resultant amplification is up to ~10^3, which compares favorably to APDs and ICCDs but with a much-reduced readout noise. EMCCDs are currently the photon detection tool of choice for low-light microscopy investigations in biology, having up to ~95% photon detection efficiency, for example, applied to single-molecule fluorescence detection, and have a reasonable sampling speed equivalent to ~1 ms per image frame for small pixel arrays of ~100 pixels of edge length, relevant to many fast biological processes.

3.4 NONFLUORESCENCE MICROSCOPY

Basic light microscopy is invaluable as a biophysical tool. However, its biggest weakness is poor image contrast, since most of the material in living organisms is water, on average ~60%. Since cells are surrounded by a fluid environment, which is largely water, the signal obtained from VIS light scattered from cellular object is small. However, there are several adaptations to basic light microscopy that can be applied to enhance image contrast.

3.4.1 BRIGHT-FIELD AND DARK-FIELD MICROSCOPY

Bright-field microscopy relies on measuring the differences in the absorbed or scattered intensity of light as it passes through different features of the sample. Incident light is generated usually from either a tungsten halogen filament broadband source or a bright LED, which is captured by short focal length collector lens. Light is then directed through a *condenser* lens using a *Köhler illumination* design that involves forming an image of the light source in the back focal plane of the condenser (Figure 3.2b). This results in a collimated beam incident on the sample and a uniform illumination intensity in the focal plane of the microscope.

A cell on a microscope coverslip/slide whose shape is broadly symmetrical on either side of a plane parallel to a focal plane taken through its midheight will exhibit minimal bright-field image contrast between the foreground cellular material and the background cell media. A simple approach to increase the contrast for the outline of the cell is to use *defocusing microscopy*. Negative defocusing, for which the focal plane is moved below the midheight level of the cell closer to the object lens, generates an image of a dark cell body with a higher-intensity cell perimeter, while positive defocusing generates the inverse of this, due to light interference at the image plane between undeviated and transmitted light beams whose optical path length (OPL), which is the product of the geometrical path length of the light beam with the index of refraction in that optical media, depends upon the extent of defocus. The contrast at the cell perimeter is a function of the radius of curvature of the cell, and for cells of a micron length scale a defocus value of a few hundred nanometers generates optimum perimeter contrast.

Another approach to improve contrast is to tag a biomolecule with a reporter probe designed to generate a high local signal upon excitation with light, for example, a probe coated in a high atomic number metal such as gold generates a high scatter signal for certain wavelengths of VIS light. Here, photon scattering is *elastic* and so the wavelength of scattered light is the same as the incident light; thus, any scattered light not from the labeled biomolecules must be blocked. These include back reflections from the glass microscope coverslip/slide. The regions of the sample not containing tagged biomolecules appear dark on the camera detector, hence the name *dark-field microscopy*. In *transmitted light dark field*, a modified condenser lens blocks out the central aperture resulting in highly oblique illumination on the sample; nonscattered light will emerge at too steep an angle to be captured by the objective lens, whereas light diffracted by the sample will be forward scattered at small angles and can be captured.

A similar approach can be used with reflected light, in general using a laser source (hence *laser dark field*) in which an oblique angled laser beam incident on the sample emerging from the objective lens is either transmitted through the coverslip in the absence of any sample, or back scattered by the sample back into the objective lens (Figure 3.2c). An additional enhancement of contrast can be achieved by the generation of *surface plasmons*, whose intensity is a function of the particle size (a few tens of nanometers) and the laser wavelength. This can generate very high SNRs on *in vitro* samples facilitating extremely high time resolutions of $\sim 10^{-6}$ s.

Dark-field microscopy has more limited use with living samples than bright-field microscopy because the relatively large size of a cell compared to the scatter signal either from a native unlabeled biomolecule or from a biomolecule that has been labeled using a dark-field probe (e.g., a gold-coated bead of tens of nanometers in diameter) can result in significant scatter from the cell body itself, which can swamp the probe signal. The scatter signal from unlabeled biomolecules can be prohibitively small, but using a scatter label can also present technical challenges; it is not easy to specifically label biomolecules inside living cells with, for example, a gold nanoscale bead without nonspecific labeling of other cellular structures.

For certain cell types (e.g., prokaryotes), it is also difficult to introduce such a large scatter probe while still keeping the cell intact, limiting its application to accessible surface features. The practical lower size limit to detect a reproducible, measurable signal from the scattered light is a few tens of nanometers; the size of the probe is large compared to single biomolecules implying some steric hindrance effects with impairment of normal biological operation. Note that there are advanced new techniques such as interferometric scattering microscopy, which use interferometric methods of scattered light detection that can be used to detect the scattered signal directly from unlabeled biomolecules themselves (see Chapter 4).

3.4.2 CONTRAST ENHANCEMENT USING OPTICAL INTERFERENCE

Optical interference techniques can be used to modify a light microscope to increase image contrast. *Phase contrast microscopy* utilizes differences in *refractive index* inside a biological sample. Optically, transparent cells and tissues exhibit a range of refractive indices n_t as a function of spatial localization across the tissue, 1.35–1.45 being typical for cellular material, which compares with that of water, $n_w \sim 1.33$. The result is that the phase of light propagating through a region of sample, which has a length parallel to the optic axis Δz of a few microns, will be retarded. Such objects are examples of *phase objects*. The retardation for many cells is roughly around a *quarter of a wavelength* λ relative to that of light passing through the largely aqueous environment in between cells (Figure 3.2d). It is trivial to derive the following relationship by comparing the speed of light in water to that inside a cell or tissue on the basis of differences in refractive indices:

(3.22)
$$\Delta z = \frac{\lambda n_w}{4\left(n_t - n_w\right)}$$

An annulus aperture in the front focal plane of the condenser lens, similar to that used for dark-field forward scatter microscopy in blocking out the central aperture of illumination, generates a cone of collimated light onto the sample. Emergent light transmitted through the sample is collected by an objective lens consisting of both undeviated light (since the angle of the cone of light is not as oblique as that used in dark-field microscopy) that has not encountered any biological material and diffracted (forwarded scattered) light that has exhibited a relative phase retardation to the undeviated light.

A *phase ring* in the back focal plane of the objective lens, in a conjugate image plane to the condenser annulus, converts this retardation into a *half wavelength phase shift*, a condition for *destructive* interference, either by introducing a half wavelength phase increase in the ring (*positive phase contrast microscopy*) by having an extra thickness of glass, for example, in which case the background appears darker relative to the foreground sample, or more commonly by introducing a further half wavelength phase retardation in the ring (*negative phase contrast microscopy*) by indenting the glass in that region, in which case the sample appears brighter relative to the background, or by coating the ring in a thin layer of aluminum.

In other words, this process transforms phase information at the sample into amplitude contrast in the intensity of the final image. The length scale of a few microns over which the retardation of the light is typically a quarter of a wavelength is comparable to some small cells in tissues, as well as cellular organelle features such as the nucleus and mitochondria. It is therefore ideal for enhancing the image contrast of cellular components.

Polarized light microscopy can increase the relative contrast of birefringent samples. *Birefringence*, as discussed for polarization spectroscopy techniques in Section 3.2.4, occurs when a sample has a refractive index which is dependent upon the orientation of the polarization E-field vector of the incident light. This is often due to repeating structural features in a sample, which have a spatial periodicity over a length scale comparable to, or less than, the wavelength of the light, which is true for several biological structures. In other words, this is a characteristic of certain crystals or more relevant for biological samples due to the fluidity of the water-solvent environment and other fluidic structures such as phospholipid bilayers, *liquid crystals*.

There are several examples of birefringent biological liquid crystals. These include fibrous proteins with well-defined spatial periodicity between bundles of smaller *fibrils* such as collagen in the extracellular matrix, cell membranes and certain proteins in the cell membranes, cytoskeletal proteins, structural proteins in the cell walls of plants (e.g., cellulose) and certain bacteria (e.g., proteoglycans), and the highly periodic protein capsid coats of viruses. Polarization microscopy is an excellent tool for generating images of these biological liquid crystal features, and there are also examples of nonliquid crystalline biomolecule samples that can be investigated similarly (e.g., crystalline arrays of certain vitamins).

For polarization microscopy, a *polarizer* is positioned in the illumination path between the VIS light source and a condenser lens, before the sample, and a second polarizer described as an *analyzer* is positioned after the transmitted light has emerged from the sample, close to the back aperture of the objective lens. The transmitted light through a birefringent sample can be split into two orthogonally polarized light components of p and s, which are either parallel to the plane of the optic axis or perpendicular to it, respectively. The speed of the light in each of these separate components is different due to the polarization dependence of the refractive index in the sample. These components therefore become out of phase with each other but are recombined with various combinations of constructive and destructive interference during their passage through the analyzer, depending upon the relative position on the sample, which is then imaged onto a camera (or viewed through eyepieces) in the normal way for basic light microscopy. Polarized light microscopy can quantify the precise amount of retardation that occurs in each polarization direction and thus generates information about the relative orientation of spatially periodic molecular structures of the birefringent sample. Some of these structures are involved in mechanical features of tissues and cells, and thus polarization microscopy can be used to probe biomechanics (see Chapter 6).

Differential interference contrast (*DIC*) microscopy is a related technique to polarization microscopy, in using a similar interference method utilizing polarized light illumination on the sample. A polarizer again is placed between the VIS light source and condenser lens, which is set at 45° relative to the optical axis of an additional birefringent optical component of either a *Wollaston prism* or a *Nomarski compound prism*, which is positioned in the beam path in the front focal plane of the condenser lens. These prisms both generate two transmitted orthogonally polarized rays of light, referred to as the "ordinary ray" (polarized parallel to the prism optical axis) and the "extraordinary ray," thus with polarization E-field vectors at 0°/90° relative to the prism optic axis. These two rays emerge at different angles relative to the incident light (which is thus said to be *sheared*), their relative angular separation called the "shear angle," due to their different respective speeds of propagation through the prism (Figure 3.3a).

These sheared rays are used to form a separate *sample ray* and *reference ray*, which are both then transmitted through the sample. After passing through the sample and objective lens, both transmitted sheared rays are recombined by a second matched Wollaston/Nomarski prism positioned in a conjugate image plane to the first, with the recombined beam then transmitted through a second polarizer analyzer oriented to transmit light polarized at 135°.

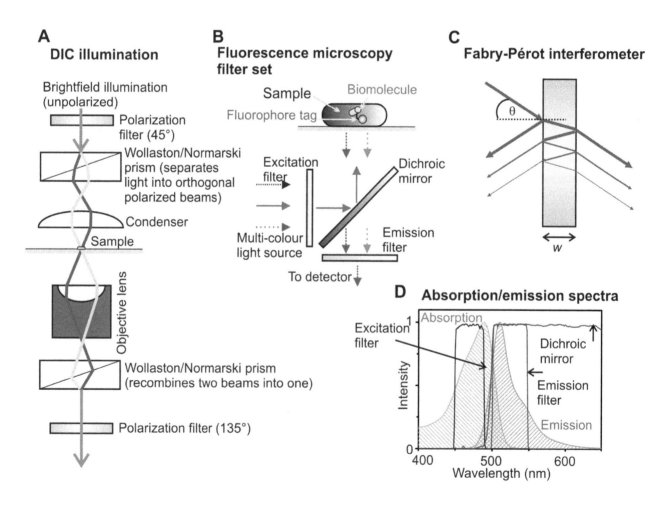

FIGURE 3.3   Generating image contrast in light microscopy. (a) Schematic of differential interference contrast illumination. (b) Typical fluorescence microscopy filter set. (c) Wavelength selection using a Fabry–Perot interferometer design. (d) Absorption and emission spectra for GFP, with overlaid excitation and emission filters and dichroic mirror in GFP filter set.

The difference in OPLs of the sample and reference beam results in an interference pattern at the image plane (normally a camera detector), and it is this wave interference that creates contrast.

Since the sample and reference beams emerge at different angles from the first Wollaston/ Nomarski prism, they generate two bright-field images of orthogonal polarization that are laterally displaced from each other by typically a few hundred nanometers, with corresponding regions of the two images resulting from different OPLs, or phases. Thus, the resultant interference pattern depends on the variations of phase between lateral displacements of the sample, in other words with the spatial *gradient* of refractive index of across a biological sample. It is therefore an excellent technique for identifying the boundaries of cells and also of cell organelles.

A related technique to DIC using polarized illumination is *Hoffmann modulation contrast* (*HMC*) microscopy. HMC systems consist of a condenser and objective lens, which have a slit aperture and two coupled polarizers instead of the first Wollaston/Nomarski prism and polarizer of DIC, and a *modulator* filter in place of the second Wollaston/Nomarski prism, which has a spatial dependence on the attenuation of transmitted light. This modulator filter has usually three distinct regions of different attenuation, with typical transmittance values of $T = 100\%$ (light), 15% (gray), and 1% (dark). The condenser slit is imaged onto the gray zone of the modulator. In regions of the sample where there is a rapid spatial change of sample optical path, refraction occurs, which deviates the transmitted light path. The refracted light will be attenuated either more or less in passing through the modulator filter, resulting in an image whose intensity values are dependent on the spatial gradient of the refractive index of the sample, similar to DIC. HMC has an advantage over DIC in that it can be used with birefringent specimens, which would otherwise result in confusing images in DIC, but has a disadvantage in that DIC can utilize the whole aperture of the condenser resulting in higher spatial resolution information from the transmitted light.

Quantitative phase imaging (QPI) (Popescu, 2011) utilizes the same core physics principles as phase microscopy but renders a quantitative image in which each pixel intensity is a measure of the absolute phase difference between the scattered light from a sample relative to a reference laser beam and has the same advantages of being label-free and thus less prone to potential physiological artifacts due to the presence of a contrast-enhancing label such as a fluorescent dye. It can thus in effect create a map of the variation of the refraction index across a sample, which is a proxy for local biomolecular concentration—for example, as a metric for the spatial variation of biomolecular concentration across a tissue or in a single cell. 2D and 3D imaging modalities exist, with the latter also referred to as *holotomograpy*. The main drawback of QPI is the lack of specificity since it is non-trivial to deconvolve the respective contributions of different cellular biomolecules to the measured refractive index. To mitigate this issue, QPI can be also combined with other forms of microscopy such as fluorescence microscopy in which specific fluorescent dye labeling can be used with multi-color microscopy to map out the spatial distribution of several different components (see section 3.5.3), while QPI can be used to generate a correlated image of the total biomolecular concentration in the same region of the cell or tissue sample. Similarly, QPI can be correlated with several other light microscopy techniques, for example including optical tweezers (see Chapter 6).

3.4.3 DIGITAL HOLOGRAPHIC MICROSCOPY

Digital holographic microscopy is emerging as a valuable tool for obtaining 3D spatial information for the localization of swimming cells, for example, growing cultures of bacteria, as well as rendering time-resolved data for changes to cellular structures involved in cell motility during their normal modes of action, for example, *flagella* of bacteria that rotate to enable cells to swim by using a propeller type action, and similarly *cilia* structures of certain eukaryotic cells. The basic physics of hologram formation involves an interference pattern between a laser beam, which passes through (or some variant of the technique is reflected from) the sample, and a reference beam split from the same coherent source that does not pass through the sample. The sample beam experiences phase changes due to the different range of refractive indices inside a cell compared to the culture medium outside the cell, similar to phase contrast microscopy. The interference pattern formed is a Fourier transform of the 3D variation of relative phase changes in the sample and therefore the inverse Fourier transform can render 3D cellular localization information.

Several bespoke digital holographic systems have been developed for studying dense cultures of swimming cells in particular to permit the investigation of biological physics features such as *hydro-dynamic coupling* effects between cells as well as more physical biology effects such as signaling features of swimming including *bacterial chemotaxis*, which is the method by which bacteria use a *biased random walk* to swim toward a source of nutrients and away from potential toxins. Some systems do not require laser illumination but can function using a relatively cheap LED light source, though they require in general an expensive camera that can sample at several thousand image frames per second in order to obtain the time-resolved information required for fast swimming cells and rapid structural transitions of flagella or cilia.

KEY BIOLOGICAL APPLICATIONS: NONFLUORESCENCE MICROSCOPY

Basic cell biology and cell organelle imaging; Monitoring cell motility dynamics.

3.5 FLUORESCENCE MICROSCOPY: THE BASICS

Fluorescence microscopy is an invaluable biophysical tool for probing biological processes *in vitro*, in live cells, and in cellular populations such as tissues. Although there may be potential issues of *phototoxicity* as well as impairment of biological processes due to the size of fluorescent "reporter" tags, fluorescence microscopy is the biophysical tool of choice for investigating native cellular phenomena in particular, since it provides exceptional detection contrast for relatively minimal physiological perturbation compared to other biophysical techniques. It is no surprise that the number of biophysical techniques discussed in this book is biased toward fluorescence microscopy.

3.5.1 EXCITATION SOURCES

The power of the excitation light from either a broadband or narrow bandwidth source may first require attenuation to avoid prohibitive photobleaching, and related photodamage, of the sample. *Neutral density (ND) filters* are often used to achieve this. These can be either absorptive or reflective in design, which attenuate uniformly across the VIS light spectrum, with the attenuation power of 10ND where *ND* is the neutral density value of the filter. Broadband sources, emitting across the VIS light spectrum, commonly include the *mercury arc lamp, xenon arc lamp*, and *metal–halide lamp*. These are all used in conjunction with narrow bandwidth *excitation filters* (typically 10–20 nm bandwidth spectral window), which select specific regions of the light source spectrum to match the absorption peak of particular fluorophores to be used in a given sample.

Narrow bandwidth sources include laser excitation, with an emission bandwidth of around a nanometer. Bright LEDs can be used as intermediate bandwidth fluorescence excitation source (~20–30 nm spectral width). *Broadband lasers*, the so-called white-light supercontinuum lasers, are becoming increasingly common as fluorescence excitation sources in research laboratories due to reductions in cost coupled with improvements in power output across the VIS light spectrum. These require either spectral excitation filters to select different colors or a more dynamic method of color selection such as an *acousto-optic tunable filter (AOTF)*.

The physics of AOTFs is similar to those of the *acousto-optic deflector (AOD)* used, for example, in many optical tweezers (*OT*) devices to position laser traps and are discussed in Chapter 5. Suffice to say here that an AOTF is an optically transparent crystal in which a standing wave can be generated by the application of radio frequency oscillations across the crystal surface. These periodic features generate a predictable steady-state spatial variation of refractive index in the crystal, which can act in effect as a diffraction grating. The diffraction angle is a function of light's wavelength; therefore, different colors are spatially split.

The maximum switching frequency for an AOTF (i.e., to switch between an "off" state in which the incident light is not deviated, and an "on" state in which it is deviated) is several tens of MHz; thus, an AOTF can select different colors dynamically, more than four orders of magnitude faster than the sampling time of a typical fluorescence imaging experiment, though the principal issues with an AOTF is a drop in output power of >30% in passing through the device and the often prohibitive cost of the device.

3.5.2 FLUORESCENCE EMISSION

The difference in wavelength between absorbed and emitted light is called the Stokes shift. The full spectral emission profile of a particular fluorophore, $\phi_{EM}(\lambda)$, is the relation between the intensity of fluorescence emission as a function of emission wavelength normalized such that the integrated area under the curve is 1. Similarly the spectral excitation profile of a particular fluorophore, $\phi_{EX}(\lambda)$, represents the variation of excitation absorption as a function of incident wavelength, which looks similar to a mirror image of the $\phi_{EM}(\lambda)$ profile offset by the Stokes shift.

A typical fluorescence microscope will utilize the Stokes shift by using a specially coated filter called a dichroic mirror, usually positioned near the back aperture of the objective lens in a *filter set* consisting of a dichroic mirror, an emission filter, and, if appropriate, an excitation filter (Figure 3.3b). The dichroic mirror reflects incident excitation light but transmits higher wavelength light, such as that from fluorescence emissions from the sample. All samples also generate elastically scattered light, whose wavelength is identical to the incident light. The largest source of elastic back scatter is usually from the interface between the glass coverslip/slide on which the sample is positioned and the water-based solution of the tissue often resulting in up to ~4% of the incident excitation light being scattered back from this interface. Typical fluorescent samples have a ratio of emitted fluorescence intensity to total back scattered excitation light of 10^{-4} to 10^{-6}. Therefore, the dichroic mirror ideally transmits less than a millionth of the incident wavelength light.

Most modern dichroic mirrors operate as *interference filters* by using multiple *etalon* layers of thin films of dielectric or metal of different refractive indices to generate spectral selectivity in reflectance and transmission. A single etalon consists of a thin, optically transparent, refractive medium, whose thickness w is less than the wavelength of light, which therefore results in interference between the transmitted and reflected beams from each optical surface (a *Fabry–Pérot interferometer* operates using similar principles). With reference to Figure 3.3c, the phase difference $\Delta\varphi$ between a pair of successive transmitted beams is

$$(3.23) \qquad \Delta\varphi = \frac{4\pi n w \cos\theta}{\lambda}$$

where
  λ is the free-space wavelength
  n is the refractive index of the etalon material

The finesse *coefficient F* is often used to characterize the spectral selectivity of an etalon, defined as

$$(3.24) \qquad F = \frac{4R}{\left(1 - R^2\right)}$$

where R is the reflectance, which is also given by $1 - T$ where T is the transmittance, assuming no absorption losses. By rearrangement

$$(3.25) \qquad T = \frac{1}{1 + F\sin^2\left(\Delta\varphi/2\right)}$$

A maximum T of 1 occurs when the *OPL* difference between successive transmitted beams, $2nw \cos \theta$, is an integer number of wavelengths; similarly the maximum R occurs when the OPL equals half integer multiples of wavelength. The peaks in T are separated by a width $\Delta\lambda$ known as the *free spectral range*. Using Equations 3.23 through 3.25, $\Delta\lambda$ is approximated as

$$(3.26) \qquad \Delta\lambda \approx \frac{\lambda_{peak}^2}{2nl \cos\theta}$$

where λ_{peak} is the wavelength of the central T peak. The sharpness of each peak in T is measured by the full width at half maximum, $\delta\lambda$, which can be approximated as

$$(3.27) \qquad \delta\lambda \approx \frac{\lambda_{peak}^2}{\pi w \sqrt{F}}$$

Typical dichroic mirrors may have three or four different thin film layers that are generated by either evaporation or sputtering methods in a vacuum (see Chapter 7) and are optimized to work at $\theta = 45°$, the usual orientation in a fluorescence microscope.

The transmission function of a typical VIS light dichroic mirror, $T_D(\lambda)$ is thus typically $<10^{-6}$ for $\lambda < (\lambda_{cut\text{-}off} - \Delta\lambda_{cut\text{-}off}/2)$ and more likely 0.90–0.99 for $\lambda > (\lambda_{cut\text{-}off} + \Delta\lambda_{cut\text{-}off}/2)$, up until using the VIS light maximum wavelength of ~750 nm, where $\lambda_{cut\text{-}off}$ is the characteristic cutoff wavelength between lower wavelength high attenuation and higher wavelength high transmission, which is usually optimized against the emission spectra of a particular fluorescent dye in question. The value $\Delta\lambda_{cut\text{-}off}$ is a measurement of the sharpness of this transition in going from very high to very low attenuation of the light, typically ~10 nm.

In practice, an additional fluorescence emission filter is applied to transmitted light, *band-pass filters* such that their transmission function $T_{EM}(\lambda)$ is $<10^{-8}$ for $\lambda < (\lambda_{midpoint} - \Delta\lambda_{midpoint}/2)$ and for $\lambda > (\lambda_{midpoint} + \Delta\lambda_{midpoint}/2)$ and for λ between these boundaries is more likely 0.90–0.99, where $\lambda_{midpoint}$ is the midpoint wavelength of the band-pass window and $\Delta\lambda_{midpoint}$ is the *bandwidth* of the window, ~10–50 nm depending on the fluorophore and imaging application (Figure 3.3d).

The fluorescence quantum yield (Φ) gives a measure of the efficiency of the fluorescence process as the ratio of emitted photons to photons absorbed, given by

$$(3.28) \qquad \Phi = \frac{k_s}{\sum_{i=1}^{n} \Phi_i k_i}$$

where

   k_s is the spontaneous rate of radiative emission
   Φ_i and k_i are the individual efficiencies and rates, respectively, for the various decay processes of the excited state (internal conversion, intersystem crossing, phosphorescence)

The fluorescence emission intensity I_{EM} from a given fluorophore emitting isotropically (i.e., with equal probability in all directions), which is detected by a camera, can be calculated as

$$(3.29) \qquad I_{EM}(\lambda) = I_{abs}\left(\frac{\Omega}{4\pi}\right)\Phi(\lambda)\phi(\lambda)T_D(\lambda)T_{EM}(\lambda)T_{OTHER}(\lambda)E_{camera}(\lambda)$$

where

   I_{abs} is the total absorbed light power integrated over all wavelengths
   Ω is the collection angle for photons of the objective lens

T_{OTHER} is a combination of all of the other transmission spectra of the other optical components on the emission path of the microscope

E_{camera} is the efficiency of photon detection of the camera

The total *SNR* for fluorescence emission detection is then

$$(3.30) \qquad SNR = \frac{G\sum_{i=1}^{n} N_{EM}}{n\sqrt{N_{EX}^2 + N_{AF}^2 + N_{CAM}^2}}$$

where N_{EM} is the number of detected fluorescence emission photons per pixel, with their summation being over the extent of a fluorophore image consisting of n pixels in total (i.e., after several capture and transmission losses, through the microscope) from a camera whose gain is G (i.e., for every photon detected the number of electron counts generated per pixel will be G) and readout noise per pixel is N_{CAM}, with N_{EX} photons transmitted over an equivalent region of the camera over which the fluorophore is detected.

3.5.3 MULTICOLOR FLUORESCENCE MICROSCOPY

A useful extension of using a single type of fluorophore for fluorescence microscopy is to use two or more different fluorophore types that are excited by, and which emit, different characteristic ranges of wavelength. If each different type of fluorophore can be tagged onto a different type of biomolecule in an organism, then it is possible to monitor the effects of interaction between these different molecular components and to see where each is expressed in the organism at what characteristic stages in the lifecycle and how different effects from the external environment influence the spatial distributions of the different molecular components. To achieve this requires splitting the fluorescence emission signal from each different type of fluorophore onto a separate detector channel.

The simplest way to achieve this is to mechanically switch between different fluorescence filter sets catered for the different respective fluorophore types and acquire different images using the same region of sample. One disadvantage with this is that the mechanical switching of filter sets can judder the sample, and this coupled to the different filter set components being very slightly out alignment with each other can make it more of a challenge to correctly coalign the different color channel images with high accuracy, necessitating acquiring separate bright-field images of the sample for each different filter set to facilitate correct alignment (see Chapter 8).

A more challenging issue is that there is a time delay between mechanically switching filter sets, at least around a second, which sets an upper limit on the biological dynamics that can be explored using multiple fluorophore types. One way round this problem is to use a specialized multiple band-pass dichroic mirror in the filter set, which permits excitation and transmission of multiple fluorophore types, and then using one more additional standard dichroic mirrors and single band-pass emission filters downstream from the filter set to then split the mixed color fluorescence signal, steering each different color channel to a different camera, or onto different regions of the same camera pixel array (Figure 3.4a). Dual and sometimes triple-band dichroic mirrors are often used. The main issue with having more bands is that since the emission spectrum of a typical fluorophore is often broad, each additional color band results in losing some photons to avoid cross talk between different bands by *bleed-through* of the fluorescence signal from one fluorophore type into the detection channel of another fluorophore type. Having sufficient brightness in all color channels sets a practical limit on the number of channels permitted, though quantum dots (QDs) have much sharper emission spectra compared to other types of fluorophores and investigations can be performed potentially using up to seven detection bands across the VIS and near IR light spectrum.

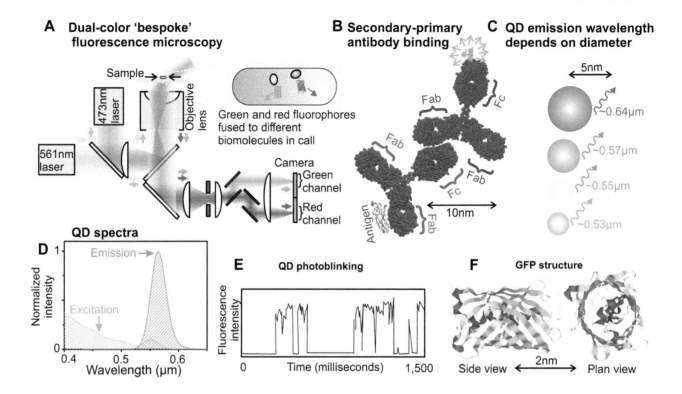

A Dual-color 'bespoke' fluorescence microscopy

B Secondary-primary antibody binding

C QD emission wavelength depends on diameter

D QD spectra

E QD photoblinking

F GFP structure

FIGURE 3.4   Fluorophores in biophysics. (a) Bespoke (i.e., homebuilt) dual-color fluorescence microscope used in imaging two fluorophores of different color in a live cell simultaneously. (b) Fluorescently labeled secondary antibody binding to a specific primary antibody. (c) Dependence of QD fluorescence emission wavelength on diameter. (d) Normalized excitation (green) and emission (orange) spectra for a typical QD (peak emission 0.57 μm). (e) Example fluorescence intensity time trace for a QD exhibiting stochastic photoblinking. (f) Structure of GFP showing beta strands (yellow), alpha helices (magenta), and random coil regions (gray).

KEY POINT 3.2

An ideal fluorescent probe binds in a highly specific way to the biological component under investigation, is small so as to cause a minimal steric interference with native biological processes, has a high quantum yield, and is relatively photostable (i.e., minimal irreversible photobleaching and/or reversible photoblinking, but note that some techniques can utilize both to generate stoichiometric and structural information). Currently, there is no single type of fluorescent probe that satisfies all such ideal criteria, and so the choice of probe inevitably results in some level of compromise.

3.5.4 PHOTOBLEACHING OF FLUOROPHORES

Single-photon excitation of a bulk population of photoactive fluorophores of concentration $C(t)$ at time t after the start of the photon absorption process follows the first-order bleaching kinetics, which is trivial to demonstrate and results in

$$(3.31) \qquad\qquad C(t) = C(0)\exp\!\left(\frac{-t}{t_b}\right)$$

where t_b is a characteristic *photobleach time*, equivalent to the lifetime of the fluorescence excited state. This photobleach time is the sum of an equivalent radiative decay time, t_{rad}

(processes involving direct emission of photon radiation) and nonradiative decay time, $t_{\text{non-rad}}$ (processes not involving the direct emission of photon radiation, such as molecular orbital resonance effects). In general, if there are a total of n fluorescence decay mechanisms, then

(3.32)
$$t_b = \frac{1}{\sum_{i=1}^{n} k_i}$$

where k_i is the rate constant of the ith fluorescence decay mechanism. *Photobleaching* of each single fluorophore molecule is a stochastic Poisson process such that its photoactive lifetime is an exponential distribution of mean time t_b. The principal cause of *irreversible photobleaching* of a fluorophore is light-dependent free radical formation in the surrounding water solvent, especially from molecular oxygen (under normal conditions, the concentration of dissolved oxygen in biological media is relatively high at ~0.5 mM, unless efforts are made to remove it). Free radicals are highly reactive chemicals containing an unpaired electron, which can combine with a fluorophore to destroy its ability to fluoresce. Many fluorophores also exhibit *reversible photobleaching* (or *blinking*), often under conditions of high excitation intensity, in which the excited state is transiently quenched to generate a stochastic dark "off" state as well as the bright "on" state.

Blinking is also known as *fluorescence intermittency* and is related to the competition between radiative and nonradiative *relaxation pathways* for the excited electron state (i.e., an excited state electron can return to its ground state via more than just a single energy transition pathway). The blinking phenomenon is exhibited by many fluorophores, especially semiconductor-based systems such as *quantum dots*, and also *organic dyes* and *fluorescent proteins* (FPs) (see the following sections in this chapter). Blinking often appears to obey a power-law distribution of on and off times with dark states in some systems lasting for tens of seconds, which is enormous on the quantum time scale, but remarkably a dark blinker will recover its fluorescence state after such a huge dark period and start emitting once again. The underlying specific physical mechanisms for blinking are largely unresolved but appear to be very specific for the fluorophore type.

3.5.5 ORGANIC DYE FLUOROPHORES

There are a large range of different organic dyes, for example, *cyanines* and *xanthenes*, whose chemical structures facilitate *electron delocalization* through a so-called π-electron system. A π *bond* is a covalent molecular orbital formed from the overlap of two *p atomic orbitals*; multiple π bonds in close proximity in a molecular structure can form a pool of spatially extended, delocalized electron density over a portion of the molecule through orbital resonance. This enables a large portion of the molecule to operate as an efficient electric dipole.

Historically, such dyes were first used to specifically label single biomolecules using *immunofluorescence*. Here, a *primary antibody* binds with high specificity to the biomolecule of interest, while a *secondary antibody*, which is chemically labeled with one or more fluorophores, then binds to the primary antibody (Figure 3.4b). The main issues with this technique concern the size of the probe and how to deliver it into a cell. The effective size of the whole reporter probe is ~20 nm, since each antibody has an effective viscous drag radius (the *Stokes radius*) of ~10 nm, which is an order of magnitude larger than some of the biomolecules being labeled. This can impair their biological functions. Second, introducing the antibody labels into living tissue is often difficult without significantly impairing the physiological functions, for example, permeabilizing the tissue using harsh detergents. With this caveat, this can result in very informative fluorescence images *in vivo*.

Fluorescence in situ hybridization (*FISH*) is a valuable labeling technique using organic dyes for probing specific regions of nucleic acids. A probe consists of a ~10 nucleotide base sequence either of singled-stranded DNA or RNA, which binds to a specific sequence of nucleic acid from a cell extract or a thin, fixed (i.e., dead) tissue sample via complementary base pairing following suitable incubation protocols normally >10 h, sometimes a few days.

A fluorophore either is chemically attached to the FISH probe via one or more bases via a fluorescent secondary antibody or is a chemical tag that binds to the probe. FISH can isolate the position of individual genes and can be used clinically in probing a range of disorders in a developing fetus in the womb by testing extracts of amniotic fluid.

Fluorophores can also use covalent chemical conjugation to attach to specific biological molecules. For example, a common strategy is to use the reactive sulfhydryl (–SH) side group of cysteine amino acids in proteins to conjugate to a fluorophore a dye molecule to generate a new *thiol* covalent bond (–S–R where R is the chemical group of the dye molecule). Proteins that contain several cysteine amino acid groups can lead to problems due to multiple labeling, and genetic modifications are often performed to knock out some of these additional cysteine residues. In addition, proteins that contain no native cysteines can be genetically modified to introduce an additional foreign cysteine. This general technique is called "site-specific cysteine mutagenesis."

Other less specific chemical conjugation methods exist. These methods target more general reactive chemical groups, including amino (–NH$_2$) and carboxyl (–COOH) groups, especially those present in the substituent group of amino acids. The principal issue is that the binding target is not specific and so dye molecules can bind to several different parts of a biomolecule, which makes interpretation of imaging data more challenging, in addition to potentially affecting biological function of the molecule in unpredictable ways. *Click chemistry* is an alternative method of conjugation. The "click" is meant to convey a convenience of simply snapping objects together. The most utilized type of click chemistry relies on strong covalent bonds formed between a reacting *azide* and a carbon–carbon *alkyne* triple bond. The main challenge is to introduce the foreign chemical group, either the azide or alkyne, into the biomolecule. With DNA, a variety of foreign chemical conjugating groups can be introduced using *oligo inserts*. These are short ~10 bp sections of DNA whose sequence is designed to insert at specific regions of the large DNA molecule used under *in vitro* investigation, and each oligo insert can be manufactured to be bound to a specific chemical group for conjugation (e.g., azide and biotin). All chemical conjugation techniques are discussed in more detail in Chapter 7.

Organic dye fluorophores may also be conjugated to small latex beads that have a typical diameter of a few hundred nanometers, small microspheres or large nanospheres depending on which way you look at it, which may then be conjugated to a biological substructure to yield a brighter probe that permits faster imaging. The biggest disadvantage is the size of these sphere probes, which is large enough to impair the normal biological processes in some way.

Most organic dye fluorophores emit VIS light. However, VIS light results in greater elastic scattering from unlabeled surrounding tissue, manifested as noise in fluorescence microscopy. Progress in developing fluorophores that are excited at longer wavelengths, some even in the IR region, is of assistance here.

3.5.6 FLASH/REASH PROBES

Improvements in minimizing biological process impairments are offered through the use of the green *fluorescein* arsenical helix (*FlAsH*) binder and the pink organic dye *resorufin* (*ReAsH*). This technology utilizes a genetically encoded arrangement of four cysteine amino acid residues in a specific protein under study inside a cell. Cells are incubated with membrane-permeable FlAsH/ReAsH reagents, which can then bind to the four cysteine residues inside the cell, which convert the dyes into a fluorescent form. FlAsH/ReAsH reagents have an effective diameter of 1–2 nm with a binding site in a protein requiring as few as six amino acids (four for the cysteine residues plus two more to generate the 3D shape for a binding pocket) and therefore exhibit minimal steric hindrance effects. But it is technically difficult to introduce the reagents, and the nonspecificity of binding combined with cellular toxicity has limited its use.

3.5.7 SEMICONDUCTORS, METAL-BASED FLUOROPHORES, AND NANODIAMONDS

Other fluorescent probes that have been applied to populations of cells include semiconductor-based fluorescent nanocrystals that generate quantized excitation energy states that relate not only to the material properties of their crystals but also to their physical dimensions and shapes. These include primarily *quantum rods* and QDs. The physics of their fluorescence properties are similar, and we use QDs as an exemplar here since these, in particular, have found significant applications in biophysics. They have advantages of being relatively bright, but their diameter is roughly an order of magnitude smaller than micro-/nanospheres. QDs have a photostability >100 times that of most organic dyes and so can be considered not to undergo significant irreversible photobleaching for most experimental applications. They thus have many advantages for monitoring single biomolecules. They are made from nanocrystal alloy spheres typically of two to three components (*cadmium selenide* (CdSe) and *cadmium telluride* (CdTe) are the most common) containing ~100 atoms. They are ~3–5 nm in core diameter (Figure 3.4c) and have semiconductor properties, which can undergo fluorescence due to an *exciton* resonance effect within the whole nanocrystal, with the energy of fluorescence relating to their precise length scale dimensions. An exciton is a correlated particle pairing composed of an electron and electron hole. It is analogous to the excited electron state of traditional fluorophores but has a significantly longer lifetime of ~10^{-6} s.

The fluorescence emission spectrum of a QD is dependent on its size. A QD is an example of quantum confinement of a *particle in a box* in all three spatial dimensions, where the particle in question is an exciton. In other words, QDs have size-dependent optical properties. The 1D case of a particle in a box can be solved as follows. The *Schrödinger equation* can be written as

$$(3.33) \qquad \left(\frac{-h^2}{8\pi^2 m} \frac{d^2}{dx^2} + V(x) \right) \psi(x) = E\psi(x)$$

where
   $\psi(x)$ is the wave function of a particle of mass m at distance x
   V is the potential energy
   E is the total energy
   h is Planck's constant

For a "free" particle, V is zero, and it is trivial to show that a sinusoidal solution exists, such that if the probability of being at the ends of the "1D box" (i.e., a line) of length a is zero, this leads to allowed energies of the form

$$(3.34) \qquad E_n = \frac{n^2 h^2}{8ma^2}$$

where n is a positive integer (1, 2, 3, …); hence, the energy levels of the particle are discrete or quantized. This formulation can be generalized to a 3D Cartesian box of dimensions a, b, and c parallel to the x, y, and z axes, respectively, which yields solutions of the form

$$(3.35) \qquad E_n = \frac{h_2}{8m} \left(\frac{n_x^2}{a^2} + \frac{n_y^2}{b^2} + \frac{n_z^2}{c^2} \right)$$

QDs, however, has a spherical geometry, and so Schrödinger's equation must be solved in spherical polar coordinates. Also, the different effective masses of the electron m_e and hole m_H ($m_H > m_e$ in general) need to be considered as do the electrostatic ground state energy and the bulk energy of the semiconductor, which leads to

$$(3.36) \qquad E = E_{bulk} + \frac{h^2}{8} \left(\frac{1}{m_e} + \frac{1}{m_H} \right) - \frac{1.8q^2}{4\pi\varepsilon_r \varepsilon_0 a}$$

where
   q is the unitary electron charge
   E_{bulk} is the bulk semiconductor energy

ε_r and ε_0 are the relative electrical permittivity and absolute electrical permittivity in a vacuum

a is the QD radius

When light interacts with a QD, an electron–hole exciton pair is created. An exciton has an associated length scale called the Bohr radius, such that beyond this length scale, the probability of exciton occurrence is very low. For CdSe QDs, the Bohr radius is ~5.6 nm, and thus quantum confinement effects occur at QD diameters that are less than ~11.2 nm. The aforementioned equation predicts that the energy state of the confined exciton decreases with increasing QD radius. In other words, smaller QDs are blue, while larger QDs are red.

QDs are characterized by a broad absorption spectrum and narrow emission spectrum—this means that they can be excited using a range of different lasers whose wavelength of emission does not necessarily correspond to an absorption peak in the fluorophore as is the case for organic dyes, with the tightness of the spectral emission meaning that emissions can be relatively easily filtered without incurring significant loss of signal (Figure 3.4d).

This narrowness of spectral emission means that several different colored QDs can be discriminated in the same sample on the basis of their spectral emission, which is useful if each different colored QD tags a different biomolecule of interest. QDs are brighter than their corresponding organic dyes at similar peak emission wavelength; however, their relative brightness is often overstated (e.g., a single QD emitting in the orange-red region of the VIS light spectrum at the corresponding excitation wavelengths and powers is typically only six to seven times brighter than a single molecule of the organic dye rhodamine).

QDs undergo a photophysical phenomenon of *blinking* (Figure 3.4e). Many different types of fluorophores also undergo blinking. Blinking is a reversible transition between a photo-active (*light*) and an inactive (*dark*) state; the dye appears to be bright and then momentarily dark in a *stochastic* manner (i.e., random with respect to time), but, in general, these are prevalent more at excitation intensities higher than would normally be used for fluorescence imaging with dark state dwell times <10 ms, which is often sufficiently fast to be averaged out during typical fluorescence microscopy sampling time windows of ~100 ms or more. QDs blink more appreciably at lower excitation intensities with longer dwell times of dark states more comparable to the time scale of typical fluorescence imaging, which can make it difficult to assess if what you see from one image to a consecutive image in a kinetic series of images is a continuous acquisition of the same tagged biomolecule or a different one that has diffused in to the field of view from elsewhere.

The actual functional diameter of a QD can be more like 15–20 nm since the core is further coated with a solvent-protective shell (typically *zinc sulfide*) and a polymer matrix for chemical functionalization. The size can increase further to ~30 nm if an antibody is also attached to the QD to allow more specific binding to a given biological molecule. A diameter of 30 nm is an order of magnitude larger than a typical biomolecule, which can result in significant steric hindrance effects, which can inhibit native biological processes and also make it challenging to deliver the QD into a cell. However, there have been several applications of cellular imaging developed using QD fluorophores, going back to the turn of the century (see Michalet et al., 2005).

An alternative metal-based fluorophore involves *lanthanides*. The lanthanides are a group of 15 *4f-orbital* metals in the periodic table between atomic number element 57 (*lanthanum*) to 71 (*lutetium*). They form unique fluorescent complexes when bound via their 3+ ion states with an organic *chelating* agent, such as a short random coil sequence of a given protein to be labeled. This can confer significant stability to the fluorophore state, since the chelation complex is protected from water and so exhibits limited free radical photobleach damage (Allen and Imperiali, 2010).

Fluorescent nanodiamond (FND) is emerging as a valuable probe for biological samples since it has a high photostability manifest as no photobleaching or blinking, is biocompatible, and has spectral properties that are relatively insensitive to the surrounding fluid environment and are spectrally distinct from the normal range of cellular autofluorescence. A nanodiamond is a synthetic nanoscale-sized particle composed of carbon atoms bound in the tetrahedral sp3 arrangement (see Chapter 2), identical to that of natural diamond. The fluorescence comes from doping the center of the nanodiamond with a high density of negatively charged nitrogen vacancy (NV$^-$) atoms. NV$^-$ is a fluorophore with peak absorption wavelength of roughly 550 nm, emitting at ~700 nm, and so not only has a wide Stokes shift facilitating fluorescence detection, but also emits at wavelengths which are 200–300 nm higher than typical contributors of native cellular autofluorescence, hence the level of background fluorescence "noise" is relatively low.

The primary issues with FNDs are that they require technically demanding high pressure and temperature conditions to manufacture and that there is currently no easy way to chemically functionalize their surface to facilitate specific labeling to biological structures; instead, they are typically embedded into a larger latex bead matrix, whose surface can be derivatized, setting a typical lower limit on their overall diameter of a few tens of nm which is sufficiently large to inhibit many cellular processes. However, their potential for enabling long-duration fluorescence imaging studies is significant so it likely that future technical developments to address their current limitations will see greater uptake of FNDs for bioimaging.

3.5.8 FLUORESCENT PROTEINS AND AMINO ACIDS

A fluorescent protein (*FP*) is the most useful fluorophore for *in vivo* fluorescence microscopy, that is, imaging of living cells. They are photophysically poor choices for a fluorophore (compared to other types of fluorophores discussed previously in this chapter; they are dim, have smaller photon absorption cross-sectional areas, and are less photostable and thus photobleach after emitting fewer photons). Despite this, significant insight has been gained from using FPs into the behavior of proteins inside living cells since the early 1990s (see the issue in *Chem. Soc. Rev.*, 2009, listed in the references).

FPs were discovered in the 1960s when it was found that a species of jellyfish called *Aequorea victoria* produced a naturally fluorescent molecule called "green fluorescent protein" (*GFP*). A breakthrough came when the GFP gene was sequenced in the early 1990s and researchers could use genetics techniques to introduce its DNA code into organisms from different species. GFP has two peak excitation absorption wavelengths at ~395 and ~475 nm and peak emission wavelength of ~509 nm. Using further molecular biology techniques, the GFP gene has been modified to make it brighter and to emit fluorescence over different regions of the VIS light spectrum, and variants of FP from other classes of organisms including corals and crustaceans are also used now.

The FP gene is fused directly to the DNA of a gene encoding a completely different protein of interest and when the genetic code is read off during transcription (see Chapter 2), the protein encoded by this gene will be fused to a single FP molecule. They are widely used as noninvasive probes to study different biological systems, from the level of whole organism tissue patterning down to single individual cells, including monitoring of protein–protein interactions and measurement of a cell's internal environment such as the concentration of protons (i.e., pH) as well as ion-sensing and local voltage measurements inside a cell.

FPs have a β-barrel-type structure (see Chapter 2) of mean diameter ~3 nm, with molecular weight ~28 kDa. The electric dipole moment of the fluorophore is formed from three neighboring amino acids that generate a cyclic *chromophore* enclosed by 11 β-strands (Figure 3.4f). Genetic modification of the chromophore groups and the charged amino acid residues inside the core of the protein has resulted in a wide range of synthetic variants having different absorption and emission peak wavelengths, with the excitation wavelength spanning not only the long UV and the VIS light spectrum but now extending into the IR. Mutation of some of the surface residues of the barrel has resulted in variants that fold into a fully functional shape faster in the living cell and have less risk of aggregating together via hydrophobic forces.

The size of an FP is larger than an organic dye molecule, resulting in more steric hindrance effects. DNA coding for a short linker region of a few amino acid residues is often inserted between the FP gene and that of the protein under investigation to allow for more rotational flexibility. In many biological systems, the FP can be inserted at the same location as the original protein gene, deleting the native gene itself, and thus the tagged protein is manufactured by the cell at roughly native concentration levels. The FP is fused at the level of the original DNA code meaning that the labeling efficiency is 100% efficient, which is a significant advantage over other fluorophores previously discussed in this section.

However, FPs are relatively dim and photobleach quickly. For example, GFP is more than two times dimmer (a measure of a relatively small absorption cross-sectional area) and photobleaches after emitting ~10 times fewer photons compared to equivalent organic dyes excited using similar light wavelengths and powers. However, the quantum yield Q_Y of GFP is actually reasonably high at ~0.79 (i.e., out of every 10 photons absorbed, ~8 are emitted in fluorescence).

Also, when the FP–protein fusion is transcribed from the modified DNA to make mRNA, which is then translated to make the protein fusion, the FP still needs to fold into its functional 3D shape and undergo chemical modifications until it is photoactive. This maturation process at best takes still several minutes, meaning that that there is always a small proportion of dark FP present in a cell during a fluorescence microscopy investigation. Also, in some cells, it is not possible to delete the native gene under investigation and still maintain its biological function, and instead the FP is expressed off a separate plasmid (see Chapter 7)—the effect of this is to generate a mix of labeled and unlabeled protein in the cell and also an overexpression of the protein in question that could affect biological processes.

Despite the flaws in FP technology, their application has dramatically increased the understanding of several fundamental biological processes in living cells. Also, highly pH-sensitive FP variants have been developed, for example, *pHlourin*, which have increased brightness sensitivity to pH at long excitation wavelengths but is insensitive to pH change if excited at shorter wavelengths. This can therefore be used as a *ratiometric* pH indicator in live cells (the fluorescence emission signal at shorter wavelengths can be used to normalize the measured signal at the longer wavelength against the total concentration of FP).

The three natural aromatic amino acid residues (which contain a benzene ring type structure as a side group, see Chapter 2) of tryptophan (Trp), tyrosine (Tyr), and phenylalanine (Phe) all exhibit low-level fluorescence, with Trp having the highest quantum yield with a peak excitation wavelength of ~280 nm and peak emission at ~340 nm. The brightness of individual Trp is not sufficient to enable single residue detection with current detector technologies. Trp fluorescence is widely used in ensemble average measurements to monitor changes in protein conformation due to fluorescent properties being solvent dependent; Trp residues are very hydrophobic due to the aromatic side group and so generally found at the core of folded protein where the polar water solvent cannot reach them. However, if the protein opens up, for example, due to unfolding, then water can access the Trp manifest as typically a ~5% increase in peak fluorescence emission wavelengths and a two-fold decrease in intensity. This effect can therefore be used as a metric for dynamic protein unfolding in a sample. Synthetic fluorescent analogs of Trp can also be manufactured with higher quantum yields than natural Trp fluorescence.

Similarly, it is also possible to generate a range of synthetic fluorescent amino acid analogs that are not directly based on the natural aromatic acids. Good examples are fluorescent D-amino acids (FDAAs)—most amino acids are L-optical isomers (see Chapter 2) but some D-amino acids are found in nature in the bacterial cell wall. It is possible to make chemical derivatives whose side-chain terminal is covalently linked to a fluorescent organic dye molecule. FDAAs will incorporate into the bacterial peptidoglycan, which is a key structural component of bacterial cell walls, and a range of fluorescence detection tools including light microscopy can be used to investigate how the cell wall assembles and how it can be disrupted using antibiotics.

3.5.9 SNAP- AND CLIP-TAGS

Some of the disadvantages of FPs are overcome in *CLIP-tag* or closely related *SNAP-tag* technology. Here, a protein probe is first encoded at the level of the DNA next to the protein under investigation inside a living cell, which in most applications consists of a modified DNA repair protein called "O6-alkylguanine-DNA alkyltransferase" (*AGT*). The cell is then incubated with a secondary probe, which is labeled with a bright organic dye that will bind with very high specificity to the primary probe. SNAP/CLIP-tags have the same advantage of FPs in specificity of labeling since the primary probe is generated at the level of the encoding DNA. The primary AGT probe has a molecular weight of 18 kDa, which is smaller than FPs and results in marginally less steric hindrance. A key advantage over FPs, however, is that the secondary probe is labeled with a bright organic dye, which has significantly superior photophysical properties.

3.5.10 OVERCOMING CELLULAR AUTOFLUORESCENCE

A strong additional source of background noise is *autofluorescence*. This is the natural fluorescence that occurs from a wide range of biomolecules in cells, especially molecules such as flavins, used in the electron transport chain, and those containing pyrimidines, for example, one of the two chemical categories for nucleotides in nucleic acids but also a component of NAD^+ and NADH also in the electron transport chain (see Chapter 2). Typically, autofluorescence is more prevalent at lower wavelengths of excitation, in the blue or long UV (~300–500 nm).

Many autofluorescent molecules have low photostability as fluorophores and so will photobleach irreversibly quickly. This trick can be used in fluorescence imaging to *prebleach* a sample with a rapid pulse of excitation light prior to data acquisition. Some autofluorescent components, however, have a longer photoactive lifetime. Certain metabolic tricks can be applied to minimize the cellular concentration of these components, for example, to culture tissues in nutrient medium that is designed to reduce the expression of flavins and pyrimidines, but this runs a risk of adjusting the physiology of the tissue unnaturally. A better approach is to avoid using fluorescent dyes that are excited by blue wavelengths in preference for longer excitation wavelength, or *red-shifted*, dyes.

The contrast of fluorescence microscopy images obtained on cellular samples can be enhanced using a technique called "optical lock-in detection" (*OLID*) that facilitates discrimination between noise components in living cells and the desired specific fluorescence signal from fluorophores inside the cell by utilizing detection of the correct time signature of an imposed periodic laser excitation function (Marriott et al., 2008). This can be particularly useful in the case of FP imaging in living cells that are often expressed in small copy numbers per cell but still need to be detected over often large levels of noise from both camera readout and native autofluorescence from the cell.

OLID implements the detection of modulated fluorescence excitation on a class of specific OLID dyes that are optimized for optical switching. The relative population of the excited and inactive fluorescence states in a dye is controlled by periodic cycles of laser activation and deactivation, which is dependent on wavelength. Neither camera noise nor autofluorescence are so sensitively dependent on wavelength, and therefore only specific biomolecules that are labeled with an OLID dye molecule will register a fluorescence emission signal of the same characteristic modulation driving frequency as the laser excitation light. Software can then lock-in onto true signal data over noise by applying cross-correlation analysis to the individual pixel data to a time series of acquired images from a sensitive camera detector pixel array, referenced against the driving excitation waveform of the laser. This allows a precise correlation coefficient to be assigned for each pixel in each image. Thus, a pixel-by-pixel 2D map of correlation coefficients is generated, which provides a high-contrast image of the localization of the specific switchable OLID dye molecules, largely uncontaminated by background noise.

The first OLID fluorophores developed were organic dyes (e.g., probes based on a dye called "nitrospirobenzopyran"). The main issues with their use were challenges associated with how to deliver them into living cells and also nonspecificity of dye binding. Recent cellular studies exploit an FP called "Dronpa" and have generated high-contrast images of cellular structures in live, cultured nerve cells, as well as in small *in vivo* samples including mammalian and fish embryos.

3.6 BASIC FLUORESCENCE MICROSCOPY ILLUMINATION MODES

There are several fluorescence microscopy methods available that allow fluorophores in labeled biomolecules to be excited and detected. These not only include camera-imaging methods of *wide-field* illumination modes comprising approaches such as *epifluorescence* and *oblique epifluorescence*, as well as narrower illumination modes such as *Slimfield* and *narrow-field*, used normally in combination with a high-quantum-efficiency EMCCD camera detector, but also include spectroscopic approaches such as fluorescence correlation spectroscopy and scanning confocal microscopy.

3.6.1 WIDE-FIELD MODES OF EPIFLUORESCENCE AND OBLIQUE EPIFLUORESCENCE

Wide-field microscopy is so called because it excites a laterally "wide" field of view of the sample (Figure 3.5a). Epifluorescence is the most standard form of fluorescence microscopy illumination and involves focusing a light beam into the back focal plane of an objective lens centered on its optical axis. This generates an excitation field that is uniform with height z into the sample from the microscope slide/coverslip surface, but with a radially symmetrical 2D Gaussian profile in the xy lateral plane. The full width at half maximum of this intensity field in xy is 30–60 μm. *Epi* refers to excitation and emission light being routed through the same objective lens in opposite directions. *Trans* fluorescence illumination, in which the excitation and emission beam travel in the same direction, is possible but not used in practice since the amount of unblocked excitation light entering the objective lens is significantly higher, thus reducing the contrast.

A laser beam can be tightly focused into the back aperture of an objective lens, unlike spatially extended sources such as arc lamps or LEDs, and so there is a room to translate the focus laterally away from the optic axis, allowing the angle of incidence to be adjusted. Beyond the critical angle, TIRF excitation is generated. However, an angle of incidence between zero and the critical angle results in *oblique epifluorescence* (Figure 3.5b), also known as *variable-angle epifluorescence, oblique epi illumination, pseudo-TIRF, quasi-TIRF, near-TIRF, leaky TIRF,* and *highly inclined and laminated optical sheet illumination* (HILO) (see Tokunaga et al., 2008). Oblique epifluorescence results in uniform excitation intensity parallel to the excitation field wave vector but has lower back scatter from cellular samples and from the surface of the microscope slide/coverslip, which can increase the contrast compared to standard epifluorescence by almost an order of magnitude.

3.6.2 TOTAL INTERNAL REFLECTION FLUORESCENCE

TIRF microscopy (for a comprehensive discussion, see Axelrod et al., 1984) generates wide-field excitation laterally in the focal plane of a few tens of microns diameter but utilizes a *near-field* effect (i.e., an optical phenomenon over a length scale of less than a few wavelengths of light) parallel to the optic axis of the microscope objective lens to generate a fluorescence excitation field very close to a glass microscope slide/coverslip of ~100 nm characteristic depth (Figure 3.5c). TIRF is an enormously powerful and common biophysical technique, and so we cover several technical aspects of its use here. TIRF usually utilizes laser excitation, such that a beam of wavelength λ is directed at an oblique supercritical angle θ to the interface between a glass microscope coverslip and a water-based solution surrounding a biological sample. Total internal reflection of the incidence beam occurs at angles of incidence, which are greater than a critical angle θ_c:

(3.37)
$$\theta_c = \sin^{-1}\left(\frac{n_w}{n_g}\right)$$

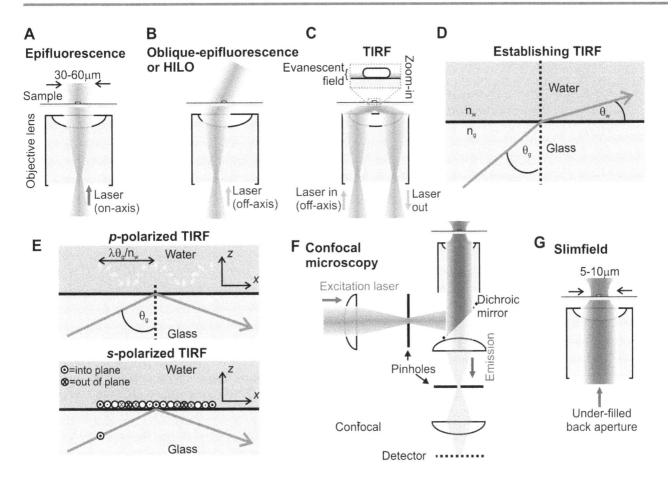

FIGURE 3.5 Different fluorescence illumination modes. (a) Wide-field epifluorescence using laser excitation. (b) Oblique epifluorescence or highly inclined and laminated optical sheet illumination. (c) Total internal reflection fluorescence (TIRF) using the objective lens method. (d) Schematic for refraction of a laser beam incident on a glass–water interface. (e) The *E*-field polarization vectors in the TIRF evanescent field generated during (A) *p*-TIRF excitation (incident polarization vector shown by small light-shaded arrow is in the plane of the page and the evanescent field polarization cartwheels across the surface) and (B) *s*-TIRF excitation (the incident polarization vector shown by small circle is *into* the plane of the page, and the axis of this polarization vector in the evanescent field remains parallel to this incident polarization, that is, conserved, though the polarization vector will also periodically extend both into and out of the plane of the page). (f) Confocal microscope schematic arrangement. (g) *Slimfield* illumination.

where n_w and n_g refer to the refractive indices of water (~1.33) and glass (~1.515, for commonly used *BK7* low-fluorescence borosilicate glass), respectively, which indicates a critical angle of ~61°. These refractive index values relate to the wavelength-dependent *dispersion relation* of light, and across the VIS light spectrum of 400–700 nm result in a range of 1.344–1.342 (n_w) and 1.529–1.514 (n_g). The simplest empirical model for the dispersion relation is embodied in the *Cauchy equation*:

$$(3.38) \qquad n(\lambda) = c_0 + \frac{c_2}{\lambda^2} + \frac{c_4}{\lambda^4} + \cdots \approx c_0 + \frac{c_2}{\lambda^2}$$

KEY BIOLOGICAL APPLICATIONS: BASIC FLUORESCENCE MICROSCOPY

General cell biology imaging; Imaging specific subcellular architectures in cells with diffraction-limited spatial resolution; Probing thin fluorescently stained tissue samples.

The agreement with experiment in the VIS range is very good (within a few %), but larger inaccuracies exist beyond this range. A more accurate approximation comes directly from the classical solution to Maxwell's equations for damped Lorentz oscillators, resulting in the *Sellmeier equation*:

$$(3.39) \qquad n(\lambda) \approx 1 + \lambda^2 \left(\frac{b_1}{\lambda^2 - c_1} + \frac{b_2}{\lambda^2 - c_2} + \frac{b_3}{\lambda^2 - c_3} \right)$$

To understand how the evanescent field is generated, we can first apply *Snell's law* of refraction, from glass to water (Figure 3.5d):

$$(3.40) \qquad \frac{\sin\theta_w}{\sin\theta_g} = \frac{ng}{n_w} (> 1 \text{ for TIRF})$$

where
   θ_g is the angle of incidence through the glass
   θ_w is the angle of refraction through the water

Thus, by rearrangement

$$(3.41) \qquad \cos\theta_w = \sqrt{1 - \left(\frac{n_g}{n_w} \right) \sin^2\theta_w} = ib$$

where *b* is a real number in the case of TIRF; thus the angle of refraction into the water is a purely imaginary number. The *E*-field in water in the 2D cross-section of Figure 3.3d can be modeled as a traveling wave with distance and wave vectors of magnitude *r(x, z)* and wave vector k_w, respectively, with angular frequency ω after a time *t*, with *z* parallel to the optic axis of the objective lens and *xy* parallel to the glass–water interface plane:

$$(3.42) \qquad E_{evanescent} = E_0 \exp i(\vec{k}_w \cdot \vec{r} - \omega t) = E_0 \exp i(k_w x \sin\theta_w + k_w z \cos\theta_w - \omega t)$$

By substitution from Equation 3.41, we get the *evanescent wave equation*:

$$(3.43) \qquad E_{evanescent} = E_0 \exp(-bk_w z)\exp i\left(k_w \frac{n_w}{n_g} x\sin\theta_w - \omega t \right) = E_0 \exp(-bk_w z)\exp i(k^* x - \omega t)$$

The intensity *I(z)* of the evanescent field at *x* = 0 as a function of *z* (i.e., as it penetrates deeper into the solution) decays exponentially from the glass–water interface, characterized by the *depth of penetration* factor *d*:

$$(3.44) \qquad I(z) = I(0)\exp\left(\frac{-z}{d} \right)$$

where

$$(3.45) \qquad d = \frac{\lambda}{4\pi\sqrt{n_g^2 \sin^2\theta_g - n_w^2}}$$

A typical range of values for *d* is ~50–150 nm. Thus, after ~100 nm, the evanescent field intensity is ~1/*e* of the value at the microscope glass coverslip surface, whereas at 1 μm depth from the coverslip surface (e.g., the width bacteria), the intensity is just a few thousandths of a percent of the surface value. Thus, only fluorophores very close to the slide are excited into significant fluorescence, but those in the rest of the sample or any present in the surrounding solution are not. In effect, this creates an *optical slice* of ~100 nm thickness, which increases the contrast dramatically above background noise to permit single-molecule detection. It is possible to adjust the angle of incidence θ_g to yield d values smaller than 100 nm, limited by the numerical aperture of the objective lens (see Worked Case Example 3.1) and also larger values, such that $d \rightarrow \infty$ as $\theta_g \rightarrow \theta_c$.

The intensity of the evanescent field is proportional to the square of the E-field amplitude, but also depends on the polarization of the incidence E-field in the glass and has different values for the two orthogonal polarization components in water. Incident light that is polarized parallel (p) to the xz plane of incidence as depicted in Figure 3.5e generates an elliptically polarized evanescent field consisting of both parallel and perpendicular (s) polarized components. Solving Maxwell's equations generates the full solutions for the electric field vector components of the evanescent field as follows:

$$(3.46) \quad E_{evanescent,x} = E_{p,0}\exp\left[-i\left(\delta_p + \frac{\pi}{2}\right)\right]\frac{2\cos^2\theta_g\sqrt{\left(2\sin^2\theta_g - \left(n_w/n_g\right)^2\right)}}{\sqrt{\left(n_w/n_g\right)^4\cos^2\theta_g + \sin^2\theta_g - \left(n_w/n_g\right)^2}}$$

$$(3.47) \quad E_{evanescent,y} = E_{s,0}\exp\left[-i\delta_s\right]\frac{2\cos^2\theta_g}{\sqrt{1 - \left(n_w/n_g\right)^2}}$$

$$(3.48) \quad E_{evanescent,z} = E_{p,0}\exp\left[-i\delta_p\right]\frac{2\cos^2\theta_g\sin\theta_g}{\sqrt{\left(n_w/n_g\right)^4\cos^2\theta_g + \sin^2\theta_g - \left(n_w/n_g\right)^2}}$$

where $E_{p,0}$ and $E_{s,0}$ are the incident light E-field amplitudes parallel and perpendicular to the xz plane. As these equations suggest, there are different phase changes between the incident and evanescent E-fields parallel and perpendicular to the xz plane:

$$(3.49) \quad \delta_p = \tan^{-1}\frac{\sqrt{\sin^2\theta_g - \left(n_w/n_g\right)^2}}{\left(n_w/n_g\right)\cos\theta_g}$$

$$(3.50) \quad \delta_s = \tan^{-1}\frac{\sqrt{\sin^2\theta_g - \left(n_w/n_g\right)^2}}{\cos\theta_g}$$

As these equations depict, the orientation of the polarization vector for s-polarized incident light is preserved in the evanescent field as the E_y component; as the supercritical angle of incidence gets closer to the critical angle, the E_x component in the evanescent field converges to zero and thus the p-polarized evanescent field converges to being purely the E_z component. This is utilized in the *p-TIRF* technique, which uses incident pure p-polarized light close to, but just above, the critical angle, to generate an evanescent field, which is polarized predominantly normal to the glass coverslip–water interface. This has an important advantage over subcritical angle excitation in standard *epifluorescence* illumination, for which the angle of incidence is zero, since in p-TIRF the polarization of the excitation field is purely parallel to the glass–water interface and is unable to excite a fluorophore whose electric dipole axis is normal to this interface and so can be used to infer orientation information for the fluorophore electric dipole axis. In the general case of supercritical angles not close to the critical angle, the *p*-polarized component in the evanescent field spirals elliptically across the spatial extent of the glass–water interface in a *cartwheel* fashion with a sinusoidal spatial periodicity of $\lambda/n_w \sin \theta_g$ (Figure 3.5f).

The intensity of both the *p* and *s* components in the evanescent field can both be several times greater than the incident intensity for values of θ_g between θ_c and ~75°–80°, with the *p* component marginally greater than the *s* component and then both tailing off to zero as $\theta_c \to 90°$:

(3.51)
$$I_{evanescent,p,0} = I_{incident,p,0} \frac{4\cos^2\theta_g \left(2\sin^2\theta_g - \left(n_w/n_g\right)^2\right)}{\left(n_w/n_g\right)^4 \cos^2\theta_g + \sin^2\theta_g - \left(n_w/n_g\right)^2}$$

(3.52)
$$I_{evanescent,s,0} = I_{incident,s,0} \frac{4\cos^2\theta_g}{1 - \left(n_w/n_g\right)^2}$$

Often, the incident *E-field* polarization will be *circularized* by a quarter-wave plate. The effect of a quarter-wave plate, similar to those utilized in phase contrast microscopy, is to retard the phase of any light whose polarization vector is aligned to the plate's *slow axis* by one quarter of a wavelength relative to incident light whose polarization vector is aligned to the plate's *fast axis* (90° rotated from the fast axis). If linearly polarized light is incident on the plate oriented at 45° to both the fast and slow axes, then circularly polarized light is generated such that the polarization vector of the light after propagating through the plate rotates around the wave vector itself with a spatial periodicity of one wavelength. The effect is to minimize any bias resulting from preferred linear polarization orientations in the absorption of the incident light from the relative orientation of the electric dipole moment of the fluorescent dye tag, but it should be noted that this does not result in a complete randomization of the polarization vector.

TIRF, in modern microscope systems, is generated either using a *prism method* or an *objective lens method*. The prism method results in marginally less undesirable incident light scattering than the objective lens method, since the light does not need to propagate across as many optical surfaces *en route* to the sample. However, fluorescence emissions need to be collected through the thickness of a microscope flow cell, ~100 μm depth filled with water; to avoid aberration effects normally requires the use of a special *water-immersion* objective lens to image through the bulk of the sample solution, which have a marginally lower numerical aperture (~1.2) than used for the objective lens method, and therefore the photon collection efficiency is lower. In Equation 3.45, the term $n_g \sin \theta_g$ is identical to the numerical aperture of the objective lens, and therefore to generate TIRF using the objective lens method requires an objective lens whose numerical aperture is greater than the refractive index of water, or ~1.33 (values of 1.4–1.5 in practice are typical).

The first application of TIRF to cellular investigation was to study epidermal growth factor (*EGF*) receptors in the cell membrane whose biological function is related to cell growth and development in the presence of other nearby cells (e.g., in a developing tissue) in which the EGF receptors were tagged with the cyanine dye Cy3 (Sako et al., 2000). Many biophysical investigations that use TIRF also utilize *Förster resonance energy transfer (FRET)*. This is a nonradiative technique occurring over a nanometer length scale between two different dye molecules, and thus is a technique for investigating putative interaction of different biomolecules (discussed fully in Chapter 4). TIRF can also be combined with fluorescence polarization microscopy measurements for *in vitro* and cellular samples.

Several surface-based *in vitro* assays benefited from the enhancement in contrast using TIRF illumination. A good historical example in biophysics is the *in vitro motility assay* used to study molecular motors. This assay was designed to monitor the interaction of molecular motors that run on molecule-specific tracks, originally developed for observing muscle-based molecular motors. Here, F-actin filaments of several microns in length are conjugated with the fluorescent dye rhodamine and can be observed using TIRF in exceptional detail to undergo active diffusion on a microscope coverslip surface coated with the protein myosin in the presence of ATP due to the interaction between the molecular motor region of the myosin head domain with its F-actin track, as occurs *in vivo* in muscle, fueled by chemical energy from the hydrolysis of ATP (the theoretical model of this motor translocation behavior is discussed in Chapter 8).

KEY POINT 3.3

TIRF is one of the most widely utilized biophysical tools for studying dynamic biological processes on, or near to, surfaces. It offers exceptional contrast for single-molecule detection and can be combined with a variety of different biophysical techniques, such as electrical measurements.

Delimitation of light excitation volumes can also be achieved through the use of waveguides. Here, light can be guided through a fabricated optically transparent material, such as glass, to generate a supercritical angle of incidence between the waveguide surface and a physiological buffer containing a fluorescently labeled biological sample. This type of approach can be used to generate an evanescent field at the tip of an optical fiber, thus allowing fluorescence detection from the end of the fiber. This approach is used in nanophotonics, which enables complex shaped evanescent fields to be generated from fabricated waveguides, and has relevance toward enhancing the contrast of fluorescence detection in microfluidics-based biosensing devices (see Chapter 7).

3.6.3 FLUORESCENCE POLARIZATION MICROSCOPY

Fluorescence polarization measurements can be performed by adapting a standard fluorescence microscope to split the orthogonal emission polarization onto either two separate cameras, or potentially onto two halves of the same camera pixel array in a similar manner to splitting the fluorescence emissions on the basis of wavelength except using a polarization splitter optic instead of a dichroic mirror. Typically, the incident E-field polarization is fixed and linear, but as discussed in the previous section, standard epifluorescence illumination is suitable since it results in excitation polarization parallel to the microscope coverslip or slide.

It is also possible to apply polarization microscopy using TIRF illumination. If the incident light onto the glass–water interface is purely s-polarized, then the polarization orientation will be conserved in the evanescent excitation. However, as discussed, useful information can also be obtained by using p-polarized excitation light for TIRF, that is, p-TIRF. Here, the polarization vector is predominantly normal to the glass–water interface, and so has application in monitoring fluorophores whose electric dipoles axes are constrained to be normal to a microscope coverslip; an example of such is for voltage-sensitive membrane-integrated dyes.

The cartwheeling polarization vector of the p component of the evanescent excitation field in the general case of supercritical angle TIRF results in a well-characterized spatial periodicity of a few hundred nanometers. This is a comparable length scale to some localized *invaginations* in the cell membrane called "caveolae" that may be involved in several different biological processes, such as environment signal detection, and how large particles enter eukaryotic cells including food particles through the process of endocytosis, how certain viruses infect cells, as well as how particles are secreted from eukaryotes through the process of exocytosis (see Chapter 2). Using specialized membrane-permeable fluorescent dyes that orient their electric dipole axis perpendicular to the phospholipid bilayer plane of the cell membrane, p-polarization excitation TIRF microscopy can be used to image the spatial architecture of such localized membrane invaginations (Hinterdorfer et al., 1994; Sund et al., 1999).

Worked Case Example 3.1: TIRF Microscopy

Video-rate fluorescence microscopy was performed using objective lens TIRF on single living Saccharomyces cerevisiae cells ("budding yeast"), spherical cells of mean diameter 5 µm, immobilized onto a glass coverslip in a low-fluorescence water-based growth medium. An orange dye was used to stain the cell membrane that aligns its electric dipole axis parallel to the hydrophobic tail of phospholipid groups in the membrane. The dye was excited into fluorescence using a 561 nm wavelength 20 mW laser with Gaussian beam profile of full width at half maximum 1.0 mm as it comes out of the laser head, which was then expanded by a factor of 3 prior to propagating through a neutral density filter of ND = 0.6 then through a +200 mm focal length lens 200 mm downstream of the back focal plane of an objective lens of effective focal length +2 mm and NA 1.49, in between which was positioned a dichroic mirror that reflected 90% of the laser light onto the sample, with net transmission power losses from all lenses between the laser and the sample being ~20%. Initially, the angle of incidence of the laser was set at 0° relative with the normal of the coverslip–water interface and was purely p-polarized. The focal plane was set to the coverslip surface and a cell positioned to the center of the camera field of view, with the measured intensity values on the camera detector at the center of the camera during continuous illumination being [112, 108, 111, 114, 112, 109, 111, 112, 107, 110, 113] counts per pixel measured using 10 consecutive image frames from the start of laser illumination. Similar values were obtained in a separate experiment performed in the absence of any dye.

a   *Calculate the approximate mean intensity of the laser at the sample in units of $W\,cm^{-2}$. Comment on the results in regard to the dye and autofluorescence and estimate the camera readout noise.*
b   *The angle of incidence of the laser was then increased to the critical angle so that the illumination was only just in TIRF mode. With the focal plane set again to the coverslip surface and a new cell not previously exposed to the laser positioned to the center of the camera field of view, the measured intensity values on the camera detector at the center of the camera during continuous illumination were [411, 335, 274, 229, 199, 176, 154, 141, 133, 122, 118] counts per pixel. Comment on the difference with the intensity data values of (a) and estimate the photobleach time of the dye under these conditions.*
c   *The angle of incidence was then increased further to generate the thinnest evanescent field possible for TIRF excitation. Assuming the photon absorption of the dye is not saturated, make a quantitative sketch of the variation in expected initial fluorescence intensity per pixel if a 1D intensity profile is plotted parallel to the focal plane through the center of the cell image.*

Answers

a   The power incident on the sample after known attenuation and transmission losses is given by

$$P = 20 \times 10^{-0.6} \times 0.9 \times (1-0.2) = 3.6 \text{ mW}$$

The width *w* of the excitation field at the sample is given by

$$1.0 \times 3 \times (2/200) = 0.03\,\text{mm} = 30\,\mu m = 3 \times 10^{-3}\,\text{cm}$$

For a rough approximation, the area of the excitation field is $\sim\pi w^2$. Therefore, the mean excitation intensity in units of $W\,cm^{-2}$ is

$$I_0 \approx P/\pi w^2 = \left(4.5 \times 10^{-13}\right)/\left(\pi \times \left(3 \times 10^{-3}\right)^2\right) \approx 130\ \text{Wcm}^{-2}$$

Since the system is in a standard epifluorescence mode with zero angle of incidence, the polarization vector will be parallel to the glass–water interface; at the point of attachment of a spherical cell to the coverslip, the cell membrane will lie parallel to the coverslip, and so the electric dipole axis of the dye will be oriented normal to this and will thus not be excited into fluorescence, explaining why the intensity data look similar when the dye is removed. A straight-line plot of the intensity points indicates no significant photobleaching above the level of noise, thus no significant autofluorescence under these conditions, with a mean camera readout noise of ~110 counts per pixel.

b   Approaching the critical angle, the system is in p-TIRF mode and so the polarization of the evanescent E-field will be predominantly normal to glass–water interface and will therefore excite the dye into fluorescence. The intensity values decrease due to stochastic dye photobleaching, converging on the camera noise level when all dyes are irreversibly bleached. Subtracting the mean camera noise value from the intensity data and doing a straight-line fit on a semilog plot, or equivalent fit, indicates a $1/e$ photobleach time of ~115 ms, assuming 33 fps on a North American model video-rate camera.

c   The devil is in the detail here! The thinnest evanescent field equates to the minimum possible depth of penetration that occurs at the highest achievable angle of incidence, θ_{max}, set by the NA of the objective lens such that $NA = n_g \sin \theta_{max}$; therefore

$$\theta_{max} = \sin^{-1}(1.49/1.515) = 79.6°, \text{ resulting in a depth of penetration}$$
$$d = 561/\left(4 \times \pi \times \sqrt{(1.49^2 - 1.33^2)}\right) = 66 \text{ nm}$$

Thus, the intensity at different x for a 1D (i.e., line) profile will change since z changes for the height of the membrane above the coverslip surface due to spherical shape of cell by a factor of $\exp(-z/d)$ where from trigonometry $z = r(1 - \sqrt{(1^2 - (x/r)^2)}$ for a cell radius r assumed to be ~5 µm, but also the dipole axis of the orientation changes. However, the orientation θ_d of the electric dipole axis relative to the normal with the coverslip surface rotates with increasing x between $x = 0 - r$ such that $\cos \theta_d = 1 - \sqrt{(1^2 - (x/r)^2)}$ and the absorption by the dye will thus be attenuated by a further factor $[1 - \sqrt{(1^2 - (x/r)^2)}]^2$. Also, in the supercritical regime, the excitation p-polarized vector cartwheels in x with a sinusoidal spatial periodicity of $\lambda_p = \lambda/n_w \sin \theta_g = 561/(1.33 \times 1.49) = 283$ nm. In a nonsaturating regime for photon absorption by the dye, the fluorescence emission intensity is proportional to the incident excitation intensity, which is proportional to the square of E-field excitation multiplied by all of the aforementioned attenuation factors. Thus, a line profile intensity plot in x should look like

$$I_0 \exp\left[-r\left(1 - \sqrt{\left(1^2 - (x/r)^2\right)}\right)/d\right]\left(1 - \sqrt{\left(1^2 - (x/r)^2\right)}\right)^2 \sin^2\left(2\pi x / \lambda_p\right) + 110 \text{ counts}$$

where 110 comes from the camera readout noise and assuming that the cell is in contact with the glass coverslip at $x = 0$, and there is negligible outer material on the surface such as a thick cell wall and that changes in intensity due to out-of-focus effects are negligible since the depth of field is at least ~400 nm, greater than the depth of penetration, namely, a damped $\sin^2 x$ function. A sketch for a line profile from a "typical" cell should ideally have an x range from −2.5 to +2.5 µm. The fluorescence intensity at $x = 0$, I_0, assuming nonsaturation photon absorption of the dye molecules, which is ~0.8/5.3 of the fluorescence intensity obtained in part (a) at the critical angle of incidence, multiplied by a factor $\cos \theta_g / \cos \theta_c \approx 0.46$, since at oblique nonzero angles of incidence the excitation field projects onto the focal plane as an ellipse whose major axis is longer than w by a factor $\cos\theta_g$. From part (b), the initial fluorescence intensity of the dye is $\sim 411 - 110 = 301$ counts, but the intensity due to the camera noise will be insensitive x; thus,

$$I_0 \approx \left((0.8 / 5.3) \times 0.46 \times 301\right) \approx 21 \text{ counts per pixel}$$

3.6.4 CONFOCAL MICROSCOPY

An *in vivo* sample in a light microscope can often encapsulate a height equivalent to tens of equivalent depth of field layers, which can generate significant background noise on the image. The most robust standard biophysical tool to limit this effect is that of *confocal micros-copy*, which uses a combination of two pinholes, in front of the sample and the detector (Figure 3.5g) to delimit the detected intensity to that emerging from the focal plane, resulting in significant increases in fluorescence imaging contrast. Laser light is focused to a volume of just 1 femtoliter (fL), 10^{-18} m^3, onto the sample that is either raster scanned across the sample or the sample stage raster scanned relative to the laser focus. Fluorescence emissions acquired during the analog raster scanning are then digitized during software reconstruction to create a 2D pixel array image.

The *confocal volume* can be approximated as a 3D Gaussian shape, roughly like an egg, with its long vertical axis parallel to the microscope optic axis that is longer than the lateral width w in the focal plane by a factor of a of typically ~ 2.5, giving a volume V:

(3.53)
$$V = a\pi^{3/2}w^3$$

Photon emissions are ultimately focused onto a sensitive detector, typically a PMT, which can then be reconstituted from the raster scan to form the 2D image. Slow speed is the primary disadvantage, limited to ~ 100 fps. Improvements have involved *high-speed spinning disk* (or *Nipkow disk*) confocal microscopy comprising two coupled spinning disks scanning ~ 1000 focused laser spots onto a sample at the same time allowing imaging of ~ 1000 fps. The prin-cipal issue with such fast confocal imaging methods is that the extra exposure to light can result in significant photodamage effects on living biological samples.

The lateral width w is determined by the PSF of the microscope. Note that this value is identical to the *optical resolution limit* in diffraction-limited light microscopy, discussed fully in Chapter 4. For a circular aperture objective lens of numerical aperture NA:

(3.54)
$$w = \frac{0.61\lambda}{NA}$$

For determining the excitation volume size in confocal imaging, this formula can be also used with the radius of the confocal volume in the focal plane equal to w. For determining the resolution of the fluorescence emission images for VIS light fluorescence, the wavelength λ is normally in the range 500–700 nm, low-magnification light microscopy allows fields of view of several hundred microns or more in diameter to be visualized in a single image frame, w can be as low as 1–2 μm (essentially the length scale of subcellular organelles in a eukaryotic tissue, or single bacterial cells within a biofilm), whereas the highest magnifica-tion light microscopes have the smallest values of w of ~ 250–300 nm. The *Nyquist criterion* indicates that pixel edge length in a digitized confocal microscopy image should be less than half the size of the smallest resolvable length scale, that is, $w/2$, in the sample to overcome undersampling, and in practice, pixel edge lengths equivalent to 50–100 nm in the sample plane are typical.

The *axial spatial resolution* in confocal microscopy is worse than the lateral resolution w by the same factor of ~2.5 for the relative stretching of the PSF parallel to optic axis to generate a roughly ellipsoidal volume. A modification to standard confocal imaging is used in *confocal theta microscopy* in which juxtaposed confocal volumes are angled obliquely to each other (ideally perpendicularly) to interfere at the level of the sample that generates a reduced size of confocal volume with a resultant small improvement to axial resolution.

The z range of confocal microscopy illumination is sometimes limited by diffractive effects of the confocal beam. The standard illumination intensity of a confocal microscope laser beam incident on the back aperture of the objective lens has a Gaussian profile, resulting in a divergent confocal volume that, as we have seen, has an aspect ratio parallel to the optic axis of ~1:3, implying that a wide area of sample can obstruct the incident beam *en route* to the focal waist if imaging reasonably deep into a large cell or multicellular tissue sample. The use of Bessel beam illumination can circumvent this problem. A Bessel beam is nondiffractive (discussed fully, in the context of their application in OTs, in Chapter 6), which means that it is relatively nondivergent and insensitive to minor obstructions in the beam profile. A Bessel beam can be used as an alternative source of scanning illumination for exciting fluorophores in the sample (see Planchon et al., 2011), but note that since the beam is nondivergent the ability to *optically section* in z is much reduced compared to standard confocal microscopy.

Another application of confocal illumination is to monitor the diffusion of biomolecules in tissues, or even single cells. This can be achieved using the technique of *fluorescence recovery after photobleaching* (FRAP). Here, a relatively intense confocal excitation volume is generated for the purpose of photobleaching dye molecules of a specific fluorescently labeled biomolecule in that region of space. If the laser intensity is high enough, then relatively little diffusion will occur during the rapid photobleach process. Before and after imaging in fluorescence then shows a dark region indicative of this photobleaching. However, if subsequent fluorescence images are acquired, then fluorescence intensity may recover in this bleached region, which is indicative of diffusion of photoactive dye-labeled biomolecules back into this bleached area. This can be used to determine rates of diffusion of the biomolecule, but also rates of biomolecule turnover in distinct molecular complexes (see Chapter 8).

A related technique to FRAP is *fluorescence loss in photobleaching* (FLIP). This experimental photobleaching method is similar, but fluorescence intensity measurements are instead made at positions outside of the original bleach zone. Here, the diffusion of photobleach dye-labeled biomolecules results in a decrease in fluorescence intensity in surrounding areas. FLIP gives similar information to FRAP but can also yield more complex features of heterogeneous diffusion between the point of photobleaching and the physically distant point of fluorescent intensity measurement.

Worked Case Example 3.2: Using Confocal Microscopy

A fluorescence microscopy experiment was performed on a live Caenorhabditis elegans embryo (see Chapter 7 for a discussion on the use of the C. elegans worm as a model organism) at room temperature to investigate stem cells near the worm's outer surface, whose nuclei are ~2 μm in diameter using a confocal microscope with high numerical aperture objective lens of NA 1.4. A bacterial protein called "LacI" was tagged with GFP and inserted into the worm, and the C. elegans DNA sequence was modified to create a LacO binding site for LacI inside the nucleus of the stem cells that could be switched on (i.e., binding site accessible by LacI) or off (i.e., binding site not accessible by LacI) by external chemical control. With the binding site switched off, a laser at a wavelength of 473 nm was focused on the sample at the same height as the center of the nucleus to generate a confocal excitation volume, and consecutive images were acquired at a rate of 1000 fps without moving the sample relative to the confocal volume, with an observation of single fluorescent particles diffusing through the confocal volume whose brightness was consistent with single molecules of GFP.

a   What % proportion of the nucleus is excited by the laser?
b   If a single particle consistent with one GFP molecule is observed for roughly 60% of consecutive image frames in a sequence, prior to any GFP irreversibly photobleaching, what is the LacI nuclear molarity in units of nM?
c   The range of time taken to traverse the confocal volume in the sample plane was estimated at ~20 ms. Assuming that the frictional drag coefficient of the LacI tagged with GFP can be approximated from Stokes law as $6\pi\eta r$ where the mean viscosity η of the nucleoplasm is roughly that of water at ~0.001 Pa·s, and r is the effective radius of the GFP-tagged LacI, which can be approximated as roughly twice that of GFP, estimate the effective Brownian diffusion coefficient in $\mu m^2 s^{-1}$ and comment on the observed traversal time.
d   If the binding site is chemically induced to be on, what might you expect to observe?

Answers

a   The excitation confocal volume can be estimated:

$$w = 0.61 \times (473 \times 10^{-9}) / 1.4 = 206 \times 10^{-9}\,m$$
$$V = 2.5 \times \pi^{3/2} \times (206 = 1.2 \times 10^{-19}\,m^3$$
$$\text{Volume of nucleus, } V_n = 4 \times \pi \times (^2/3 = 4.2 \times 10^{-18} = 3\%$$
$$\text{Proportion of nucleus excited as } \% = 100V/V_n = 100 \times 1.2 \times 10^{-19}/4.2 \times 10^{-18} = 3\%$$

b   If the observed "occupancy" of the confocal volume is 60% by time for a single diffusing molecule of GFP, then on average, there will be a single molecule of GFP that occupies a characteristic mean volume $V_G = $ of $V / 0.6 = 1.2 \times 10^{-19} / 0.6 = 2.0 \times 10^{-19}\,m^3 = 2.0 \times 10^{-16}\,L$
1 molecule is equivalent to $1/A_v$ moles where A_v is Avogadro's number = $1/(6.022 \times 10^{23}) = 1.7 \times 10^{-24}$ mol
Therefore, the molarity, $m = (1.7 \times 10^{-24})/(2.0 \times 10^{-16})\,M = 8.5 \times 10^{-9}\,M = 8.5$ nM

c   The Stokes radius r_L of the LacI molecule tagged to GFP $\approx 2 \times 2$ nm = 4×10^{-9} m. Therefore, the theoretical diffusion coefficient D_{th} assuming a room temperature of ~20°C is

$$D_{th} = (1.38 \times 10^{-23} \times (20 + 273.15))/(6 \times \pi \times 1 \times 10^{-3} \times 4 \times 10^{-9}$$
$$= 5.4 \times 10^{-11}\,m^{-2}s^{-1} = 54\ \mu m^2 s^{-1}$$

However, the experimental observations indicate that the traversal time in the (2D) focal plane is 20 ms, which is the time taken on average to diffuse in 2D across the confocal volume *waist* (a distance of 2w); thus from rearranging Equation 2.12 assuming n = 2, the observed diffusion coefficient D_{obs} is

$$D_{obs} = (2 \times 206 \times 10^{-9})^2 / (2 \times 2 \times 20 \times 10^{-3}) = 2.1 \times 10^{-12}\,m^2 s^{-1} = 2.1\ \mu m^2 s^{-1}$$

The observed rate of diffusion is smaller by a factor of ~25 compared to theoretical Brownian diffusion. Thus, the diffusion is likely to be non-Brownian (one explanation might be that the LacI diffusion is being impeded by the presence of the tightly packed DNA in the cell nucleus).

d   If the LacI binding sites are accessible to LacI, then there is an increase likelihood that LacI might bind to these sites, in which case one would expect to see diffusing spots suddenly stop moving. However, note that the DNA itself has some mobility, for example, diffusion coefficients in the range $1-2 \times 10^{-3}$ μm^2 s^{-1} have been measured, so, for example, during the ~20 ms "binding site off" traversal time, a DNA-bound LacI might have a root mean squared displacement of up to ~90 nm, which is then roughly a half of the confocal volume focal waist radius—so knowing whether a molecule is bound or not is not a trivial task (see Chapter 7 for analytical methods to assist here).

3.6.5 ENVIRONMENTAL FLUORESCENCE MICROSCOPY

Fluorescence microscopy can be adapted to provide information about the local physical and chemical environment in the vicinity of a fluorescent dye label. For example, confocal microscopy can be utilized to probe the intracellular environment by precise measurement of the lifetime of a fluorophore attached to a specific biological molecule using a technique called FLIM, which is spatially dependent upon physical and chemical parameters such as the local pH and the concentration of certain ions, especially chloride (Cl$^-$). The lifetime of the excited fluorescence state typically varies in the range 10^{-9} to 10^{-7}s and can be measured by synchronizing the fluorescence detection to the excitation using a rapidly pulsed laser at least tens of megahertz frequency. Usual confocal-type imaging is applied with lateral xy scanning of the sample through the confocal volume resulting in a 2D map of fluorescence lifetimes across the extent of the cell, unlike conventional fluorescence imaging that generates maps of fluorescence intensity. Confocal microscopy allows images at different values of height z to be generated, thus in principle enabling the fluorescence lifetime volume in a cell to be reconstructed. The same principle can also be used applying multiphoton excitation to further improve the spatial resolution.

The fluorescence lifetime can be expressed as the reciprocal of the sum of all the rate constants for all the different fluorescence decay processes present. It can be measured using either a frequency or time domain method. For the *time-domain* method, a high-bandwidth detector such as an APD or PMT is used to perform *time-correlated single-photon counting* (*TCSPC*) to count the arrival time of a photon after an initial laser excitation pulse. To improve sampling speed, multiple detection (usually in the range 16–64) can be employed.

The arrival times are modeled as a *Poisson distribution*, and after this process is repeated several times, sufficient statistics are acquired to estimate the fluorescence lifetime from binned arrival time histograms using an exponential fit. Some older FLIM equipment still in operation use instead a *gating optical intensifier* that is only activated after a small proportion of the fluorescence light is directed onto a photodiode via a dichroic mirror, such that photon detection is only possible after a programmed electronic delay. By performing detection across a range of delay times thus similarly allows the fluorescence lifetime to be estimated, but with poorer time resolution compared to TCSPC and lower photon collection efficiency.

For the *frequency domain* approach, the fluorescence lifetime is estimated from the phase delay between a high-frequency modulated light source such as an LED or laser modulated using an AOD (see Chapter 6), coupled to a fast intensified camera detector. An independent estimate of the lifetime may also be made from the modulation ratio of the y components of the excitation and fluorescence. If these values do not agree within experimental error, it may indicate the presence of more than one lifetime component. Frequency domain methods are faster than time domain approaches due to the use of camera detection over slow lateral scanning, and thus are the most commonly used for dynamic cellular studies.

Molecular interactions may also be monitored by using a variant of the technique called FLIM–FRET. This technique utilizes the fact that the lifetime of a fluorophore will change if energy is non-radiatively transferred to, or from it, from another fluorophore in a very close proximity via FRET. Separate lifetime measurements can be made on both the FRET *donor* dye at shorter wavelengths and FRET *acceptor* dye at longer wavelengths (FRET effects are discussed fully in Chapter 4) to infer the nanoscale separation of two different biological molecules separately labeled with these two different dyes. FLIM–FRET therefore can generate cellular *molecular interaction* maps.

Also, a range of fluorophores are now available whose fluorescence lifetime is dependent on the local viscosity of their cellular environment (Kuimova, 2008). These dye molecules typically operate by undergoing periodic mechanical transitions as nanoscale rotors in forming a transient electrical dipole that can absorb excitation light. As with all dyes, each molecule will emit a characteristic approximate number of photons prior to irreversible photobleaching most likely due to free radical chemical damage of the dye. Since the frequency of rotation of the dye is a function of local viscosity, the dye fluorescence lifetime is therefore a metric for viscosity, and thus FLIM measurements using such dyes can map out viscosity over single cells. This is important since local cellular viscosity is a manifestation of the underlying subcellular architecture in that specific region of the cell and thus gives us insight into these different biological features at the nanoscale.

There are also several fluorophores available whose fluorescence emission output is particularly sensitive to specific chemical and physical environmental conditions. Into this category can be included voltage-sensitive dyes and probes which can measure molecular crowding (for example, a FRET pair of fluorescent protein molecules attached by a lever arm which closes to give high FRET efficiency at high molecular crowding conditions but opens out to give a lower FRET efficiency at low molecular crowding. But other dyes also exist, which have been chemically optimized to be highly sensitive to local pH or the binding of ions such calcium (Ca^{2+}), whose fluorescence intensity and fluorescence lifetime change in response to binding. These dyes therefore act as nanoscale environmental sensors, and FILM can map out the absolute values of these environmental parameters in the cell. Many of these dyes operate through having specific regions of their emission spectra, which are sensitive to environmental change, whereas other regions of the emission spectrum may be relatively insensitive. Usually, therefore, a ratiometric approach is taken to measure the relative ratio of emission intensity change at the sensitive and insensitive regions of the emission spectrum, since this ratio will no longer be sensitive to absolute concentrations of the dye in a given localization of the cell.

Direct measurement of the integrated fluorescence intensity of individual dye molecules can also be used as a metric for the physical and chemical environment, that is, the total brightness of a dye molecule is a function of several different environment factors, depending upon the specifics of the dye. A more precise metric is to perform *spectral imaging* of the dye molecule. Here, the fluorescence emission signal can be directed through a transmission diffraction grating, such that the zeroth order (undeviated light) can be imaged onto one-half of a camera detector, while the first order (deviated light) is imaged onto the other half. The zeroth order can be used to determine precisely where the molecule is by using localization fitting algorithms (discussed in Chapter 4) while the first order is a measurement of the transmission spectrum of that dye molecule, since the diffraction angle is wavelength dependent. Thus, the 1D profile of this spectral image can therefore be used as a very precise indicator for local environmental parameters.

KEY POINT 3.4

Fluorescent dyes are often sensitive to many local physical and chemical parameters. Such dyes can be optimized so that they can be used as direct reporters for the output of these physicochemical parameters in live cells, observing changes to fluorescence lifetimes, through a direct or ratiometric intensity approach of fluorescence emissions or through direct spectral imaging.

3.6.6 SLIMFIELD AND NARROW-FIELD EPIFLUORESCENCE MICROSCOPY

Biological molecules typically diffuse much faster inside the watery innards of cells than in membrane surfaces due to lower effective viscosities in the cell cytoplasm by two to three orders of magnitude, and they have a high likelihood of moving during each sampling time window for slow scanning techniques and thus are too blurred to monitor their localization in a time-resolved fashion (i.e., it is not possible to *track* them). To beat the blur time of biomolecules requires imaging faster than their characteristic diffusional time.

In a sampling time window Δt, a molecule with effective diffusion coefficient D will diffuse a root mean squared displacement $\sqrt{\langle R^2 \rangle}$ of $\sqrt{(2Dn\Delta t)}$ (see Equation 2.12) in n-dimensional space. To estimate what maximum value of Δt we can use in order to see a fluorescently labeled molecule unblurred, we set $\sqrt{\langle R^2 \rangle}$ equal to the PSF width. Using this simple assumption in conjunction with the Stokes–Einstein relation (Chapter 2) it is trivial to derive:

$$(3.55) \qquad \Delta t_{max} \approx \frac{1.17 \eta \lambda^2_{peak} \pi r_{Stokes}}{k_g nNA^2 T}$$

where λ_{peak} is the peak fluorescence emission wavelength through an objective lens of numerical aperture NA of a fluorescent-labeled molecule of effective Stokes radius r_{Stokes} diffusing in a medium of viscosity η. For typical nanoscale globular biomolecules for a high-magnification fluorescence microscope, this indicates a maximum sampling time window of a few hundred milliseconds for diffusion in cell membranes (2D diffusion), and more like a few milliseconds in the cytoplasm (3D diffusion). Thus, to image mobile molecular components inside cells requires millisecond time resolution. Note that if the fluorophore-labeled biomolecule exhibits ballistic motion as opposed to diffusive motion (e.g., in the extreme of very short time intervals that are less than the mean collision time of molecules in the sample), then the root mean squared displacement will scale *linearly* with Δt as opposed to having a square root dependence, thus requiring a shorter camera sampling time window than Equation 3.55 suggests.

However, the normal excitation intensities used for conventional epifluorescence or oblique epifluorescence generate too low a fluorescence emission signal for millisecond sampling, which is swamped in camera readout noise. This is because there is a limited photon budget for fluorescence emission and carving this budget into smaller and smaller time windows reduces the effective signal until it is hidden in the noise. To overcome this, the simplest approach is to shrink the area of the excitation field, while retaining the same incident laser power, resulting in substantial increases in excitation intensity. *Narrow-field* epifluorescence shrinks the excitation intensity field to generate a lateral full width at half maximum of ~5–15 µm, which has been used to monitor the diffusion of single lipid molecules with millisecond time resolution (Schmidt et al., 1996), while a variant of the technique delimits the excitation field by imaging a narrow pinhole into the sample (Yang and Musser, 2006).

A related technique of *Slimfield* microscopy generates a similar width excitation field in the sample but achieves this by propagating a narrow collimated laser of width w_s (typically <1 mm diameter) into the back aperture of a high NA objective lens resulting in an expanded confocal volume of lateral width w_c, since there is a reciprocal relation in Gaussian optics between the input beam width and output diffraction pattern width (see Self, 1983):

$$(3.56) \qquad w_c = \frac{f\lambda}{\pi w_s}$$

where f is the focal length of the objective lens (typically 1–3 mm). Since the Slimfield excitation is a confocal volume and therefore divergent with z away from the laser focus, there is some improvement in imaging contrast over narrow field in reducing scattering from out-of-focus image planes.

The large effective excitation intensities used in narrow-field and Slimfield approaches result in smaller photobleach times for the excited fluorophores. For example, if the GFP fluorophore is excited, then it may irreversibly photobleach after less than a few tens of milliseconds, equivalent to only 5–10 consecutive image frames. This potentially presents a problem since although the diffusional time scale of biomolecules in the cytoplasm is at the millisecond level, many biological processes will typically consist of *reaction–diffusion* events,

**KEY BIOLOGICAL
APPLICATIONS:
DIFFERENT
FLUORESCENCE
MICROSCOPY MODES**

Investigating cell membrane
processes in live cells; Probing
rapid processes in the cell cytoplasm of live cells; Rendering
axial as well as lateral localization
information for processes and
components inside cells.

characterized by fast diffusion of one more molecular component but often slow reaction events with other components (see Chapter 8). Thus, the time scale for the entire process may be substantially longer than the millisecond diffusional time scale. To overcome this issue, individual millisecond image frames may be obtained *discontinuously*, that is, using *strobing*. In this way, the limited fluorescence emission photon budget of a single fluorescent protein molecule may be used over substantially longer time scales than just a few tens of milliseconds to gain insight into the dynamics of several different processes inside the cell while still enabling unblurred fluorophore detection on individual images frames. A technical issue in regard to strobing is the bandwidth of the shuttering mechanism used to turn the laser excitation on and off. Typical mechanical shutters are limited to a bandwidth of ~100 Hz, and so the minimum sampling time window that can be accommodated is ~10 ms, which is too high for millisecond imaging. Alternative faster shuttering can be implemented directly through electronic modulation of the laser power output on some devices or utilizing acousto-optic-based technology for shuttering, for example, an *acousto-optic modulator* that can be shuttered at >MHz bandwidth using similar physics principles to AODs (see Chapter 6).

KEY POINT 3.5

In fluorescence imaging at the molecular scale, many single-particle tracks are truncated, often because of rapid fluorophore photobleaching. Strobing can permit the limited photon emission budget to be stretched out over larger time scales to probe slower biological processes.

Worked Case Example 3.3: Slimfield

A Slimfield microscope comprised a 488 nm wavelength laser of beam width 0.1 mm directed into the back aperture of a high numerical aperture objective lens of NA 1.45, focal length 2 mm, to image a single GFP molecule at room temperature, peak fluorescence wavelength ~505 nm, tagged to a cell membrane protein such that the protein is integrated into the phospholipid membrane. The membrane viscosity was measured separately to be ~100 cP, but the GFP molecule itself sits just outside the membrane in the cytoplasm, whose viscosity was closer to 1 cP. The cell membrane is in contact with the microscope coverslip and can be considered reasonably planar over a circle of diameter ~5 μm.

a If the frictional drag coefficient of the membrane integrated protein is 1.1 × 10⁻⁷ kg/s, what is the maximum sampling time per image frame which will allow you to observe the GFP unblurred? Before the protein integrates into the membrane, it must first diffuse through the 3D volume of the cytoplasm. If you wish to monitor this cytoplasmic diffusion as molecules come into focus before they integrate into the membrane, what is the maximum sampling time that could be used?

b In a separate experiment, the average time that a single GFP molecule fluoresces before photobleaching using identical Slimfield imaging conditions as that for measuring the cytoplasmic diffusion was measured to be 30 ms. If the membrane protein is integrated into a lipid raft region of the cell membrane whose effective diameter is 200 nm, do you think you will be able to also monitor the diffusion of the membrane integrated protein as it moves from the center to the edge of a raft?

Answers

a The PSF width for imaging a GFP molecule is given by Equation 3.54 of $w = 0.61\lambda/NA$. Here λ relates to the fluorescence emission, not the excitation, so using the characteristic peak value given indicates that:

$$w = 0.61 \times (505 \times 10^{-9})/1.45 = 212 \times 10^{-9} \text{ m (i.e. 212 nm)}$$

The lateral width of the Slimfield excitation field at the sample is given by Equation 3.56 of $w_c = f\lambda/\pi w_s$. Here, $w_s = 0.1$ mm, $f = 2$ mm, but now λ refers instead to the excitation wavelength, so:

$w_c = (2 \times 10^{-3}) \times (488 \times 10^{-9})/(\pi \times 0.1 \times 10^{-3}) = 3.1 \times 10^{-6}$, or ~3 µm. Therefore, since this width is < 10 µm, we can assume the cell membrane is flat over this Slimfield excitation field and thus diffusion within its membrane in that region is in the 2D focal plane of the microscope. To calculate the diffusion coefficient D in the membrane we use the Stokes–Einstein relation of Equation 2.11 $D = k_B T/\gamma$ where k_B is the Boltzmann constant (1.38×10^{-23} J/k), T the absolute temperature (~300 K for room temperature), and γ is the frictional drag coefficient (such that the drag force = speed multiplied by this frictional drag coefficient), thus:

$$D = (1.38 \times 10^{-23} \times 300)/(1.1 \times 10^{-7}) = 3.7 \times 10^{-14} \text{ m}^2/\text{s}$$

The maximum sampling time Δt to avoid blurring of the PSF image for GFP (see Equation 2.12) occurs roughly when the mean squared displacement of the diffusing integrated membrane protein is the same as the PSF width w for the GFP molecule, so $w^2 = 2Dn\Delta t$ where the spatial dimensionality n here is 2. So:

$$\Delta t = (212 \times 10^{-9})^2/(4 \times 3.7 \times 10^{-14}) = 0.3 \text{ s or 300 ms}$$

The frictional drag coefficient scales linearly with viscosity, so the equivalent Δt in the cytoplasm will be lower by a factor of ~100cp/1cp, or 100, so the equivalent value of D in the cytoplasm will be higher by a factor of 100 compared to the membrane. However, the cytoplasmic diffusion is in 3D not 2D, so $n = 3$, thus the equivalent Δt will be smaller by a factor of $100 \times 3/2$ or 150, giving $\Delta t = 2$ ms.

b   To diffuse from the center of a lipid raft to the edge is a distance d of 100 nm. If we equate this to the root mean squared distance for diffusion in 2D this implies $d^2 = 2Dn\Delta t_{tot}$ so the total time taken for this to occur is:

$\Delta t_{tot} = (100 \times 10^{-9})^2/(2 \times 3.7 \times 10^{-14} \times 2) = 6.8 \times 10^{-5}$ s, or 680 ms. So, a single GFP molecule is very likely to have bleached before it reaches the raft edge as this is more than 20-fold larger than the typical 30 ms photobleaching time indicated. This illustrates one issue with using Slimfield, or indeed any high intensity fluorescence microscopy, in that the high laser excitation intensities required mean that the total time before photobleaching occurs is sometimes too short to enable monitoring of longer duration biological processes. One way to address this issue here could be to use the longer sampling time of 300 ms we estimated for the membrane diffusion, since to retain the same brightness of a GFP molecule might then require 100-fold less excitation intensity of the laser, assuming the fluorescence emission output scales linearly with excitation intensity, and so the GFP might photobleach after more like ~100 × 0.3 s = 30 s. But then we would struggle to be able to observe the initial more rapid 3D diffusion unblurred in the cytoplasm with this much longer sampling time. Another alternative approach could be to use the rapid 2 ms sampling time throughout, but when the protein is integrated into the membrane to *stroboscopically* illuminate it, so to space out the 30 ms/2 ms or ~15 image frames of fluorescence data that we have before photobleaching occurs over the ~680 ms required for the diffusion process to the edge of the raft. But one issue here would be synchronization of the software in real time to the strobing control for the laser excitation so that the strobing starts automatically only when the protein integrates into the membrane. Technically non-trivial to achieve, as they say, but it illustrates that measuring biological processes across multiple time and length scales does still present demanding instrumentation and analysis challenges for biophysics!

3.7 SUMMARY POINTS

- Elastic scattering light spectroscopy—using VIS light, UV, and IR—is a robust tool to determine the concentration of biological scattering particles in solution.
- Fluorescence spectroscopy and FACS can characterize and help isolate different cell types.
- Image contrast can be improved in bright-field microscopy using a range of tools, especially those involving optical interference, which includes phase contrast and DIC microscopy.
- Single-photon excitation fluorescence microscopy is one of the most widely used and valuable biophysical tools to investigate functional biological material, especially when combined with multiple color dye tags.
- Of the many different types of dye tags used in fluorescence microscopy, FPs offer the greatest physiological insight but have suboptimal photophysical features and often cause steric hindrance of native biological functions.
- There are several different modes of illumination for fluorescence microscopy, from which TIRF offers huge enhancements in contrast for monitoring processes in cell membranes in particular.

QUESTIONS

3.1   Give an example of how a biological process spans multiple length and time scales and crosses over and feedbacks at several different levels of length and time scale. Describe an experimental biophysical technique that can be used to generate information potentially at the whole organism, single-cell, and single-molecule levels simultaneously. Should we try to study even broader length and time scales regimes, for example, at the level of ecosystems at one end of the spectrum or quantum biology at the other? Where should we stop and why?

3.2   Transmission electron microscopy (see Chapter 5) on a layer of cells in a tissue suggested their nuclei had mean diameters of 10.2 ± 0.6 µm (± standard deviation). Negative-phase contrast microscopy images from this tissue suggested that the nuclei were the brightest features in the image when the nuclei were most in focus.
   a   Derive a relation between the length through a cell over which the phase of propagating light is retarded by one quarter of a wavelength, stating any assumptions.
   b   Estimate the range of refractive index for the nuclei.

3.3   What do we mean by an *isotropic emitter* in the context of a fluorescent dye molecule? Derive an expression relating the geometrical efficiency of photon capture of an objective lens of numerical aperture (NA) (i.e., what is the maximum proportion of light emitted from an isotropic emitter, neglecting any transmission losses through the lens). What factors may result in fluorescence emission not being isotropic?

3.4   Fluorescence anisotropy experiments were performed on a GFP-tagged protein in aqueous solution, whose effective Stokes radius was roughly twice that of a single GFP molecule. Estimate what the minimum sampling frequency in GHz on the photon detector needs to be to detect anisotropic emission effects. (Assume that the viscosity of water is ~0.001 Pa·s.)

3.5   The same protein of Question 3.3 under certain conditions binds to another protein in the cell membrane with the same diameter, whose length spans the full width of the membrane. If this GFP-tagged protein is imaged using a rapid sampling fluorescence microscope, which can acquire at a maximum sampling time of 1000 image frames per second, comment on whether it will be possible to use fluorescence polarization images to determine the state of the membrane protein's angular rotation in the membrane.

3.6   A yellow FP called "YPet," at peak emission wavelength ~530 nm, was used to tag a low copy number cytoplasmic protein in a spherical bacterial cell of radius ~1 µm. A slimfield microscope, using an objective lens of NA 1.45 with EMCCD camera detector with pixel edge length 18.4 μm and a magnification of 300 between the sample and the camera, was used to monitor the localization of the protein with millisecond sampling, suggesting the protein was freely diffusing with a mean of two fluorescent spots detected at the start of each acquisition if the focal plane of the microscope was set to the midpoint of each cell. Estimate the intracellular concentration in nanomolar of the protein.

3.7 What do we mean by the *Stokes shift*, and where does it originate from? Why does the normalized excitation spectrum of a fluorophore look like a rough mirror image of the normalized emission spectrum?

3.8 A fluorescence microscopy experiment using an objective lens of numerical aperture 0.9 was performed on a zebrafish embryo (a "model organism" used to investigate multicellular tissues, see Chapter 7) to investigate a single layer of GFP-labeled cells of 10 μm diameter, focusing at the midheight level of the cells. These cells were expected to express a mean number of 900 GFP-tagged protein molecules per cell with a standard deviation of 500 molecules per cell. Typical cells had a measured intensity during fluorescence microscopy of ~10^7 counts per cell integrated over the whole of the data acquisition prior to cells being completely photobleached. Data were acquired from a high-efficiency camera detector whose magnification per pixel was equivalent to 200 nm at the sample, gain was 300. Similar cells from another zebrafish in which there was no GFP had a total measured intensity during the same fluorescence microscopy imaging of ~2×10^6 counts.

a Assuming that captured fluorescence emissions come from a cell slice whose depth is equivalent to the depth of field of the objective lens, estimate the number of fluorescence emission photons detected from a single molecule of GFP.

b The quantum efficiency of the camera was ~90%, the transmission function of the dichroic mirror transmits a mean of 85% of all GFP fluorescence emissions, and an emission filter between the dichroic mirror and camera transmits 50% of all GFP fluorescence emissions. The transmission losses due to other optical components on the emission pathway of the microscope resulted in ~25% of all light not being transmitted. Estimate the mean number of photons emitted in total per GFP molecule, stating any assumptions.

3.9 A similar experiment was performed on the cells from Question 3.5, but on a thicker tissue section 80 μm thick, consisting of a 70 μm layer of similar cells not labeled with GFP close to the microscope's coverslip surface, above which are the single layer of GFP-labeled cells.

a Assuming that the rate at which light is absorbed when it propagates through a homogeneous tissue is proportional to its intensity and to the total number of molecular absorption event, derive the Beer–Lambert law, stating any assumptions.

b If each layer of cells attenuates, the excitation beam by 4% calculates the total integrated emission signal due to GFP from a single cell using the same microscope and camera detector and of focusing at mid-cell height for the GFP-labeled cells, acquiring data until the cells are completely photobleached as before.

c Estimate the total noise per cell detected during these data acquisitions.

d To detect a given cell reliably above the level of noise, the effective SNR needs to be above ~2. If the tissue area is ~100 × 100 μm, estimate how many cells you would expect to detect, stating any assumptions you make.

3.10 Describe the technique of TIRF microscopy and give an example of its application in biophysics. Most TIRF *in vivo* studies investigate membrane complexes; why is this? Would TIRF still be effective if there was a high concentration of auto-FPs in the cytoplasm? Can TIRF be applied to monitoring the nucleus of cells?

3.11 The wave equation for a plane wave of light has solutions of the form

$$E = E_0\exp\left\{i\left[kx\sin\theta + ky\cos\theta - \omega t\right]\right\}$$

For real angles θ, this represents a traveling wave in the focal xy plane of the sample.

a   Show that the same solution but with complex θ describes the electric field in water near a water–glass interface in the case where plane wave illumination is totally internally reflected within the glass at the interface.

b   Obtain an expression in terms of the angle of incidence in glass, angular frequency ω, and the refractive indices of glass and water, for the electric field in water for the aforementioned case. Describe using a sketch the wave that this represents.

3.12 Does TIRF illumination require a coherent laser source, or would a noncoherent white-light source be okay?

3.13 A fluorescence imaging experiment was performed at video rate on a bespoke inverted microscope using *Escherichia coli* bacteria that have a shape close to a cylinder of length 2 μm capped by hemispheres of diameter 1 μm, in which a diffused cell membrane protein was tagged using GFP, with biochemical assays suggesting 200 proteins per cell, using a simple flow cell that consists of a glass coverslip in optical contact with a high numerical aperture objective lens via immersion oil, with a cell stuck to the coverslip surface in a water-based buffer, with the walls of the flow cell being 120 μm high stuck on the upper side to a glass microscope slide, above which was then simply air. The cells were first imaged using epifluorescence in which the emergent 473 nm laser excitation beam traveled from below and then straight up through the sample, resulting in a halolike appearance in fluorescence to cells that were stuck with their long axis parallel to the glass coverslip surface (a bright fluorescence at the perimeter of the cell when setting the focal plane to be at the midpoint of the cylinder).

a   Explain these observations. The angle of incidence of the excitation beam was then increased from zero (epifluorescence), which resulted in the beam emerging from the top of the microscope slide at increasingly shallow angles. Eventually, the emergent beam angle was shallow enough that it just dipped below the horizon and could not be seen exiting the microscope slide. At that point, the experimentalists concluded that the system was set for TIRF imaging. However, they were surprised to still see a halolike appearance to the fluorescence images of the cells when they had expected to see a brighter region that marked the cell body.

b   Explain these observations. (*Hint*: think about the microscope slide as well as the coverslip.)

3.14 A confocal microscopy experiment is performed where the focused laser volume had a lateral width measured at 230 nm and an axial width measured at 620 nm. Bacteria in the sample were rodlike with end-to-end micron length and 1 μm width, shaped like a cylinder capped at either end by a hemisphere, and contained a GFP-tagged protein in their cytoplasm.

a   How much of a typical cell is excited by the confocal volume?

b   If the focused laser beam is occupied by a single GFP-tagged protein in the cytoplasm of the cell for 50% of the time, what is the molarity of that protein?

The laser beam is focused on to the cell's midpoint in the cytoplasm. From measurements of GFP-tagged proteins diffusing through the confocal volume of fluorescence pulses from a few fluorescently tagged molecules, the range of time taken to traverse the confocal spot in the sample plane was estimated at 1.7 ± 0.7 ms.

c   Estimate the effective diameter of the protein, stating any assumptions you make. (Assume that the viscosity of the cytoplasm in this case is ~0.002 Pa · s.)

3.15 An epifluorescence microscope with oil immersion (refractive index =1.515) objective lens NA 1.4, focal length 2 mm, was used to monitor a dye called Hoerchst, fluorescence emission peak wavelength 454 nm, which labels the DNA in a single rod-like *E. coli* bacterium lying flat with its long axis (length 3 μm, width 1 μm) on a coverslip. The downstream imaging path after the objective lens comprised a focal length 200 mm lens that makes a telescope with the objective lens, and another telescope with lenses of focal lengths 50 mm and 200 mm, which images onto a CMOS camera with pixel size 20 μm. 3D fluorescence imaging suggested that the width of the nucleoid was 0.48 μm. If a fluorescently tagged DNA binding protein bound transiently to the "bottom" region of the nucleoid closest to the coverslip but then diffused through the nucleoid and subsequently emerged from the 'top' nucleoid region furthest away in right from the coverslip, would you be able to track the protein for the entirety of this journey? (see Worked Case Example 8.4 for a related question.)

REFERENCES

KEY REFERENCE

Axelrod, D. et al. (1984). Total internal reflection fluorescence. *Annu. Rev. Biophys. Bioeng.* *13*:247–268.

MORE NICHE REFERENCES

Allen, K.N., and Imperiali, B. (2010). Lanthanide-tagged proteins—An illuminating partnership. *Curr. Opin. Chem. Biol. 14*:247–254.

Corry, B. et al. (2006). Determination of the orientational distribution and orientation factor for transfer between membrane-bound fluorophores using a confocal microscope. *Biophys. J. 91(3)*:1032–1045.

Fara, P. (2009). A microscopic reality tale. *Nature 459*:642–644.

Hinterdorfer, P. et al. (1994). Reconstitution of membrane fusion sites. A total internal reflection fluorescence microscopy study of influenza hemagglutinin-mediated membrane fusion. *J. Biol. Chem.* 269:20360–20368.

Hooke, R. (1665). *Micrographia: Or Some Physiological Descriptions of Minute Bodies Made by Magnifying Glasses with Observations and Inquiries Thereupon.* Royal Society, London, U.K. Scanned manuscript from 1754 available at http://lhldigital.lindahall.org/cdm/ref/collection/nat_hist/id/0.

Hughes, B.D., Pailthorpe, B.A., and White, L.R. (1981). The translational and rotational drag on a cylinder moving in a membrane. *J. Fluid. Mech. 110*:349–372.

Kuimova, M.K. et al. (2008). Molecular rotor measures viscosity of live cells via fluorescence lifetime imaging. *J. Am. Chem. Soc. 130(21)*:6672–6673.

Marriott, G. et al. (2008). Optical lock-in detection imaging microscopy for contrast-enhanced imaging in living cells. *Proc. Natl. Acad. Sci. USA 105*:17789–17794.

Michalet, X. et al. (2005). Quantum dots for live cells, in vivo imaging, and diagnostics. *Science 307*:538–544.

Nasse, M.J. et al. (2011). High-resolution Fourier-transform infrared chemical imaging with multiple synchrotron beams. *Nat. Methods 8*:413–416.

Piston, D.W. (2010). Fluorescence anisotropy of protein complexes in living cells. *Biophys. J. 99(6)*:1685–1686.

Planchon, T.A. et al. (2011). Rapid three-dimensional isotropic imaging of living cells using Bessel beam plane illumination. *Nat. Methods 8*:417–423.

Popescu, G. (2011) *Quantitative Phase Imaging of Cells and Tissues*, McGraw-Hill, New York.

Saffman, P.G. and Delbrück, M. (1975). Brownian motion in biological membranes. *Proc. Natl. Acad. Sci. USA 72(8)*:3111–3113.

Sako, Y. et al. (2000). Single-molecule imaging of EGFR signalling on the surface of living cells. *Nat. Cell Biol. 2*:168–172.

Schmidt, T. et al. (1996). Imaging of single molecule diffusion. *Proc. Natl. Acad. Sci. USA 93*:2926–2929.

Self, S.A. (1983). Focusing of spherical Gaussian beams. *Appl. Opt. 22*:658–661.

Sund, S.E. et al. (1999). Cell membrane orientation visualized by polarized total internal reflection fluorescence. *Biophys. J. 77*:2266–2283.

Tokunaga, M. et al. (2008). Highly inclined thin illumination enables clear single molecule imaging in cells. *Nat. Methods* 5:159–161.

van Leeuwenhoek, A. (1702). Part of a letter from Mr Anthony van Leewenhoek, F.R.S., concerning green weeds growing in water, and some animalcula found about them. *Philos. Trans.* 23:1304–1311.

Yang, W. and Musser, S.M. (2006). Visualizing single molecules interacting with nuclear pore complexes by narrowfield epifluorescence microscopy. *Methods* 39:316–328.

Making Light Work Harder in Biology

Advanced, Frontier UV–VIS–IR Spectroscopy and Microscopy for Detection and Imaging

4

The dream of every cell is to become two cells.

—**Francois Jacob (Nobel laureate Prize for Physiology or Medicine, 1965, From Monod (1971))**

General Idea: The use of visible light and "near" visible light in the form of ultraviolet and infrared to detect/sense, characterize, and image biological material is invaluable. Here, we discuss advanced biophysical techniques that use visible and near visible light, including microscopy methods, which beat the optical resolution limit, nonlinear visible and near visible light tools, and light methods to probe deeper into biological material than basic microscopy permits.

4.1 INTRODUCTION

The United Nations Educational, Scientific and Cultural Organization announced that 2015 was the International Year of Light, highlighting the enormous achievements of light science and its applications. It is no surprise that there are several biophysical tools developed that use light directly to facilitate detection, sensing, and imaging of biological material. Many of these go far beyond the basic methods of light microscopy and optical spectroscopy we discussed in Chapter 3.

For example, up until the end of the twentieth century, light microscopy was still constrained by the optical resolution limit set by the diffraction of light. However, now we have a plethora of the so-called super-resolution techniques that can probe biological samples using advanced fluorescence microscopy to a spatial precision that is better than this optical resolution limit. An illustration of the key importance of these methods was marked by the award of a Nobel Prize in 2014 to Eric Betzig, Stephan Hell, and William Moerner for the development of "super-resolved" fluorescence microscopy. The fact that they won their Nobel Prize in Chemistry is indicative of the pervasive interdisciplinary nature of these tools.

There are other advanced methods of optical spectroscopy and light microscopy that have been developed, which can tackle very complex questions in biology, methods that can use light to probe deep into tissues and *label-free* tools that do not require a potentially invasive fluorescent probe but instead utilize advanced optical technologies to extract key signatures from native biological material.

The macroscopic length scale of whole organisms presents several challenges for light microscopy. The most significant of these is that of sample heterogeneity, since the larger the sample, the more heterogeneous it is likely to be, with a greater likelihood of being composed

DOI: 10.1201/9781003336433-4

of a greater number of soft matter materials each with potentially different optical properties. Not only that, but larger samples encapsulate a greater range of biological processes that may be manifest over multiple lengths and time scales, making biological interpretation of visible light microscopy images more challenging.

Experimental biophysical techniques are sometimes optimized toward particular niches of detection in length and time scale, and so trying to capture several of these in effect in one go will inevitably present potential technical issues. However, there is good justification for attempting to monitor biological processes in the whole organism, or at least in a population of many cells, for example, in a functional tissue, because many of the processes in biology are not confined to specific niches of length and time scale but instead crossover into several different length–time regimes via complex feedback loops. In other words, when one monitors a particular biological process in its native context, it is done so in the milieu of a whole load of other processes that potentially interact with it. It demands, in effect, a level of *holistic biology* investigation. Methods of advanced optical microscopy offer excellent tools for achieving this objective.

KEY POINT 4.1

The processes of life operate at multiple lengths and time scales that, in general, may crossover and feedback in complex ways. There is no "unique" level to characterize any native process, and so ideally each has to be interpreted in the context of all levels. This is technically demanding, but potentially makes "*in vivo*" experiments the most relevant in terms of genuine biological insight, provided a broad range of length and time scales can be explored.

4.2 SUPER-RESOLUTION MICROSCOPY

Light microscopy techniques, which can resolve features in a sample better than the standard optical resolution limit, are called "*super-resolution*" methods (sometimes written as *superresolution* or *super resolution*). Although super-resolution microscopy is not in the exclusive domain of cellular biology investigations, there has been a significant number of pioneering cellular studies since the mid-1990s.

4.2.1 ABBE OPTICAL RESOLUTION LIMIT

Objects that are visualized using scattered/emitted light at a distance greater than ~10 wavelengths are described as being viewed in the *far-field regime*. Here, the optical diffraction of light is a significant effect. As a result, the light intensity from *point source emitters* (e.g., approximated by a nanoscale fluorescent dye molecule, or even quantum dots [QDs] and fluorescent nano-spheres of a few tens of nanometers in diameter) blurs out in space due to a convolution with the imaging system's *point spread function* (*PSF*). The analytical form of the PSF, when the highest numerical aperture component on the imaging path has a circular aperture (normally the objective lens), is that of an *Airy ring* or *Airy disk* (Figure 4.1a). This is the *Fraunhofer diffraction pattern* given by the squared modulus of the Fourier transform of the intensity after propagating through a circular aperture. Mathematically, the intensity I at a diffraction angle a through a circular aperture of radius r given a wavenumber k ($= 2\pi/\lambda_m$ for wavelength λ_m of the light propagating through a given optical medium) can be described by a first-order *Bessel function* J_1:

$$(4.1) \qquad I(\alpha) = I(0)\left(\frac{2J_1(kr\sin\alpha)}{kr\sin\alpha}\right)^2$$

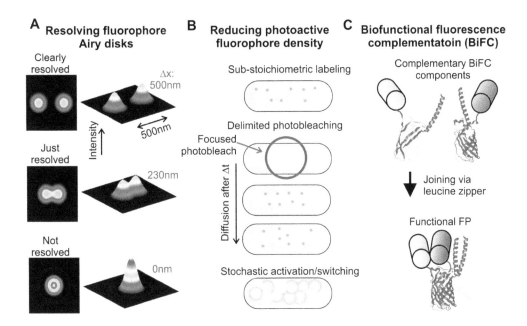

A Resolving fluorophore Airy disks

Clearly resolved

Just resolved

Not resolved

Intensity

Δx: 500nm

500nm

230nm

0nm

B Reducing photoactive fluorophore density

Sub-stoichiometric labeling

Delimited photobleaching

Focused photobleach

Diffusion after Δt

Stochastic activation/switching

C Biofunctional fluorescence complementatoin (BiFC)

Complementary BiFC components

Joining via leucine zipper

Functional FP

FIGURE 4.1 Resolving fluorophores. (a) Airy disk intensity functions displayed as false-color heat maps corresponding to two identical fluorophores separated by 500 nm (clearly resolved), 230 nm (just resolved), and 0 nm (not resolved—in fact, located on top of each other). (b) Methods to reduce the density of photoactive fluorophores inside a cell to ensure that the concentration is less than C_{lim}, at which the nearest-neighbor photoactive fluorophore separation is equal to the optical resolution limit. (c) BiFC that uses genetic engineering technology (see Chapter 7) to generate separate halves of a fluorescent protein that only becomes a single-photoactive fluorescence protein when the separate halves are within a few nanometers of each other to permit binding via a leucine zipper.

This indicates that the first-order intensity minimum at an angle a satisfies

$$(4.2) \qquad \sin\alpha \approx 1.22\frac{\lambda_m}{2r}$$

The *Rayleigh criterion* for optical resolution states that two PSF images can just be resolved if the peak intensity distribution of one falls on the first-order minimum of the other. The factor of 1.22 comes from the circularity of the aperture; the wavelength λ_m in propagating through an optical medium of refractive index n is λ/n where λ is the wavelength in a vacuum. Therefore, if f is the focal length of the objective lens, then the distance the optical resolution in terms of displacement along the x axis in the focal plane Δx is

$$(4.3) \qquad \Delta x = f\sin\theta = \frac{1.22 f \lambda_m}{2r} = \frac{0.16\lambda}{n\sin\theta_{max}} = \frac{0.61\lambda}{NA}$$

This is a modern version of the *Abbe equation*, with the optical resolution referred to as the *Abbe limit*. Abbe defined this limit first in 1873 assuming an angular resolution of $\sin\theta = \lambda_m/d$ for a standard rectangular diffraction grating of width d, with the later factor of 1.22 added to account for a circular aperture. An Airy ring diffraction pattern is a circularly symmetrical intensity function with central peak containing ~84% of the intensity, such that multiple outer rings contain the remaining intensity interspaced by zero minima.

The Rayleigh criterion, from the eponymous astronomer, based on observations of stars, which appear to be so close together that they are difficult to resolve by optical imaging, is semiarbitrary in suggesting that two objects could be resolved using circular optics if the central maximum of the Airy ring pattern of one object is in the same position as the first-order minimum of the other, implying the Abbe limit of $\sim0.61\lambda/NA$. However, there are other suggested resolution criteria that are perfectly reasonable, in that whether or not objects are resolved is as much to do with neural processing of this intensity information in the observer's brain as it is with physics. One such alternative criterion calls for there being no local dip in intensity between the neighboring Airy disks of two close objects, leading to the *Sparrow limit* of $\sim0.47\lambda/NA$. The experimental limit was first estimated quantitatively by the astronomer Dawes, concluding that two similarly bright stars could be just resolved if the dip in intensity between their images was no less than 5%, which is in fact closer to the Sparrow limit than the Rayleigh criterion by $\sim20\%$. Note that there is in fact a hard upper limit to optical resolution beyond which no spatial frequencies are propagated (discussed later in this chapter).

In other words, the optical resolution is roughly identical to the PSF width in a vacuum. For visible light emission, the optical resolution limit is thus in the range $\sim250–300$ nm for a high-magnification microscope system, two orders of magnitude larger than the length scale of a typical biomolecule. The z resolution is worse still; the PSF parallel to the optic axis is stretched out by a factor of ~2.5 for high-magnification objective lenses compared to its width in x and y (see Chapter 3). Thus, the optical resolution limits in x, y, and z are at least several hundred nanometers, which could present a problem since many biological processes are characterized by the nanoscale length scale of interacting biomolecules.

4.2.2 LOCALIZATION MICROSCOPY: THE BASICS

The most straightforward super-resolution techniques are *localization microscopy* methods. These involve mathematical fitting of the theoretical PSF, or a sensible approximation to it, to the experimentally measured diffraction-limited intensity obtained from a pixel array detector on a high-sensitivity camera. The *intensity centroid* (see Chapter 8 for details of the computational algorithms used) is generally the best estimate for the location of the point source emitter (the method is analogous to pinpointing with reasonable confidence where the peak of a mountain is even though the mountain itself is very large). In doing so, the center of the light intensity distribution can be estimated to a very high spatial precision, σ_x, which is superior to the optical resolution limit. This is often performed using a Gaussian approximation to the analytical PSF (see Thompson et al., 2002):

$$(4.4) \qquad \sigma_x = \sqrt{\frac{s^2 + a^2/12}{N} + \frac{4\sqrt{\pi}s^2 b^2}{aN^2}}$$

where
  s is the width of the experimentally measured PSF when approximated by a Gaussian function
  a is the edge length of a single pixel on the camera detector multiplied by the magnification between the sample and the image on the camera
  b is the camera detector *dark noise* (i.e., noise due to the intensity readout process and/or to thermal noise from electrons on the pixel sensor)
  N is the number of photons sampled from the PSF

The distance w between the peak of the Airy ring pattern and a first-order minimum is related to the Gaussian width s by $s \approx 0.34w$.

The Gaussian approximation has merits of computational efficiency compared to using the real Airy ring analytical formation for a PSF, but even so is still an excellent model for 2D PSF images in widefield imaging. The aforementioned Thompson approximation results in marginally overoptimistic estimates for localization precision, and interested readers are directed to later modeling (Mortensen et al., 2010), which includes a more complex but realistic treatment.

The spatial precision is dependent on three principal factors: Poisson sampling of photons from the underlying PSF distribution, a *pixelation noise* due to an observational uncertainty as to where inside a given pixel a detected photon actually arrived, and noise associated with the actual camera detection process. It illustrates that if N is relatively large, then σ_x varies roughly as $s/\sqrt{N}$. Under these conditions, σ_x is clearly less than w (a condition for super-resolution). Including the effects of pixelation and dark noise indicates that if N is greater than $\sim 10^6$ photons, then the spatial precision can in principle be at the level of 1 nm to a few tens of nanometers. A popular application of this method has been called "fluorescence imaging with one nanometer accuracy" (see Park, 2007).

KEY POINT 4.2

In localization microscopy, the spatial precision scales roughly with the reciprocal of the square root of the detected integrated intensity from a diffraction-limited fluorescent spot.

To avoid aliasing due to undersampling of the intensity distribution by the camera, *Nyquist theory* (also known as *Nyquist–Shannon information theory*) indicates that the pixel edge length multiplied by the image magnification must be less than w. Equation 4.4 can be used with physically sensible values of s, N, and b to estimate the optimal value of a to minimize σ_x in other words, to optimize the image magnification on the camera to generate the best spatial precision. Typical optimized values of pixel magnification are in the range 50–100 nm of the sample plane imaged onto each pixel.

The effective photon collection efficiency of a typical high-magnification microscope used for localization microscopy is at best $\sim 10\%$. Therefore, if one were to achieve a theoretical precision as good as 1 nm, then a fluorescence point source emitter must emit at least $\sim 10^7$ photons. A bright single organic dye molecule imaged under typical conditions of epifluorescence microscopy will emit this number of photons after a duration of ~ 1 s. This sets a limit on the speed of biological processes, which can be probed at a precision of 1 nm to a few tens of nanometers, typical of super-resolution microscopy. Note, however, that in practice, there is often a short linker between the dye tag and the biomolecule being tracked, so the true spatial precision for the location of the biomolecule is a little worse than that expected from localization fitting theory since it needs to include the additional flexibility of the linker also.

4.2.3 MAKING THE MOST OUT OF A LIMITED PHOTON BUDGET

To investigate faster processes requires dividing up the *photon budget* of fluorescence emission into smaller sampling windows, which therefore implies a poorer spatial precision unless the photon emission flux is increased. For many fluorophores such as bright organic dyes and QDs, this is feasible, since they operate in a subsaturation regime for photon absorption and therefore their fluorescence output can be increased by simply increasing the excitation intensity. However, several less stable fluorophores are used in biophysical investigations, including fluorescent proteins (FPs), which undergo irreversible photobleaching after emitting at least an order of magnitude fewer photons compared to organic dyes.

For example, green fluorescent protein (GFP) emits only $\sim 10^6$ photons prior to irreversible photobleaching for GFP and so can never achieve a spatial precision to the level of 1 nm in localization microscopy in the typical high-efficiency photon collection microscopes currently available. Some variants of FP, such as *yellow* FP, (YFP), emit in excess of 10^7 photons prior to irreversible photobleaching and therefore have potential application for nanoscale imaging. But similarly, there are less stable FPs (a good example is *cyan* FP [CFP]) that only emit $\sim 10^5$ photons before irreversible photobleaching.

Irreversible photobleaching, primarily due to free radical formation, can be suppressed by chemical means using quenchers. In essence, these are chemicals that mop up free radicals and prevent them from binding to fluorophores and inactivating their ability to fluorescence. The most widely used is based on a combination of the sugar glucose (G) and two enzymes called "glucose oxidase" (*GOD*) and "catalase" (*CAT*). The *G/GOD/CAT freeradical quencher* works through a reaction with molecular oxygen (O·):

(4.5)
$$G + O^{\cdot} \xrightarrow{\quad GOD \quad} GA + H_2O_2$$
$$2H_2O_2 \xrightarrow{\quad CAT \quad} 2H_2O + O_2$$

where glucose is a *substrate* for GOD, which transfers electrons in glucose to form a product *glucuronic acid* (*GA*), and hydrogen peroxide (H_2O_2). In a second reaction, CAT transfers electrons from two molecules of hydrogen peroxide to form water and oxygen gas. The downside is that as GA accumulates, the pH of the solution potentially drops, and so strong *pH buffers* are required to prevent this, though there are some newer quencher systems available that have less effect on pH.

As a rough guide, at video-rate imaging of a typical FP such as GFP, a high-magnification fluorescence microscope optimized for single-molecule localization microscopy can achieve a localization precision in the lateral *xy* focal plane of a few tens of nanometers, with irreversible photobleaching occurring after 5–10 image frames per GFP, or ~200–400 ms at video-rate sampling. If faster sampling time is required, for example, to overcome motion blurring of the fluorophore in cytoplasmic environments, then detection may need to be more in the range of 1–5 ms per image frame, and so the total duration that a typical FP can be imaged is in the range ~5–50 ms. However, as discussed in the previous section, strobing can be implemented to space out this limited photon emission budget to access longer time scales where appropriate for the biological process under investigation.

4.2.4 ADVANCED APPLICATIONS OF LOCALIZATION MICROSCOPY

Localization microscopy super-resolution approaches have been successively applied to multicolor fluorescence imaging in cells, especially *dual-color imaging*, also known as *colocalization microscopy*, where one biomolecule of interest is labeled with one-color fluorophore, while a different protein in the same cell is labeled with a different color fluorophore, and the two emission signals from each are split optically on the basis of wavelength to be detected in two separate channels (see Chapter 8 for robust computational methods to determine if two fluorophores are colocalized or not). This has led to a surplus of acronyms for techniques that essentially have the same core physical basis. These include *single-molecule high-resolution colocalization* that can estimate separations of different colored fluorophores larger than ~10 nm (Warshaw et al., 2005, for the technique's invention; Churchman et al., 2005, for invention of the acronym). Also, techniques called "single-molecule high-resolution imaging with photobleaching" (Gordon et al., 2004) and "nanometer-localized multiple single-molecule fluorescence microscopy" (Qu et al., 2004) both use photobleaching to localize two nearby fluorophores to a precision of a few nanometers up to a few tens of nanometers. *Single-particle tracking localization microscopy* (*TALM*) uses localization microscopy of specifically mobile-tagged proteins (Appelhans et al., 2012).

4.2.5 LIMITING CONCENTRATIONS FOR LOCALIZATION MICROSCOPY

Localization microscopy super-resolution techniques are effective if the mean nearest-neighbor separation of fluorophores in the sample is greater than the optical resolution limit, permitting the PSF associated with a single fluorophore to be discriminated from others in solution. Therefore, there is a limiting concentration of fluorescently tagged molecules in a cell that will satisfy this condition. This depends upon the spatial dimensionality of the localization of the biomolecule. For example, it might be 3D in the cell cytoplasm, 2D confined to the cell membrane, or even 1D delimited to a filamentous molecular track. Also, it depends upon the mobility of the molecule in question.

The distribution of nearest-neighbor distances can be modeled precisely mathematically (see Chapter 8); however, to obtain a rough idea of the limiting concentration C_{lim}, we can use the simple arguments previously in this chapter indicating that in the cytoplasm, the mean fluorophore concentration in a typical bacterial cell such as *Escherichia coli* used is equivalent to ~50–350 molecules per cell, depending on whether they are free to diffuse (low end of the range) or immobile but randomly distributed (high end of the range). In practice, much of a cell contains excluded volumes (such as due to the presence of DNA genetic material), and/or biomolecules may group together in a nonrandom way, so in reality, there may be nontrivial differences from cell to cell and molecule to molecule (see Worked Case Example 4.1).

There are several different types of biomolecules that are expressed in low copy numbers in the cell, some of which, such as transcription factors, regulate the on/off switching of genes, down to only 1–10 per cell in *E. coli* at any one time, which therefore satisfy the mean nearest-neighbor distance condition to be distinctly detected. However, there are similarly other types of molecules that are expressed at effective mean concentration levels per cell of four or more orders of magnitude beyond this (see Chapter 1), whose concentration therefore results in typical nearest-neighbor separations that are less than the optical resolution limit.

In practice, what often happens is that single fluorescently tagged molecules, often FPs, integrate into molecular machines in living cells. These often have characteristic *modular molecular architecture*, meaning that a given molecule may be present in multiple copies in a given molecular complex in the machine. These machines have a characteristic length scale of ~5–50 nm, much less than the optical resolution limit, and since the image is a convolution of the PSF for a single fluorophore with the spatial probability function for all such fluorophores in the machine, this results in a very similar albeit marginally wider PSF as a single fluorophore but with an amplitude greater by a factor equal to the number of copies of that fluorophore in the machine.

4.2.6 SUBSTOICHIOMETRIC LABELING AND DELIMITED PHOTOBLEACHING

There are several techniques to overcome the nearest-neighbor problem. One of the simplest is to substoichiometrically label the molecular population of interest, for example, adjusting the concentration of fluorescent dyes relative to the biomolecule of interest and reducing the incubation time. This involves labeling just a subpopulation of molecules of a specific type such that the cellular concentration of fluorophore is below C_{lim}. Irreversibly, photobleaching a proportion of fluorophores in a cell with excitation light for a given duration prior to normal localization microscopy analysis can also reduce the concentration of photoactive fluorophore in cell to below C_{lim} (Figure 4.1b).

This method is superior to substoichiometric labeling in that there are not significant numbers of unlabeled molecules of interest in the cell, which would potentially have different physical properties to the tagged molecule such as mobility and rates of insertion into a complex so forth, and also has the advantage of being applicable to cases of genomic FP-fusion labeling. This method has been used to monitor the diffusion of fluorescently labeled proteins in the cell membrane of bacteria using a high-intensity focused laser bleach at one end of the cell to locally bleach a ~1 μm diameter region and then observe fluorescence subsequently at a lower laser intensity. The main issue with both of these approaches is that they produce a *dark population* of the particular biomolecule that is under investigation, which may well be affecting the experimental measurements but which we cannot detect.

Dynamic biological processes may also be studied using substoichiometric labeling in combination with *fluorescent speckle microscopy* (see Waterman-Storer et al., 1998). Here, a cellular substructure is substoichiometrically labeled with fluorescent dye (i.e., meaning that not all of the molecules of a particular type being investigated are fluorescently labeled). This results in a speckled appearance in conventional fluorescence imaging, and it has been employed to monitor the kinetics and mobility of individual protein molecules in large molecular structures. The fluorescent speckle generates an identifiable pattern, and movement of the protein assembly as a whole results in the pattern image translating. This can be measured accurately without the need for any computationally intensive fitting algorithms and has been applied to the study of microtubular structures in living cells.

4.2.7 GENETIC ENGINEERING APPROACHES TO INCREASE THE NEAREST-NEIGHBOR DISTANCE

For FP-labeling experiments, it may be possible to control concentration levels of the fluorophore through the application of inducer chemicals in the cell (see Chapter 7). This is technically challenging to optimize predictably, however. Also, there are issues of deviations from native biological conditions since the concentration of the molecules observed may, in general, be different from their natural levels.

Pairs of putatively interacting proteins can satisfy the C_{lim} condition using a technique called "bifunctional fluorescence complementation" (*BiFC*). Here, one of the proteins in the pair is labeled with a truncated nonfluorescent part of a FP structure using the same type of genetics technology as for conventional FP labeling. The other protein in the pair is labeled with the complementary remaining part of the FP structure. When the two molecules are within less than roughly a nanometer of each other, the complementary parts of the FP structure can bind together facilitated by short alpha helical attachment made from leucine amino acids that interact strongly to form a *leucine zipper* motif. In doing so, a fully functional FP is then formed (Figure 4.1c), with a cellular concentration, which may be below C_{lim} even though those of the individual proteins themselves may be above this threshold.

4.2.8 STOCHASTIC ACTIVATION AND SWITCHING OF FLUOROPHORES

Ensuring that the photoactive fluorophore concentration is below C_{lim} can also be achieved through *stochastic activation, photoswitching*, and *blinking* of specialized fluorophores.

The techniques of *photoactivatable localization microscopy* (*PALM*) (Betzig et al., 2006) are essentially the same in terms of core physics principles as the ones described for *fluorescence photoactivatable localization microscopy* and *stochastic optical reconstruction microscopy* (*STORM*) (Rust et al., 2006). They use *photoactivatable* or *photoswitchable* fluorophores to allow a high density of target molecules to be labeled and tracked. Ultraviolet (UV) light is utilized to stochastically either activate a fluorophore from an inactive into a photoactive form, which can be subsequently excited into fluorescence at longer visible light wavelengths, or to switch a fluorophore from, usually, green color emission to red.

Both approaches have been implemented with organic dyes as well as FPs (e.g., photoactivatable GFP [*paGFP*], and PAmCherry in particular, and photoswitchable proteins such as *Eos* and variants and mMaple). Both techniques rely on photoconversion to the ultimate fluorescent state being stochastic in nature, allowing only a subpopulation to be present in any given image and therefore increasing the typical nearest-neighbor separation of photoactive fluorophores to above the optical resolution threshold. Over many ($>10^4$) repeated activation/imaging cycles, the intensity centroid can be determined to reconstruct the localization of the majority of fluorescently labeled molecules. This generates a super-resolution reconstructed image of a spatially extended subcellular structure.

The principal problems with PALM/STORM techniques are the relatively slow image acquisition time and photodamage effects. Recent faster STORM methods have been developed, which utilize bright organic dyes attached via genetically encoded SNAP-Tags. These permit dual-color 3D dynamic live-cell STORM imaging up to two image frames per second (Jones et al., 2011), but this is still two to three orders of magnitude slower than many dynamic biological processes at the molecular scale. Most samples in PALM/STORM investigations are chemically fixed to minimize sample movement, and therefore, the study of dynamic processes, and of potential photodamage effects, is not relevant. However, the use of UV light to activate and/or switch fluorophores, and of visible excitation light for fluorescence imaging, over several thousand cycles substantially increases concentration of free radicals in cellular samples. This potentially impairs the viability of non-fixed cells.

The phrase *time-correlated single-molecule localization microscopy (tcSMLM)* is sometimes used to describe the subset of localization microscopy techniques, which render time-resolved information. This is generally from using tracking of positional data, which estimate the peak of the fluorophore's 2D intensity profile, but other methods also utilize time-dependent differences in photophysical fluorophore properties such fluctuations of brightness and fluorescence lifetimes, which may be applied to multiple fluorescent emitters in a field of view, including *super-resolution optical fluctuation imaging (SOFI)* (Dertinger et al, 2009) which uses temporal intensity statistics to generate super-resolved image data, and a range of recent algorithms which use statistical analysis of dye brightness fluctuations, for example, Bayesian analysis of bleaching and blinking (3B) (Cox et al, 2012), which can overcome several of the issues associated with high fluorophore density, which straightforward tracking methods can be limited by.

The most widely applied tcSMLM approach uses PALM instrumentation, called time-correlated PALM (*tcPALM*), in which PALM is used to provided time-resolved information from individual tracks (Cissé et al. 2013). As with normal PALM, only a fraction of the tagged molecules present can be localized so this does not render a definitive estimate for the total number of molecules present of that particular type but does yield quantitative details of their dynamics from the subset that are labeled and detected once stochastically activated. Here, for example, PALM can image a region of a cell expected to have a high concentration of molecular clusters comprising a specific protein labeled with a photoactivatable dye, and then a time-series is acquired recording the time from the start of activation at which every track is subsequently detected. The peak time from this distribution, inferred from many such fields of view form different cells, is then used as a characteristic arrival time for that protein into a molecular complex, which can then be compared against other estimates in which the system is perturbed in some way at which characteristic time for fluorophore to be detected time, and which can be a very tool to understand the reactions that occur in the cluster assembly process.

4.2.9 STOCHASTIC BLINKING

Reversible photobleaching of fluorophores, or blinking, can also be utilized (for a good review of the fluorophore photophysics, see Ha and Tinnefeld, 2012). The physical mechanism of blinking is heterogeneous, in that several potential photophysical mechanisms can lead to the appearance of reversible photobleaching. Occupancy of the triplet state, or *triplet blinking*, is one of such; however, the triplet state lifetime is $\sim 10^{-6}$ s, which is too small to account for observed blinking in fluorescence imaging with a sampling time window of $\sim 10^{-3}$ s and above. *Redox blinking* is another possible mechanism in that an excited electron is removed (one of the definitions of oxidation), which induces a dark state that is transient up until the time that a new electron is elevated to the excited state energy level. However, many different fluorophores also appear to have nonredox photochemical mechanisms to generate blinking.

The stochastic nature of *photoblinking* can be carefully selected using different chemical redox conditions but also, in some cases, through a dependence of the blinking kinetics on the excitation light intensity. High-intensity light, in excess of several kW cm^{-2}, can give rise to several reversible blinking cycles before succumbing to irreversible photobleaching. This reduces the local concentration of photoactive fluorophores in any given image frame, facilitating super-resolution localization microscopy. This technique has been applied to living bacterial cells to map out DNA binding proteins using YFP (Lee et al., 2011).

An alternative stochastic super-resolution imaging method is called "point accumulation for imaging in nanoscale topography." Here, fluorescence imaging is performed using diffusing fluorophore-tagged biomolecules, which are known to interact only transiently with the sample. This method is relatively straightforward to implement compared to PALM/STORM. This method has several variants, for example, it has also been adapted to a membrane-localized protein to generate super-resolution cell membrane features in a technique described as super-resolution by power-dependent active intermittency and points accumulation for imaging in nanoscale topography (SPRAIPAINT) (Lew et al., 2011).

Dye blinking of organic dyes has also been utilized in a technique called "blinking assisted localization microscopy" (Burnette et al., 2011). This should not be confused with *binding-activated localization microscopy*, which utilizes a fluorescence enhancement of a dye when bound to certain cellular structures such as nucleic acids compared to being free in solution, which can be optimized such that the typical nearest-neighbor distance is greater than the optical resolution limit (Schoen et al., 2011). Potential advantages over PALM/STORM of blinking localization microscopy are that the sampling time scales are faster and also that there is less photodamage to living cells in avoiding the more damaging shorter wavelength used in UV-based activation.

Improvements to localization microscopy precision can be made using prior information concerning the photophysics of the dyes, resulting in a hybrid technique of analytical inference with standard localization tools such as PALM/STORM. Bayesian analysis is an ideal approach in this regard (discussed fully in Chapter 8). This can be applied to photoblinking and photobleaching observations trained on prior knowledge of both. Bayesian blinking and bleaching microscopy (3B *microscopy*) analyzes data in which many overlapping fluorophores undergo both bleaching and blinking events to generate spatial localization information at enhanced resolution. It uses a *hidden Markov model* (*HMM*). An HMM assumes that the underlying process is a *Markov process* (meaning future states in the system depend only on the present state and not on the sequence of events that preceded it, i.e., there is no memory effect) but with unobserved (*hidden*) states and is often used in Bayesian statistical analysis (see Chapter 8). It enables information to be obtained that would be impossible to extract with standard localization microscopy methods.

The general issue of photodamage with fluorescence imaging techniques should be viewed in the following context:

1 All imaging of live cells with fluorescence (and many other modalities) is potentially damaging, for example, fast confocal scanners often kill a muscle cell in seconds. That is, however, acceptable, for certain experiments where we are looking at fast biological processes (e.g., a few milliseconds) and the mindful biologist builds in careful control experiments to make sure they can put limits on these effects.
2 The degree of damage seen is likely closely connected to the amount of information derived but can be reduced if light dosage is reduced using hybrid approaches such as 3B microscopy. Some spatial resolution is inevitably sacrificed for time resolution and damage reduction—nothing is for free.
3 Many dark molecules in photoactivating super-resolution methods greatly reduce absorption in the sample, and the UV exposure to photoactivate is generally very low.

4.2.10 RESHAPING THE PSF

The Abbe diffraction limit for optical resolution can also be broken using techniques that reduce the size of the PSF. One of these is *4Pi microscopy* (Hell and Stelzer, 1992). Here, the sample is illuminated with excitation light from above and below using matched high *NA* objective lenses, and the name of the technique suggests an aspiration to capture all photons emitted from all directions (i.e., 4π steradians). However, in reality, the capture solid angle is less than this. The technique improves the axial resolution by a factor of ~5 to more 100–150 nm, generating an almost spherical focal volume six times smaller than confocal imaging.

Stimulated-emission depletion microscopy (*STED*) (see Hell and Wichmann, 1994) and adapted techniques called "ground state depletion," "saturated structured illumination microscopy" (*SSIM*), and "reversible saturable optical fluorescence transitions" microscopy all reduce the size of the excitation volume by causing depletion of fluorescence emissions from the outer regions of the usual Airy ring PSF pattern. Although these techniques began as *in vitro* super-resolution methods, typically to investigate the aspect of cytoskeletal structure

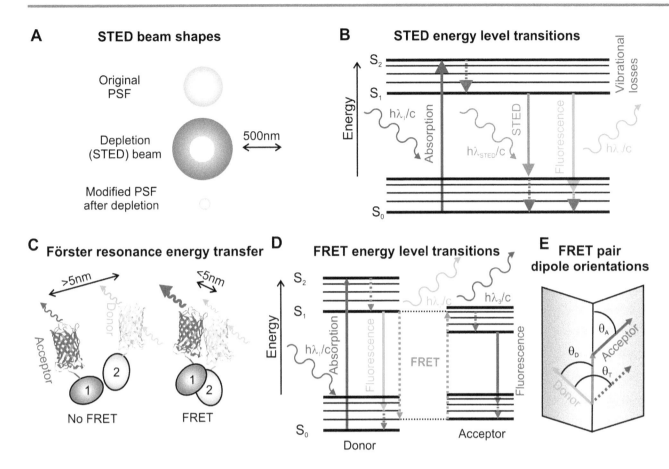

FIGURE 4.2 Stimulated-emission depletion microscopy (STED) and Förster resonance energy transfer (FRET). (a) Relative sizes and shapes of STED depletion beam and original PSF of a fluorophore; the donut-shaped depletion beam stimulates emission depletion from the fluorophore to generate a much smaller point spread function intensity volume, with (b) the associated Jablonski energy level diagram indicated. (c) Schematic depiction of FRET, here, indicated between a donor and acceptor fluorescent protein, which are excited by short and long light wavelengths respectively, with (d) associated Jablonski energy level diagram and (e) schematic indicating the relative orientation of the donor–acceptor fluorophore electric dipole moment axes; each respective electric dipole axis lies in a one of two planes separated by angle θ_T, and these two planes intersect at a line defined by arrow indicated that makes angles θ_A and θ_D with the acceptor and donor electric dipole axes, respectively.

(to this day, the ability to resolve single-microtubule filaments of a few tens of nanometer width in a tightly packed filament is treated as a benchmark of the technique), they have also now developed into powerful cellular imaging tools.

In STED, this reduction in laser excitation volume is achieved using a second *stimulated emission* laser beam in addition to an excitation beam, which is shaped like a donut in the focal plane with a central intensity minimum of width ~200 nm. This annulus intensity function can be generated using two offset beams or by phase modulation optics (Figure 4.2a).

This beam has a longer wavelength than the laser excitation beam and stimulates emission from the fluorescence excited state (Figure 4.2b), resulting in the *depletion* of the excited state in the high-intensity region of the donut, but a nondepleted central region whose volume is much smaller than the original PSF. In STED, it is now standard to reduce the width w of the excitation volume in the sample plane to <100 nm, which, assuming an objective lens of numerical aperture NA, is given by

(4.6)
$$w = \frac{\lambda}{2NA\sqrt{1 + I/I_s}}$$

where

λ is the depletion beam wavelength with *saturating intensity* I_s (intensity needed to reduce fluorescence of the excited state by a factor of 2)

I is the excitation intensity at the center of the donut

For large I, there is an $\sim 1/\sqrt{I}$, dependence on w, so w could in principle be made arbitrarily small. A limiting factor here is irreversible photobleaching of the fluorophores used, resulting in a few tens of nanometers for most applications at present. Due to the absence of shorter wavelength UV activation, STED excitation can penetrate deeper with less scatter into large cells, which has some advantages over PALM/STORM. Also, maximum image sampling rates are typically higher than for PALM/STORM at a few tens of frames per second currently higher for STED compared to a few per second for PALM/STORM. However, there are potentially greater issues of photodamage with STED due to the high intensity of the depletion beam.

Minimal photon FLUX (MinFlux) microscopy (Balzarotti et al 2016) is a related super-resolution tool that combines SMLM and STED while using fewer fluorescence photons but enabling higher spatial and time resolution. In MinFlux, the STED donut-shaped bead is steered to map onto the molecular position itself while eliciting minimum fluorenscent photon emissions from the dye molecule. In Minflux, the donut beam is scanned across a sample to minimally acquire emission data sufficient to estimate roughly where a dye molecule is by using probabilistic triangulation criteria based on the brightness of the fluorescence and the spatial position of the donut beam. This estimate is then used to fine-tune the position of the donut beam to center it over the dye by shifting the beam over an area of length scale, $L \sim 50$ nm, and then STED as normal is performed. However, since the center of the donut beam, the zero-excitation intensity region, is now roughly colocalized with the dye position, then the dye molecule subsequently emits relatively low numbers of fluorescent photons.

The spatial precision, instead of scaling with $\sim \lambda/(NA\sqrt{N})$ as suggested by Equation 4.6 scales as $\sim \sim L/\sqrt{N}$. This results in a spatial precision of 1–3 nm for as few as a ~ 500 emitted photons but can be made significantly smaller by reducing L to nanoscale levels, thus allowing for true nanoscale spatial resolution but with substantively longer duration acquisitions while minimizing photobleaching of the dyes. Also, since steering of the donut beam uses rapid piezoelectric and electro-optical control, the time resolution for 2D imaging can be as low as a few hundred microseconds, hence rapid enough to enable single-molecule dye diffusion to be tracked unblurred in the cytoplasm of live cells, with tracking really then limited only by photoblinking of the dyes themselves.

Variants of the technique enabling multicolor 3D MinFlux imaging now exist (e.g. using a "z-donut," i.e., a 3D shell-intensity depletion beam volume). At the time of writing, basic SMLM bespoke microscopy can be implemented for as a little as few tens of thousands of USD with higher throughput compared to MinFlux, whereas the equivalent cost for a basic MinFlux system is roughly an order of magnitude greater. Although promising developments are being made with structured illumination to increase MinFlux throughput, the main barrier to its more widespread application is arguably cost.

4.2.11 PATTERNED ILLUMINATION MICROSCOPY

Structured illumination microscopy (SIM), also known as *patterned illumination microscopy*, is a super-resolution method that utilizes the *Moiré* pattern interference fringes generated in the focal plane using a spatially patterned illumination (Gustafsson, 2000). Moiré fringes are equivalent to a beat pattern. When measurements are made in the so-called reciprocal space or frequency space in the Fourier transform of the image, smaller length scale features in the sample are manifested at higher spatial frequencies, such that the smallest resolvable feature using conventional diffraction-limited microscopy has the highest spatial frequency. In generating a beat pattern, spatial frequencies above this resolution threshold are translated to lower values in frequency space. On performing an inverse Fourier transform, this therefore reveals spatial features that would normally be smaller than the optical resolution limit. In practice, the fringe pattern is rotated in the focal plane at multiple orientations (three orientations separated by 120° is typical) to obtain resolution enhancement across the full lateral plane. The actual pattern itself is removed from the imaging by filtering in frequency space; however, unavoidable artifacts of the pattern lines do occur, which can result in embarrassing overinterpretation of cellular data if careful controls are not performed.

The spatial resolution enhancement in standard SIM relies on a linear increase in spatial frequency due to the sum of spatial frequencies from the sample and pattern illumination. The latter is diffraction-limited and so the maximum possible enhancement factor for spatial resolution is 2. But, if the rate of fluorescence emission is nonlinear with excitation intensity (e.g., approaching very high intensities close to photon absorption saturation of the fluorophore), then the effective illumination pattern may contain harmonics with spatial frequencies that are integer multiples of the fundamental spatial frequency from the pattern illumination and can therefore generate greater enhancement in spatial resolution. This has been utilized in nonlinear SIM techniques called "saturated pattern excitation microscopy" and *SSIM*, which can generate a spatial resolution of a few tens of nanometers. The laser excitation intensities required are high, and therefore, sample photodamage is an issue, and the imaging speeds are currently still low at a maximum of tens of frames per second.

KEY POINT 4.3

Most super-resolution techniques suffer issues of cellular photodamage to differing extents. For example, PALM/STORM uses harmful UV light of several thousand activation cycles, and photoblinking methods also use very high-excitation intensities of visible light, STED using a damaging high-intensity depletion beam. Caution should be applied when interpreting any study, which purports to perform "live-cell" studies super-resolution techniques. However, a key here is the use of appropriate biological control experiments—any live-cell imaging requires a large number of careful control experiments.

4.2.12 NEAR-FIELD EXCITATION

Optical effects that occur over distances less than a few wavelengths are described as *near-field*, which means that the light does not encounter significant diffraction effects and so the optical resolution is better than that suggested by the Abbe diffraction limit. This is utilized in *scanning near-field optical microscopy* (*SNOM* or *NSOM*) (Hecht et al., 2000). This often involves scanning a thin optical fiber across a fluorescently labeled sample with excitation and emission light conveyed via the same fiber. The vertical distance from sample to fiber tip is kept constant at less than a wavelength of the emitted light. The lateral spatial resolution is limited by the diameter of the optical fiber itself (~20 nm), but the axial resolution is limited by scanning reliability (~5 nm). Scanning is generally slow (several seconds to acquire an image), and imaging is limited to topographically accessible features on the sample (i.e., surfaces).

However, samples can also be imaged in SNOM using other modes beyond simply capturing reflected and/or emitted light. For example, many of SNOM's applications use transmission mode with the illumination external to the fiber. These are either oblique above the sample for reflection, or from underneath a thin transparent sample for transmission. The fiber then collects the transmitted/reflected light after it interacts with the sample.

An important point to consider is the long time taken to acquire data using SNOM. SNOM takes several tens of minutes to acquire a single image at high pixel density, an order of magnitude longer than alternative scanning probe methods such as atomic force microscopy (AFM) for the equivalent sample area (see Chapter 6). The high spatial resolution of ~20 nm that results is a great advantage with the technique, though the poor time resolution is a significant drawback with regard to monitoring the dynamic biological processes. Fluorescent labeling of samples is not always necessary for SNOM, for example, the latest developments use label-free methods employing an infrared (IR)-SNOM with tuneable IR light source.

Near-field fluorescence excitation fields can be generated from *photonic waveguides*. Narrow waveguides are typically manufactured out of etched silicon to generate channels of width ~100 nm. A laser beam propagated through the silicon generates an evanescent excitation field in much the same wave as for total internal reflection fluorescence (TIRF) microscopy (see Chapter 3). Solutions containing fluorescently labeled biomolecules can be flowed through a channel and excited by the evanescent near-field. Many flow channels can be manufactured in parallel, with surfaces precoated by antibodies, which then recognize different biomolecules, and this therefore is a mechanism to enable biosensing. Recent improvements to the sensitivity of these optical microcavities utilize the *whispering gallery mode* in which an optically guided wave is recirculated in silicon crystal of a circular shape to enhance the sensitivity of detection in the evanescent near-field to the single-molecule level (Vollmer and Arnold, 2008).

4.2.13 SUPER-RESOLUTION IN 3D AND 4D

Localization microscopy methods are routinely applied to 2D tracking of a variety of fluorophore-labeled biomolecules in live cells. However, to obtain 3D tracking information presents more challenges, due in part to mobile particles diffusing from a standard microscope's depth of field faster than refocusing can be applied, and so they simply go out of focus and cannot be tracked further. This, coupled with the normal PSF image of a fluorophore within the depth of field, results in relative insensitivity to z displacement. With improvements in cytoplasmic imaging techniques of cellular samples, there is a motivation to probe biological processes deeper inside relatively large cells and thus a requirement for developing 3D tracking methods. (Note that these techniques are sometimes referred to as "4D," since they generate information from three orthogonal spatial dimensions, in addition to the dimension of time.)

There are three broad categories of 3D tracking techniques. *Multiplane imaging*, most commonly manifested as *biplane imaging*, splits fluorescence emissions from a tracked particle to two or more different image planes in which each has a slight focal displacement offset. This means that the particle will come into sharp focus on each plane at different relative distances from the lateral focal plane of the sample. In its most common configuration of just two separate image planes, this can be achieved by forming the two displaced images onto separate halves of the same camera detector. Standard 2D localization fitting algorithms can then be applied to each image to measure the xy localizations of the particle as usual and also estimate the z value from extrapolation of the pixel intensity information compared to a reference of an in-focus image of, for example, a surface-immobilized fluorophore. This technique works well for bright fluorophores, but in having to split the fluorescence photon budget between different images, the localization precision is accordingly reduced by a factor of $\sim\sqrt{2}$. Also, although the real PSF of a fluorescence imaging system shows in general some asymmetry in z, this asymmetry is normally only noticeable at a few hundred nanometers or more away from the focal plane. Therefore, unless the different focal planes are configured to be separated by at least this threshold distance, there is some uncertainty as to whether a tracked particle very close to the focal plane is diffusing above or below it or, in having to separate the focal planes by relatively large distances inevitably reduces the sensitivity in z at intermediate smaller z values. Similarly, there can be reductions in z sensitivity with this method if separate tracked spots are relatively close to each other in z so as to be difficult to distinguish (in practice, the threshold separation is the axial optical resolution limit that is ~2.5 times that of the lateral optical resolution limit, close to 1 μm, which is the length scale of some small cells such as bacteria and cell organelles in eukaryotes such as nuclei).

Astigmatism imaging is a popular alternative method to multiplane microscopy, which is relatively easy to implement. Here, a long focal length cylindrical lens is introduced in the optical pathway between the sample and the camera detector to generate an image of tracked particles on the camera. The cylindrical lens has intrinsic astigmatism, meaning that it has marginally different focal lengths corresponding to the x-axis and y-axis. This results in fluorophores above or below the xy focal plane having an asymmetric PSF image on the camera, such that fluorescence intensity appears to be stretched parallel to either the x- or y-axis depending upon whether the fluorophore is above or below the focal plane and the relative geometry of the cylindrical lens and the camera.

Measuring the separate Gaussian widths in x and y of such a fluorophore image can thus be used as a sensitive metric for z, if employed in combination with prior calibration data from surface-immobilized fluorophores at well-defined heights from the focal plane. The rate of change of each Gaussian width with respect to changes in z when the Gaussian width is minimum is zero (this is the condition when the fluorophore image is in focus with respect that the appropriate axis of x or y). What is normally done therefore is to toggle between using the x and y widths for the best metric of z, at different z, in order to span the largest possible z range for an accurate output prediction of z. The main issues with the astigmatism method are that the localization precision in z is worse than that in x and y by a factor of ~1.5 and also that the maximum range in z, when using a typical high-magnification microscope optimized for localization microscopy, is roughly ±1 μm.

Corkscrew PSF methods, the most common of which is the *double-helical PSF (DH-PSF)* approach, can be used to generate z information for fluorophore localization. These techniques use phase modulation optics to generate helical, or in the case of the DH-PSF method, double-helical-shaped PSFvolumes in the vicinity of the sample plane. The helical axis is set parallel to the optic (z) axis such that when a fluorophore is above or below the focal plane, the fluorophore image rotates around this central axis. In the case of DH-PSF imaging, there appear to be two fluorescent spots per fluorophore, which rotate around the central axis with changes in z. In this instance, x and y can also be determined for the fluorophore localization as the mean from the two separate intensity centroid values determined for each separate spot in a pair.

This method has a downside of requiring more expensive phase modulation optics in the form of either a fixed *phase modulation plate* placed in a conjugate image plane to the back aperture of the objective lens (conjugate to the Fourier transformation plane of the sample image) or a *spatial light modulator* (*SLM*) consisting of an array of electrically programmable LCD crystals, which can induce controllable levels of phase retardation across a beam profile. However, the precision in z localization is more than twice as good as the other two competing methods of multiplane and astigmatism imaging (see Badieirostami et al., 2010). A downside is that all multilobe-type methods have a larger in-plane extent, which further reduces the density of active markers that can be imaged without overlap, which can present a real issue in the case of intermediate-high copy number systems (i.e., where the concentrations of fluorescently labeled biomolecules is reasonably high).

KEY BIOLOGICAL APPLICATIONS: SUPER-RESOLUTION METHODS

Imaging subcellular architectures with nanoscale spatial precision; Quantifying molecular mobility; Determining molecular colocalization.

Worked Case Example 4.1: Super-Resolution Imaging

Video-rate imaging epifluorescence microscopy with a water immersion objective lens NA 1.2 was performed on a low density of surface-immobilized GFP using a 473 nm laser to reveal fluorescent spots that bleached after an average of ~30 consecutive image frames. Spherical bacterial cells of diameter ~2 μm were engineered to contain a GFP-labeled protein X that under the conditions of the experiment was known not to aggregate or oligomerize and was expressed in the cytoplasm with typically ~120 molecules present per cell at any one time. Narrow-field epifluorescence was used on these cells immobilized to a microscope coverslip to sample at 3 ms per frame, which initially indicated bright slightly grainy fluorescence intensity toward the outer perimeter of the cell when focusing at a midcell height, much dimmer toward the cell center, but after some continuous illumination, the outer region also became dimmer until eventually distinct diffusing fluorescent spots could be seen, ~15% of which had similar integrated intensity to the epifluorescence data, with the remaining spots having integrated intensity values of ~2, ~3, or more rarely ~4+ times within relative proportions of ~50%, ~25%, and ~10% that of the epifluorescence data. Preliminary tracking data suggest that a typical spot diffused its own PSF width in the microscope focal plane during ~1 image frame, with the lateral spatial resolution using super-resolution localization microscopy algorithms being ~50 nm.

a   *Estimate the 1/e photobleaching time for GFP-X during the narrow-field experiments, stating any assumptions.*

b   *Explain what the observation of a dimmer central region might imply with regard to the bacterial nucleoid. How many image frames would be acquired using narrow field before you might expect to see distinct fluorescent spots?*

c   *Explain with quantitative reasoning the observed distribution of fluorescent spots in narrow-field.*

d   *The conditions of the experiment were changed such that all of the protein X molecules in the cell can oligomerize to form a globular chain, which can translocate across the cell membrane. Discuss if it might be possible to observe the translocation process directly.*

Answers

a   Assuming that photon absorption is linear in the range of excitation intensity, and since the measure of typical spot emission intensity is the same, the excitation intensity of narrow-field must be greater than that of standard epifluorescence by a factor equal to the reciprocal of the ratio of the exposure times used, that is, ~(33 ms)/(3 ms) = 11. The photobleach 1/e time will also be inversely proportional under these conditions to excitation intensity, so under epifluorescence this time is ~10 × 33 ms; therefore, in narrow-field the photobleach time t_b is $t_b = (10 \times 33)/11 = 10 \times 3$ ms = 30 ms (in other words, the total number of consecutive image frames before a typical GFP molecule bleaches is the same as it was for standard epifluorescence).

b   The dim central region of the cell is consistent with the nucleoid being an excluded volume to GFP-X. The typical volume of this for bacteria is ~1/3 of the total cell volume (see Chapter 2); thus, the volume accessible to GFP-X may only be ~2/3 of the total cell volume. To just resolve distinct fluorescent spots, assuming the Rayleigh criterion, requires the nearest-neighbor separation to be such that a first-order minimum of the Airy pattern of a fluorescent spot coincides with the peak of another. A simple approximation suggests that this distance equates to the PSF width w plus the typical distance diffused in one image frame, which in this instance is also w. For a more robust analytical treatment of this, see Chapter 7. For simplicity, one possible approximation is then to say that at this limiting density of photoactive GFP-X, the effective radius of the equivalent sphere is $(w + w) = 2w$ such that a total of N spheres are all tightly packed to occupy the accessible volume of the cell (assumed to be ~2/3 total cell volume):

$$N \times \left(4\pi \times (2w)^3 / 3\right) \approx (2/3) \times 4\pi \times (1)^3 / 3$$

Using the Abbe limit and assuming the characteristic emission, wavelength is given roughly by the emission peak for GFP of ~509 nm (see Chapter 3):

$$w = 0.61 \times (0.509)/1.2 = 0.26 \ \mu m$$

Therefore,

$$N \approx 2/3 \times (1/0.52)^3 \approx \text{photoactive molecules of GFP-X per cell}$$

Assuming that there are initially ~120 photoactive GFP-X molecules per cell, to end up with just 5 per cell requires a continuous photobleach of duration t such that

$$5 \approx 120 \times exp\left[-t/t_b\right], \text{ therefore}$$

$$t \approx 30 \times ln(120 / 2) \approx 95 \text{ ms. This equates to } \sim(95/3) \approx 32 \text{ frames}$$

c   The distribution of integrated spot intensity values cannot be explained here by aggregation/oligomerization. A possible explanation is that two or more fluorescent spots are detected when they are closer than the limiting distance of ~2w in this instance. The robust calculation of the probability of this occurring is in Chapter 7, but for a simple approximation, we could say that the probability that a second spot is within ~2w of a first is $p_1 = V_{spot}/V_{cell}$ where V_{spot} is the volume of a sphere of radius 2w and V_{cell} is the total accessible volume. With $(N - 1)$ such spot, the overall probability p_2 for such a double is ~$(N - 1)p_1$ and similarly for a triple spot is $p_3 \sim (N-2)p_1 = (N-1)(N-2)p_1^2$, and $p_4 \sim (N-2)p_1 = (N-1)(N-2)(N-3)p_1^3$

Setting $N \sim 5$ and using $V_{spot} = 4\pi \times (2w)^3/3$ and $V_{cell'} = (2/3) \times 4\pi \times (1)^3/3$ indicates

$$p_1 \approx 0.14, p_2 \approx 0.56, p_3 \approx 0.24, p_{4+1} \approx 1-\left(p_1 + p_2 + p_3\right) \approx 0.06$$

These probability predictions are close to the observed intensity distributions.

d   If all protein X molecules in a typical cell oligomerize, then the stoichiometry will be equivalent to ~120 molecules of GFP. Thus, the lateral resolution might be ~50/V(120) ≈ 4.6 nm. The width of the cell membrane is ~5 nm (see Chapter 2), which is marginally higher than the localization lateral resolution, so we might *just* be able to observe the translocation process.

4.3 FÖRSTER RESONANCE ENERGY TRANSFER

Förster resonance energy transfer (FRET) is a nonlinear optical technique that operates over length scales, which are approximately two orders of magnitude smaller than the optical resolution limit. Thus it be considered a super-resolution technique, but is discussed as a separate section due to its specific utility in probing molecular interactions in biology. Although there is a significant body of literature now concerning the application of FRET in light microscopy investigations, the experimental technique was developed originally from bulk ensemble *in vitro* assays not using light microscopy. FRET still has enormous applications in that context. Changes to FRET efficiency values can be measured in a suitable fluorimeter, which contains two-color detector channels, one for the so-called donor and the other for acceptor fluorescence emissions. However, the cutting edge of FRET technology uses optical microscopy to probe putative molecular interactions at a single-molecule level.

4.3.1 EFFICIENCY OF FRET

This *is* a nonradiative energy transfer between a *donor* and *acceptor* molecule over a length scale of ~1–10 nm, which occurs due to overlapping of the electronic molecular orbitals in both spatial localization and in transition energy level gaps. Often, in practice, as an experimental technique, FRET utilizes fluorescent molecules for donor and acceptor whose electronic energy levels for excitation and emission overlap significantly, and so the term *fluorescent energy resonance transfer* is sometimes applied, though the physical process of FRET in itself does not necessarily require fluorescence. The length scale of operation of FRET is comparable to that of many biomolecules and molecular machines and so can be used as a metric for molecular interaction between two different molecules if one is labeled with a donor and the other with an appropriate acceptor molecule.

In FRET, the donor fluorophore emits at a lower peak wavelength than the acceptor fluorophore. Resonance transfer of energy between the two is therefore manifest as a small reduction in fluorescence emission intensity from the donor, and a small increase in fluorescence emission intensity of the acceptor (Figure 4.2c and d). The length scale of energy transfer is embodied in the *Förster radius*, R_0, which is the distance separation that yields a FRET efficiency of exactly 0.5. The efficiency ε of the energy transfer as a function of the length separation R of the donor–acceptor pair is characterized by

$$(4.7) \qquad \varepsilon = \frac{k_{FRET}}{\sum_{i=1}^{donors} k_i} = \frac{k_{FRET}}{k_{FRET} + k_{radiative} + \sum_{j=1}^{other\ donors} k_j} = \frac{1}{1 + (R/R_0)^6}$$

The constant k_{FRET} is the rate of energy transfer from donor to acceptor by FRET, whereas the summed parameters Σk_i are the energy transfer rates from the donor of all energy transfer processes, which include FRET and radiative processes plus various non-FRET and nonradiative processes (Σk_j). With no acceptor, a donor transfers energy at rate ($k_{radiative}$ + Σk_j), and so the mean donor lifetime T_D is equal to $1/(k_{radiative} + \Sigma k_j)$. With an acceptor present, FRET occurs at a rate k_{FRET} such that the donor lifetime τ_{DA} is then equal to $(R_0/R)^6/k_{FRET}$, indicating that $\varepsilon = 1 - \tau_{DA}/\tau_D$.

We can also write $\varepsilon = 1 - I_{DA}/I_D$ where I_{DA} and I_D are the *total* fluorescence emission intensities of the donor in the presence and the absence of the acceptor, respectively; in practice, the intensity values are those measured through an emission filter window close to the emission peak of the donor fluorophore in question. Similarly, we can say that $\varepsilon = (I_{AD} - I_A)/I_A$ where I_{AD} and I_A are the *total* fluorescence emission intensities of the acceptor in the presence and the absence of the donor, respectively. These formulations assume that there is minimal fluorophore cross talk between the two excitation lasers used for the acceptor and donor (i.e., that the donor is not significantly excited by the acceptor laser, and the acceptor is not significantly excited by the donor laser). Also, that there is minimal bleed-through between the fluorescence emissions of each fluorophore between the two detector emission channels. A simpler formulation involves the *relative FRET efficiency* used in *ratiometric FRET*, of $\varepsilon_{rel} = I_A/(I_A + I_D)$ with I_A and I_D being the total fluorescence intensities for acceptor and donor, respectively, following excitation of just the donor. However, if the acceptor and donor emission spectra overlap, then this mixed spectrum must be decomposed into the separate component spectra to accurately measure I_A and I_D, which is often nontrivial. Rarely, one can equate ε_{rel} to the actual FRET efficiency (ε) in the case of minimal laser/fluorophore cross talk, in practice, though converting ε_{rel} to the actual FRET efficiency (ε) usually requires two correction factors of the contribution from direct acceptor excitation to I_A and the ratio between the donor and the acceptor fluorescence emission quantum yields. Additionally, corrections may be needed to account for any fluorescence bleed-through between the acceptor and donor detector channels.

Note that sometimes a FRET pair can actually consist of a dye molecule and a nearby quencher molecule, instead of a donor and acceptor molecule. Here, the distance dependence between the dye and quencher is the same as that of a donor and acceptor molecule since the mechanism of nonradiative energy transfer is the same. However, the quencher does not emit fluorescence, and so the drop in normalized intensity of a dye molecule undergoing quenching is $1 - \varepsilon$.

4.3.2 FÖRSTER RADIUS AND THE KAPPA-SQUARED ORIENTATION FACTOR

The Förster radius R_0 is given by a complex relation of photophysical factors:

(4.8)
$$R_0 \approx \left(\frac{0.529 \kappa^2 Q_D \int_0^\infty \varepsilon_A f_D \lambda^4 \, d\lambda}{n^4 N_A \int_0^\infty f_D \, d\lambda} \right)^{1/6}$$

where
  Q_D is the quantum yield of the donor in the absence of the acceptor
  κ^2 is the *dipole orientation factor*
  n is the refractive index of the medium (usually of water, ~1.33)
  N_A is the Avogadro's number
  the integral term in the numerator is for the spectral overlap integral such that f_D is the
    donor emission spectrum (with the integral in the denominator normalizing this)
  ε_A is the wavelength-dependent *molar extinction coefficient* or *molar absorptivity* of the
    acceptor

Typical values of R_0 for FRET pairs are in the range 3–6 nm. The R^6 dependence on ε results in a highly sensitive response with distance changes. For example, for $R < 5$ nm, the FRET efficiency ε is typically 0.5–1, but for $R > 5$ nm, ε falls steeply toward zero. Thus, the technique is very good for determining putative molecular interaction. The κ^2 factor is given by

(4.9)
$$\kappa^2 = \left(\cos\theta_T - 3\cos\theta_A \cos\theta_D \right)^2$$

where angles θ_T, θ_A, and θ_D are relative orientation angles between the acceptor and donor, defined in Figure 4.2e. The κ^2 factor can in theory vary from 0 (transition dipole moments are *perpendicular*) to 4 (transition dipole moments are *collinear*), whereas *parallel* transition dipole moments generate a κ^2 of exactly 1. FRET donor–acceptor fluorophore pairs that rotate purely isotropically have an expected κ^2 of precisely 2/3. However, care must be taken not to simply assume this isotropic condition. The condition is only true if the rotational correlation time for both the donor and acceptor fluorophores is significantly less than the sampling time scale in a given experiment. Typical rotational correlation times scales are ~10^{-9} s, and so for fluorescence imaging experiments where the sampling times are 10^{-2} to 10^{-3} s, the assumption is valid, though for fast nonimaging methods such as confocal fluorescence detection and fluorescence correlation spectroscopy (FCS) sampling may be over a ~10^{-6} time scale or faster and then the assumption may no longer be valid. An implication of anisotropic fluorophore behavior is that an observed change in FRET efficiency could be erroneously interpreted as a change in donor–acceptor distance when in fact it might just be a relative orientation change between their respective dipole moments.

4.3.3 SINGLE-MOLECULE FRET

FRET is an enormously valuable tool for identifying putative molecular interactions between biomolecules, as we discussed for the FLIM–FRET technique previously (see Chapter 3). But using light microscopy directly in nonscanning imaging methods enables powerful *single-molecule FRET* (*smFRET*) techniques to be applied to addressing several biological questions *in vitro* (for a practical overview see Roy et al., 2008). The first smFRET biophysical investigation actually used near-field excitation (Ha et al., 1996), discussed later in this chapter. However, more frequently today, smFRET involves diffraction-limited far-field light microscopy with fluorescence detection of both the donor and accept or fluorophore in separate color channels of high-sensitivity fluorescence microscope (see Worked Case Example 4.2). The majority of smFRET studies to date use organic dyes, for example, a common FRET pair being variants of a Cy3 (green) donor and Cy5 (red) acceptor, but the technique has also been applied to QD and FP pairs (see Miyawaki et al., 1997).

The use of organic fluorophore FRET pairs comes with the problem that the chemical binding efficiency to a biomolecule is never 100%, and so there is a subpopulation of unlabeled "dark" molecules. Also, even when both fluorophores in the FRET pair have bound successively, there may be a subpopulation that are photoinactive, for example, due to free radical damage. Again, these "dark" molecules will not generate a FRET response and may falsely indicate the absence of molecular interaction.

Since the emission spectrum of organic dye fluorophores is a continuum, there is a risk of bleed-through of each dye signal into the other's respective detector channel, which is difficult to distinguish from genuine FRET unless meticulous control experiments are performed. These issues are largely overcome by *alternating laser excitation* (*ALEX*) (Kapanidis et al., 2005). Here, donor and acceptor fluorophores are excited in alternation with respective fluorescence emission detection synchronized to excitation.

One of the most common approaches for using smFRET is with confocal microscope excitation *in vitro*. Here, interacting components are free to diffuse in aqueous solution in and out of the confocal excitation volume. The $\sim 10^{-6}$ m length scale of the confocal volume sets a low upper limit on the time scale for observing a molecular interaction by FRET since the interacting pair diffuses over this length scale in typically a few tens of milliseconds.

An approach taken to increase the measurement time is to confine interacting molecules either through tethering to a surface (Ha et al., 2002) or confinement inside a lipid *nanovesicle* immobilized to a microscope slide (Benitez et al., 2002). This latter method exhibits less interaction with surface forces from the slide. These methods enable continuous smFRET observations to be made over a time scale of tens of seconds.

A significant disadvantage of smFRET is its very limited application for FP fusion systems. Although certain paired combinations of FPs have reasonable spectral overlap (e.g., CFP/YFP for blue/yellow and GFP/mCherry for green/red), R_0 values are typically ~ 6 nm, but since the FPs themselves have a length scale of a few nanometers, this means that only FRET efficiency values of ~ 0.5 or less can be measured since the FPs cannot get any closer to each other due to their β-barrel structure. In this regime, it is less sensitive as a *molecular ruler* compared to using smaller, brighter organic dye pairs, which can monitor nanoscale conformational changes.

A promising development in smFRET has been its application in structural determination of biomolecules. Two-color FRET can be used to monitor the displacement changes involved between two sites of a molecule in conformational changes, for example, during *power stroke* mechanisms of several molecular machines or the dynamics of protein binding and folding. It is also possible to use more than two FRET dyes in the same sample to permit FRET efficiency measurements to be made between three or more different types of dye molecule. This permits *triangulation* of the 3D position of the dye molecule. These data can be mapped onto atomic level structural information where available to provide a complementary picture of *time-resolved* changes to molecular structures.

KEY BIOLOGICAL APPLICATIONS: FRET

Determining molecular interactions over a ~0–5 nm length scale.

Worked Case Example 4.2: FRET

A single-molecule FRET experiment was performed on a short linear synthetic DNA construct composed of 18 nucleotide base pairs, which was surface immobilized to a glass coverslip in an aqueous flow-cell, to which was attached FRET donor Cy3 (green) fluorescent dye molecule at one end and FRET acceptor Cy5 (red) fluorescent dye molecule at the other end. The construct was excited into fluorescence using ALEX of a 532 and 640 nm laser at 1 kHz modulation for each laser using TIRF excitation for which the polarization had been circularized. In one experiment, the average brightness of the green and red dye molecules was measured at ~7320 and ~4780 counts on an EMCCD detector when both molecules were emitting at the same time. When a red molecule photobleached during an acquisition in which both green and red molecules had been emitting, the average green molecule brightness then changed to ~5380 counts. Similarly, when a green molecule photobleached during an acquisition in which both green and red molecules had been emitting, the average red molecule brightness then changed to ~6450 counts.

a   *Explain these observations and estimate the Förster radius for Cy3–Cy5, stating any assumptions you make.*
A molecular machine that was known to undergo a conformational change by a "lever arm" component of the whole molecular complex upon hydrolyzing ATP had two protein subunits X and Y labeled with Cy3 (green) and Cy5 (red) fluorescent dyes, respectively, in a single-molecule ALEX FRET experiment in vitro, one attached to the lever arm and the other attached on a nonlever component of the complex, here, involving diffusion of the molecular machine complex through a confocal excitation volume. In one experiment, a "loose" short linker was used between the dyes and the proteins allowing free rotation of the dyes. For this, addition of ATP resulted in changing the average green and red channels' signals from ~5560 and ~6270 counts, respectively, before the addition of ATP and ~7040 and ~5130 counts in the first ~5 ms after the addition of ATP.

b   *Calculate the average speed in microns per second of the molecular conformational change due to ATP hydrolysis stating any assumptions you make.*
In a second experiment, a "tight" linker has prevented any free rotation between the dyes and the protein subunits, and a chemical cross-linker was added to prevent any relative translational change between the two dye loci. For this, addition of ATP resulted in changing the average green and red channels' signals from roughly the same values, respectively, as for the part (b) experiment earlier before the addition of ATP, but ~6100 and ~5800 counts in the first ~5 ms after the addition of ATP.

c   *Explain these observations, and comment on previous speculation that the molecular machine might operate via either a "pincer" or a "twister" type of action.*

Answers

a   The average distance between Cy3 and Cy5 molecules is ~18 × 0.34 = 6.12 nm (see Chapter 2), which is within the typical "FRET ruler" scale of ~0–10 nm; thus, one might expect FRET to be occurring when both molecules are emitting. When the donor (green) molecule bleaches, the FRET transfer from green to red is stopped and so the acceptor (red) molecule brightness goes down. Similarly, if the acceptor molecule (red) bleaches first, then no energy is subsequently transferred from the green to the red molecule by FRET, and so the green molecule brightness goes up. Assuming isotropic emissions of the dyes in the time scale of sampling, no polarization dependence of fluorescence excitation for TIRF, that the height of the dyes molecules are the same in evanescent field (i.e., construct is sitting "flat"), and that there is minimal bleed-through between the acceptor and donor detector channels, the FRET efficiency can then be calculated using either the donor intensity as

$$1-(4780 \text{ counts})/6450 \text{ counts} = 0.36$$

or the acceptor intensity as

$$(7320-5380 \text{ counts})/(5380 \text{ counts}) = 0.36$$

Thus, from Equation 4.7 the Förster radius is

$$(6.12\text{nm})/(1/0.36-1)^{1/6} = 5.6 \text{ nm}$$

b   Using the same assumptions as for part (a), then, from the donor intensity values, the FRET efficiency before adding ATP is

$$1-6270\,\text{counts}/6450\,\text{counts}=0.03$$

Similarly, using the acceptor intensity values,

$$(5560-5380\,\text{counts})/(5380\,\text{counts})=0.03$$

This is significantly less than 0.5 and therefore "low FRET." Upon adding ATP, there is a molecular conformational change, and FRET efficiency can be similarly calculated using either the donor or acceptor intensity values to be ~0.20, which is "high" FRET.

A "loose" linkage implies isotropic excitation absorption and emission of the dye molecules; thus, the size of the displacement during the conformational change of the FRET pair dye molecules is given by

$$(5.6\,\text{nm})\times\left((1/0.03-1)^{1/6}-(1/0.20-1)^{1/6}\right)=2.9\ \text{nm}$$

This conformational change is accomplished in 5 ms; therefore, the average speed is given by

$$(2.9\times10^{-9}\,\text{m})/\left(5\times10^{-3}\text{s}\right)=5.8\times10^{-7}\,\text{ms}^{-1}=58\ \mu\text{ms}^{-1}$$

c   A "tight" linkage prevents free rotation of the dye molecules; therefore, the brightness in the absence of FRET (no ATP added) of the molecule is a function of the projection of the polarization excitation vector onto the respective dye dipole axis. The fact that the dye brightness values are similar to part (b) is consistent with a random orientation of the construct on the coverslip surface, since the dyes themselves cannot undergo free rotation relative to the molecular complex. Since there is no translational movement, the change in FRET efficiency must be due to relative rotational movement of the dipole axes of two dye molecules due to relative rotation between the whole lever arm and the rest of the molecular complex. If θ is the relative angle between the acceptor and donor dipole axes, then the E-field from the donor projected onto the acceptor scales as $\cos\theta$; the FRET efficiency scales as the local intensity of the donor E-field, which scales as the square of the amplitude of the f-field and thus as $\cos^2\theta$. Thus,

$$\cos^2\theta=0.03/(1-5800/6450)=0.30$$

indicating $\theta\approx57°$.

Thus, the molecular conformational change upon ATP hydrolysis is comprised of both a translation component and a rotation component of the lever arm, suggesting a combination of both a "pincer" and a "twister" model.

4.4 FLUORESCENCE CORRELATION SPECTROSCOPY

Fluorescence correlation spectroscopy (FCS) is a technique in which fluorescently labeled molecules are detected as they diffuse through a confocal laser excitation volume, which generates a pulse of fluorescence emission prior to diffusing out of the confocal volume. The time correlation in detected emission pulses is a measure of fluorophore concentration and rate of diffusion (Magde et al., 1972). FCS is used mainly *in vitro* but has been recently also applied to generate fluorescence correlation maps of single cells.

4.4.1 DETERMINING THE AUTOCORRELATION OF FLUORESCENCE DATA

FCS is a hybrid technique between ensemble averaging and single-molecule detection. In principle, the method is an ensemble average tool since the analysis requires a distribution of dwell times to be measured from the diffusion of many molecules through the confocal volume. However, each individual pulse of fluorescence intensity is in general due to a single molecule. Therefore, FCS is also a single-molecule technique.

The optical setup is essentially identical to that for confocal microscopy. However, there is an additional fast real-time acquisition card attached to the fluorescence detector output that can sample intensity fluctuation data at tens of MHz to calculate an *autocorrelation* function, I_{Auto}. This is a measure of the correlation in time t of the pulses with intensity I:

$$(4.10) \qquad I_{Auto}\left(t'\right) = \frac{\left(I(t) - \langle I(t)\rangle\right)\left(I\left(t+t'\right) - \langle I\left(t+t'\right)\rangle\right)}{\langle I(t)\rangle^2}$$

where
   the parameter t' is an equivalent time interval value
   $\langle I(t)\rangle$ is the time-averaged intensity signal over some time T of experimental observation

$$(4.11) \qquad \langle I(t)\rangle = \frac{1}{T}\int_0^T I(t)\,\mathrm{d}t$$

If the intensity fluctuations all arise solely from local concentration fluctuations δC that are within the volume V of the confocal laser excitation volume, then

$$(4.12) \qquad \delta I = \int_V I_1 P(\vec{r})\,\delta C(\vec{r})\,\mathrm{d}V$$

where
   r is the displacement of a given fluorophore from the center of the confocal volume
   P is the PSF
   I_1 is the effective intensity due to just a single fluorophore

For normal confocal illumination FCS, the PSF can be modeled as a 3D Gaussian volume (see Chapter 3):

$$(4.13) \qquad P(x,y,z) = \exp\left[-\frac{1}{2}\left(\frac{\left(x^2+y^2\right)}{w_{xy}} + \frac{z^2}{w_z}\right)\right]$$

where w_{xy} and w_z are the standard deviation widths in the xy plane and parallel to the optical axis (z), respectively. The normalized autocorrelation function can be written then as

$$(4.14) \qquad I_{Auto}\left(t'\right) = \frac{\langle \delta I(t)\,\delta I\left(t+t'\right)\rangle}{\langle \delta I(t)\rangle^2}$$

This therefore can be rewritten as

(4.15)
$$I_{Auto}(t') = \frac{\iint P(\vec{r})P(\vec{r}')\langle \delta C(\vec{r},0)\delta C(\vec{r}',t')\rangle \mathrm{d}V \mathrm{d}V'}{\left(\langle C\rangle \int P(\vec{r})\mathrm{d}V\right)}$$

The displacement of a fluorophore as a function of time can be modeled easily for the case of Brownian diffusion (see Equation 2.12), to generate an estimate for the *number density auto-correlation term* $\langle \delta C(r,0)\delta C(r',\tau)\rangle$:

(4.16)
$$\langle \delta C(\vec{r},0)\delta C(\vec{r}',t')\rangle = \langle C\rangle \frac{1}{(4\pi Dt')^{3/2}}\exp\left[-\frac{|\vec{r}-\vec{r}'|^2}{4Dt'}\right]$$

An important result from this emerges at the zero time interval value for the autocorrelation intensity function, which then approximates to $1/V\langle C\rangle$, or $1/\langle N\rangle$, where $\langle N\rangle$ is the mean (time averaged) number of fluorophores in the confocal volume. The full form of the autocorrelation function for one type of molecule diffusing in three spatial dimensions through a roughly Gaussian confocal volume with anomalous diffusion can be modeled as I_m:

(4.17)
$$I_m(t') = I_m(\infty) + \frac{I_m(0)}{\left(1+(t'/\tau)^\alpha\right)\sqrt{1+(t'/\tau)/\alpha^2}}$$

Fitting experimental data I_{Auto} with model I_m yields estimates for parameters $I_m(0)$ (simply the intensity due to the mean number of diffusing molecules inside the confocal volume), $I(\infty)$ (which is often equated to zero), τ, and α. The parameter a is the *anomalous diffusion coefficient*. For diffusion in n spatial dimensions with effective diffusion coefficient D, the general equation relating the mean squared displacement $\langle R^2\rangle$ after a time t for a particle exhibiting normal or Brownian diffusion is given by Equation 2.12, namely, $\langle R^2\rangle = 2nDt$. However, in the more general case of *anomalous diffusion*, the relation is

(4.18)
$$\langle R^2\rangle = 2nDt^\alpha$$

The anomalous diffusion coefficient varies in the range 0–1 such that 1 represents *free* Brownian diffusion. The microenvironment inside a cell is often crowded (certain parts of the cell membrane have a protein crowding density up to ~40%), which results in hindered mobility termed anomalous or subdiffusion. A "typical" mean value of a inside a cell is 0.7–0.8, but there is significant local variability across different regions of the cell.

The time parameter τ in Equation 4.17 is the mean "on" time for a detected pulse. This can be approximated as the time taken to diffuse in the 2D focal plane, a mean squared distance, which is equivalent to the lateral width w of the confocal volume (the full PSF width equivalent to twice the Abbe limit of Equation 4.3, or ~400–600 nm), indicating

(4.19)
$$\tau \approx \left(\frac{w^2}{4D}\right)^{1/\alpha}$$

Thus, by using the value of τ determined from the autocorrelation fit to the experimental data, the translational diffusion coefficient D can be calculated.

4.4.2 FCS ON MIXED MOLECULE SAMPLES

If more than one type of diffusing molecule is present (*polydisperse diffusion*), then the autocorrelation function is the sum of the individual autocorrelation functions for the separate diffusing molecule types. However, the main weakness of FCS is its relative insensitivity to changes in molecular weight, M_w. Different types of biomolecules can differ relatively marginally in terms of M_w; however, the "on" time τ scales approximately with the frictional drag of the molecule, roughly as the effective Stokes radius, which scales broadly as $\sim M_w^{1/3}$. Therefore, FCS is poor at discriminating different types of molecules unless the difference in M_w is at least a factor of ~4.

FCS can also be used to measure molecular interactions between molecules. Putatively, interacting molecules are labeled using different colored fluorophores, mostly dual-color labeling with two-color detector channels to monitor interacting pairs of molecules. A variant of standard FCS called "fluorescence cross-correlation spectroscopy" can then be applied. A modification of this technique uses dual-color labeling but employing just one detector channel, which captures intensity only when the two separately labeled molecules are close enough to be interacting, known as *FRET-FCS*.

4.4.3 FCS ON MORE COMPLEX SAMPLES

FCS can also be performed on live-cell samples. By scanning the sample through the confocal volume, FCS can generate a 2D image map of mobility parameters across a sample. This has been utilized to measure the variation in diffusion coefficient across different regions of large living cells. As with scanning confocal microscopy, the scanning speed is a limiting factor. However, these constraints can be overcome significantly by using a spinning-disk system. FCS measurements can also be combined with simultaneous topography imaging using AFM (see Chapter 6). For example, it is possible to monitor the formation and dynamics of putative lipid rafts (see Chapter 2) in artificial lipid bilayers using such approaches (Chiantia, 2007).

KEY BIOLOGICAL APPLICATIONS: FCS

Quantifying molecular mobility; Determining molecular interactions.

Worked Case Example 4.3: Localization Microscopy

A time-correlated PALM experiment was performed on live E. coli bacteria stuck to coverslip to track a protein known to incorporate into liquid–liquid phased separated droplets in the cytoplasm whose formation was triggered by stressing the cell using a toxin. The droplet protein was tagged with a red photoactivatable fluorescent protein called PAmCherry, which was stochastically activated using low intensity 405 nm wavelength laser excitation and imaged using a high intensity 561 nm wavelength laser excitation in narrow-field mode, peak fluorescence wavelength 595 nm, using a 1.4 NA oil immersion (refractive index = 1.515) objective lens to track individual molecules with sampling time of 5 ms per consecutive image frame. Once fluorescence light from the PAmCherry had been captured by the objective lens, only ~30% was transmitted through all of the imaging system to the entrance of the camera detector. The camera detector had a quantum efficiency of 80% at a wavelength of 595 nm. The average effector diameter of the droplet protein tagged with PAmCherry is ~2–3 nm.

a *If the PAmCherry emits ~10^5 fluorescent photons prior photobleaching on average resulting in single-molecule tracks of average duration ~50 ms under these narrow-field conditions, estimate the localization precision when using a 2D Gaussian fitting to track single molecules in the microscope focal plane, if you assume that the effect from non-photon sources on localization precision is negligible.*

b *By increasing the intensity of the 405 nm wavelength activation laser the number of tracks detected increased, but also in some cases the brightness of tracks was twice that detected at the lower excitation intensity. Offer an explanation for these observations and estimate the smallest diameter of droplet that could be detected under these conditions. What does that imply in terms of the smallest droplet we can detect and the maximum number of proteins within it?*

c *Under the lower intensity activation laser setting, a plot of the average mean squared displacement versus time interval for all detected tracks was initially a straight line with a positive gradient at low time interval values, but at higher time interval values, the gradient would decrease until reaching a plateau at the very highest time intervals. What do you think could explain this?*

d   *With increasing incubation of the cells in the toxin, it was found that the frequency of detection of the tracks in the droplet increased, that the initial gradient of the average mean squared displacement versus time interval plot was lower, and the plateau was reached at a lower level of mean squared displacement. Explain these observations.*

Answers

a   The localization precision in 1D for a single photon is given simply by the optical resolution limit, which is the PSF width, assuming that the characteristic wavelength can be best approximated by the peak emission wavelength here, using Equation 4.3:

$$\Delta x = 0.61 \times 595 \times 10^{-9}/1.4 = 2.59 \times 10^{-7} \text{ or } 259 \text{ nm}$$

However, with Gaussian fitting there is a better estimate for the center of the dye fluorescence intensity profile, such that the localization precision improves with the number N of fluorescence photons sampled (Equation 4.4). If the non-photon associated localization error (i.e., dark noise and localization uncertainty error associated with finite pixel size on the camera) is negligible, then the 1D localization precision σ_x from Gaussian fitting scales as $\sim s/\sqrt{N}$ where s approximates to the single photon localization precision, or $\sim\Delta x$. To determine N, we need to estimate the number of fluorescence photons captured in total from a single PAmCherry molecule and detected by the camera. This *total* efficiency ε of photon detection equals 0.50 multiplied by 0.88 multiplied by the geometrical photon capture efficiency (the solid angle Ω subtended at the objective lens for detecting photons divided by all solids angle of 4π steradians). Ω is given by Equation 3.21 $\Omega = 2\pi(1 - \cos\theta)$, where $NA = n\sin\theta$, assuming the refractive index $n = 1.515$. Thus:

$$\theta = \sin^{-1}(1.4/1.515) = 67.5°$$

$$\varepsilon = (2\pi \times (1 - \cos(67.5))/4\pi) \times 0.30 \times 0.80 = 0.07 \text{ or } 7\%$$

The total number of photons therefore detected by the camera from one PAmCherry molecule prior to photobleaching is $10^5 \times 0.07 = 7000$. However, these photons are detected over a total of 50 ms on average, with a sampling time per image frame of 5 ms, so the number of photons detected by the camera per image frame is $N = 7000 \times (5/50) = 700$. Thus:

$$\sigma_x = 259/\sqrt{(700)} = 9.8 \text{ nm}$$

But by combining the total sigma x and y values and by symmetry the 2D localization is:

$$\sigma_{2D} = \sqrt{(\sigma_x^2 + \sigma_y^2)} = \sigma_x\sqrt{2} = 13.8 \text{ nm}$$

b   The increase in the number of detected tracks is expected since the intensity of activation increases the activation probability. The observation of two-fold increases in brightness of the tracks may indicate that two photoactive PAmCherry molecules are colocalized to within the localization precision, which might mean that the protein in not a single monomer but is a dimer. This will at most increase the number of detected photons by a factor of two, thus will improve (i.e., reduce) σ by a factor of √2, i.e., 9.4 nm. To determine if a protein is localized into a droplet would ideally require us to track it in the droplet at least over a distance greater than its 2D localization precision, so the minimum droplet size by this logic that we could observe this would have a diameter of ~14 nm. We are told that the protein plus PAmCherry tag has an effective diameter of 2–3 nm, so very roughly say that a dimer has an effective diameter of more like ~5 nm. The maximum number of dimers n_d that could be present in the droplet would involve tight-packing of the dimers, so estimate that under this condition n_d multiplied by the volume of one dimer then equals the total volume of one droplet. Assuming spherical volumes:

$n_d \times 4\pi.5^3/3 = 4\pi.14^3/3$ ∴ $n_d = (14/5)^3 \approx 22$, or 11 dimers, so although not at the very first stages in the nucleation process of droplet formation (i.e., two dimer interacting together presumably), is still at a relatively early formation stage beyond this.

c   The initial straight line indicates Brownian diffusion whose diffusion coefficient is proportional to the gradient. The decrease in gradient *could* indicate a decrease in mobility toward the edges of the droplet, so some subdiffusive or anomalous diffusion behavior, but with the presence of a plateau more likely indicates that diffusion is confined and the plateau is the boundary of the confinement (i.e., the edge of the droplet). An increase in tracks detected in the droplet indicates that more proteins are likely to be present inside the droplet. A decrease in the plateau height indicates a smaller effective confinement diameter. So, the concentration of protein inside the droplet will increase. This molecular crowding could potentially result in an increase in viscosity for the diffusion of any given protein in the droplet, thus explain the smaller observed initial gradient.

4.5 LIGHT MICROSCOPY OF DEEP OR THICK SAMPLES

Although much insight can be gained from light microscopy investigations *in vitro*, and on single cells or thin multicellular samples, ultimately certain biological questions can only be addressed inside thicker tissues, for example, to explore specific features of human biology. The biophysical challenges to deep tissue light microscopy are the attenuation of the optical signal combined with an increase in background noise as it passes through multiple layers of cells in a tissue and the optical inhomogeneity of deep tissues distorting the optical wave front of light.

Some nonlinear optics methods have proved particularly useful for minimizing the background noise. Nonlinear optics involve properties of light in a given optical medium for which the dielectric polarization vector has a nonlinear dependence on the electric field vector of the incident light, typically observed at high light intensities comparable to interatomic electric fields (~10^8 V m^{-1}) requiring pulsed laser sources.

4.5.1 DECONVOLUTION ANALYSIS

For a hypothetically homogeneous thick tissue sample, the final image obtained from fluorescence microscopy is the convolution of the spatial localization function of all of the fluorophores in the sample (in essence, approximating each fluorophore as a point source using a delta function at its specific location in the sample) with the 3D PSF of the imaging system. Therefore, to recover the true position of all fluorophores requires the reverse process of *deconvolution* of the final image. The way this is performed in practice is to generate *z-stacks* through the sample using confocal microscopy; this means generating multiple images through the sample at different focal heights, so in effect *optically sectioning* the sample.

Since the height parameter z is known for each image in the stack, deconvolution algorithms (discussed in Chapter 7) can attempt to reconstruct the true positions of the fluorophores in the sample providing the 3D PSF is known. The 3D PSF can be estimated separately by immobilizing the sparse population of purified fluorophore onto a glass microscope coverslip and then imaging these at different incremental heights from the focal plane to generate a 3D look-up table for the PSF, which can be interpolated for arbitrary value of z during the *in vivo* sample imaging.

The main issues with this approach are the slowness of imaging and the lack of sample homogeneity. The slowness of the often intensive computational component of conventional deconvolution microscopy in general prevents real-time imaging of fast dynamic biological processes from being monitored. However, data can of course be acquired using fast confocal Nipkow disk approaches and deconvolved offline later.

A significant improvement in imaging speed can be made using a relatively new technique of *light-field microscopy* (see Levoy et al., 2009). It employs an array of *microlenses* to produce an image of the sample, instead of requiring scanning of the sample relative to the confocal volume of the focused laser beam. This results in a reduced effective spatial resolution, but with a much enhanced angular resolution, that can then be combined with deconvolution analysis offline to render more detailed in-depth information in only a single image frame (Broxton et al., 2013), thus with a time resolution that is limited only by the camera exposure time. It has been applied to investigating the dynamic neurological behavior of the small flatworm *model organism* of *Caenorhabditis elegans* (see Section 7.3).

4.5.2 ADAPTIVE OPTICS FOR CORRECTING OPTICAL INHOMOGENEITY

However, deconvolution analysis in itself does not overcome the problems associated with the degradation of image quality with deep tissue light microscopy due to heterogeneity in the refraction index. This results from imaging through multiple layers of cells, which causes local variations in phase across the wave front through the sample, with consequent interference effects distorting the final image, which are difficult to predict and correct analytically and which can be rate limiting in terms of acquiring images of sufficient quality to be meaningful in terms of biological interpretation.

Variations in refractive index across the spatial extent of a biological sample can introduce optical aberrations, especially for relatively thick tissue samples. Such optical aberrations reduce both the image contrast and effective optical resolution. They thus set a limit for practical imaging depths in real tissues. *Adaptive optics* (AO) is a technology that can correct for much of this image distortion.

In AO, a reference light beam is first transmitted through the sample to estimate the local variations of phase due to the refractive index variation throughout the sample. The phase variations can be empirically estimated and expressed as a 2D matrix. These values can then be inputted into a 2D phase modulator in a separate experiment. Phase modulators can take the form either of a *deformable mirror, microlens array,* or an SLM. These components can all modulate the phase of the incident light wave front before it reaches the sample to then correct for the phase distortion as the light passes through the sample (Figure 4.3a).

The end result is a reflattened, corrected wave front emerging from the sample (for a recent review, see Booth, 2014). AO has been applied to image to tissue depths of up to several hundred microns. It is also compatible with several different forms of light microscopy imaging techniques.

4.5.3 PTYCHOGRAPHY METHODS FOR NUMERICAL FOCUSING

Another emerging light microscopy technique that shows promise for imaging at tissue depths in excess of 100 µm is *ptychography* (also referred to as *Fourier ptychography*). Ptychography

A   **Normal image distortion**   **Adaptive optics correction**   **B**   **Optical ptychography**

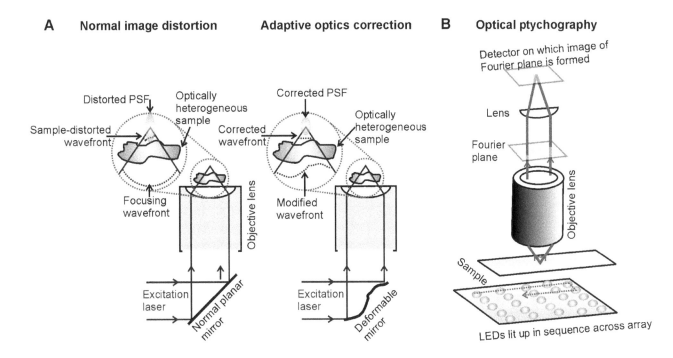

FIGURE 4.3 Methods to correct for image distortions and numerically refocus. (a) Uncorrected illumination (left panel) through an optically heterogeneous sample can result in image distortion during fluorescence excitation, which can be corrected by using adaptive optics (right panel, here shown with a deformable mirror on the excitation path, but a similar optical component can also be placed in the imaging path for fluorescence emissions). (b) Optical ptychography generates sample images by Fourier transforming the Fourier plane image of the sample, which permits a greater effective numerical aperture for imaging compared to the physical objective lens and also enabling numerical refocusing of the sample.

is a *label-free method*, which uses advanced computational algorithms to *numerically focus* a sample, in effect using a *virtual objective lens*, and has been applied to investigating various aspects of cell cycle changes (see Marrison et al., 2013).

Ptychography was originally developed for the analysis of x-ray microscopy scattering data (see Chapter 5). With x-rays, there is no equivalent "lens" to form an image, but ptychography was implemented to allow numerical focusing from computational reconstructions of the x-ray diffraction pattern. These methods have now been implemented in light microscopy systems.

In a common design, the specimen and a coherent illuminating beam are moved relative to one another to sequentially illuminate the sample with overlapping areas (Figure 4.3b). Another method achieves a similar effect but using a 2D array of LED light sources to sequentially illuminate the sample, which circumvents the requirement for relatively slow scanning. The diffracted light pattern is then detected by a 2D pixel photodetector array, such as simple CCD camera. The spatial extent of this detector array can give access to a far greater region of reciprocal (i.e., frequency) space than is available to a physical objective lens, which is limited by its numerical aperture.

By utilizing different sequences of illumination areas, different conventional contrast enhancement modes of light microscopy can be replicated. For example, illumination in the center of the sample produces brightfield images, whereas illuminating the outer regions (equivalent to obtaining data from higher diffraction angles than those in principle obtainable from the finite small numerical aperture of a typical objective lens) can generate equivalent dark-field images. Similarly, sequentially taking pairs of images with alternating halves of the sample illuminated allows the reconstruction of phase contrast images. Performing a full sequential illumination scan over the whole extent of the sample allows accurate recovery of the phase of the wave as it travels through the specimen. This has great potential for rectifying the aberration due to optical inhomogeneity in thick samples, since corrections for the phase variations can be made numerically in reciprocal space.

One potential problem is that the detected diffraction pattern consists of just intensity data and does not contain information concerning the relative phase of a scattered beam from a particular part of the sample. However, this phase information can be recovered since the illuminated area moves over the sample to generate redundancy in the data since there is always some overlap in the sampled regions, which can be used to retrieve the phase from the scattered object using an algorithm called the "pytchographic iterative engine" (Faulkner and Rodenburg, 2004).

The main issues with ptychography are the huge volumes of data captured (a gigapixel image for each illuminated area on the sample) and a requirement for potentially very long acquisition time scales. A typical single dataset from a static biological sample contains hundreds of images to obtain sufficient information from different diffraction angles. The LED array approach improves the time resolution issue to some extent; however, to monitor any time-resolved process potentially involves datasets that would fill a normal computer hard drive very quickly.

4.5.4 MULTIPHOTON EXCITATION

Multiphoton excitation (*MPE*) is a nonlinear optical effect. In *MPE microscopy*, the transition energy required to excite a ground state electron to a higher level during fluorescence excitation in a fluorophore can in principle be contributed from the summation of the equivalent quantum energies of several photons, provided these photons are all absorbed within a suitably narrow time window. In *two-photon excitation microscopy* (or *2PE microscopy*), the initial excitation of a ground state electron is made following the absorption of two photons of the same wavelength λ during a time window of only $\sim 10^{-18}$ s, since this is the lifetime of a *virtual state* halfway between the excited and ground states (Figure 4.4a). This means that λ is twice that of the required for the equivalent single-photon excitation process, and so for visible light, two-photon excitation fluorescence detection *near IR* (NIR) incident wavelengths ($\sim$ a micron) are typically used.

Two-photon absorption, also known as the *Kerr effect*, is described as a *third-order* nonlinear effect because of the dependence of the complex polarization parameter of the optical medium on the cubic term of the electrical susceptibility. Since two photons are required, the rate of two-photon absorption at a depth z depends on the square of the incident photon intensity I, whereas for one-photon absorption, the dependence is linear, such that the overall rate has a quadratic dependence:

$$(4.20) \qquad \frac{\mathrm{d}I}{\mathrm{d}z} = -\left(\alpha I + \beta I^2 \right)$$

where α and β are the one- and two-photon absorption coefficients, respectively.

The longer wavelengths required result is less scattering from biological tissue, for example, *Rayleigh scattering*, for which the length scale of the scattering objects is much smaller than the incident wavelength, has a very sensitive $1/\lambda^4$ dependence. Much of the scattering in tissue is also due to *Mie scattering*, that is, from objects of size comparable to or greater than λ, for which there is a less sensitive dependence of λ than for Rayleigh scattering, but still a reduction at higher wavelengths. This is significant since at depths greater than a few hundred microns, tissues are essentially opaque at visible light wavelengths due to scattering, whereas they are still optically transparent at the NIR wavelength used in two-photon microscopy. The *geometrical scattering* regime applies to scattering objects whose effective radius r is at least an order of magnitude greater than the wavelength of light.

The measure of the ability of an object to scatter can be characterized by its scattering cross-section. The cross-section σ can be deduced from the Mie scattering model for any general

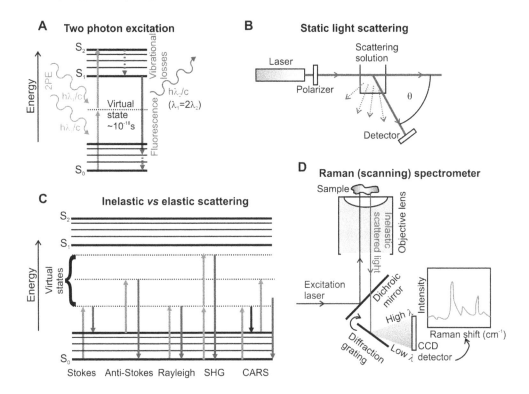

FIGURE 4.4 Nonlinear excitation and inelastic scattering: (a) Jablonski diagram for two-photon excitation. (b) Schematic of static light scattering apparatus. (c) Jablonski diagram for several inelastic light scattering modes compared against elastic Rayleigh scattering. (d) Schematic of Raman imaging spectrophotometer (typically based on a scanning confocal microscope core design).

wavelength though it has a complex formulation, but the two extremes of this at very short (Rayleigh) and very high (geometrical) wavelength generate the following approximations, first for Rayleigh, σ_R,

$$(4.21) \qquad \sigma_R = \frac{128\pi r^6}{3\lambda^4}\left(\frac{n_r^2 - 1}{n_r^2 + 2}\right)^2$$

and for geometrical scattering, σ_G,

$$(4.22) \qquad \sigma_G = \pi r^2$$

where n_r is the ratio of refractive indices for the scattering particle and its surrounding media, n_b/n_w, where n_b is the refractive index of the biological scattering object and n_w is the refractive index of water. In other words, the cross-section for Rayleigh scatterers scale with $\sim V^2$ where V is their volume, whereas for geometrical scatterers this dependence with volume is much less sensitive at $\sim V^{2/3}$.

KEY POINT 4.4

"Geometrical" (i.e., ballistic type) scattering applies for effective diameter $r > 10\lambda$, Mie/Tyndall scattering for $\sim 0.1\lambda < r < 10\lambda$ and Rayleigh scattering for $\sim r > 0.1\lambda$. For an electric dipole, the polarizability p is proportional to an incident sinusoidal E-field plane wave; the scattered E-field is proportional to the second derivative of p, hence $1/\lambda^2$ dependence, and the scattered intensity is proportional to the square of the scattered E-field, hence $1/\lambda^4$ dependence.

Another advantage of two-photon microscopy is the greater localization precision of the *excitation volume*. The effective cross-section for two-photon absorption is very small compared to the single-photon absorption process due to the narrow time window required for the absorption process. This requires a high incident flux of photons from a focused laser source with the two-photon absorption only probably very close to the center of the focal volume. This means that the effective focal excitation volume is an order of magnitude smaller than that of single-photon confocal microscopy, or ~0.1 fl. Thus, the excitation is significantly more localized, resulting in far less contamination in the images from out of focal plane emissions.

The spectral emission peak of QDs is temperature sensitive since the population of high-energy excitons is governed by the Boltzmann factor, which is temperature sensitive, but this sensitivity is significantly more for two-photon laser excitation compared to the standard one-photon excitation process (see Chapter 3), and this has been exploited in using QDs as *nanother-mometers* (see Maestro et al., 2010). This is manifest as a drop in QD brightness when measuring over a wavelength window close to the peak of a factor of ~2 when changing the local temperature from 30°C to 50°C and thus potentially is a good probe for investigating temperature changes relevant to biological samples, which has been tested as a proof-of-principle to measure the local temperatures inside human cancer cells.

A significant disadvantage with 2PE microscopy is that the incident light intensity needs to be so high that photodamage/phototoxicity becomes problematic; this can be seen clearly by using *death sensors*, in the form of individual muscle cells whose speed of contraction is inversely proportional to the extent of their photodamage. Also, the technique requires raster scanning technology before images can be reconstructed, and so it is slower than camera detector pixel array–based imaging, which requires no scanning, limiting the range of dynamic biological processes that can be investigated. However, tissue depths of up to 1.6 mm have been imaged using 2PE, with investigation of gray brain matter in mice (Kobat, 2011).

The limit beyond this depth using 2PE is again due to scatter resulting from unavoidable single-photon scattering. To counteract this, researchers have developed *three-photon excitation microscopy*, such that the wavelength of the incident photons is three times that of the required for the equivalent single-photon absorption event, which reduces the single-photon scattering even more. 3PE has enabled brain imaging in mice to be extended to ~2.5 mm depth (Horton et al., 2013).

KEY POINT 4.5

MPE fluorescence microscopy is emerging as a tool of choice for imaging deep into tissues, resulting in significantly reduced back scatter and a spatially more confined excitation volume compared to single-photon fluorescence microscopy and is especially powerful when combined with AO to correct for optical inhomogeneity in thick tissues.

Developments in *optogenetics* technologies (Chapter 7) have also benefited from MPE microscopy. Optogenetics is a method that uses light to control nerve tissue by genetically inserting lightsensitive proteins into nerve cells that open or close ion channels in response to the absorption of light at specific wavelengths. In combining this approach with 2PE, it is now possible to control the operation of multiple specific nerve fibers relatively deep in living tissue (Prakash et al., 2012).

4.5.5 SECOND-HARMONIC IMAGING

While conventional light microscopes obtain image contrast largely from the spatial differences either in optical density or refractive index in the sample, *second-harmonic imaging (SHI) microscopy* utilizes the generation of second harmonics from the incident light in the sample. *Second-harmonic generation (SHG)*, or *frequency doubling*, involves two photons of the same frequency interacting to generate a single photon with twice the frequency and half the wavelength (an example of *sum frequency generation*). Biological matter capable of SHG requires periodic structural features with chiral molecular components, which result in *birefringence*, an optical feature in which the refractive index of a medium depends upon the wavelength of incident light (see Chapter 3). Good examples of this are the extracellular matrix protein collagen, well-ordered myosin protein filaments in muscle tissue, microtubules from the cytoskeleton, and structurally ordered features of cell membranes. SHI can also be used in monitoring the formation of crystals (see Chapter 7) for use in x-ray crystal diffraction, which can be used for determining the structure of biomolecules to an atomic level precision (see Chapter 5).

SHI microscopy offers many advantages for *in vivo* imaging. It is a label-free method and so does not impair biological function due to the presence of a potentially bulky fluorophore probe. Also, since it requires no fluorescence excitation, there is less likelihood from phototoxicity effects due to free radical formation. Typically, SHI microscopy is utilized with NIR incident light and so has much reduced scattering effects compared to visible light methods and can be used to reconstruct 3D images of deep tissue samples. Similarly, *third-harmonic imaging* microscopy, in which three incident photons interact with the sample to generate a single photon of one-third the original wavelength, has been utilized in some *in vivo* investigations (see Friedl, 2007).

4.5.6 LIGHT SHEET MICROSCOPY

Light sheet microscopy, also known as *selective plane illumination microscopy*, is a promising biophysical tool that bridges the length scales between single-cell imaging and multicellular sample imaging (see Swoger et al., 2014, for a modern review). It evolved from confocal theta microscopy and involves orthogonal-plane fluorescence optical sectioning; illuminating a sample from the side typically via a thin sheet of light was generated using a cylindrical lens onto a single plane of a transparent tissue sample, which has been fluorescently labeled. Since just one plane of the sample is illuminated, there is minimal out-of-plane fluorescence emission contamination of images, permitting high-contrast 3D reconstruction of several diverse *in vivo* features.

The development of live fruit fly embryos has been investigated with this technique (Huisken et al., 2004), as well as tracking of nuclei in live zebrafish embryos (Keller et al., 2008), growing roots tissue in developing plants, developing gut tissue, and monitoring down to subcellular levels in functional salivary glands to imaging depths of ~200 µm (Ritter et al., 2010). Variants of this technique have now been applied to monitor single cells.

A recent *tissue decolorization* method has been used on live *mice* in combination with light sheet fluorescence microscopy (Tainaka et al., 2014). This involves using a specific chemical treatment involving *aminoalcohol*, which results in removing the normal pink color associated with oxygen-carrying *heme* chemical groups in the hemoglobin of the red blood cells, thereby decolorizing any tissue that contains blood. Decolorization thus reduces the absorption of excitation light and improves its depth of penetration in live tissues and its consequent signal-to-noise ratio, facilitating single-cell resolution while performing whole-body imaging. This approach shows significant promise in determining the functional interactions between cells in living tissue, to the so-called cellular circuits of organisms.

4.5.7 OPTICAL COHERENCE TOMOGRAPHY

Optical coherence tomography (OCT) is based on *low-coherence interferometry*. It uses the long coherence length of light sources to act as a *coherence rejection filter* to reduce the detection of multiple scattering events. In conventional interference techniques, for example, single-wavelength laser interferometry, interference occurs over a distance of a few meters. In OCT, a less coherent light source, for example, an LED, might be used, which exhibits shorter *coherence lengths* over a few tens of microns, which is useful for biophysics in corresponding approximately to the length scale of a few layers of cells in a tissue.

The incident light in an OCT system is normally divided into two beams to form a sample path and a reference path. A confocal light volume is typically used as a mode of illumination onto a sample. After both beams are scattered from the confocal volume, they are recombined and imaged onto one or more photodiode detectors. The two beams will generate an interference pattern on the detector if the *optical path length* from both beams is less than the coherence length of the light source.

In conventional light microscopy, the majority of scattered light generates background noise. This is especially prevalent with deep tissue imaging due to multiple scattering events through several layers of cells. However, in OCT, multiple scatter events can be rejected on the basis of them having a longer optical path length than the optical coherence length, since these events do not form an interference pattern. Thus, an accepted scattered photon will have arrived typically from just a single back reflection event from a cellular structure.

This rejection of multiple scatter noise permits a 3D tomographic image to be reconstructed down to a depth of a several tens of microns. OCT is now a standard technique in biomedical imaging for *ophthalmology*, for example, to generate 3D details of the retina of the eye but is emerging as a useful biophysical tool in research labs for imaging deep tissues and bacterial biofilms.

A variant of OCT is *angle-resolved low-coherence interferometry (a/LCI)*. This is a relatively new light-scatting tool, which can obtain information about the size of cellular structures, including organelles such as cell nuclei. It combines the depth resolution of OCT with angle-resolved elastic light-scattering measurements (see section in the following text) to obtain in-depth information on the shape and optical properties of cellular organelles. In a/LCI, the light scattered by a sample at different angles is mixed with a reference beam to produce an inference pattern. This pattern can then be analyzed to generate the spatial distribution of scattering objects in the sample using inverse light-scattering analysis based on *Mie scattering theory*, which assumes spherical scattering objects (or the equivalent *T-matrix theory*, which is computationally more expensive but can be applied to nonspherical particles). Since the interference pattern is a measure of differences in optical path length of the scale of less than the wavelength of light, this approach can generate very precise estimates of the size and shape of intracellular-scattering objects like nuclei. This biophysical technology also shows promise as a clinical tool for detecting cancerous cells.

4.5.8 REMOVING THE DEEP TISSUE BARRIER

Arguably, the simplest approach to overcoming the issues of optical heterogeneity and signal and the attenuation effect of excitation and signal intensity of light when imaging through relatively thick sections of biological tissue is to remove that barrier of thick tissue. For example, this approach has been used in experiments on nerve cells in the brains of living rodents and primates using an optogenetics approach. To excite the proteins in individual nerve cells using light, it is often easiest to remove a small section of the bone from the skull. Superficial areas of the brain (i.e., relatively close to the skull), which include the cerebral cortex responsible for voluntary control of muscles, can then be activated by light using either an optical fiber or LED directly mounted to the skull of the animal, so that the light does not have to propagate through the bony tissue. Other methods transect out a portion of bone from the skull but replace it using a *zirconium dioxide* substrate (also known as *zirconia*), which is mechanically strong but optically transparent. Areas of the brain far from the surface can, in principle, be accessed using implanted optical fibers to deliver and receive light as appropriate.

Similar *tissue resection* methods can be applied for imaging several types of biological tissues that are close to the surface. Optical fibers can also be used more generally to access

KEY BIOLOGICAL APPLICATIONS: DEEP IMAGING

Monitoring processes at a depth of at least several cells, *for example*, in tissues and biofilms.

deeper into tissue, for example, by directing an optical fiber through natural channels in an animal whose diameter is larger than that of a cladded optical fiber. A multimodal cladded fiber has a diameter of a few hundred microns, which is small enough to be directed through the gut, large blood vessels, and lymphatic vessels. This at least allows the light source/detector to be brought close enough to the internal surface of many organs in the body within significant disruption to the native physiology.

4.6 ADVANCED BIOPHYSICAL TECHNIQUES USING ELASTIC LIGHT SCATTERING

Scattering of light, as with all electromagnetic or matter waves, through biological matter is primarily due to linear optical processes of two types, either elastic or inelastic. Rayleigh and Mie/Tyndall scattering (see Chapter 3) are elastic processes in which an emergent scattered photon has the same wavelength as the incident photon. In the previous chapter, we encountered Mie/Tyndall scattering used in simple optical density measurements in a visible light spectrophotometer to determine the concentration of scattering particles such as cells in a sample. More advanced applications of elastic light scattering, which can reveal molecular level details, include specific techniques called "static light scattering" (*SLS*) and "dynamic light scattering" (*DLS*).

Rayleigh scattering occurs when the scattering particle has a length scale at least an order of magnitude less than the incident light (a semiarbitrary condition often used is that the length scale is less than ~1/20 of the light wavelength), such that the entire surface of the particle will in effect scatter roughly with the same phase. This is the length scale regime of many small biomolecules and molecular complexes. If molecules are randomly positioned, the arrival of a photon at any molecular surface will be random, resulting in incoherent scattered light whose intensity is just the sum of squares of the amplitudes from all particles. Using a simple harmonic dipole oscillator model for electromagnetic radiation scattering leads to the *Rayleigh equation* (or *scattering formula*):

$$(4.23) \qquad I_s(\theta) = I_s(0)C\left(\frac{1}{2d^2}\right)\left(\frac{2\pi}{\lambda}\right)^4\left(\frac{n^2-1}{n^2+1}\right)r^6(1+\cos^2\theta)$$

where I_s is the Rayleigh scattered light of wavelength λ when measured at a distance d and angle θ from the incident light direction from a sample composed of C scattering particles per unit volume of refractive index n and effective radius r. Thus, the scattered intensity is proportional to the reciprocal of the fourth power of the light wavelength and the sixth power of its radius.

4.6.1 STATIC LIGHT SCATTERING

SLS can be used to obtain estimates for the molecular weight M_w of an *in vitro* biological sample in the Rayleigh scattering regime, and for larger molecules in the Mie scattering regime can estimate M_w as well as generate a measure of the length scale of such *macromolecules* given by their root-mean-squared radius, denoted as the radius of gyration, R_G.

Typically, the scattered intensity from a visible laser light beam incident on an *in vitro* solution of a particular type of biomolecule at high concentration (equivalent to ~1 mg mL^{-1}), or a mixture of molecule types, is measured as a function of the scatter angle θ (Figure 4.4b), either by rotating the same detector or by using multiple fixed detectors located at different angles (*multiangle light scattering*), often using 10 different values of θ spanning a typical range ~30°–120°. A *time-averaged* scattered signal intensity is obtained at each angle; hence, there is no time-resolved information and so the technique is described as "static."

One analytical approach to understand these data is to apply continuum modeling (i.e., to derive an analytical formulation) to Rayleigh scattering (which is clearly from discrete particles) to approximate the scattering as emerging from *scattering centers* that are local fluctuations δC in the continuum concentration function for molecules in solution. This fluctuation approach is a common theoretical biophysical tool used to interface continuum and discrete mathematical regimes (see Chapter 8). Here, δC can be estimated from the Boltzmann distribution through a fluctuation in *chemical potential energy* $\Delta\mu$, which relates to the *osmotic pressure* of the solution Π in the sample volume V, through a simple Taylor expansion of the natural logarithm of $(1 - \delta C)$, as

$$(4.24) \qquad \Delta\mu = RT\ln(1 - \delta C) \approx RT\,\delta C = -\Pi V$$

where
  R is the molar gas constant
  T is the absolute temperature

But Π can also be related to C through a *virial expansion* that is in essence an adaptation of the *ideal gas law* but taking into account real effects of interaction forces between molecules and the solvent (usually an organic solvent such as toluene or benzene). This indicates

$$(4.25) \qquad \frac{\Pi}{C} \approx RT\left(\frac{1}{M_w} + BC + O(C)^2\right)$$

where B is the second *virial coefficient*. Combining Equations 4.23 through 4.25 gives the *Debye equation*:

$$(4.26) \qquad \frac{KC}{R(\theta)} \approx \frac{1}{M_w} + 2BC$$

where R is the *Rayleigh ratio* given by:

$$(4.27) \qquad R(\theta) = \frac{d^2 I(\theta)}{I(0)}$$

And the factor K depends on the parameters of the SLS instrument and the molecular solution:

$$(4.28) \qquad K = \left(\frac{\partial n}{\partial C}\right)_{P,T}\left(\frac{2n^2\pi^2}{\lambda^4}\right)(1 + \cos^2\theta)$$

Therefore, a linear plot of KC/R *versus* C has its intercept at $1/M_w$. Thus, M_w can be determined.

For larger molecules in the Mie/Tyndall scattering regime, the Debye equation can be modified to introduce a form or shape factor $P(Q)$, where Q is the magnitude of *scattering vector* (the change in the photon wave vector upon scattering with matter):

$$(4.29) \qquad \frac{KC}{R(\theta)} \approx \left(\frac{1}{M_w} + 2BC\right)\frac{1}{P(Q)}$$

The exact formulation of P depends on the 3D shape and extension of the molecule, embodied in the R_G parameter. The most general approximation is called the "Guinier model," which can calculate R_G from any shape. A specific application of the Guinier model is the *Zimm model* that assumes that each molecule in solution is approximated as a long random coil whose end-to-end length distribution is dependent on a Gaussian probability function, a *Gaussian polymer coil* (see Chapter 8), and can be used to approximate P for relatively small Q:

$$(4.30) \qquad \frac{KC}{R(\theta)} \approx \left(\frac{1}{M_w} + 2BC \right)\left(1 + \frac{R_G^2}{3Q^2} \right)$$

The scattering vector in the case of elastic scattering can be calculated precisely as

$$(4.31) \qquad Q = \frac{4\pi}{\lambda}\sin\left(\frac{\theta}{2} \right)$$

Therefore, a plot of *KC/R versus C* at high values of θ would have a gradient approaching $\sim 2B$, which would allow the second virial coefficient to be estimated. Similarly, the gradient at small values of θ approaches $\sim 2B\left(1 + R_G^2/3Q^2\right)$ and therefore the radius of gyration can be estimated (in practice, no gradients are manually determined as such since the B and R_G parameters are outputted directly from least-squares fitting analysis). Typically, the range of C explored varies from the equivalent of ~0.1 up to a few mg mL^{-1}.

A mixed/polydisperse population of different types of molecules can also be monitored using SLS. The results of M_w and R_G estimates from SLS will be manifested with either multimodal distributions or apparent large widths to unimodal distributions (which hides underlying multimodality). This can be characterized by comparison with definitively pure samples. The technique is commonly applied as a purity check in advance of other more involved biophysical techniques, which require ultrahigh purity of samples, for example, the formation of crystals for use in x-ray crystallography (see Chapter 7).

4.6.2 DYNAMIC LIGHT SCATTERING

Dynamic light scattering (DLS), also referred to as *photon correlation spectroscopy* (or *quasielastic light scattering*), is a complementary technique to SLS, which uses the time-resolved fluctuations of scattered intensity signals, and is therefore described as "dynamic." These fluctuations result from molecular diffusion, which is dependent on molecular size. It can therefore be used to determine the characteristic *hydrodynamic radius*, also known as the *Stokes radius R_S*, using an *in vitro* solution of biomolecules, as well as estimating the distribution of the molecular sizes in a polydisperse solution.

As incident light is scattered from diffusing biomolecules in solution, the motion results in randomizing the phase of the scattered light. Therefore, the scattered light from a population of molecules will interfere, both destructively and constructively, leading to fluctuations in measured intensity at a given scatter angle θ as a function of time t. Fluctuations are usually quantified by a normalized second-order autocorrelation function g, similar to the analysis performed in FCS discussed previously in this chapter:

$$(4.32) \qquad g(\tau,\theta) = \frac{\langle I(\tau,\theta)I(\tau+t,\theta)\rangle}{\langle I(t,\theta)\rangle^2}$$

A monodispersed molecular solution can be modeled as g_m:

$$(4.33) \qquad g_m(\tau,\theta) = g_m(\infty,\theta) + \beta\exp(-2DQ^2\tau)$$

where $g_m(\infty, \theta)$ is the baseline of the autocorrelation function at "infinite" time delay. In practice, autocorrelations are performed using time delays τ in the range $\sim 10^{-6}$ s^{-1}, and so $g_m(\infty,$

θ) would be approximated as ~g_m using a value of ~1 s for τ. However, in many experiments, there is often very little observable change in g at values of τ above a few milliseconds, since this is the typical diffusion time scale of a single molecule across the profile of the laser beam in the sample.

Q in Equation 4.30 is the θ-dependent scattering vector as described previously in Equation 4.28. Many conventional DLS machines monitor exclusively at a fixed angle θ = 90°; however, some modern devices allow variable θ measurement. In conventional DLS, the sample often has to be diluted down to concentration levels equivalent to ~0.1 mg mL^{-1} to minimize stochastic noise on the autocorrelation function from scattered events through the entire sample cuvette. "Near" backscatter measurements (e.g., at θ ≈ 170°) have some advantages in that they allow focusing of the incident laser beam to sample scattered signals just from the front side of sample cuvette, which reduces the need to dilute the sample, thus increasing the total scattered intensity signal.

D is the translational diffusion coefficient for the biomolecule. This is related to the drag coefficient by the Stokes–Einstein relation (Equation 2.11), which, for a perfect sphere, is given by

$$(4.34) \qquad D = \frac{k_B T}{6\pi\eta R_s}$$

where
k_B is the Boltzmann constant
T is the absolute temperature
η is the viscosity of the solvent

Thus, by fitting g_m to the experimental autocorrelation data, the diffusion coefficient, and hence the Stokes radius of the molecule, can be determined. A polydisperse system of N different biomolecule types generates an N-modal autocorrelation response, which can be approximated by a more general model of

$$(4.35) \qquad g_m(\tau,\theta) = g_m(\infty,\theta) + \sum_{i=1}^{N} \beta_i \exp\left(-2D_i Q^2 \tau\right)$$

Thus, in principle, this allows estimation of the Stokes radii of several different components present in solution, though in practice separating out more than two different components in this way can be nontrivial unless they have distinctly different sizes.

4.6.3 ELECTROPHORETIC LIGHT SCATTERING

A modification of DLS is *electrophoretic light scattering*. Here, an oscillating electric E-field is applied across the sample during DLS measurements, usually parallel to the incident laser light. This results in biased electrophoretic velocity of the molecules in solution, v, determined by the molecules *electrophoretic mobility* μ_E, which depends on their net surface charge:

$$(4.36) \qquad v = \mu_E E$$

A laser beam is first split between a reference and a sample path, which are subsequently recombined to generate an interference pattern at the photodetector. The molecular motion from electrophoresis results in a *Doppler shift* (v_D) on the distribution of fluctuation frequencies observed from the scattered signal, manifested as a phase shift between the sample and reference beams, which can therefore be measured as a change in the interference pattern at the photodetector. On simple geometrical considerations

$$(4.37) \quad \begin{aligned} \upsilon_D &= \frac{\mu_E E \sin\theta}{\lambda} \\ \therefore \mu_E &= \frac{\upsilon_D \lambda}{E \sin\theta} \end{aligned}$$

If we consider a molecule as an ideal sphere, and balance the electrophoretic and drag forces, this indicates that

$$(4.38) \quad \begin{aligned} qE &= \gamma v \\ \therefore \mu_E &= \frac{v}{E} = \frac{q}{6\pi\eta R_s} \end{aligned}$$

where γ is the viscous drag coefficient on the sphere of Stokes radius (R_s) and net surface charge q in a solution of viscosity η. Thus, the net surface charge on a molecule can be estimated as

$$(4.39) \quad q = \frac{6\pi\eta\upsilon_D \lambda R_s}{E \sin\theta}$$

The Stokes radius can be estimated using similar autocorrelation analysis to that of DLS earlier. The net surface charge can be related to other useful electrical parameters of a molecule, such as the *zeta potential*. For biological colloidal dispersions such as large biomolecules in water, the zeta potential is the voltage difference between the water in the bulk of the bulk liquid and the *electrical double layer* (*EDL*) of ions and counterions held by electrostatic forces to the molecule surface. The EDL is an important parameter in determining the extent of aggregation between biomolecules in solution.

4.6.4 INTERFEROMETRIC ELASTIC LIGHT SCATTERING FOR MOLECULAR IMAGING

Interferometric light scattering microscopy (a common method used is known as *iSCAT*) has sufficiently high contrast to enable imaging of single protein molecules without the need for any fluorescent labels, for example, demonstrated with the observation of nanoscale molecular conformational changes of the protein myosin used in the contraction of muscle tissue. Here, the sample is illuminated using coherent laser light, such that the sample consists of weakly scattering objects localized on a microscope coverslip at the glass–water interface. The detected light intensity (I_d) from a fast camera detector is the sum of reflected light from this interface and that scattered from the proteins on the coverslip surface:

$$(4.40) \quad I_d = \left| E_{ref}^2 + E_{scat}^2 \right| = \left| E_i^2 \right| \left(R^2 + |S|^2 - 2R|s|\sin\phi \right)$$

where
  E_i, E_{ref}, and E_{scat} are the incident, reflected, and scattered light E-field amplitudes
  R and s are the reflected and scattering amplitudes
  ϕ is the phase between the scattered and reflected light

For small scattering objects, the value of $|s|^2$ is close to zero. This is because the Rayleigh scattering cross-section, and hence the scattering amplitude $|s|^2$, scales with V^2 for a small scattering particle whose radius is much less than the wavelength of light (see Equation 4.21), for example, the scattering cross-section of a 40 nm gold nanoparticle is ~10^7 that of a typical globular protein of a few nanometers in effective diameter; a few tens of nanometers is the practical lower limit for reproducible detection of scattered light from the laser dark-field technique (see Chapter 3), which demonstrates the clear problem with attempting to detect the scatter signal from small biomolecules directly. However, the *interference term* $2R|s|\sin\phi$ only scales with V and so is far less sensitive to changes in scatterer size, and the detection of this term is the physical basis of iSCAT.

An iSCAT microscope setup is similar to a standard confocal microscope, in terms of generating a confocal laser illumination volume, which is laterally scanned across the sample, though instead of detecting fluorescence emissions, the interference term intensity is extracted by combining a quarter wave plate with a polarizing beamsplitter. This utilizes the phase difference between the interference term with respect to the incident illumination and rotates this phase to enable highly efficient reflection of just this component at the polarizing beamsplitter, which is directed not through a pinhole as for the case of traditional confocal microscopy but rather onto a fast CCD camera, such as a CMOS camera.

An enormous advantage of iSCAT, and similar interferometric imaging methods, over fluorescence imaging is speed. Fluorescence imaging can achieve a significant imaging contrast but to do so ultimately requires a sufficiently large sampling time window to collect fluorescence emission photons. Fluorophores, as we have seen, are ultimately limited by the number of photons that they can emit before irreversible photobleaching. Interferometric scattering is not limited in this manner; in fact, the background signal in iSCAT scales with $\sim\sqrt{N}$ from Poisson sampling statistics, where N is the number of scattered photons detected; therefore, since the signal scales with $\sim N$, then the imaging contrast, which is a measure of the signal-to-noise ratio, itself scales with $\sqrt{N}$. That is, a larger contrast is achievable by simply increasing the power of laser illumination. There is no photon-related physical limit, rather a biological one in increased sample damage at high laser powers.

KEY BIOLOGICAL APPLICATIONS: ELASTIC SCATTERING TOOLS

Estimating molecular shapes and concentrations *in vitro*; label-free monitoring of biomolecules.

4.7 TOOLS USING THE INELASTIC SCATTERING OF LIGHT

Scattering of light, as with all electromagnetic or matter waves, through biological matter is primarily due to linear optical processes of two types, either elastic or inelastic. Rayleigh and Mie scattering (see Chapter 3) are both elastic processes in which the emergent scattered photon has the same wavelength as the incident photon. One of the key inelastic processes with regard to biophysical techniques is *Raman scattering*. This results in the incident photon either losing energy prior to scattering (*Stokes scattering*) or gaining energy (*anti-Stokes scattering*). For most biophysical applications, this energy shift is due to vibrational and rotational energy changes in a scattering molecule in the biological sample (Figure 4.4c), though in principle the Raman effect can also be due to interaction between the incident light and to a variety of quasiparticles in the system, for example, *acoustic matter waves* (*phonons*). There are also other useful inelastic light scattering processes that can also be applied to biophysical techniques.

4.7.1 RAMAN SPECTROSCOPY

Raman scattering is actually one of the major sources of bleed-through noise in fluorescence imaging experiments, which comes mainly from anti-Stokes scattering of the incident excitation light from water molecules. A Raman peak position is normally described in terms of wavenumbers ($2\pi/\lambda$ with typical units of cm^{-1}), and in water, this is generally ~ 3400 cm^{-1} lower/higher than the equivalent excitation photon wavenumber depending on whether the peak is Stokes or anti-Stokes (typically higher/lower wavelengths by ~ 20 nm for visible light excitation).

Some dim fluorophores can have comparable Raman scatter amplitudes to the fluorescence emission peak itself (i.e., this Raman peak is then the limiting noise factor). However, in general, the Raman signal is much smaller than the fluorescence emission signal from typical fluorophores, and only 1 in $\sim 10^{6}$ incident photons will be scattered by the Raman effect. But a *Raman spectrum*, although weak, is a unique signature of a biomolecule with a big potential advantage over fluorescence detection in not requiring the addition of an artificial label to the biomolecule.

The Raman effect can be utilized in biophysics techniques across several regions of the electromagnetic spectrum (including x-rays, see Chapter 5), but most typically, a *near IR* laser (wavelength ~1 μm) is used as the source for generating the incident photons. The shift in NIR Raman scattered energy for biomolecules is typically measured in the range ~200–3500 cm^{-1}. Lower energy scattering effects in principle occur in the range ~10–200 cm^{-1}; however, the signal from these typically gets swamped by that due to Rayleigh scattering.

A Raman spectrometer consists of a laser, which illuminates the sample, with scattered signals detected at 90° from the incident beam (Figure 4.4d). A notch rejection filter, which attenuates the incident laser in excess of 10^6 over a narrow bandwidth of a few nanometers, eliminates the bulk of the elastic Rayleigh scattering component, leaving the inelastic Raman scattered light. This is imaged by a lens and then spatially split into different color components using a diffraction grating, which is projected onto a CCD detector array such that different pixel positions correspond to different wavenumber shift values. Thus, the distribution of pixel intensities corresponds to the Raman spectrum.

In principle, Raman spectroscopy has some similarities to IR spectroscopy discussed in Chapter 3. However, there are key differences to IR spectroscopy, for example, the Raman effect is scattering as opposed to absorption, and also although the Raman effect can cause a change in electrical polarizability in a given chemical bond, it does not rely on exciting a different bond vibrational mode, which has a distinctly different electrical dipole moment. Key biomolecule features that generate prominent Raman scattering signatures include many of the bonds present in nucleic acids, proteins, lipids, and many sugars. The weakness of the Raman scatter signal can be enhanced by introducing small *Raman tags* into specific molecular locations in a sample, for example, alkyne groups, which give a strong Raman scatter signal, though this arguably works against the primary advantage of conventional Raman spectroscopy over fluorescence-based techniques in being label-free.

4.7.2 RESONANCE RAMAN SPECTROSCOPY

When the incident laser wavelength is close to the energy required to excite an electronic transition in the sample, then *Raman resonance* can occur. This can be especially useful in enhancing the normally weak Raman scattering effect. The most common method of Raman resonance enhancement as a biophysical tool involves surface-enhanced Raman spectroscopy (SERS), which can achieve molecular level sensitivity in biological samples *in vitro* (see Kneipp et al., 1997).

With SERS, the sample is placed in an aqueous colloid of gold or silver nanoparticles, typically a few tens of nanometers in diameter. Incident light can induce surface plasmons in the metallic particles in much the same way as they do in surface plasmon resonance (see Chapter 3). In the vicinity of the surface, the photon electric field E is enhanced by a factor ~E^4. This enhancement effect depends sensitively on the size and shape of the nanoparticles. For spherical particles, the enhancement factor falls by 50% over a length scale of a few nanometers.

Heuristic power-law dependence is often used to model this behavior:

$$(4.41) \qquad I(z) = I(0)\frac{R}{(R+z)^a}$$

where $I(z)$ is the Raman scatter intensity at a distance z from surface of a spherical particle of radius R. Although different experimental studies suggest that the parameter a varies broadly in the range ~3–6, with ~4.6 being given consensus by many.

A typical enhancement in measurement sensitivity, however, is >10^5, with values up to ~10^{14} being reported. Therefore, if a biomolecule is bound to the surface of a nanoparticle, then an enhanced Raman spectrum can be generated for that molecule. The enhancement allows sample volumes of $\sim 10^{-11}$ L to be probed at concentrations of $\sim 10^{-14}$ M, sufficiently dilute to permit single-molecule detection (for a review, see Kneipp et al., 2010).

SERS has been applied to detecting nucleotide bases relevant to DNA/RNA sequencing, amino acids, and large protein structures such as hemoglobin, in some cases pushing the sample detection volume down to $\sim 10^{-13}$ L. It has also been used for living cell samples, for example, to investigate the process of internalization of external particles in eukaryotic cells of endocytosis (see Chapter 2).

SERS can also be used in conjunction with microscale tips used in AFM (see Chapter 6). These tips are pyramidal in shape and have a height and base length scale of typically several microns. However, the radius of curvature is more like ~ 10 nm, and so if this is coated in gold or silver, there will be a similar SERS effect, referred to as *tip-enhanced Raman spectroscopy* (*TERS*), which can be used in combination with AFM imaging. Recently, carbon nanotubes have also been used as being mechanically strong, electrically conductive extensions to AFM tips.

SERS has also been performed on 2D arrays of silver holes nanofabricated to have diameters of a few hundred nanometers. The ability to controllably nanofabricate a 2D pattern of holes has advantages in increased throughput for the detection of biological particles (e.g., of a population of cells in a culture, or a solution of biomolecules), which facilitates miniaturization and coupling to microfluidics technologies for biosensing application. Also, although still in its infancy, the technology is compatible with rendering angle-resolved Raman scattering signals in using a polarized light source, which offers the potential for monitoring molecular orientation effects.

4.7.3 RAMAN MICROSCOPY

A Raman microscope can perform Raman spectroscopy across a spatially extended sample to generate a spatially resolved Raman spectral image. Raman microscopy has been used to investigate several different, diverse types of cells grown in culture. For example, these include spores of certain types of bacteria, sperm cells, and cells that produce bone tissue (*osteocytes*). In its simplest form, a Raman microscope is a modified confocal microscope whose scattered light output captured by a high *NA* objective lens is then routed via an optical fiber to a Raman spectrometer. Devices typically use standard confocal microscope scanning methods.

This is an example of *hyperspectral imaging* or *chemical imaging*. In the case of Raman microscopy, it can generate thousands of individual Raman spectra across the whole of the field of view. The molecular signatures from these data can then, in principle, be extracted computationally and used to generate a 2D map showing the spatial localization and concentration of different biochemical components in cellular samples. In practice, however, it is challenging to extract the individual signature from a complex mix of anything more than a handful of different biochemical components due to the overlap between Raman scatter peaks, and so the method is largely limited to extracting strong signals from a few key biochemical components.

Hyperspectral imaging is a slow technique, limited by the scanning of the sample but also in the required integration time for a complete Raman spectrum to be generated for each pixel in the digitized confocal Raman map. A typical scan for a small pixel array can take several minutes. This increased exposure to incident light increases the risk of sample photodamage and limits the utility of the technique for monitoring dynamic biological processes.

Improvements in sampling speed can be made using direct Raman imaging. Here, only a very narrow range of wavenumbers corresponding to a small Raman scattering bandwidth is sampled to allow a standard 2D photodetector array system, such as an EMCCD camera, to be used to circumvent the requirement for mechanical scanning and full Raman spectrum acquisition, for example, to monitor the spatial localization of just a single biochemical component such as cholesterol in a cell. Also, the temporal resolution is improved using related coherent Raman spectroscopy methods. This has significant advantages for this emerging field of *chemical imaging;* this is an example of a biophysical imaging technique used to investigate the *chemical environment* of biological samples, and improving the time resolution enables better insight into the underlying dynamics of this chemistry.

4.7.4 COHERENT RAMAN SPECTROSCOPY METHODS

Two related Raman techniques that specifically use coherent light sources are *coherent anti-Stokes Raman scattering* (*CARS*) and *coherent Stokes Raman scattering* (*CSRS*, pronounced "scissors"). However, CSRS is rarely favored over CARS in practice because its scattering output is at higher wavelengths than the incident light and is therefore more likely to be contaminated by autofluorescence emissions. CARS, in particular, has promising applications for *in vivo* imaging. Both CARS and Raman spectroscopy use the same Raman active vibration modes of molecular bonds, though the enhancement of the CARS signal compared to conventional Raman spectroscopy is manifested in an improvement in sampling time by a factor of ~10^5. In other words, CARS can be operated as a real-time video-rate technique.

CARS is a third-order nonlinear optical effect, which uses three lasers, one to pump at frequency ω_{pump}, a Stokes laser of frequency ω_{Stokes}, and a probe laser at frequency ω_{probe}. All these interact with the sample to produce a coherent light output with anti-Stokes shifted frequency of ($\omega_{pump} + \omega_{probe} - \omega_{Stokes}$). When the frequency difference between the pump and the Stokes lasers (i.e., $\omega_{pump} - \omega_{Stokes}$) coincides with the Raman resonance frequency of a specific vibrational mode of a molecular bond in the sample, there is a significant enhancement of the output. Molecules such as lipids and fatty acids, in particular, have strong Raman resonances. When combined with microscopic laser scanning, 2D images of *in vivo* samples can be reconstructed to reveal high-resolution details for cellular structures that have high fat content, a good example being the fatty *myelin sheath*, which acts as a dielectric insulator around nerve fibers, with some CARS imaging systems now capable of video-rate time resolution. This technology not only gives access to dynamic imaging of biochemical components in cell populations that are difficult to image using other techniques but like conventional Raman spectroscopy is also a label-free approach with consequent advantages of reduced impairment to biological functions compared to technologies that require components to be specifically labeled.

4.7.5 BRILLOUIN LIGHT SCATTERING

When light is transmitted through a sample that contains periodic variation in refractive index, *Brillouin light scattering* (*BLS*) can occur. In a biological sample, the most likely cause of Brillouin inelastic scattering is the interaction between acoustic oscillation of the sample matter, in the form of phonons, and the incident light. Phonons can establish standing waves in a sample, which causes the periodic changes to the molecular structure correlated over an extended region of the sample, resulting in periodicity in refractive index over the length scale of phonon wavelengths.

Similar to Raman scattering, an incident photon of light can gain energy from a phonon (Stokes process) or lose energy in creating a phonon in the material (anti-Stokes process), resulting in a small *Brillouin shift* to the scattered light's wavelength. The shift can be quantified by a Brillouin spectrometer, which works on principles similar to a *Fabry–Pérot interferometer* (or *etalon*), as discussed in Chapter 3. Although not a unique *fingerprint* in the way that a Raman spectrum is, a Brillouin spectrum can render structural details of optically transparent biological materials, for example, to investigate the different optical materials in an eye. The largest technical drawbacks with *Brillouin spectroscopy* are the slowness of sampling and the often limited spatial information, in that most devices perform essentially point measurements, which require several minutes to acquire. However, recent developments have coupled Brillouin spectroscopy to confocal imaging to generate a *Brillouin microscope*, with improvements in parallelizing of scatter signal sampling to permit sampling rates of

**KEY BIOLOGICAL
APPLICATIONS:
I NELASTIC
SCATTERING TOOLS**

Label-free molecular identification; Monitoring mechanical
stress in tissues.

~1 Hz per acquired spectrum (Scarcelli and Yun, 2007). BLS has been used in a number of studies to investigate the biomechanical properties of a variety of tissues, which are discussed in Chapter 6.

4.8 SUMMARY POINTS

- There are several super-resolution methods, characterized by techniques that use light microscopy to determine the location of a single dye molecule better than the diffraction-determined optical resolution limit.
- An array of nonlinear methods is useful for studying biological processes deep into tissues, especially two-photon excitation fluorescence imaging.
- Tools to correct for optical inhomogeneity in deep tissues exist, such as AO.
- Interactions between two different biomolecules can be monitored using FRET, down to the level of single molecules.
- FCS is a robust tool to determine diffusion rates, which complements single-particle tracking methods.
- Inelastic scattering methods such as Raman and Brillouin scattering are "label-free" and facilitate monitoring of a range of biomolecule types beyond just the primary focus of fluorescence microscopy, which is fluorescently labeled proteins.

QUESTIONS

4.1 A high-power oil-immersion objective lens has an NA or 1.45, transmitting 80% of visible light, total magnification 300.
   a What is the photon capture efficiency of the lens?
   In a TIRF experiment using the objective lens method with a laser excitation of wavelength 561 nm, a protein is labeled with the FP mCherry.
   b Using PubSpectra (http://www.pubspectra.org/) or equivalent, select a suitable dichroic mirror and emission filter for exciting and visualizing mCherry molecules and estimates the maximum proportion of fluorescence photons that can in principle reach a camera detector.
   c If the camera detector converts 90% of all incident visible light photons into a signal and has a pixel size of 18 μm and a background noise level (b) equivalent to ~200 counts, what is the best spatial precision with which one can estimate the position of a single mCherry molecule stuck to the surface of a glass coverslip? In an actual experiment, the best precision was measured to be approximately 10 times worse than the theoretical estimate.
   d What is the most likely cause of this discrepancy?
4.2 Derive formulas for the relation between sigma width and w for the full width at half maximum value to a Gaussian approximation to the PSF.
4.3 X-ray diffraction data (see Chapter 5) suggest that a certain protein found in the cell membrane can be present as both a monomer and a dimer. A single-molecule yellow-color FP called YPet was fused to the end of this protein to perform live-cell fluorescence imaging. TIRF microscopy was used on live cells indicating typically six detected fluorescent spots of full width at half maximum intensity profile ~180 nm per cell in each illuminated section of cell membrane. TIRF microscopy here was found to illuminate only ~15% of the total extent of the cell membrane per cell, with the total membrane area being ~30 μm² per cell. On the basis of intensity measurements of the fluorescent spots, this indicated that ~75% of the spots were monomers, ~15% were dimers, and the remaining 10% were consistent with being oligomers. Explain with reasoning if you think the oligomers are "real" or simply monomers or dimers, which are very close to each other. (*Hint:* if two fluorescent molecules observed using

diffraction-limited microscopy are separated by less than the optical resolution limit, we may interpret them as a single "spot.")

4.4   A synthetic molecular construct composed of DNA was labeled with a bright organic "intercalating" PALM dye, which decorated the DNA by binding between every other nucleotide base pair. Each organic dye molecule, once stochastically activated during PALM imaging, could fluorescently emit an average of $\sim10^7$ photons of peak wavelength at ~550 nm, before irreversibly photobleaching, and could emit these at a maximum flux rate of $\sim10^8$ photons per second. If the *total* detection efficiency for all photons emitted is 10% at the EMCCD camera of the PALM imaging device, calculate, stating any assumptions you make, what the maximum theoretical imaging frame rate on the PALM imaging device would be to permit two neighboring dye molecules on the DNA to be resolved in the focal plane of the microscope.

4.5   The PALM imaging device of Question 4.4 was modified to permit 3D localization of the organic dye using astigmatism imaging. If the DNA construct was aligned with the optic axis, what then is the maximum frame rate that would permit two neighboring dye molecules on the DNA to be resolved?

4.6   An *in vitro* fluorescence imaging experiment involving GFP was performed with and without the presence of the G/GOD/CAT free radical quencher. When performed in pure water without any quencher the GFP bleached rapidly, whereas upon adding the quencher the rate of bleaching was lower. If the same experiment was performed with the quencher but in a solution of PBS (phosphate-buffered saline, a simple pH buffer) the rate of bleaching was the same, but the brightness of single GFP molecules appeared to be greater than before. Explain these observations.

4.7   A protein is labeled with a donor and acceptor fluorophore to study a conformational change from state 1 to 2 by using single-molecule FRET. The FRET acceptor–donor pair has a known Förster radius of 4.9 nm, and the measured fluorescence lifetimes of the isolated donor and the acceptor fluorophores are 2.8 and 0.9 ns, respectively.
   a   Show that the FRET efficiency is given by $(1 - \tau_{DA}/\tau_D)$ where τ_{DA} and τ_D are the fluorescence lifetimes of the donor in the presence and absence of acceptor, respectively.
   b   What is the distance between the donor and acceptor if the measured donor lifetime in conformation 1 is 38 ps? Structural data from x-ray crystallography (see Chapter 5) suggest that the fluorophore separation may increase in a distinct step by 12 Å when the protein makes a transition between states 1 and 2.
   c   Calculate the donor fluorescence lifetime of conformation 2. How small does the measurement error of the FRET efficiency need to be to experimentally observe? See the changes of state from 1 to 2.
   d   What is the maximum change in FRET efficiency that could be measured here?

4.8   A FLIM–FRET experiment using GFP and mCherry FPs as the donor and acceptor FRET fluorophores, respectively, indicated that the fluorescence lifetimes of both proteins were 5 ns or less. Both proteins have a molecular weight of ~28 kDa and an effective diameter of ~3 nm.
   a   What, with reasoning, is the typical rotational time scale of these FPs if their motions are unconstrained?
       The researchers performing the experiment assumed that any dynamic changes observed in the estimated FRET efficiency were due to relative separation changes of the GFP and mCherry.
   b   Discuss, with reasoning, if this is valid or not.
   c   Why is this experiment technically challenging when attempted at the single-molecule level?

4.9   A 4π microscope was composed of two matched oil-immersion (refractive index $\approx$ 1.52) objective lenses of *NA* 1.49. In an exam answer, a student suggested that this should be better described as a 3.2π microscope. Discuss.

4.10 A bespoke STED microscope used a white-light laser emitting at wavelengths from ~450 to 2000 nm of total power 6 W, generated in regular pulse bursts at a frequency of 60 MHz such that in each cycle, the laser produces a square-wave pulse of 10 ps duration.

a   What is the mean power density in mW per nm from the laser?

b   How does this compare with the peak power output?

The excitation laser was split by a dichroic mirror and filtered to generate a blue beam that would transmit 90% of light over a 20 nm range centered on 488 nm wavelength, and zero elsewhere, before being attenuated by an ND1 filter and focused by a high NA objective lens onto a sample over an area equivalent to a diffraction-limited circle of effective diameter equal to the PSF width. A red beam was directed through a loss-free prism that would select wavelengths over a range 600–680 nm, before being attenuated by an ND4 filter, phase modified to generate a donut shape and focused by the same objective lens in the sample of area ~0.1 μm^2. The oil-immersion objective lens had an NA of 1.49 and a mean visible light transmittance of 80%. A biological sample in the STED microscope consisted of fibers composed of microtubules (see Chapter 2) or diameter 24 nm, which were labeled on each tubulin monomer subunit with a fluorescent STED dye. Assume here that the saturating intensity of STED is given by the mean in the ring of the donut.

c   Discuss, with reasoning, if this microscope can help to address the question of whether fibers in the sample consist of more than one individual microtubule.

4.11 A typical FCS experiment is performed for a single type of protein molecule labeled by GFP, which exhibits Brownian diffusion through a confocal laser excitation volume generated by focusing a 473 nm laser into the sample solution to generate a diffraction-limited intensity distribution.

a   Show, stating any assumptions you make, that the number density autocorrelation function is given by $\langle C \rangle \exp\left[-(r-r')^2/(4Dt')\right]/(1/4\pi Dt')^{3/2}$ and r is positional vector of a fluorophore relative to the center of the confocal excitation volume of light, r' is the equivalent autocorrelation positional vector interval, and t' is the autocorrelation time interval value.

b   The effective confocal volume V can be defined as $\left(\int P(r)\mathrm{d}V\right)^2 / \left(\int P(r)\mathrm{d}V\right)$ where P is the associated PSF. Demonstrate that V can be approximated by Equation 3.53. (You may assume the Gaussian integral formula of $\int \exp(-r^2)\mathrm{d}r = \sqrt{(\pi/2)}$ if the integral limits in r are from zero to infinity.)

c   In one experiment, the autocorrelation function at small time interval values converged to ~2.6 and dropped to ~50% of this level at a time interval value of ~8 ms. Estimate with reason the diffusion coefficient of the protein and the number of protein molecules that would be present in a 1 μL drop of that solution.

4.12 A protein is labeled with a donor–acceptor FRET pair whose Förster radius is 5.1 nm to investigate a molecular conformational change from state I to II. The fluorescence lifetimes of the donor and the acceptor fluorophores are known to be 5.3 and 1.1 ns, respectively.

a   The donor lifetime for state I has a measured value of 54 ps. Calculate the FRET efficiency in this state and the distance between the FRET dye pair.

b   There is a decrease in this distance of 0.9 nm due to the conformational change of state I–II. Estimate the fluorescence lifetime of the donor in this state, and comment on what the error of the FRET efficiency measurement must be below in order to observe this distance change.

4.13 A gold-coated AFM tip was used as a fluorescence excitation source for TERS by focusing a laser beam on to the tip and then scanning the tip over a fluorescently labeled sample. The fluorescence intensity measured from the sample when the tip was close to the surface was ~10^6 counts, which decreased to only ~10^2 counts when the tip was moved back from the sample by just 1 nm. Estimate the radius of curvature of the AFM tip.

4.14 If a 50 nm diameter nanobead can just be detected using laser dark-field, which relies on Rayleigh scattering, with a maximum sampling frequency of 1 kHz, estimate the minimum time scale of a biological process that could in principle be investigated using the same instrument but to monitor a biomolecular complex of effective diameter of 10 nm, assuming the nanobead and biomolecules have similar refractive indices.

REFERENCES

KEY REFERENCE

Thompson, R.E., Larson, D.R., and Webb, W. (2002). Precise nanometer localization analysis for individual fluorescent probes. *Biophys. J.* 82:2775.

MORE NICHE REFERENCES

Appelhans, T. et al. (2012). Nanoscale organization of mitochondrial microcompartments revealed by combining tracking and localization microscopy. *Nano Lett.* 12:610–616.

Badieirostami, M. et al. (2010). Three-dimensional localization precision of the double-helix point spread function versus astigmatism and biplane. *Appl. Phys. Lett.* 97:161103.

Balzarotti, F. et al. (2016) Nanometer resolution imaging and tracking of fluorescent molecules with minimal photon fluxes. *Science* 355:606–612.

Benitez, J.J., Keller, A.M., and Chen, P. (2002). Nanovesicle trapping for studying weak protein interactions by single-molecule FRET. *Methods Enzymol.* 472:41–60.

Betzig, E. et al. (2006). Imaging intracellular fluorescent proteins at nanometer resolution. *Science* 313:1642–1645.

Booth, M. (2014). Adaptive optical microscopy: The ongoing quest for a perfect image. *Sci. Appl.* 3:e165.

Broxton, M. et al. (2013). Wave optics theory and 3-D deconvolution for the light field microscope. *Opt. Exp.* 21:25418–25439.

Burnette, D.T. et al. (2011). Bleaching/blinking assisted localization microscopy for superresolution imaging using standard fluorescent molecules. *Proc. Natl. Acad. Sci. USA* 108:21081–21086.

Chiantia, S., Kahya, N., and Schwille, P. (2007). Raft domain reorganization driven by short- and long-chain ceramide: A combined AFM and FCS study. *Langmuir* 23:7659–7665.

Churchman, L.S. et al. (2005). Single molecule high-resolution colocalization of Cy3 and Cy5 attached to macromolecules measures intramolecular distances through time. *Proc. Natl. Acad. Sci. USA* 102:1419–1423.

Cissé, I. I. et al. (2013). Real-time dynamics of RNA polymerase II clustering in live human cells. *Science.* 341:664–667.

Cox, S. et al. (2012). Bayesian localization microscopy reveals nanoscale podosome dynamics. *Nat. Methods* 9:195–200.

Dertinger, T. et al. (2009). 3D super-resolution optical fluctuation imaging (SOFI). *Proc. Natl. Acad. Sci. USA* 106:22287–22292.

Faulkner, H.M. and Rodenburg, J.M. (2004). A phase retrieval algorithm for shifting illumination. *Appl. Phys. Lett.* 85:4795–4797.

Friedl, P. et al. (2007). Biological second and third harmonic generation microscopy. *Curr. Protoc. Cell Biol.* 4:Unit 4.15.

Gordon, M.P., Ha, T., and Selvin, P.R. (2004). Single-molecule high-resolution imaging with photobleaching. *Proc. Natl. Acad. Sci. USA* 101:6462–6465.

Gustafsson, M.G. (2000). Surpassing the lateral resolution limit by a factor of two using structured illumination microscopy. *J. Microsc.* 198:82–87.

Ha, T. and Tinnefeld, P. (2012). Photophysics of fluorescence probes for single molecule biophysics and super-resolution imaging. *Annu. Rev. Phys. Chem. 63*:595–617.

Ha, T. et al. (1996). Probing the interaction between two single molecules: Fluorescence resonance energy transfer between a single donor and a single acceptor. *Proc. Natl. Acad. Sci. USA 93*:6264–6268.

Ha, T. et al. (2002). Initiation and re-initiation of DNA unwinding by the *Escherichia coli* Rep helicase. *Nature 419*:638–641.

Hecht, B. et al. (2000). Scanning near-field optical microscopy with aperture probes: Fundamentals and applications. *J. Chem. Phys. 112*:7761–7774.

Hell, S.W. and Stelzer, E.H.K. (1992). Fundamental improvement of resolution with a 4Pi-confocal fluorescence microscope using two-photon excitation. *Opt. Commun. 93*:277–282.

Hell, S.W. and Wichmann, J. (1994). Breaking the diffraction resolution limit by stimulated emission: Stimulated-emission-depletion fluorescence microscopy. *Opt. Lett. 19*:780–782.

Horton, N.G. et al. (2013). In vivo three-photon microscopy of subcortical structures within an intact mouse brain. *Nat. Photon. 7*:205–209.

Huisken, J. et al. (2004). Optical sectioning deep inside live embryos by selective plane illumination microscopy. *Science 305*:1007–1009.

Jones, S.A. et al. (2011). Fast, three-dimensional super-resolution imaging of live cells. *Nat. Methods 8*:499–508.

Kapanidis, A.N. et al. (2005). Alternating-laser excitation of single molecules. *Acc. Chem. Res. 38*:523–533.

Keller, P.J. et al. (2008). Reconstruction of zebrafish early embryonic development by scanned light sheet microscopy. *Science 322*:1065–1069.

Kneipp, K. et al. (1997). Single molecule detection using surface-enhanced Raman scattering (SERS). *Phys. Rev. Lett. 56*:1667–1670.

Kneipp, J. et al. (2010). Novel optical nanosensors for probing and imaging live cells. *Nanomed. Nanotechnol. Biol. Med. 6*:214–226.

Kobat, D. et al. (2011). In vivo two-photon microscopy to 1.6-mm depth in mouse cortex. *J. Biomed. Opt. 16*:106014.

Lee, S.F. et al. (2011). Super-resolution imaging of the nucleoid-associated protein HU in *Caulobacter crescentus*. *Biophys. J. 100*:L31–L33.

Levoy, M. et al. (2009). Recording and controlling the 4D light field in a microscope using microlens arrays. *J. Microsc. 235*:144–162.

Lew, M.D. et al. (2011). Three-dimensional superresolution colocalization of intracellular protein superstructures and the cell surface in live *Caulobacter crescentus*. *Proc. Natl. Acad. Sci. USA 108*:E1102–E1110.

Maestro, L.M. et al. (2010). CdSe quantum dots for two-photon fluorescence thermal imaging. *Nano Letters 10*:5109–5115.

Magde, D., Elson, E.L., and Webb, W.W. (1972). Thermodynamic fluctuations in a reacting system: Measurement by fluorescence correlation spectroscopy. *Phys. Rev. Lett. 29*:705–708.

Marrison, J. et al. (2013). Ptychography—A label free, high-contrast imaging technique for live cells using quantitative phase information. *Sci. Rep. 3*:2369.

Miyawaki, A. et al. (1997). Fluorescent indicators for Ca^{2+} based on green fluorescent proteins and calmodulin. *Nature 388*:882–887.

Monod, J. (1971). *Chance and Necessity*. Vintage Books, New York.

Mortensen, K.I. et al. (2010). Optimized localization analysis for single-molecule tracking and super-resolution microscopy. *Nat. Methods 7*:377–381.

Park, H., Toprak, E., and Selvin, P.R. (2007). Single-molecule fluorescence to study molecular motors. *Quart. Rev. Biophys. 40*:87–111.

Prakash, R. et al. (2012). Two-photon optogenetic toolbox for fast inhibition, excitation and bistable modulation. *Nat. Methods 9*:1171–1179.

Qu, X., Wu, D., Mets, L., and Scherer, N.F. (2004). Nanometer-localized multiple single-molecule fluorescence microscopy. *Proc. Natl. Acad. Sci. USA 101*:11298–11303.

Ritter, J.G. et al. (2010). Light sheet microscopy for single molecule tracking in living tissue. *PLoS One 5*:e11639.

Roy, R., Hohng, S., and Ha, T. (2008). A practical guide to single-molecule FRET. *Nat. Methods* 5:507–516.

Rust, M.J., Bates, M., and Zhuang, X. (2006). Sub-diffraction-limit imaging by stochastic optical reconstruction microscopy (STORM). *Nat. Methods* 3:793–795.

Scarcelli, G. and Yun, S.H. (2007). Confocal Brillouin microscopy for three-dimensional mechanical imaging. *Nat. Photon.* 2:39–43.

Schoen, I. et al. (2011). Binding-activated localization microscopy of DNA structures. *Nano Letters* 11:4008–4011.

Swoger, J. et al. (2014). Light sheet-based fluorescence microscopy for three-dimensional imaging of biological samples. *Cold Spring Harb. Protoc.* 2014:1–8. http://www.nature.com/nmeth/coll ections/lightsheet-microscopy/index.html.

Tainaka, K. et al. (2014). Whole-body imaging with single-cell resolution by tissue decolarization. *Cell* 159:911–924.

Vollmer, F. and Arnold, S. (2008). Whispering-gallery-mode biosensing: Label-free detection down to single molecules. *Nat. Methods* 5:591–596.

Warshaw, D.M. et al. (2005). Differential labeling of myosin V heads with quantum dots allows direct visualization of hand-over-hand processivity. *Biophys. J.* 88:L30–L32.

Waterman-Storer, C.M. et al. (1998). Fluorescent speckle microscopy, a method to visualize the dynamics of protein assemblies in living cells. *Curr. Biol.* 8:1227–1230.

Detection and Imaging Tools that Use Nonoptical Waves

Radio and Microwaves, Gamma and X-Rays,

and Various High-Energy Particle Techniques

5

It must be admitted that science has its castes. The man whose chief apparatus is the differential equation looks down upon one who uses a galvanometer, and he in turn upon those who putter about with sticky and smelly things in test tubes.

—Gilbert Newton Lewis, *The Anatomy of Science* **(1926)**

General Idea: In this chapter, we explore many of the detection and sensing tools of biophysics, which primarily use physical phenomena of high-energy particles or electromagnetic radiation that does not involve visible or near-visible light. These include, in particular, several methods that allow the structures of biological molecules to be determined.

5.1 INTRODUCTION

Modern biophysics has grown following exceptional advances in *in vitro* and *ex vivo* (i.e., experiments on tissues extracted from the native source) physical science techniques. At one end, these encapsulate several of what are now standard characterization tools in a biochemistry laboratory of biological samples. These methods typically focus on one, or sometimes more than one, physical parameter, which can be quantified from a biological sample that either has been extracted from the native source and isolated and purified in some way or has involved a bottom-up combination of key components of a particular biological process, which can then be investigated in a controlled test tube level environment. The quantification of these relevant physical parameters can then be used as a metric for the type of biological component present in a given sample, its purity, and its abundance.

Several of these *in vitro* and *ex vivo* techniques utilize detection methods that do not primarily use visible, or near-visible, light. As we shall encounter in this chapter, there are electromagnetic radiation probes that utilize radio waves and microwaves, such as nuclear magnetic resonance (NMR) as well as related techniques of electron spin resonance (ESR) and electron paramagnetic resonance (EPR), and terahertz spectroscopy, while at the more energetic end of the spectrum, there are several x-ray tools and also some gamma ray methods. High-energy particle techniques are also very important, including various forms of accelerated electron beams and also neutron probes and radioisotopes.

Many of these methods come into the category of *structural biology tools*. As was discussed in Chapter 1, historical developments in structural biology methods have generated enormous insight into different areas of biology. Structural biology was one of the key drivers in the formation of modern biophysics. Many expert-level textbooks are available, which are dedicated to advanced methods of structural biology techniques, but here we discuss the core

DOI: 10.1201/9781003336433-5

physics and the essential details of the methods in common use and of their applications in biophysical research laboratories.

5.2 ELECTRON MICROSCOPY

Electron microscopy (EM) is one of the most established of the modern biophysical technologies. It can generate precise information of biological structures extending from the level of small but whole organisms down through to tissues and then all the way through to remarkable details at the molecular length scale. Biological samples are fixed (i.e., dead), and so one cannot explore functional dynamic processes directly, although it is possible in some cases to generate snapshots of different states of a dynamic process, which gives us indirect insight into time-resolved behavior. In essence, EM is useful as a biophysical tool because the spatial resolution of the technique, which is limited by the wavelength of electrons, in much the same way as that of light microscopy is limited by the wavelength of light. The electron wavelength is of the same order of magnitude as the length scale of individual biomolecules and complexes, which makes it one of the key tools of structural biology.

5.2.1 ELECTRON MATTER WAVES

Thermionic emission from a hot electrode source, typically from a tungsten filament that forms part of an *electron gun*, generates an accelerated electron beam in an electron microscope. Absorption and scattering of an electron beam in air is worse at high pressures, and so conventional electron microscopes normally use high-vacuum pressures <10^{-3} Pa and in the highest voltage devices as low as ~10^{-9} Pa. Speeds v up to ~70% that of light c in a vacuum (3 × 10^8 m s^{-1}) can be achieved and are focused by either electromagnetic or electrostatic lenses onto a thin sample, analogous to photons in light microscopy (Figure 5.1a). However, the effective wavelength λ is smaller by nearly five orders of magnitude. The difference between an electron's rest ($E(0)$) and accelerated ($E(v)$) energy is provided by the electrostatic potential

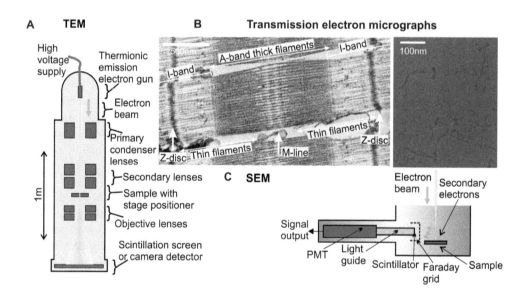

FIGURE 5.1 Electron microscopy. (a) Schematic of a transmission electron microscope. (b) Typical electron micrograph of a negatively stained section of the muscle tissue (left panel) showing a single myofibril unit in addition to several filamentous structural features of myofibrils and a positively shadowed sample of purified molecules of the molecular motor myosin, also extracted from muscle tissue. (Both from Leake (2001).) (c) Scanning electron microscope (SEM) module schematic.

energy qV, where q is the magnitude of the unitary charge on the electron ($\sim1.6 \times 10^{-19}$ C) being accelerated through a voltage potential difference V (a broad range of ~0.2–200 kV depending on the specific mode of EM employed):

$$(5.1) \qquad E(v) - E(0) = qV$$

The *relativistic relation* between an electron's rest mass m_0, $\sim9.1 \times 10^{-31}$ kg, its momentum p, and its energy given by the energy-momentum equation,

$$(5.2) \qquad p^2 c^2 = E(v^2) - \left(m_0 c^2\right)^2$$

But the wavelength of the accelerated electron can be determined from the *de Broglie relation* that embodies the duality of waves and particles of matter, which on rearranging yields

$$(5.3) \qquad \lambda = \frac{h}{p} = \frac{h}{\sqrt{2m_0 qV}} - \frac{1}{\sqrt{1 + \left(qV/2m_0 c^2\right)}}$$

where h is *Planck's constant* $\sim6.62 \times 10^{-34}$ m^2 kg s^{-1}. Thus, the usual classical approximation (cited in many textbooks)

$$(5.4) \qquad \lambda \approx \frac{h}{\sqrt{2m_0 qV}}$$

still holds to within $\sim10\%$. Typical accelerated electrons have such *matter wave* wavelengths of 10^{-12} to 10^{-11} m, and waves will exhibit wavelike phenomena such as reflection and diffraction.

The hypothetical spatial resolution Δx of an electron beam probe is diffraction limited in the same sense as discussed previously in Chapter 4 for a visible light photon beam probe, which is determined by the Abbe diffraction limit for circularly symmetrical imaging apertures of $\sim0.61\lambda/NA$. For a high-resolution (i.e., short wavelength) electron microscope, which might accelerate electrons with ~100 kV, the wavelength λ is $\sim4 \times 10^{-12}$ m, whereas the effective numerical aperture, NA, is ~0.01. This would imply an Abbe limit of ~0.2 nm for spatial resolution. However, in practice, the experimental spatial resolution is an order of magnitude worse than would be expected from the Abbe limit at a given wavelength, which is more like 1–2 nm in this instance, mainly due to the limitations of spherical aberration on the electron beam, but also compounded by the finite size of scattering objects used as typical contrast reagents and of spatial distortions to the sample cause by the method of fixation.

5.2.2 FIXING A SAMPLE FOR ELECTRON MICROSCOPY AND GENERATING CONTRAST

The ultralow pressures used in standard electron microscopes would result in rapid, uncontrolled vaporization of water from wet biological samples, which would result in sample degradation. The specific methods of sample preparation differ depending on the type of EM imaging employed. For example, *cryo-EM* (discussed in detail later in this chapter) has distinctly different preparation methods compared to *transmission EM* and *scanning EM* techniques. Also, the method of preparation depends of the length scale of the sample—whether one is fixing an entire insect, a cell, a subcellular cell compartment, a macromolecular complex.

Tissue samples prepared for EM are fixed so as to prevent uncontrollable water loss, through either dehydration or freezing, and are often fixed to lock the movement of the biological components in the sample. *Chemical fixation* is a gradual multistage process of sample dehydration with organic solvents such as ethanol and acetone; incubation with a bivalent *aldehyde* chemical, typically *glutaraldehyde* or a modified variant, generates chemical cross-links that are relatively indiscriminate between different biomolecular structures in the sample. The dehydrated, cross-linked sample is then embedded in *paraffin wax*, which is sliced with a *microtome* to generate sections of a just a few tens of nanometers of thickness.

The most significant disadvantage with this multistage stage chemical preparation is that it often generates considerable, and sometimes inconsistent, experimental artifacts. Not least of which are volume changes in the sample during dehydration, which potentially affect different parts of a tissue to different extents and therefore lead to sample distortion. *Cryofixation* (also referred to as "snap freezing") rapidly cools the sample using a cryogen such as liquid nitrogen or liquid propane instead of chemical fixation, which eliminates some of these problems. Common methods to achieve this include slam freezing, in which the sample is mechanically positioned rapidly against a cold, flat metallic surface, and high-pressure freezing, which is normally achieved at a pressure of ~2000 atm.

A general method to minimize experimental artifacts is to at least aim for *robustness* in the sample preparation conditions. By this, we mean that the various steps of the sample preparation procedure should be optimized so that the appearance of the ultimate EM images becomes relatively insensitive to small changes in sample preparation, for example, to select a choice of dehydrating reagent that does not result in markedly different images to many other reagents. In other words, this is to optimize the chemical and incubation conditions of sample preparation to be relatively insensitive to their being perturbed.

The key aim of all sample freezing techniques is to vitrify the liquid phases of a biological matter, principally water, to solid to minimize motion of the internal components and to ensure that an *amorphous*, as opposed to a crystalline, vitreous solid results. The biggest problem is the formation of ice crystals, which occurs if the rate of drop in temperature is less than ~10^4 K s^{-1}, which in practice means that freezing needs to occur within a few milliseconds. Slam freezing can achieve this on samples, provided they are less than ~10 μm in thickness, while high-pressure freezing can achieve this on larger samples for up to ~200 μm thick.

Cryosubstitution can then be performed on the frozen sample, which involves low-temperature dehydration by substitution of the water components with organic chemical solvents. In essence, the sample temperature is raised very slowly (over a period of a few days typically), and as it melts, the liquid phase water becomes substituted with organic solvents; this can facilitate stable cross-links between large biomolecules driven by hydrophobic forces in the absence of covalent bond cross-links, so eliminating the need for a specific chemical fixation step. *Cryoembedding* is then performed at temperatures less than −10°C, and samples can be sectioned using a cooled microtome.

Take the example of large protein complexes in the cell membrane. These include membrane-based molecular machines such as the flagellar motor in bacteria that rotates to drive the swimming of bacteria and the ATP synthase molecular machine that generates molecules of ATP (see Chapter 2). Cryofixation is an invaluable preparation approach for these, especially when coupled to a method called "freeze-fracture" or "freeze-etch electron microscopy," which has been used to gain insight into several structural features of cells and subcellular architectures. Here, the surface of the frozen sample is fractured using the tip of a microtome, which can reveal a random fracture picture of the structural makeup immediately beneath the surface, yielding structural details of the cell membrane and the pattern of integrated membrane proteins.

Aficionados of both cryofixation and chemical fixation in EM report a variety of pros and cons for both methods, for example, on the different respective abilities of each to stabilize the motions of certain cellular components during sample fixation. However, one should be mindful of the fact that although EM has excellent spatial resolution and imaging contrast, all sample preparation methods generate distortions when compared against the relatively less invasive biophysical imaging technique of light microscopy.

5.2.3 GENERATING CONTRAST IN ELECTRON MICROSCOPY

Biological matter is mostly water and carbon, comprising relatively low-atomic-number elements. This results in a far greater *mean free collision path* for electrons in carbon compared to high-atomic-number metals. For example, at a 100 kV accelerating voltage, an electron in carbon has a mean free collision path of ~150 nm, whereas in gold, it is ~5 nm. To visualize biological material therefore, a high-atomic-number metal *contrast reagent* is applied.

Negative staining can be applied on both chemically and cryofixed samples, usually by including a heavy metal contrast reagent such as *osmium tetroxide* or *uranyl acetate* dissolved in an organic solvent such as acetone. The contrast reagent preferentially fills the most accessible volumes in the sample (those least occupied by the densest biological matter). This therefore results in a *negative* image of the sample if the electrons are transmitted onto a suitable detector. This technique can generate excellent contrast between heterogeneous biological matter found *in vivo* (e.g., illustrated in the case of muscle tissue in Figure 5.1b).

Another contrast reagent incorporation method involves positive staining via *metallic shadowing*, typically of evaporated platinum. This not only can be applied to relatively large length scale samples (e.g., small whole organisms such as insects) to coat the surface for visualization of backscattered electrons reflected from the metallic coat but is also a common approach applied to visualizing single molecules from visualization of transmitted electrons through the sample. Here, a dilute purified solution of the biomolecules is first sprayed onto a thin sheet of evaporated carbon, which is supported from an EM-grid sample holder. The sample aqueous medium is then dried in a vacuum and platinum is evaporated onto the sample from a low angle <10° as the sample is rotated laterally.

This creates a uniform metallic shadow of topographical features of any single molecules stuck to the surface of the carbon, which are electron dense, generating a high scatter signal from an electron beam, whereas the supporting thin carbon sheet is relatively transparent to electrons and thus results in a "positive" image. Single gold or platinum atoms have a diameter of ~0.3–0.4 nm, but typically, a minimum-sized cluster of atoms in a shadowed single region in a sample might consist of 5–10 such atoms, in which case the real spatial resolution may be worse than expected, after the effects of diffraction and spherical aberration, by a factor of ~2–4.

Metallic shadowing is also used for generating contrast in freeze-fracture samples. Here, larger angles of ~45° are applied in order to reach more recessed surface features compared to single-molecule samples. This method can generate excellent images of the phospholipid bilayer architecture of cell membranes, down to a precision sufficient to visualize single polar head groups of a phospholipid molecule.

Both tissue-/cellular-level samples and single-molecule samples can also be visualized using *immunostaining* techniques. These involve incubating the sample with a specific antibody, which contains a heavy metal tag of just a few gold atoms. The antibodies will then bind with high affinity to specific molecular features of the sample, thus generating high electron beam attenuation contrast for those regions, often used to complement negative staining. However, a single antibody has a Stokes radius of ~10 nm, which reduces the effective spatial resolution of this method. Recent improvements have involved the development of genetically encoded EM labels for use in *correlative light and electron microscopy* (*CLEM*) techniques (discussed later in this chapter).

5.2.4 TRANSMISSION ELECTRON MICROSCOPY

Biological samples can be imaged by detecting the intensity of transmitted electrons, in *transmission electron microscopy* (*TEM*), or by the backscattered secondary electrons, in *scanning electron microscopy* (*SEM*). TEM is valuable for probing cellular morphology in tissues, subcellular architectures, and a range of molecular-level samples. In TEM, the accelerating voltage is ~80–200 kV, capable of generating a wide-field electron beam at the sample of up to several tens of microns in diameter. Contrast reagents are normally used in the form of negative staining, metallic shadowing, or immunostaining. *Low-voltage electron microscopy (LVEM)* in the range ~0.2–10 kV can also be used in transmission mode. The electron wavelength is larger by a factor of 3–4, which therefore reduces the spatial resolution by the same factor. Also, the mean collision path at these lower electron energies in carbon is more like ~15 nm. This means that the biological sample must be of comparable thickness, that is, sectioned very thinly and consistently; otherwise, insufficient electrons will be transmitted. However, a by-product of this is that additional contrast reagents are not required and thus the data are potentially more physiologically relevant.

Some old machines are still in operation in which transmitted electrons are detected via a phosphor screen of typically zinc sulfide, which can then be imaged onto a CCD camera (in fact some machines in operation still use photographic emulsion film). As time advances, many of these older machines will inevitably become obsolete, though a significant minority are still being used in research laboratories. Most modern machines detect the transmitted electrons directly using optimized CCD pixel arrays, which offer some improvement in avoiding secondary scatter effects of emitted light from a phosphor.

A useful variant of TEM is *electron tomography (ET)*. This involves tilting the biological sample stage over a range of ±60° from the horizontal around the *x* and *y* axes of the *xy* sample plane. This generates different projections of the same sample, which can be reconstructed to generate 3D information. The reconstruction is usually performed in reciprocal space; though there are missing angles due to the finite range of stage tilt permitted, there is a *missing wedge* of data in the Fourier plane corresponding to these unsampled orientations. There is a reduction in spatial resolution by factor of ~10 compared to conventional TEM at comparable electron energies, but the insight into molecular structures, especially when combined with cryogenic sample conditions (often referred to as *cryo-ET*, discussed later in this chapter), can be significant.

In principle, 3D information can also be generated through electron holography. Some working designs that utilize adaptations to transmission mode LVEM using electron energies can generate an electron holograph (also known as a *Gabor hologram*, a *Ronchigram*) or a nonbiological sample, using, in essence, the same physical principles as those for digital holography in light microscopy discussed previously (Chapter 3). These techniques have yet to find important applications in biophysics, which is ironic since the original concept of holography developed by Dennis Gabor was to improve the spatial resolution achievable in EM by dispensing with the need for electron lenses to focus the beam, which result in the resolution-limiting spherical aberration (Gabor, 1948). The conceived instrument was to be called the "electron interference microscope," though the practical implementation at the time was not possible since it required a point source of electrons that was technically not achievable with existing technologies. However, a variation of this technique is ptychography, which has made promising progress discussed later in this section.

5.2.5 SCANNING ELECTRON MICROSCOPY

SEM is a lower magnification technique compared to TEM and can generate important structural details on the surface of tissues and small organisms at a length scale of more like several tens to hundreds of microns (Figure 5.1c). It uses a lower range of accelerating voltage of ~10–40 kV compared to TEM. The beam is focused onto the sample to generate a confocal volume, similar in egg shape to that of light microscopy but with a lateral diameter of typically only a few nanometers. The beam passes through pairs of scanning electromagnetic coils or paired electrostatic deflector plates, which displace it laterally to scan the confocal electron volume over the sample surface in a raster fashion.

Electrons from this confocal volume lose energy due to scattering and absorption, which extends to a larger interaction volume whose length scale is greater than that of the confocal volume by at least an order of magnitude. Detected electrons from the sample are either those due backscattered/reflected electrons via elastic scattering, or more likely due to secondary electrons due to inelastic scattering. These have relatively low energies <50 eV and result from the absorption and then ejection from a K-shell electron in a scattering atom from the sample. This low energy manifests as a small mean collision path in the sample of only a few nanometers, and so any secondary electrons that are detected ultimately originate very close from the sample surface. Thus, SEM using secondary electron detection generates just a topographical detail of the sample.

Such surface secondary electrons are first accelerated toward an electrically biased grid at ~90° to the electron beam by a few hundred volts and then further toward a phosphor scintillator inside a Faraday cage (also known as a *Everhart–Thornley detector*), coupled to a photomultiplier tube (PMT) with a higher *E*-field of ~2 kV potential difference to energize the electrons sufficiently to allow scintillation in the phosphor. The resulting PMT electric current is then used as a metric for the secondary electron intensity. Although SEM in itself is not a 3D technique, the same stage tilting and image reconstruction technology as for transmission ET can be applied to generate 3D information on topographical features.

Rarer elastically backscattered electrons are higher in energy and so can scatter at relatively high angles. The electrons can emerge from anywhere in the sample, and thus, backscattered electron detection is not a topographic determination technique. To detect backscattered electrons and not secondary electrons, similar scintillation PMT detectors can be placed in a ring around the main electron beam (i.e., at relatively high scatter angles), allowing *electron backscatter diffraction* images to be generated.

The extent of backscatter is dependent on the atomic number of the metal element in the contrast reagent. In principle, this offers the potential to apply differential imaging on the basis of different atomic number components used to stain the sample. This has been applied to a few exceptional multiple length scale investigations, for example, to probe the optic nerve tract by using a nonspecific lead metal stain, which reveals topographic information of the tract from the detected secondary electrons, while using a specific silver metal stain, which targets just the nerve fibers themselves inside the tract. Silver has a higher atomic number than lead and thus backscatter electron detection can be used to image just the localization of the nerve fibers in the same optic nerve tract.

An SEM can, in principle, be modified to operate simultaneously in the transmission mode. This involves implementing detectors below the sample to capture transmitted electrons, as for conventional TEM. Most mainstream EM machines do not operate in this hybrid manner; however, there is a benefit in using *transmission scanning electron microscopy* since, if used in conjunction with LVEM on *unstained* samples, it improves the image contrast. Thus, this may serve as a useful control at least against the presence of experimental artifacts caused through chemical staining procedures.

Some SEM machines are also equipped with an *x-ray spectrometer. X-ray spectroscopy* is discussed in more detail later in this chapter, but in essence, K-shell electron ejection also generates x-rays and their wavelength is dependent on the specific electronic energy levels of the atom involved. It can therefore be used to investigate the elemental makeup of the sample (*elemental analysis*).

Conventional SEM uses the same high vacuum as TEM. The requirement for dehydrated or frozen samples means that imaging cannot be done under normal "environmental" conditions. However, the *environmental scanning electron microscope* (*ESEM*) overcomes this limitation to a large extent. ESEM utilizes the same generic SEM design but implements a modified sample chamber, which allows a higher pressure to be maintained in a humidified environment. The electron beam attenuation in air increases exponentially with the distance as the electron beam must penetrate into the sample; therefore, the key developments in ESEM have been in miniaturization of the sample chamber. Modern ESEM devices often have variable pressure options with Peltier temperature control for the sample chamber, allowing a range of EM modes to be used, with pressures of a few kilopascals being sufficient to prevent water vaporization from wet samples.

5.2.6 CRYO-EM AND CRYOET

The term cryo-EM is often misused in any EM performed on samples, which have been prepared using cryofixation. However, a better use is for describing EM on a *native* sample involving no dehydration step at which the sample temperature throughout, not just the fixation step but the entirety of the investigation from sample preparation through to the final imaging acquisition, has been kept below 140 K, which is the *vitrification temperature* of water, or in other words biological samples with very minimal sample preparation artifacts. These investigations require a specialized cold stage, typically using liquid nitrogen (boiling point 77 K, which allows a stable cold stage of 110 K to be maintained) or, in some advanced machines, liquid helium (boiling point 4 K).

Cryo-EM is particularly useful as a structural biology tool, both using metallic shadowing and negative staining techniques, and can be applied in transmission and scanning modes. For molecular-level structural investigations, cryo-EM is used for superior spatial resolution compared to SEM. However, the absolute level of spatial resolution in raw cryo-TEM molecular reconstructions is still an order of magnitude worse than the *definitive* atomic-level resolution achievable by the techniques of *nuclear magnetic resonance (NMR)* and *x-ray crystallography*. However, improvements in the methods of image analysis in particular mean that cryo-EM in many cases rivals the traditional atomic-level structural biology methods.

For example, the inferior raw spatial resolution of EM compared to the atomistic-level structural biology techniques can be improved by *subclass averaging*. This operates by categorizing each raw image of a molecular complex into a distinct class of image type, aligning each image within that class and then generating a single average image for each subclass. In the early days of this technique, in fact, close to the turn of the twentieth century, such averaging was performed manually, in a highly precarious and potentially subjective way. However, improvements in modern subclass categorization methods involve *principal component analysis* of *eigenimages* (originally described as *eigenfaces* from its implementation in face recognition software), although there are still potential issues with user-defined thresholds for determining and recognizing subclass features (discussed in Chapter 8).

However, cryo-EM also has some important advantages over x-ray crystallography and NMR, in that it can be applied to molecular complexes that are >250 kDa in summed molecular weight, which is far greater than NMR (~90 kDa maximum) and can be applied to intact large molecular complexes unlike x-ray crystallography, which requires the formation of highly pure crystals, which are too difficult to generate either because they require the presence of a phospholipid bilayer to form stably or because they consist of multiple molecular components. These include not only the large membrane complexes of the flagellar motor and ATP synthase mentioned earlier but also certain essential macromolecular complexes in the cytoplasm such as the intact ribosome and large intact viruses. The ribosome is a particularly good example since the separate components of a ribosome can be purified and structures are determined by x-ray crystallography, whereas to visualize the entirety intact ribosome requires a technique such as cryo-EM.

Electron cryotomography (CryoET) is a specific application of cryo-EM for which 3D images can be reconstructed from multiple 2D images of a sample obtained by tilting over a range of orientations up to a limit of around 70°. Since the electron propagation distance though the sample increases during tilting, this imposes a practical sample thickness upper limit to avoid significant electron beam attenuation, typically around 0.5 μm. Many CryoET studies to date have thus focused on unicellular microbes and viruses, and macromolecular complexes, though thinning of larger samples can be performed using focused ion beam (FIB) milling, in addition to normal cryo-sectioning. CryoET may also be combined with fluorescence microscopy methods to generate more specificity for identifying cellular structures, using similar correlative approaches to those described in Section 5.2.7.

5.2.7 CORRELATIVE LIGHT AND ELECTRON MICROSCOPY

Correlative light and electron microscopy (*CLEM*) combines the advantages of the time-resolved fluorescence microscopy on live cellular material with the higher spatial resolution achievable with EM. As we discussed in Chapter 4, fluorescence microscopy offers a minimally invasive high-contrast tool, which can be used on live-cell samples to monitor dynamic biological processes to a precision of single molecules. However, the diffraction-limited lateral spatial resolution of conventional far-field fluorescence microscopes is ~200–300 nm. This can be improved by an order of magnitude by superresolution techniques but is still another order of magnitude inferior to TEM. But TEM, in turn, suffers the prime disadvantage of being a dead sample technique. CLEM has made important advances in developing methods to combine some of the advantages of both the approaches.

CLEM can utilize a variety of different stains, which can specifically label a biological structure in the sample but be visible in both fluorescence microscopy and TEM. These stains include novel hybrid probes such as fluorescent derivatives of nanogold particles and also quantum dots, since the cadmium atoms at the QD core are electron dense. FlAsH and ReAsH can also be utilized by using a specific photon-induced oxidation reaction with a chemical called "diaminobenzidine" (*DAB*), which causes the DAB to polymerize. In its polymeric state, it can react rapidly with osmium used in negative staining. Secondary antibodies used for immunofluorescence can also be labeled with a fluorophore called "eosin," which is also a substrate that is sensitive to photooxidation of DAB.

The most promising developments involve the use of cryo-EM and genetically encoded fluorescent protein labels. The use of chemical fixation affects the ability of fluorescent proteins to fluoresce; although some fixative recipes exist, which affect fluorescent proteins less, there is still a drop in fluorescence efficiency. However, the rapid freezing methods of cryofixation methods have shown promise in preserving the photophysics of fluorescent proteins. Although fluorescent proteins show no clear direct sensitivity to DAB, there have been some positive results using secondary immunolabeling of green fluorescent protein (GFP) itself. The state of the art is the *mini-singlet oxygen generator* (*miniSOG*), which is a fluorescent *flavoprotein* engineered from a *phototropin* protein from the plant of genus *Arabidopsis* (used as a common model organism, see Chapter 7). MiniSOG contains only 106 amino acids, roughly half as many as GFP, and illumination generates enough singlet oxygen to locally catalyze the polymerization of DAB, which is then resolvable by EM.

The ability to image the same region of a sample is facilitated by gridded or patterned coverslips, which aid in pattern recognition between light and electron microscopes. But a key development for CLEM has been the reduction in the time taken to transfer a sample between the two modes of microscopy. Automated fast-freezing systems can now allow samples to be imaged by fluorescence microscopy, cryofixed within ~4 s, and then imaged immediately afterward using TEM.

KEY POINT 5.1

Although EM was historically one of the pioneering modern biophysical tools, the potential for generating experimental artifacts is high, compared to many other modern techniques. However, both TEM and SEM still are powerful tools and are regularly used in biophysical research laboratories. One useful modern method to increase the confidence in interpretation of EM images is to utilize fluorescence imaging and EM on the same sample. Also, revolutionary recent advances have been made in cryo-EM to rival the spatial resolution of traditional atomistic-level tools such as NMR and x-ray crystallography.

5.2.8 ELECTRON DIFFRACTION TECHNIQUES

Electron diffraction works on the same scattering principles as for light diffraction discussed previously in Chapters 3 and 4; however, the incident beam of accelerated electrons interacts far more strongly with matter. This means that 3D crystals are largely opaque to electron beams. However, 2D spatially periodic cellular structures can generate a strong emergent scatter pattern, which can be used to determine structural details. Since electron beams can be focused using electromagnetic lenses, the diffraction pattern retains phase information from the sample in much the same way as focused rays of light in optical microscopy. This offers an advantage over x-ray diffraction for which phase information has to be inferred indirectly (discussed later in this chapter).

A key biophysical application of electron diffraction is determining structural details of lipid arrays and membrane proteins, for which 3D crystals are difficult to manufacture, which is a requirement for x-ray crystallography. Close-packed 2D lipid–protein arrays are feasible to make, to determine the spacing of periodic biological structures in the sample, using both backscattered electrons in *Bragg reflection* experiments and transmitted electrons. Electrons incident on a sample having periodic features over a characteristic length scale d_b can generate backscattered electrons (also known as "Bragg reflection" or "Bragg diffraction") by an angle θ_b from the normal. Since the backscattered electrons are coherent, they can interfere, such that the condition for constructive interference generates an nth order intensity maxima, which are given by *Bragg's law*, where n is a positive integer:

$$(5.5) \qquad \sin\theta_b = \frac{\lambda n}{2d_b}$$

Primary electrons may, of course, also be transmitted through the sample at an angle θ_t from the normal due to electron diffraction through periodic layer features of length scale d_t, with the condition for constructive interference being

$$(5.6) \qquad \sin\theta_t = \frac{\lambda n}{2d_t}$$

Selected area diffraction is often used for electron diffraction, in which a metal plate containing different aperture sizes can be moved to illuminate different sizes and regions of the sample. This is important in heterogeneous samples; these are potentially *polycrystalline*, which can result in difficult interpretations of electron diffraction patterns if more than one equivalent periodic structure is present. If it is possible to spatially delimit the area of illumination to just one diffracting periodic region, this problem can often be eradicated. However, the strong interaction with matter of the electron beam confers a significant danger of radiation damage of the sample, and consequently, samples need to be cooled using liquid nitrogen or sometimes liquid helium.

Electron diffraction can also be used in *ptychographic EM* (Humphry et al., 2012). The key physical principles of ptychographic diffractive imaging for EM are the same as those discussed previously for light microscopy in Chapter 4. In essence, physical lenses used for imaging can be replaced by an inverse Fourier transform of the diffraction data detected from the sample.

The same method has been applied in a bespoke setup using relatively low-energy 30 kV electrons to form a transmitted electron diffraction image. By modifying an SEM, the primary electron is defocused to generate a broader 20–40 nm illumination patch on the sample. The effects of spherical aberration by the objective electron lens, which normally focuses the beam onto the sample, are largely eradiated since it is used simply to concentrate the electron beam into a delimited region of the sample, as opposed to acting as an imaging component.

A CCD detector is located below the sample to detect the transmitted diffracted electrons (the diffraction pattern formed is a type of a Gabor hologram), which is combined with a much stronger signal from transmitted nonscattered electrons. The phases from a scattering object can be recovered in a similar way using the ptychographic iterative engine (PIE) algorithm as for optical ptychography, since the scanned electron beam moves over the sample to generate overlap.

Measuring the diffraction intensity with the CCD and calculating the respective phases in principle would allow 3D reconstruction of the sample. However, thus far, samples have been limited to being relatively thin. But even so, this method, in eradicating spherical aberration limits, has improved spatial resolution at 30 kV by a factor of ~5 compared to equivalent energies in a conventional TEM.

KEY BIOLOGICAL APPLICATIONS: EM

Determining molecular structures; Imaging tissue, cellular, and subcellular architectures; Imaging surface topologies.

Worked Case Example 5.1: Applying Electron Microscopy

A thin section of skin tissue was prepared to purify planar cell membrane components normal to an electron beam in a diffraction experiment in a 200 kV electron microscope. Some of the transmitted electrons were diffracted with a first-order deflection of 0.5°, while a minority were scattered back with a first-order maxima deflections of 0.015° from the axis normal to the membrane surface. Comment of the angular deflections and intensity of the scattered/diffracted electrons.

Answers

Using the nonrelativistic approximation for electron wavelength and the de Broglie relation indicates

$$\lambda = (6.62 \times 10^{-34}) / \sqrt{\left(2 \times 9.1 \times 10^{-31} \times 1.6 \times 10^{-19} \times 200 \times 10^3\right)} = 2.7 \times 10^{-12}\,\mathrm{m}$$

Using the Bragg reflection formula and rearranging indicate a periodic spacing perpendicular to the membrane of $d_b = (1 \times 2.7 \times 10^{-12})/(2 \times \sin(0.015°)) = 5.1 \times 10^{-9}$ m.

Using the Bragg transmitted diffracted beam formula and rearranging indicate a periodic spacing parallel to the membrane of $d_t = (1 \times 2.7 \times 10^{-12})/(\sin(0.5°)) = 2.4 \times 10^{-10}$ m.

The estimated value of d_b is consistent with the *width* of a cell membrane and might thus be due to interference from the polar head groups that are separated by ~5 nm at either side of the membrane (see Chapter 2). The estimated value of d_t is consistent with the *lateral* spacing of polar head groups if the phospholipid monomers are tightly packed. Constructive interference can occur between several adjacent head groups to generate a first-order diffraction peak of the transmitted beam, whereas interference can only occur between two layers for the backscattered interference (as the cell membrane is a bilayer), and thus the intensity of the first-order maxima will be much less.

5.3 X-RAY TOOLS

X-rays (originally known as "Röntgen rays" in Germany where they were first discovered) are composed of high-energy electromagnetic waves, which have a typical range of wavelength of ~0.02–10 nm. This is very similar to the length scale for the separation of individual atoms in a biological molecule and also for the size of certain larger scale periodic features at the level of molecular complexes and higher length-scale molecular structures, which makes x-rays ideal probes of biomolecular structure. X-ray diffraction, in particular, is an invaluable biophysical tool for determining molecular structures—in excess of 90% of all known molecular structures that have been determined using x-ray diffraction techniques, compared to ~10% by NMR and <1% by EM methods, at the time of writing.

5.3.1 X-RAY GENERATION

In some research laboratories in the world, x-rays are still generated from a relatively small *x-ray tube* beam generator, which can fit into a typical small research lab (Figure 5.2a). This device generates electrons from a hot filament (often made from tungsten but also *thorium* and *rhenium* compounds) in a similar way to an electron microscope using thermionic emission but accelerates these electrons using high voltages of typically ~20–150 kV to impact onto a metal target plate embedded into a *rotating anode*. Rotation, at a rate of 100–200 Hz, increases the effective surface area of the metal target to distribute the high heat generated over a greater area. The target is usually composed of either copper or molybdenum, though tungsten, chromium, and iron are also sometimes used. The high energy of the electrons can be sufficient to displace atomic electrons from their atomic orbitals resulting in x-ray emission, either through a *Bremsstrahlung* mechanism (Figure 5.2b), which results in a continuous x-ray emission spectrum, or *x-ray fluorescence*, which generates emission peaks at distinct wavelengths.

In x-ray fluorescence, incident electrons can have sufficient energy to displace ground-state electrons from the K-shell (i.e., 1s orbital) to generate metal ions (Figure 5.2c). This creates a vacancy in the K-shell, which can be filled by higher-energy electrons from the *L* (2*p* orbital) or *M* (3*p* orbital) shells, coupled to the fluorescence emission of an x-ray photon of energy equal to the energy difference between these K–L and K–M levels minus any vibrational energy losses of the excited state electron as per the fluorescence mechanism described

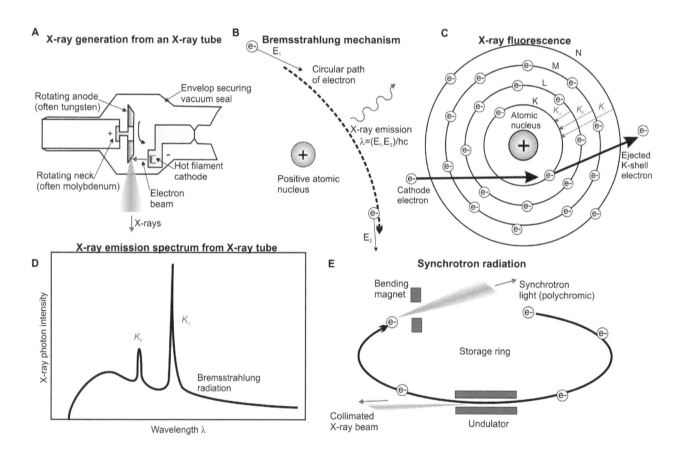

FIGURE 5.2 X-ray generation. (a) X-ray tube, with rotating anode. (b) Mechanism of x-ray generation from the Bremsstrahlung mechanism in which the energy lost by an electron accelerating round a positively charged high-atomic-number nucleus is emitted as x-rays. (c) Process of x-ray fluorescence following ejection of a core shell electron followed by higher energy shell electrons filling this vacancy, with the energy difference emitted at distinct wavelengths, which can be seen (d) overlaid as peaks on the Bremsstrahlung continuum x-ray spectrum. (e) Intense x-rays may also be generated from a synchrotron facility.

for optical microscopy (see Chapter 3). This gives rise to K_α (transition from principal quantum number $n = 2–1$) and less intense K_β (transition from principal quantum number $n = 3–1$) x-ray emission lines, respectively, at a wavelength of $\sim10^{-10}$ m (see Table 5.1 for typical wavelengths for K_α). Other shell transitions are possible to the $n = 2$ level, or L-shells are designated as L x-rays (e.g., $n = 3 \rightarrow 2$ is L_α, $n = 4 \rightarrow 2$ is L_β, etc.), but in general, all but the most intense K_α transitions are filtered out from the final emission output from an x-ray tube collected at right angles to the incident electron beam.

The choice of target in an x-ray tube is a trade-off against the x-ray emission wavelengths desired, the intensity of K_α emission lines, and the target metal having a sufficiently high melting point (since $\sim$99% of the energy from the accelerated electrons is actually converted into heat). Melting point has no clear overall trend across the periodic table, though there is some periodicity to melting point with the atomic number Z and all of the common target metals used are clustered into regions of high melting point on the periodic table. In terms of wavelength of the emission lines, this can be modeled by *Moseley's law*, which predicts that the frequency ν of emission scales closely to $\sim Z^2$:

$$(5.7) \qquad \sqrt{v} = k_1\left(Z - k_2\right)$$

where k_1 and k_2 are constants relating to the type of electron shell transition; however, for all K_α transitions $k_1 = k_2$ and the equation can be rewritten as

$$(5.8) \qquad v = \left(2.5\times10^{-15}\right)\times\left(Z-1\right)^2 \text{Hz}$$

The alternative x-ray generation mechanism to x-ray fluorescence is that which produces Bremsstrahlung radiation. Bremsstrahlung radiation is a continuum of electromagnetic wave emission output across a range of wavelengths. When a charged particle is slowed down by the effects of other nearby charged particles, some of the lost kinetic energy can be converted into an emitted photon of Bremsstrahlung radiation. In the raw output from the metal target in an x-ray tube, this emission is present as a background underlying the x-ray fluorescence emission peaks (Figure 5.2d), though in most modern biophysical applications, Bremsstrahlung radiation is filtered out.

Most x-rays generated for use in biophysical research today are generated from a *synchrotron*. The principle of generating synchrotron radiation is similar to that of a *cyclotron*; in that, it involves accelerating charged particles using radiofrequency voltages and multiple electromagnet B-field deflectors to generate circular motion, here of an electron (Figure 5.2e). These bending magnet deflectors alter the path of electrons in the storage ring. The theory of synchrotron radiation is nontrivial but is confirmed both in classical physics and at the quantum mechanical levels. In essence, a curved trajectory of a charged particle results in warping of the shape of the electric dipole force field to produce a strongly forward peaked distribution of electromagnetic radiation, which is highly collimated; this is synchrotron radiation.

However, synchrotrons use radiofrequency (*f*) values that, unlike cyclotrons, are not fixed and also operate over much larger diameters than the few tens of meters of a cyclotron, more typically a few hundred meters. The United States has several large synchrotron facilities including the *National Synchrotron Light Source* at Brookhaven, with the United Kingdom also investing substantial funds in the *DIAMOND synchrotron* national facility, with 100 other synchrotron facilities around the world, at the time of writing. Note the largest particle accelerator as such, though not explicitly designed as a synchrotron source of x-rays, is 27 km in diameter, which is the *Large Hadron Collider* near Geneva, Switzerland.

Equating magnetic and centripetal forces on an electron of mass m and charge q, traveling at speed v with kinetic energy E in a circular orbit of radius r implies simply

TABLE 5.1 **Wavelength Values of Typical K_α Lines of Common Metal Targets Used in the Generation of X-Rays**

Element	$K_\alpha \lambda$ (nm)
Mo	0.071
Cu	0.154
Co	0.179
Fe	0.194
Cr	0.229
Al	0.834

$$ r = \frac{mv}{qB} = \frac{m\omega r}{qB} = \frac{mfr}{2\pi qB} \therefore f = \frac{2\pi qB}{m} $$

(5.9)

$$ E \approx \frac{1}{2}mv^2 = \frac{q^2 B^2 r^2}{2m} $$

Thus, f is independent of v, assuming nonrelativistic effects, which is the case for cyclotrons. Synchrotrons have larger values of r than cyclotrons and therefore greater values of E, which can exceed 20 MeV after which noticeable relativistic effects occur; thus, f must be varied with v to produce a stable circular beam.

A synchrotron is a large-scale infrastructure facility but produces brighter beams than x-ray tubes, with a greater potential range of wavelength ultimately permitting greater spatial resolution. The use of major synchrotron facilities for providing dedicated x-ray beamlines for crystallography has increased enormously in recent years. In the two decades, since 1995, the number of molecular structures solved using x-ray crystallography, which were deposited each year in the *Protein Data Bank* archive (see Chapter 7) from nonsynchrotron x-ray crystallography has remained roughly constant at ~1000 structures every year, whereas those solved using synchrotron x-ray sources has increased by a factor of ~20 over the same period.

Synchrotrons can generate a continuum of highly collimated, intense radiation from lower energy infrared (~10^{-6} m wavelength) up to a much higher energy hard x-rays (10^{-12} m wavelength). Their output is thus described as *polychromatic*. The spectral output from a typical x-ray tube is narrower at a wavelength of ~10^{-11} m, but both synchrotron x-ray and x-ray tube will often propagate through a *monochromator* to select a much narrower range of wavelength from the continuum.

Monochromatic x-rays simplify data processing significance and improve the effective resolution and signal-to-noise ratio of the probe beam, as well as minimize damage to the sample from extraneous *satellite lines*. An x-ray monochromator typically consists of a quartz (SiO_2) crystal, often fashioned into a cylindrical geometry, which results in constructive interference at specific angles on for a very narrow range of wavelength due to Bragg reflection at adjacent crystal planes. For a small region of the crystal, the difference in optical path length between the backscattered rays emerging at an angle θ from two adjacent layers, which are separated by a spacing d of an x-ray scattering sample is $2d \sin \theta$, and so the condition for constructive interference is that this difference is equal to a whole integer number n of wavelengths λ, hence $2d \sin \theta = n\lambda$. Quartz has a *rhombohedral* lattice with an interlayer spacing of $d = 0.425$ nm; the K_a line of aluminum has a wavelength of $\lambda = 0.834$ nm; therefore, this specific beam can be generated at an angle of $\theta = 78.5°$. The typical bandwidth of a monochromatic beam is ~10^{-12} m.

A recent source of x-rays for biophysics research has been from the *x-ray free-electron laser* (*XFEL*). Although currently not being in sufficient mainstream use to act as a direct alternative to synchrotron-derived x-rays, the XFEL may enable a new range of experiments not possible with synchrotron beams. With x-ray tubes and conventional synchrotron radiation, the x-ray source is largely *incoherent*, that is, a random distribution of phases of the output photons. However, high-energy synchrotron electrons can be made to emit coherently

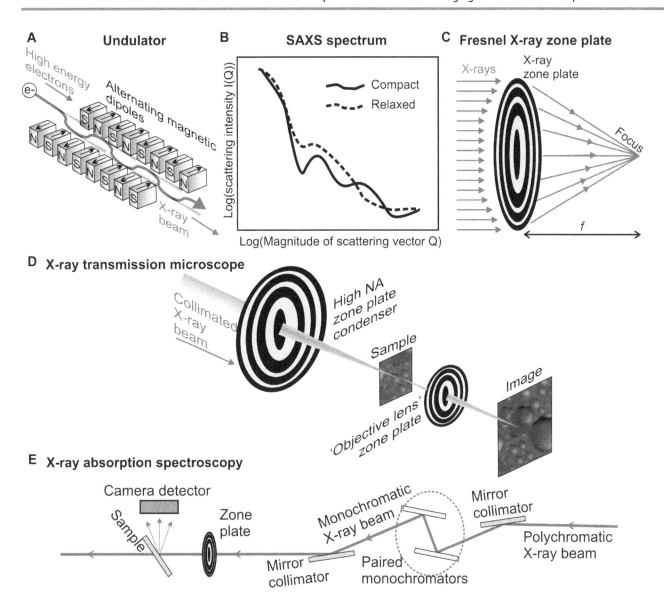

FIGURE 5.3 X-ray applications. (a) Undulator, used in a LINAC, x-ray free-electron laser, or as a module in a synchrotron, which generates a periodic wiggle in the electron beam resulting in an amplified x-ray emission. (b) Schematic of a typical SAXS spectrum of a protein complex in a solution, which allows quantitative discrimination between, for example, two different molecular conformational states. (c) Fresnel zone plate that acts as a "lens" for x-rays and can be used in (d) and x-ray transmission microscope, as well as (e) and x-ray absorption spectrometer.

either if electrons bunch together over a length scale, which is significantly shorter than the wavelength of their emitted radiation bunch that is short with respect to the radiation wavelength, or if the electron density in a given bunch of electrons is modulated with the same frequency as the emitted synchrotron radiation wavelength. For x-rays, it is too challenging currently to directly produce sufficiently small electron bunches; however, electron bunch modulation is now technically feasible and is the basis of the XFEL.

In essence, a linear electron beam is generated using high voltage to give relativistic speeds, either from an output port of a conventional synchrotron or from using a *linear accelerator* (*LINAC*) design. LINACs have a disadvantage over synchrotrons in requiring greater straight-line distances over which to operate (e.g., the Stanford LINAC, which currently operates as the world's only superconducting LINAC, is ~3 km in length), but have an advantage in that less energy from accelerated particles is unavoidably lost as synchrotron radiation. The accelerated electron beam is propagated through an *undulator* consisting of a periodic arrangement of magnets transverse to the beam, with adjacent magnets on each side of the beam arranged with alternating pole geometries (Figure 5.3a) and having a period length parallel to the beam axis of usually a few tens of millimeters, which generate a *B*-field amplitude of ~1 T. This causes a *wiggle* on the electron beam to generate a sinusoidal electron path around the main beam axis such that the high curvature at the peak sinusoidal amplitudes results in the release of synchrotron radiation generated toward the forward beam axis direction, which is highly coherent. However, unlike a visible light laser, there are no equivalent mirror for x-rays, which could be used to generate a resonant cavity (i.e., to reflect the synchrotron radiation back along the undulator thereby amplifying the x-ray laser output) so instead an extended undulator length is used up to a few meters.

The wavelength range of an XFEL currently is $\sim 10^{-10}$ to 10^{-9} m, with an average brightness ~100 times greater than the most advanced synchrotron sources. However, since the magnets in the undulator have a well-defined periodicity, the laser output is pulsed, with a pulse duration of $\sim 10^{-13}$ s, compared to an equivalent pulse duration of $\sim 10^{-11}$ s for a synchrotron source, and so the peak brightness of the XFEL can be several orders of magnitude greater. This ultrashort pulse duration is having a significant impact into conventional x-ray crystallography for determining the structure of biomolecules in reducing the sample radiation damage dramatically—the rapid pulse x-ray beam results in *diffraction before destruction.*

One such application is in *X-ray pump–probe* experiments. Here, ultrashort optical laser pulses are directed onto crystal to generate transient states of matter, which can subsequently be probed by hard x-rays. The fast pump rate of the XFEL (pulse duration of a few tens of femtoseconds) enables time-resolved investigation, that is, more than one shot to be made on the same crystal to monitor rapid structural dynamics. Also, as discussed in the following, there is significant benefit from having a coherent x-ray source in obtaining direct phase information from x-ray scattering atoms in a biomolecule sample.

5.3.2 X-RAY DIFFRACTION BY CRYSTALS

A 3D crystal is composed of a regular, periodic arrangement of several thousand individual molecules. When a beam of x-ray photons propagates through such a crystal, the beam is diffracted due to interference between backscattered x-rays from the different crystal layers. The scattering effect is due primarily to *Thompson elastic scattering*, which results from the interaction of an x-ray photon with a free outer shell *valence* electron, unlike electron scattering, which is from atomic nuclei, and is also influenced mainly by the electron orbital density. The angle of an emergent diffracted x-ray beam is inversely related to the length of separation within the periodic structures involved in the scattering, in exactly the same way that was discussed for electron diffraction, modeled by Bragg's law discussed previously for electron diffraction.

The smallest repeating structure in a crystal is called the "unit cell," and for the simplest crystal shape, which is that of an ideal cubic crystal, the unit cell can be characterized by a crystal lattice parameter a_0 and the interplanar spacings, d_{hkl} of planes, which are labeled by *Miller indices* (*h, k, l*):

(5.10)
$$d_{hkl} = \frac{a_0}{\sqrt{h^2 + k^2 + l^2}}$$

Similar relations for d_{hkl} exist for each different-shaped unit cells in a crystal (e.g., orthorhombic, tetragonal, hexagonal). As an example, for the cubic unit cell, the diffractive intensity maxima generated at angle θ_{hkl} satisfies

(5.11)
$$\theta_{hkl} = \sin^{-1}\left(\frac{\lambda\sqrt{h^2 + k^2 + l^2}}{2a_0} \right)$$

Broadly, there are three practical methods for observing clear diffraction peaks from crystalline samples. Some samples may consist of heterogeneous crystals, that is, are polycrystalline, and the *Debye–Scherrer method* uses a monochromatic source of x-rays, which can determine the distribution of interlayer spacing. The *Laue method* uses instead a polychromatic x-ray source, which produces a range of different diffraction peaks as a function of wavelength that can be used to determine the interlay spacing distribution provided the sample consists of just a single crystal (the combination of a polycrystalline sample with a polychromatic x-ray source generates a diffraction pattern, which is difficult to interpret in terms of the underlying distribution of interlayer spacings). The most useful approach is the *single-crystal monochromatic radiation method*, which generates the most easily interpreted diffraction pattern concerning interlayer spacings.

The intensity of the diffraction pattern can be modeled as the Fourier transform of a function called the "Patterson function," which characterizes the spatial distribution of electron density in the crystal. The pattern of all the scattered rays appears as periodic spots of varying intensity and may be recorded behind the crystal using a CCD. Typically, the crystal will be rotated on a stable mount so that diffraction patterns can be collated from all possible orientations. However, growing a crystal from a given type of biomolecule with minimal imperfections can be technically nontrivial (see Chapter 7). To maximize the effective signal-to-noise ratio of the scattered intensity from a crystal, there is a benefit of growing one large crystal as opposed to multiple smaller ones, and this larger scale is also a benefit due to radiation damage destroying many smaller crystals. In many examples of biomolecules, it is simply not possible to grow stable crystals.

The intensity and spacing of the spots in the diffraction patterns is the 2D projection of the Fourier transform of spatial coordinates of the scattering atoms. The coordinates can be reconstructed using intensive computational analysis, hence, to solve the molecular structure, with a typical resolution being quoted as a few angstroms (which equals 10^{-10} m, useful since it is of a comparable length scale to covalent bonds). However, an essential additional requirement in this analysis is information concerning the phase of scattered rays. For conventional x-ray crystallography, which uses either incoherent x-ray tube or synchrotron radiation, the intensity and position of the maxima in the diffraction pattern alone do not provide this, since there is no x-ray "lens" as such to form a direct image that can be done using visible light wavelengths, for example.

Crystallographers refer to this as the *phase problem*, and this phase information is often then obtained indirectly using a variety of additional methods such as doping the crystals with heavy metals at specific sites, which have known phase relationships. Phase information is normally generated by using iterative computational methods, the most common being the *hybrid input–output algorithm (HIO algorithm)*. Here, a Fourier transformation and an inverse Fourier transformation are iteratively applied to shift between real space and reciprocal space under specific boundary conditions in each. This approach is also coupled to oversampling by sampling the diffraction intensities in each dimension of reciprocal space at an interval of at least twice as fine as the *Bragg peak frequency* (the highest spatial frequency detected for a diffraction peak in reciprocal space). For the real space part of structural refinement, molecular dynamics and structural modeling/validation are also widely used (see Chapter 8).

X-ray crystallography has been at the heart of the development of modern biophysics. For example, the first biomolecule structure solved was that of cholesterol as early as 1937 by Dorothy Hodgkin, and the first protein structures solved were myoglobin in 1958 (John Kendrew and others) followed soon after by hemoglobin in 1959 (Max Perutz and others). There are important weaknesses to the method, which should be noted, however. A key disadvantage of the technique, as with all techniques of diffraction, is that it requires an often artificially tightly packed spatial ordering of molecules, which is intrinsically nonphysiological. In addition, the approach is reliant upon being able to manufacture highly pure crystals, which are often relatively large (typically a few tenths of a millimeters long, containing $\sim 10^{15}$ molecules), which thus limits the real molecular heterogeneity that can be examined since the diffraction information obtained relates to mean ensemble interference properties from a given single crystal. In some cases, smaller crystals approaching a few microns of length scale can be generated.

Also, the crystal-manufacturing process is technically nontrivial, and many important biomolecules, which are integrated into cell membranes, are difficult, if not impossible, to crystallize due to the requirements of added solvating detergents affecting the process. In addition, since the scattering is due to interaction with regions of high electron density, the positions of hydrogen atoms in a structure cannot be observed by this method directly since the electron density is too low, but rather, need to be inferred from knowledge of typical bond lengths. Just as important, however, is the lack of real time-resolved information of a biological structure—a crystal is very much a locked state. Since dynamics are essential to biological processes, this is a significant disadvantage, although efforts can be made to infer dynamics by investigating a variety of different locked intermediate states using crystallographic methods. Diffusing x-ray scattering from amorphous samples (see in the following text) can circumvent some of the issues encountered earlier regarding the use of crystals, since they can reveal some information about protein dynamics, albeit under nonphysiological conditions.

5.3.3 X-RAY DIFFRACTION BY NONCRYSTALLINE SAMPLES

X-ray diffraction can also be performed on a powder if it is not possible to grow sufficiently large 3D crystals. A suitable powder is not entirely amorphous but is composed of multiple small individual crystals with a random orientation. Therefore, all possible Bragg diffractions can be exhibited in the powder pattern. However, the relative positions and intensities of peaks in the observed diffraction pattern can be used to estimate the interplanar spacings. Similarly, biological fibers can be subjected to x-ray diffraction measurements if there is sufficient spatial periodicity. For example, muscle fibers have repeating subunits arranged periodically in one dimension parallel to the long axis of the fiber. This approach was also used to great effect in solving the double-helical structure of DNA from the work of Crick, Franklin, Wilkins, and Watson in 1953.

In *small-angle x-ray scattering* (SAXS), a 3D crystalline sample is not needed, and the technique is particularly useful for exploring the longer scale periodic features encountered in many biological fibers. The range of scattered angles explored is small (typically <10°) with a typical spatial resolution of ~1–25 nm. It is used to infer the spacing of relatively large-scale structures up to ~150 nm (e.g., to study periodic features in muscle fibers). The scatter signal is relatively weak compared to higher angle scattering methods of x-ray crystallography and so a strong synchrotron beamline is generally used. SAXS does not generate atomistic-level structural information like x-ray crystallography or NMR, but it can determine structures, which are coarser grained by an order of magnitude in a matter of days for biological structures, which span a much wider range of size and mass.

SAXS is performed using an x-ray wavelength of ~0.15 nm, directing the beam to a solution of the biomolecular structure, and the emergent scatter angle θ and beam intensity I are recorded. The magnitude of the scattering vector $Q = (4\pi/k)\sin(\theta/2)$, the formulation identical to that discussed for static light scattering previously (see Chapter 4), is normally plotted as a function of I (Figure 5.3b) and the position and sizes and the typically broad peaks in this curve are used to infer the size and extent of spatial periodicity values from the sample. The same level of analysis for determining radius of gyration can also be performed for static light scattering, also including information about the coarse shape of periodic scattering objects in the sample, but SAXS also has sufficiently high spatial resolution to investigate different molecular states of the same complexes, for example, to be able to discriminate between different conformational states of the same enzyme provided the whole sample solution is sufficiently synchronized. And, being in solution, it also offers significant potential for monitoring time-resolved changes to molecular structure, which 3D x-ray crystallography cannot. The use of coherent x-rays as available from XFEL can generate the speckled interference patterns from SAXS investigations, which can be used to generate phase information directly from the sample in much the same way as for XFEL on 2D crystal arrays.

SAXS, like 3D x-ray crystallography, utilizes elastic x-ray photon scattering. Inelastic scattering is also possible, for which the wavelength of the emergent x-rays is greater than the incident beam (i.e., the scattered beam has a lower energy). Here, some portion of the incident photon energy is transferred from the beam to energize a process in the sample, for example, to excite an inner shell electron to a higher energy level. This is not directly useful in determining atomic-level structures but has been utilized in the form of *resonant inelastic soft x-ray scattering (RIXS)*, which can be applied to a solution of biomolecules in the same way as SAXS.

However, since RIXS is often associated with changes to the energy state of atomic electrons, it is often used in biophysical investigations that involve changes to the oxidation state of transition metal atoms in electron-carrier enzymes, for example, those used in oxidative phosphorylation and photosynthesis (see Chapter 2) but has also been applied to biological questions including solvation effects in chemoreceptors and studying the dynamics of phospholipid bilayers.

5.3.4 X-RAY MICROSCOPY METHODS

X-ray microscopy methods have been developed both for transmission and scanning modes similar to the principles of EM and optical microscopy. However, the principal challenge is how to focus x-rays, since no equivalent lens as such exists as for the transparent glass lenses of optical microscopy or the electromagnetic/electrostatic lenses of EM. The solution is to use *zone plates* (Figure 5.3c), also known as *Fresnel zone plates*, which utilize diffraction for focusing instead of reflection or refraction.

Zone plates are micro- or nanofabricated concentric ring structures known as Fresnel zones, which alternate between being opaque and transparent. They can be used for focusing across the electromagnetic spectrum, and in fact for any general waveform such as sounds waves but are particularly valuable for x-ray focusing. X-rays hitting the zone plate will diffract around the opaque zones. The zone spacing between the rings is configured to allow diffracted light to constructively interfere only at a desired focus. The condition for this is

$$(5.12) \qquad r_n = \sqrt{n\lambda f - \frac{n^2\lambda^2}{4}}$$

where

  r_n is the radius of the switch position between the nth opaque and transparent zones from the center of the zone plate, such that n is a positive integer
  f is the effective focal length of the zone plate

Analogous to the diffraction resolution limit in optical microscopy (Chapter 4), the smallest resolvable object feature length Δx when using a zone plate limit is given by

$$(5.13) \qquad \Delta x = 1.22\Delta r_n$$

Therefore, the resolution limit is really determined by the precision of the micro-/nanofabrication. At the time of writing, the current reliable limit is ~12 nm.

Typical designs for a *transmission x-ray microscope (TXM)* and a *scanning transmission x-ray microscope (STXM)* are shown in Figure 5.3d. "Soft" x-rays are used typically from a collimated synchrotron source, of wavelength ~10–20 nm. The TXM uses two zone plates as equivalent condenser and objective "lenses" to form a 2D image on a camera detector, whereas the STXM typically utilizes just a single zone plate to focus the x-ray beam onto a sample. As a robust biophysical technique, x-ray microscopy is still in its infancy, but it has been tested on single-cell samples.

An alternative to using physical focusing methods of x-rays with zone plates is to perform numerical focusing through similar techniques of *coherent x-ray diffraction imaging (CXDI or CDI)* and ptychography (which was discussed previously as part of optical microscopy techniques in Chapter 4). CXDI involves a highly coherent incident beam of synchrotron x-rays, which scatter from the sample and generate a diffraction pattern, which is recorded by a camera. This raw diffraction pattern is used to reconstruct the image of the sample through a Fourier transform on the intensity data combined with computational iterative phase recovery algorithms to recover the phase information due to the lack of sufficient coherence used in synchrotron radiation. In effect, a computer performs the job of an equivalent objective lens to convert reciprocal space data into a real space image. The main advantage of CXDI is that it does not require lenses to focus the beam so that the measurements are not affected by aberrations in the zone plates but rather is only limited by diffraction and the x-ray intensity. Although not yet a mainstream biophysical technique, the superior penetration power of x-rays combined with their small wavelength and thus high spatial resolution has realistic potential for future studies of complex biological samples (see Thibault et al., 2008). A future potential for these techniques lies in time-resolved x-ray imaging.

5.3.5 X-RAY SPECTROSCOPY

An incident x-ray photon can have sufficient energy to eject a core electron through the photoelectric effect, resulting in the appearance of significant absorption edges in the spectra of transmitted photons through the sample, which correspond to the binding energies for an electron in different respective shells (K, L, M, etc.). This subatomic process can involve subsequent fluorescence emission analogous to that exhibited in light microscopy (Chapter 3); if an excited electron undergoes vibrational losses prior to returning to its ground state, it results in radiative *x-ray fluorescence* emission of a photon of slightly longer wavelength than the incident photon. Also, when the ejection of the core inner shell electrons occurs, it results in higher energy outer shell electrons dropping to these lower energy vacant states with a resultant radiative emission of a secondary x-ray photon whose energy is the difference between the binding energies of the two electronic levels. The position and intensity of these absorption and emission peak as a function of photon wavelength, constituting a unique fingerprint for the host atom in question, and thus, *x-ray absorption spectroscopy* (*XAS*) (also known variously as very similar/identical techniques of *energy-dispersive x-ray spectroscopy*, *energy-dispersive x-ray analysis*, and simply *x-ray spectroscopy*) is a useful biophysical tool for determining the makeup of individual elements in a sample, that is, performing *elemental analysis*.

X-ray absorption spectra of relevance to biological questions can be categorized into *x-ray absorption near edge structure*, which generates data concerning the electronic "oxidation state" of an atom and the spatial geometry of its molecular orbitals, and *extended x-ray absorption fine structure*, which generates information about the local environment of a metal atom's binding sites (for an accessible review, see Ortega et al., 2012). The penetration of lower energy secondary x-rays (wavelengths >1 nm) through air is significantly worse than those of higher energy secondary x-rays (wavelength <1 nm). This characteristic wavelength for K-line transitions varies as $\sim(Z-1)^2$ as predicated by Moseley's law, and the ~1 nm cutoff occurs at around $Z = 12$ for magnesium. Thus, most metals generate detectable secondary x-rays, which facilitate metal elemental analysis. Of special relevance are metal-binding proteins, or *metalloproteins*, and XAS can probe details such as the type of neighboring atoms, how many bonds are formed between them, over what distances, and others. This is a particularly attractive feature of the technique, since proteins containing metal ions actually constitute more than one-third of all known proteins.

A schematic of a typical setup is shown in Figure 5.3e, utilizing a polychromatic synchrotron x-ray source, which generates a suitably intense and collimated beam required for XAS. Normally, *hard x-rays* are used, with a monochromator then utilized to scan through a typical wavelength range of ~0.6–6 nm. Samples, which can include cultures of cells but more typically consist of high concentrations (~0.5 mM) of protein, need to be cryofixed to a glassy frozen state to stabilize thermal disorder and minimize sample radiation damage. But measurements can at least be performed in a hydrated environment, which increases its physiological relevance.

A standard XAS investigation measures the absorption coefficient as a function of incident wavelength, characterized by the simple Beer–Lambert law (see Chapter 3) from measuring the transmission of x-rays through the sample. However, this transmission mode has too low a sensitivity for the often meager concentration of metals found in many biological materials, and in this instance, x-ray fluorescence emission is a better metric, with the detector position at 90° from the incident beam. Detectors are typically based on doped semiconductor designs such that the absorption of an x-ray photon at a *p–i–n* junction of PIN diodes (where *i* is an insulating layer between positive *p* and negative *n* doped regions) creates a hotspot of electron–hole pairs, which can be detected as a voltage pulse.

X-ray photoelectron spectroscopy (*XPS*) is an alternative technique to XAS. A competing mechanism to X-fluorescence following absorption of an x-ray photon by an atom is the emission of a so-called Auger electron—the term *Auger electron spectroscopy* is synonymous with XPS, and often the technique is abbreviated simply to *electron spectroscopy*. Here, low-energy x-rays, either from an x-ray tube or synchrotron source, are used to stimulate the photoelectric effect in sample atoms, and these photoelectrons are detected directly by a high-resolution electron spectrometer, and electron intensity is determined as a function of energy. The penetration distance of photoelectrons is ~10 nm in a sample, and so XPS renders surface information from a sample, in addition to requiring high-vacuum conditions between the sample and detector. XPS is less sensitive than XAS with therefore more limited application, but as a tool potentially offers advantages over XAS in being able to utilize x-ray tube sources as opposed to requiring access to a synchrotron facility. The temporal resolution of XPS is in femtoseconds, which is ideal for probing electronic resonance effects in complex biomolecules; for example, this has been applied to investigating different forms of chlorophyll (see Chapter 9), which is the key molecule that absorbs photons coupled to the generation of high-energy electrons in the process of photosynthesis in plants and several other unicellular organisms (see Chapter 2).

In principle, it offers a similar elemental signature, sensitive enough to detect and discriminate between the energies of the photoelectric emissions from all atomic nuclei with an atomic number Z of at least 3 (i.e., lithium and above). A limitation for probing biological material is that the sample must be in a vacuum to minimize scatter of the emitted electrons; however, it is possible to keep many samples in a cold, glassy, hydrated state just up the point at which XPS is performed, before which ice sublimes off at the ultralow pressures used. XPS has been applied to quantify the affinity and geometry of metal binding in protein complexes and larger scale biological structures such as collagen fibers but is also used in elemental analysis on wood/plant matter and teeth (e.g., in *bioarcheology* investigations).

5.3.6 RADIATION DAMAGE OF BIOLOGICAL SAMPLES BY X-RAYS AND WAYS ON HOW TO MINIMIZE IT

A significant limitation to the use of x-ray photon probes in biological material is the high likelihood of stochastic damage to the sample. X-ray–associated radiation damage is primarily due to the photoelectric effect. As we have seen, the initial absorption event of an x-ray photon by an atom can result in the complete ejection of an inner shell electron. The resulting atomic orbital vacancy is filled by an outer shell electron. For high-atomic-number elements, including many metals, there is a significant likelihood of subsequent x-ray fluorescence, however, for low Z elements, many of which are biologically, highly relevant such as C, N, and O, but also S and P; the electron ejection energy is transmitted to an outer shell electron, which is ejected as an Auger electron in a process, which takes $\sim 10^{-14}$ s.

This photoelectric effect can then lead to secondary electron ionization in other nearby atoms by electron-impact ionization, resulting in the formation of chemically highly reactive free radicals. It is these free radicals that cause significant damage through indiscriminate binding to biological structures. Cooling a sample can minimize this damage simply by reducing the rate of diffusion of a free radical in the sample, and it is common to cool protein crystals in x-ray crystallography with liquid nitrogen to facilitate longer data acquisition periods.

Use of smaller crystals (e.g., down to a length scale of a few tenths of microns) also reduces the effect of x-ray radiation damage. This is because the loss of photoelectrons from a crystal scales with its surface area, whereas the number of photoelectrons produced scales with its volume. Thus, the relative probability of photoelectron-related damage scales with the effective crystal diameter. However, using small crystals reduces the x-ray diffraction signal, which reduces the effective spatial resolution of the biomolecular structure determination, but also, results in inhomogeneity in the crystal (see Chapter 7) having a more pronounced detrimental effect on the diffraction pattern relative to the signal due to homogeneous regions of the crystal.

Another strategy to reduce x-ray damage is the use of *microbeams*. Synchrotron sources have highly collimated beams, with typical diameters of a few hundreds of microns. However, the small beam divergence of ~µrad allows much narrower beams to be generated, to as low as ~1 µm. That can be employed as a much finer probe for x-ray crystallography (Schneider, 2008), reducing the effective diffraction volume in the sample exposed to the beam to just ~20 µm³. Reducing the sample volume illuminated by x-rays substantially reduces radiation damage. Also, it allows x-ray crystallography to be performed on much smaller crystals, which significantly reduces the bottleneck of requiring large and perfect crystals.

Also, the emergence of very intense, coherent x-rays from XFEL sources has allowed much shorter duration pulses for crystallography. This again reduces radiation damage to the sample and similarly permits much smaller samples to be used. As opposed to a perfect 3D crystal, 3D structural determination is now possible using x-ray diffraction from a coherent XFEL source using just a monolayer of protein generated on a surface.

KEY BIOLOGICAL APPLICATIONS: X-RAYS

Determining atomic-level precise molecular structures from crystals; Estimating elemental composition of biological samples.

5.4 NMR AND OTHER RADIO FREQUENCY AND MICROWAVE RESONANCE SPECTROSCOPIES

NMR is a powerful technique utilizing the principle that magnetic atomic nuclei will undergo resonance by absorbing and emitting electromagnetic radiation in the presence of a strong external magnetic field. The resonance frequency is a function of the type of atom undergoing resonance and of the strong external magnetic field but is also dependent on the smaller local magnetic field determined by the immediate physical and chemical environment of the atom. Each magnetic atomic nucleus in a sample potentially contributes a different relative shift in the resonance frequency, also known as the *chemical shift*, hence, the term *NMR spectroscopy*, in being a technique capable of acquiring the spectra of such chemical shifts. Put in simple terms, the spatial dependence on the chemical shift can be used to reconstruct the physical positions of atoms in a molecular structure. Other related radiowave resonance techniques include *electron spin resonance* (*ESR*) and *electron paramagnetic resonance* (*EPR*), which operate on resonance behavior in the electron cloud around atoms as opposed to their nuclei.

5.4.1 PRINCIPLES OF NMR

To have a magnetic nucleus implies a nonzero spin angular momentum. The *standard model* of particle physics proposes that atomic nuclei contain strong forces of interaction known as the *tensor interaction*, which allows neutrons and protons to be paired in an atomic nucleus in a quantum superposition of angular momentum states. These interactions can be modeled by the quantum field theory of *quantum chromodynamics*, which bind together two down (each of the charge $-e/3$ with paired 1/2 spins, where e is the magnitude of the electron charge) and one up quark (of charge $+e/3$ and 1/2 spin) in a neutron, while a proton contains one down and two up quarks. This implies that both the neutron and proton are spin-1/2 particles. Therefore, all stable isotopes whose atomic nuclei possess an odd total atomic mass number (i.e., the number of protons plus neutrons) are magnetic (and if the atomic number minus the neutron number is ±1 as is commonly the case for many stable isotopes the result are *spin-1/2 nuclei*).

The most common isotopes used for biological samples are ^1H and ^{13}C (see Table 5.2). ^1H is the most sensitive stable isotope, whereas ^{13}C has relatively low natural abundance

TABLE 5.2   Nuclear Magnetic Spin Properties of Common Half-Integer Spin Nuclei Isotopes Used in NMR (Bold) Compared Against Zero or Integer Spin Atomic Nuclei (Not Bold)

Proton Number (Z)	Neutron Number (N)	Isotope	Nuclear Spin Quantum Number (I)	Natural Abundance (%)	Magnetogyric Ratio/2π (MHz T⁻¹)	Resonance Frequency if ¹H is 400 MHz (MHz)
1 (odd)	**0 (—)**	**¹H**	**1/2 (half integer)**	**99.998**	**42.6**	**400**
1 (odd)	1 (odd)	²H (D)	1 (integer)	0.002	6.5	61.4
6 (even	6 (even)	12C	0 (zero spin)	98.89	0	—
6 (even)	**7 (odd)**	**¹³C**	**1/2 (half integer)**	**1.07**	**10.7**	**100.6**
7 (odd)	7 (odd)	¹⁴N	1 (integer)	99.63	3.1	28.9
7 (odd)	**8 (even)**	**¹⁵N**	**1/2 (half integer)**	**0.37**	**−4.4**	**40.5**
8 (even)	8 (even)	¹⁶O	0 (zero spin)	99.757	0	—
9 (odd)	**10 (even)**	**¹⁹F**	**1/2 (half integer)**	**100**	**40.1**	**376.5**
15 (odd)	**16 (even)**	**³¹P**	**1/2 (half integer)**	**100**	**17.2**	**162.1**

compared to the nonmagnetic ^{12}C and also a low sensitivity. Since carbon is a key component of all organic compounds, it is widely used in NMR, but the ^{13}C isotope has to be included in the sample preparation process due to its low natural abundance. Other lesser used isotopes include ^{15}N (which has low sensitivity but is used since nitrogen is a key component in proteins and nucleic acids), ^{19}F (which has a high sensitivity, is rarely present in natural organic compounds and thus, needs to be chemically bound into the sample in advance), and ^{31}P (which has a moderate sensitivity, and phosphorous is a key element of many biological chemicals).

For a *nuclear spin quantum number* of I, the *nuclear angular momentum L* is given by

$$(5.14) \qquad L = \frac{h}{2\pi}\sqrt{I(I+1)}$$

where h is the Planck's constant. The magnetic moment has discrete directionality such that the angular momentum parallel to an arbitrary z-axis L_z is given by

$$(5.15) \qquad L_z = \frac{hm}{2\pi}$$

where the *magnetic quantum number, m*, is allowed to take a total of $(2I + 1)$ different values of $-I, -I + 1, I - 1, +I$. In the absence of an external magnetic field, all the different orientation states have the same energy, for example, they are *degenerate*. The different spin states of the nuclei have a different magnetic moment μ whose magnitude is given by

$$(5.16) \qquad \begin{aligned} \mu &= \gamma L \\ \therefore \mu_z &= \frac{\gamma h m}{2\pi} \end{aligned}$$

where
  μ_z is the z component of μ
  γ is a constant called the magnetogyric ratio (also known as the *gyromagnetic ratio*)

Typical values of γ are equivalent to $\sim 10^7$ T^{-1} s^{-1} but are often quoted as these values divided by 2π and are given for a few atomic nuclei in Table 5.2. The bulk magnetization M of a sample is the sum of all the atomic nuclear magnetic moments, which average out to zero in the absence of an external magnetic field.

However, in the presence of an external magnetic field, there is a nonzero net magnetization, and each atomic nuclear magnetic state will also have a different energy E due to the coupling interaction between the B-field and the magnetic moment (also known as the *Zeeman interaction*), which is given by the dot product of the external magnetic field B with the atomic nucleus magnetic moment:

$$(5.17) \qquad E_m = -\vec{\mu} \cdot \vec{B} = -\mu_z B_z = -\frac{\gamma B_z h m}{2\pi}$$

Therefore, the presence of an external magnetic field splits the energy into $(2I + 1)$ discrete energy levels (*Zeeman levels*), a process known as *Zeeman splitting*, with the lower energy levels resulting from the alignment of atomic nuclear magnetic moment with the external B-field and higher energies with alignment against the B-field. The transition energy between each level is given by

$$(5.18) \qquad \Delta E = -\frac{\gamma B_z h}{2\pi}$$

If a photon of electromagnetic energy $h\nu$ matches ΔE, it can be absorbed to excite a nuclear magnetic energy level transition from a lower to a higher state; similarly, a higher energy state can drop to a lower level with consequent photon emission, with quantum selection rules permitting $\Delta m = \pm 1$, which indicates $2I$ possible reversible transitions. An absorbed photon of frequency ν can thus result in a resonance between the different spin energy states. This resonance frequency is also known as the *Larmor frequency* and is identical to the classically calculated frequency of precession of an atomic nucleus magnetic moment around the axis of the external B-field vector.

The value of ν depends on γ and on B (in most research laboratories, B is in the range of ~ 1–24 T, $\sim 10^6$ times the strength of Earth's magnetic field), but is typically $\sim 10^8$ Hz, and it is common to compare the resonance frequencies of different atomic nuclei under standard reference conditions in relation to a B-field, which would generate a resonance frequency of 400 MHz for ^{1}H ($B \sim 9.4$ T), some examples of which are shown in Table 5.1. For magnetic atomic nuclei, these are radio frequencies. For example, the resonance frequency of ^{13}C is very close to that of a common FM transmission frequency of ~ 94 MHz for New York Public Radio. A typical value of ΔE, for example, for ^{1}H in a "400 MHz NMR machine" (i.e., $B \sim 9.4$ T) is $\sim 3 \times 10^{-25}$ J. Experiments in such machines are often performed at ~ 4 K, and so $k_B T/\Delta E \sim 180$, hence, still a significant proportion of occupied lower energy states at thermal equilibrium.

The occupational probability p_m of the mth state (see Worked Case Example 5.2) is given by the normalized Boltzmann probability of

$$(5.19) \qquad p_m = \frac{\exp\left[-E_m/k_B T\right]}{\sum_{all\,m}\exp\left[-E_m/k_g T\right]} = \frac{\exp\left[\gamma B_z h m/2\pi k_B T\right]}{\sum_{all\,m}\exp\left[\gamma B_z h m/2\pi k_B T\right]}$$

The *relative* occupancy N of the different energy levels can be predicted from the Boltzmann distribution:

$$(5.20) \qquad \frac{N_{m=1}}{N_{m=I+1}} = \exp\left[\frac{-\Delta E}{k_B T}\right] = \exp\left[\frac{-\gamma B h}{2\pi k_B T}\right]$$

where

k_B is the Boltzmann constant

T is the absolute temperature

For spin-1/2 nuclei, the only photon absorption transition is thus $-1/2 \rightarrow +1/2$ (which involves *spin-flip* in going from a spin-down to a spin-up orientation). For higher-spin half-integer nuclei (e.g., ^{23}Na is a 3/2-spin nucleus), other transitions are possible; however, the $-1/2 \rightarrow +1/2$ transition, called the central transition, is most likely, whereas other transitions, known as *satellite transitions*, are less likely.

5.4.2 NMR CHEMICAL SHIFT

However, all atomic nuclei in a sample will not have exactly the same differences in spin energy states because there is a small shielding effect from the surrounding electrons, which causes subtle differences to the absolute level of the external magnetic field sensed in the nucleus. These differences are related to the physical probability distribution of the local electron cloud, which in turn is a manifestation of the local chemical environment. In other words, this shift in the resonance frequency, the chemical shift (δ), can be used to infer the chemical structure of the sample. The resulting B-field magnitude B' at the nucleus can be described in terms of a shielding constant σ:

$$(5.21) \qquad\qquad B' = (1 - \sigma) B$$

In practice, however, NMR measurements rarely refer to σ directly. Chemical shifts are typically in the range of a few parts per million ("ppm") of the nonshifted resonance frequency (so in absolute terms will correspond to a shift of $\sim$1–20 kHz):

$$(5.22) \qquad\qquad \delta = 10^6 \left(\frac{v_{samples} - v_{references}}{v_{references}} \right)$$

An NMR spectrum consists of a plot of (radio frequency) electromagnetic radiation absorption intensity in arbitrary units on the vertical axis as a function of δ in units of ppm on the horizontal axis, thus generating a series of distinct peaks of differing amplitudes, which correspond to a sample's molecular fingerprint, often called the fine structure. The most common form of NMR is performed on samples in the liquid state, and here, the chemical shift is affected by the type of solvent, so is always referred to against a standard reference.

For ^{1}H and ^{13}C NMR, the reference solvent is often *tetramethylsilane* (*TMS*) of chemical formula $Si(CH_3)_4$, though in specific NMR spectroscopy on protein samples, it is common to use the solvent *DSS* (2,2-dimethyl-2-silapentane-5-sulfonic acid). Thus, it is possible to generate both negative (*downfield shift*) and positive (*upfield shift*) values of δ, depending upon whether there is less or more nuclear screening, respectively, in the specific reference solvent. It is also common to use deuterated solvent (i.e., solvents in which ^{1}H atoms have been exchanged for ^{2}H or *deuterium*, D, usually by exchanging $\sim$99% of ^{1}H atoms, which leaves sufficient remaining to generate a detectable proton NMR reference peak) since most atomic nuclei in a solution actually belong to the solvent. The most common deuterated solvent is *deuterochloroform* ($CDCl_3$). This is a strongly hydrophobic solvent. For hydrophilic samples, *deuterated water* (D_2O) or *dimethyl sulfoxide* (*DMSO*), $(CD_3)_2SO$, are often used as an alternative.

In principle, there is an orientation dependence on the chemical shift. The strength of the shielding interaction varies in the same way as the magnetic dipolar coupling constant, which has a $(3\cos^2 \theta - 1)$ dependence where θ is the angle between the atomic nuclear magnetic dipole axis and the external B-field. However, in liquid-state NMR, more commonly applied in biophysical investigations than solid-state NMR, molecular reorientation averages out this anisotropic effect.

5.4.3 OTHER NMR ENERGY COUPLING PROCESSES

The overall NMR Hamiltonian function includes the sum of several independent Hamiltonian functions for not only the Zeeman interaction and chemical shift coupling, but also terms relating to other energy coupling factors. There are *spin–spin coupling*, which includes both *dipolar coupling* (also known as *magnetic dipole–dipole interactions*) and *J-coupling* (also known as *scalar coupling* or *indirect dipole–dipole coupling*). And there is also a nuclear *E*-field coupling called "quadrupolar coupling."

In dipolar coupling, the energy state of a nuclear magnetic dipole is affected by the magnetic field generated by the spin of other nearby magnetic atomic nuclei, since over short distances comparable to typical covalent bond lengths (but dropping off rapidly with distance r between nuclei with a $1/r^3$ dependence), nuclei experience the B-field generated from each other's spin in addition to the external (and in general shielded) magnetic field. This coupling is proportional to the product of the two associated magnetogyric ratios (whether from the same or different atoms) and can result in additional splitting of the chemical shift values depending on the nearby presence of other nuclei.

Several magnetic atomic nuclei used in NMR are not spin-1/2 nuclei, and in these cases, the charge distribution in each nucleus may be nonuniform, which results in an electrical quadrupole moment, though these have a limited application in biophysics. An *electrical quadrupole moment* may experience the *E*-field of another nearby electrical quadrupole moment, resulting in quadrupolar coupling. In liquid-state NMR, however, since molecular motions are relatively unconstrained, molecular reorientation averages out any fixed shift on resonance frequency due to dipolar or quadrupolar coupling but can result in broadening of the chemical shift peaks.

However, in solid-state NMR, and also NMR performed in solution but on liquid crystals, molecular reorientation cannot occur. Although liquid-state/solution NMR has the most utility in biophysics, solid-state NMR is useful for studying biomineral composites (e.g., bone, teeth, shells) and a variety of large membrane protein complexes (e.g., transmembrane chemoreceptors and various membrane-associated enzymes) and disease-related aggregates of proteins (e.g., *amyloid fibrils* that form in the brains of many patients suffering with various forms of *dementia*), which are inaccessible either with solution NMR or with x-ray diffraction methods. Solid-state NMR results in peak broadening and shifting of mean energy levels in an anisotropic manner, equivalent to ~10 ppm for dipolar coupling, but as high as ~10^4 ppm in the case of quadrupolar coupling. There are also significant anisotropic effects to the chemical shift. To a certain extent, anisotropic coupling interactions can be suppressed by inducing rotation of the solid sample around an axis of angle ~54.7°, known as the "magic angle" relative to the external B-field, in a process known as "magic-angle spinning" requiring a specialized rotating sample stage, which satisfies the conditions of zero angular dependence since $(3\cos^2\theta - 1) = 0$.

In liquid-state NMR, the most significant coupling interaction in addition to the Zeeman effect and the chemical shift is J-coupling. J-coupling is mediated through the covalent bond linking the atoms associated with two magnetic nuclei, arising from *hyperfine* interactions between the nuclei and the bonding electrons. This results in hyperfine structure of the NMR spectrum splitting a single chemical shift peak into multiple peaks separated by a typical amount of ~0.1 ppm given by the J-coupling constant. The multiplicity of splitting of a chemical shift peak is given by the number of equivalent magnetic nuclei in *neighboring* atoms n plus one, that is, the *n + 1 rule*.

The example of this rule often quoted is that of the ^{1}H NMR spectrum of ethanol (Figure 5.4a), which illustrates several useful features of NMR spectra. Carbon atom 1 (C1), part of a methyl group, is covalently bound to C2, which in turn is bound to two ^{1}H atoms, and the nucleus (a proton) of each has one of two possible orientations (parallel, p, or antiparallel, a, to the external B-field), indicating a total of 2^2 of 4 possible spin combinations (p–p, p–a, a–p, a–a) with a proton of an ^{1}H atom bound to C1. Low and high energy states are p–p and a–a, respectively, but p–a and a–p are energetically identical; therefore, the single chemical shift peak for the C1 protons is split into a triplet with the central peak amplitude higher by a factor of 2 compared to the smaller peaks due to the summed states of p–a and a–p together, so the amplitude ratio is 1:2:1. Similarly, the C1 atom is covalently bound to three ^{1}H atoms, which results in 2^3 or eight possible spin combinations with one of the protons of the ^{1}H atom bound to C2, which can be grouped into four energetically identical combinations as follows (from low to high energy states):

$$\{p-p-p\}_1, \{p-p-a, p-a-p, a-p-p\}_2, \{a-a-p, a-p-a, p-a-a\}_3, \{a-a-a\}_4$$

Thus, the single chemical shift peak for the C2 protons is split into a quartet of relative amplitude 1:3:3:1. The ratio of amplitudes in general is the $(n + 1)$th level of *Pascal's triangle* (so a quintet multiplicity would have relative amplitudes of 1:4:6:4:1). Note also that J-coupling can also be detected through H-bonds, indicating some covalent character of hydrogen bonding at least. Observant readers might note from Figure 5.4a that there appears to be only a single peak corresponding to the ^{1}H atom attached to the O atom of the –OH group, whereas from the simple logic earlier, one might expect a triplet O is covalently bonded to C2. However, the effects of J-coupling in this instance are largely lost and are similar for all ^{1}H atoms in general, which are bound to *heteroatoms* (specifically, –OH and –NH groups) due to a rapid chemical transfer of a proton H$^+$, allowing it to exchange with another proton from OH or NH in aqueous solution, which can occur even with tiny traces of water in a sample. This exchange process results in line broadening of all peaks in the hypothetical triplet, resulting in the appearance of a single broad peak.

5.4.4 NUCLEAR RELAXATION

The transition from a high to low magnetic nuclear spin energy state in general is not a radiative process since the probability of spontaneous photon reemission varies as v^3, which is insignificant at radio frequencies. The two major processes, which affect the lifetime of an excited state, are spin–lattice relaxation and spin–spin relaxation, which are important in practical NMR spectroscopy since the absence of relaxation mechanisms would imply rapid saturation to high energy states and thus a small equivalent resonance absorption signal in a given sample.

Spin–spin relaxation (also known as "transverse relaxation") involves coupling between nearby magnetic atomic nuclei, which have the same resonance frequency but which differ in their magnetic quantum numbers. It is also known as the "nuclear Overhauser effect" and, unlike J-coupling, is a free-space interaction not mediated through chemical bonds. A transition can occur in which the two magnetic quantum numbers are exchanged. There is therefore no change in the occupancy of energy states; however, this does decrease the "on" time probability of the excited state since an exchange in magnetic quantum number is equivalent to a transient misalignment of the magnetic dipole with the external B-field, which also broadens absorption peaks. Transient misalignment can also be caused by inhomogeneity in the B-field. The mean relaxation time associated with this process is denoted as T_2. Solids can have T_2 of a few milliseconds, while liquids more typically tens to hundreds of milliseconds.

Spin–lattice relaxation (also known as "longitudinal relaxation") is due to a coupling between a spinning magnetic atomic nucleus and its surrounding lattice, for example, to collisions between mobile sample molecules and the solvent. This results in energy loss from the magnetic spin state and a consequent rise in temperature of the lattice. The mean relaxation time taken to return from an excited state back to the thermal equilibrium state is denoted T_1. In general, $T_1 > T_2$, such that in normal nonviscous solvents at room temperature T_1 ranges from ~0.1 to 20 s.

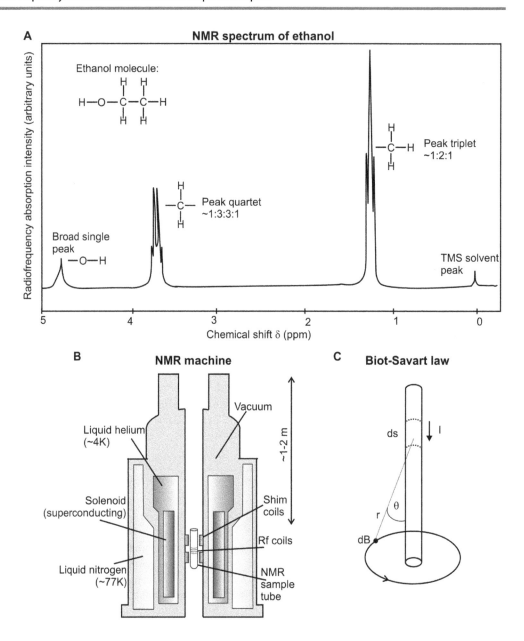

FIGURE 5.4 NMR spectroscopy. (a) Schematic of an NMR spectrum taken on with a "400 MHz" NMR machine in TMS solvent. (b) Schematic of a typical research NMR machine with double-skin Dewar. (c) Biot–Savart law: the contribution **dB** to the total circular **B**-field around an electrically conducting wire carrying current **I** can be calculated from the incremental element **ds** along the wire length of the length.

5.4.5 NMR IN PRACTICE

NMR often requires isotope enrichment, which can be technically rate limiting for investigations, in addition to typically tens of milligrams of purified sample, which can present a significant challenge in the case of many biomolecules, with liquid-state NMR requiring this to be dissolved in a few hundred microliters of pH-buffered solvent to produce high millimolar concentrations. To achieve the high B-fields required for NMR, with several machines capable of generating >20 T, magnets are based on a solenoid coil design. Early NMR magnets had an iron core and could generate fields strength up to ~5 T; however, most modern NMR machines, which can achieve maximum field strengths of ~6–24 T, utilize a superconducting solenoid, and some prototype machines using larger solenoids can operate in shorter pulses and generate higher transient B-fields, for example, up to ~100 T for millisecond-duration pulses.

Superconducting NMR solenoids of a coil length of ~100 km are composed of superconducting wire made usually from an alloy of *niobium* with *tin* and *titanium*, for example, $(NbTaTi)_3Sn$, which is embedded in copper for mechanical stability and cooled to ~4 K using a liquid helium reservoir inside a Dewar, which is in turn thermally buffered from the room temperature environment by a second outer Dewar of liquid nitrogen (Figure 5.4b). The sample is lowered into the central solenoid bore, whose a diameter and length are both typically a few centimeters, which enclose transmitter/receiver radio frequency coils that surround the sample placed inside a narrow glass tube on the central solenoid axis. The size of the Dewars required result in such machines occupying the size of a room often requiring stair access to the sample's entry port and the Dewar openings and are suitably expensive to purchase and maintain, necessitating an NMR facility infrastructure.

The B-field inside a long solenoid of length s, a coil current I, and a number of turns n can be modeled by the simple relation easily derived from the *Biot–Savart law* of (in reference to Figure 5.4c) $dB = \mu_0 I \sin \theta ds/(4\pi r^2)$:

(5.23)
$$B = \frac{n\mu_0 I}{s}$$

where μ_0 is the vacuum permeability. The signal-to-noise ratio of an NMR measurement scales roughly as $\sim B^{3/2}$ (the bulk magnetization of the sample scales as $\sim B$; see Worked Case Example 5.2, but the absorbed power also scales with ν, which scales with $\sim B$, whereas the shot noise scales with $\sim\sqrt{\nu}$) so there is a motivation to generate higher fields. Field strengths of ~12 T are possible with solenoid cooling at 4 K, which corresponds to an ~500 MHz resonance frequency for ^1H. To generate higher field strengths requires cooling lower than the ~4 K boiling point of helium, using the *Joule–Thompson effect* in a gas expansion unit to maintain solenoid temperatures as low as ~2 K, which can result in a coil current of a few hundred amperes, equivalent to a resonance frequency for ^1H of up ~900 MHz.

Older NMR machines use a continuous wave (*CW NMR*) approach to sequentially probe the sample with different radio frequencies. The primary limitation with CW NMR is one of time, since multiple repeated spectra are usually required to improve signal-to-noise ratio, which can result in experiments taking several hours. Modern NMR machines use a frequency domain method known as "Fourier transform NMR (FT NMR)," which dramatically reduces the data acquisition time. Here, a sequence of short pulses of duration τ of a carrier wave of frequency f is composed of a range of frequency components, which span $\sim f \pm 1/2\pi\tau$. The value of f used is catered to the unshielded resonance frequency of the magnetic atomic nucleus type under investigation, while τ is usually in the range 10^{-6} to 10^{-3} s to give sufficient frequency resolution to probe shifts in the resonance frequency of <0.1 ppm (typically ~0.02 ppm), with an averaged NMR spectroscopy trace typically taking less than 10 min to acquire.

As discussed previously, after the absorption of radio frequency energy, atomic nuclei relax back to a state of thermal equilibrium. This relaxation process involves the ultimate emission of tiny amounts of radio frequency energy from the high-energy-state nuclei. These tiny signals can be detected by radio frequency detector coils around the sample, and it is these that ultimately constitute the NMR signal.

5.4.6 NMR SPECTROSCOPY PULSE SEQUENCES

In practice, an NMR spectroscopy experiment is performed by using several repeated radiofrequency driving pulses, as opposed to continuous wave stimulation. However, different specific pulse sequences can generate different levels of information in regard to the spin relaxation processes. The simplest pulse sequence is just a single pulse followed by the detection of resonance signal, damped by relaxation (Figure 5.5a), known as the free induction

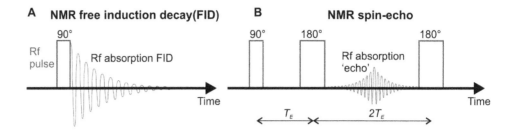

FIGURE 5.5   NMR spectroscopy pulse profiles. (a) Simple free induction decay. (b) Spin-echo pulse sequence (which is repeated multiple times on the same sample to allow averaging to improve the signal-to-noise ratio).

decay (*FID*). The signal damping in this mode is exponential with a decay time referred to as T_2^*. This basic mode is the essence of all *pulsed NMR methods*, and in a biomedical setting, single-pulse methods such as this form the basis of a common pulsing sequence used in magnetic resonance imaging (see Chapter 7) called "gradient recalled echos."

A pulse here is a superposition of oscillating radiofrequency waves (or *spin packets*) with a broad range of frequencies, which are used to rotate the bulk magnetization of the sample, which is set by the external *B*-field vector. Pulses are described as having a specific phase in terms of the angle of rotation of the bulk magnetization. So, for the simplest form of FID, a 90° pulse (referred to commonly as a "90" or "pi over two" pulse) is normally applied initially, with decay then resulting in dephasing 90° back to realignment of the bulk magnetization to its original orientation. This simple pulse sequence will normally be repeated to improve the signal-to-noise ratio. For analysis, this time-resolved repeating signal is usually Fourier transformed, resulting in a signal amplitude S, which depends on the *relaxation time* T_1 as well as the time between *pulse loops* called the "repetition time" (T_R):

$$(5.24) \qquad S = k\rho \left(1 - \exp\left[\frac{-T_R}{T_1} \right] \right)$$

where k is a constant of proportionality with a density of spin nuclei in the sample given by ρ.

The *spin-echo (SE)* pulse sequence (also known as *Hahn echo*) is another common mode. Here, a sample is stimulated with two or more radio frequency pulses with subsequent detection of an echo resonance signal at some time after these initial pulses. Usually, this involves an initial 90° pulse, a wait period known as the echo time T_E, then a 180° *refocusing pulse*, another wait period T_E, then observation of the energy peak of the SE signal (Figure 5.5b)—the 180° pulse causes the magnetization to at least partially *rephase*, which results in the echo signal. On top of this are T_1 and T_2 relaxation processes, so the Fourier transformed signal is described by

$$(5.25) \qquad S = k\rho \left(1 - \exp\left[\frac{-T_R}{T_1} \right] \right) \exp\left[\frac{-T_E}{T_2} \right]$$

The enormous advantage of SE pulsing is that normally inhomogeneous relaxation processes will ultimately cause dephasing following repeating 90° pulsing, that is, different spin nuclei from the same atom types will start to precess at noticeably different rates, whereas the 180° pulse largely resets this dephasing back to zero, allowing more pulse loops before dephasing effects dominate, resulting in large effective increases in the signal-to-noise ratio.

The *inversion recovery* sequence is similar to SE pulsing, but here a 180° radio frequency pulse is initially applied. After a given time period known as the "inversion time" T_I, during which time the bulk magnetization undergoes spin–lattice relaxation aligned 180° from the original vector, a 90° pulse is applied, which rotates the longitudinal magnetization into the

XY plane. In this example, the 90° pulse is then applied, and the magnetization dephases giving an FID response as before. Normally, an inversion recovery sequence is then repeated every T_R seconds to improve the signal-to-noise ratio, such that

$$(5.26) \qquad S = k\rho \left(1 - 2\exp\left[\frac{-T_I}{T_1} \right] \right) \exp\left[\frac{-T_R}{T_1} \right]$$

5.4.7 MULTIDIMENSIONAL NMR

For complex biomolecules, often containing hundreds of atoms, overlapping peaks in a spectrum obtained using just a single magnetic atomic nucleus type, the so-called 1D-NMR, can make interpretation of the relative spatial localization of each different atom challenging. The correct assignment of atoms for structural determination of all the major classes of biomolecules is substantially improved by acquiring NMR spectra for one and then another type of magnetic atomic nuclei simultaneously, known as "multidimensional NMR" or "NMR correlation spectroscopy" (COSY). For example, the use of *2D-NMR* with ^{13}C and ^{15}N isotopes can be used to generate a 2D heat map plot for chemical shift for each isotope plotted on each axis, with the 2D hotspots, as opposed to 1D peaks on their own, used to extract the molecular signature, which is particularly useful for identify backbone structures in proteins, a technique also referred to as "nuclear Overhauser effect spectroscopy" (NOESY). These correlative NMR approaches can be adapted in several multichannel NMR machines for *3D-NMR* and *4D-NMR*, with averaged spectra taking more like ~100 min to acquire.

Correlative NMR spectroscopy has been enormously successful in determining the structures of several types of biomolecules. These include complex lipids, carbohydrates, short nucleic acid sequences of $\lesssim 100$ nucleotides, and peptides and proteins. The upper molecular weight limit for proteins using these NMR methods is ~35 kDa, which is comparatively small (e.g., an IgG antibody has a molecular weight of ~150 kDa). Multidimensional NMR can to a great extent overcome issues of overlapping chemical shift peaks associated with larger proteins; however, a larger issue is that the sample magnetization relaxes faster in large proteins, which ultimately sets a limit on the time to detect the NMR signal. Larger proteins have longer rotational correlation times and shorter transverse (T_2) relaxation times, ultimately leading to line broadening in the NMR spectrum.

Transverse relaxation optimized spectroscopy (TROSY) has been used to overcome much of this line broadening. TROSY suppresses T_2 relaxation in multidimensional NMR spectroscopy by using constructive interference between dipole–dipole coupling and anisotropic chemical shifts to produce much sharper chemical shift peaks. TROSY can also be used in combination with deuteration of larger proteins, that is, replacing ^1H atoms with ^2H, which further suppresses T_2 relaxation. These improvements have allowed the structural determination of much larger proteins and protein complexes with nucleic acids, up to ~90 kDa.

NMR spectroscopy in its modern cutting-edge form has been used to great effect in obtaining atomic-level structures of several important biomolecules, especially of protein membranes. These are in general very difficult to crystallize, which is a requirement of the competing atomic-level structural determination technique of x-ray crystallography. A related spatially resolved technique used in biomedical *in vivo* diagnostics is *magnetic resonance imaging (MRI)*, discussed in Chapter 7.

5.4.8 ELECTRON SPIN RESONANCE AND ELECTRON PARAMAGNETIC RESONANCE

ESR, also referred to as EPR, relies on similar principles to NMR. However, here the resonance is from the absorption and emission of electromagnetic radiation due to transitions in the spin states of the electrons as opposed to magnetic atomic nuclei. This only occurs for an unpaired electron since paired electrons have a net spin of zero. ESR resonance peaks occur in the microwave range of ~10 GHz, with ESR spectrometers normally generating B-fields of ~1 T or less.

Unpaired electrons are chemically unstable, associated with highly reactive species such as free radicals. Such chemical species are short-lived, which limits the application of ESR, though this can be used to an advantage in that standard solvents do not give rise to a measurable ESR signal; therefore, the relative strength of the signal from the actual sample above this background *solvent noise* can be very high.

Site-directed spin labeling is a genetics technique that enables unpaired electron atom labels (i.e., *spin labels*) to be introduced into a protein. This uses a genetics technique called site-directed mutagenesis (discussed in more detail in Chapter 7). Here, specific labeling sites in the DNA genetic code of that protein are introduced. Once incorporated into the protein, a spin label's motions are dictated by its local physical and chemical environment and give a very sensitive metric of molecular in the vicinity of the label. A common spin label is *nitroxide*, also known as "amine oxide" or "N-oxide," which has a general chemical formula of $R_3N^+–O^-$ where R is a substituent organic chemical group, which contains an unpaired electron predominantly localized to the N–O bond, which has been used widely in the study of the structure and dynamics of large biomolecules using ESR.

5.4.9 TERAHERTZ RADIATION APPLICATIONS AND SPECTROSCOPIES

Terahertz radiation (T-rays) occupies a region of the electromagnetic spectrum between microwaves and infrared radiation, often referred to as the *terahertz gap*, where technologies for its generation and measurement are still in development. The fastest existing digital photon detectors have a bandwidth of a few tens of GHz, so the ~10^{11}–10^{13} Hz characteristic frequencies of terahertz radiation (corresponding to wavelengths of ~30–3000 μm) are too high to be measured digitally but instead must be inferred indirectly, for example, by energy absorption measurements sampled at lower frequencies. However, the energy involved in transitions between different states in several fundamental biological processes has a characteristic equivalent resonance frequency in this terahertz range.

These include, for example, the collective vibrational motions of hydrogen-bonded nucleotide base pairs along the backbone of a DNA molecule, as well as many different molecular conformational changes in proteins, especially with a very high sensitivity to water, which is a useful metric for exposure of different molecular surfaces in a protein undergoing conformational changes. *Terahertz spectroscopy*, only developed at around the turn of the twentieth century, has yet to emerge into mainstream use in addressing practical biological questions (though for a good review of emerging applications, see Weightman, 2012). However, it has significant future potential for investigating a variety of biological systems.

Terahertz spectroscopy typically utilizes a rapid pulsed Ti–sapphire laser for generation of both terahertz radiation and detection. The laser output is in the near infrared; however, the pulse widths of these NIR wave packets are ~10^{-13} to 10^{-14} s, implying a frequency range of several terahertz, though centered on an NIR wavelength output of ~800 nm (~375 THz), which thus needs to be downshifted by two orders of magnitude.

KEY POINT 5.2

A pulse of electromagnetic radiation is an example of a wave packet, and if it has a duration of Δt, it can be deconstructed into several Fourier frequency components, which span an angular frequency range $\Delta \omega$, which satisfies $\Delta \omega \Delta t \approx 1$. This can be viewed as a classical wave equivalent of the quantum uncertainty principle. This implies the range of frequencies in a wave packet $\Delta \upsilon \approx 1/2\pi\Delta t$. An alternative depiction of this is with position x and wave vector k uncertainty, $\Delta k \Delta x \approx 1$, for example, superposing several waves of different wavelengths generates an interference pattern, which results in increased spatial localization of this wave train.

A *terahertz spectrometer* is very similar in design to an FTIR spectrometer, in measuring the laser transmission in a cryofixed sample over a typical frequency range of ~0.3–10 THz. Samples are mounted in polyethylene holders, which are transparent to THz radiation, and held on an ultracooled stage at a temperature of ~4 K or less by liquid helium to minimize vibrational noise in the sample. THz spectroscopy has been applied in particular to investigating different topologies and flexibility of both DNA and RNA molecules in a cryofixed but physiologically relevant hydrated state, with an ability to detect base pair mutations in short oligonucleotide sequences from differences to the THz transmission spectra. THz spectroscopy can also be adapted to slow confocal scanning techniques across a thin sample, thus allowing *terahertz imaging*.

T-rays also have biophysical applications for tissue imaging. Intense T-rays can be controllably generated from a variety of sources, for example, both a synchrotron *and free-electron laser* (FEL) in addition to generating a continuum of x-ray radiation can be utilized to provide a stable source of T-rays, but also smaller sources that do require a very large facility such as lower power FEL sources or *free-electron masers* (which generate T-rays through cyclotron resonance of electrons in a device called a "gyrotron"), in this case due to high-frequency T-rays overlapping with low-frequency microwaves in the electromagnetic spectrum. Unlike x-rays, T-rays are nonionizing due to a lower photon energy and so do not result in the often high level of cellular damage of x-rays, especially due to damage of cellular DNA.

However, T-rays can penetrate into millimeters of biological tissues, which have low water content, such as fat, but have a high reflectivity for high water content tissues. Thus, T-rays can be used to measure tissue differences in water content, which has been used for the detection of various forms of epithelial cancer. Similarly, T-rays have been applied to generating more accurate images of teeth compared to x-rays in dentistry (see Chapter 7).

A recent application of T-rays has involved investigations of the structural states of the protein lysozyme, which was used as a model enzyme system (Lundholm et al., 2015). Here, the time-resolved structure of lysozyme was monitored at as low as ~1 K temperatures using x-ray crystallography, before and after bombarding the crystals with T-rays. Instead of being dissipated rapidly in a few nanoseconds as heat in the anticipated process of *thermalization*, the T-rays were absorbed in a spatially extended state of several coupled lysozyme molecules, extending the absorption lifetime to time scales three to six orders of magnitude longer than expected for single molecules. This coupled system is consistent with a state of condensed matter theoretically predicted in 1968 by Herbert Fröhlich (1968) as a possible theoretical mechanism for ordered energy storage in dielectric biomolecules in cell membranes, but never experimentally confirmed until now, called the Fröhlich condensate, which is the lowest order vibrational mode of condensed dielectric matter analogous to the *Bose–Einstein condensate* of a gas of bosons in quantum mechanics—in essence, long-range electrostatic Coulomb forces are coupled between molecules in a pool, resulting in coherent electrical oscillations, thus trapping absorbed energy of the right frequency (T-rays, in this case) for much longer than would be expected from individual electric dipole oscillations. The result is still being hotly debated as it could have enormous relevance to the existence of nontrivial quantum mechanical effects in many biological processes (see Chapter 9) and certainly may have implications for the real mode of operation of enzymes on their substrates, that is, potentially involving more physical-based processes cooperatively than what were imagined previously.

Worked Case Example 5.2: NMR Spectroscopy

An NMR spectrometer contains a bespoke superconducting solenoid magnet with a length of 7 cm and an inner bore diameter of 5 cm, and an outer diameter of 6 cm was composed of tightly wound, superconducting wire with a diameter of 0.85 mm, with each wire comprising ~500 individual conducting filaments. If cooled to ~4 K, a stable coil current of ~100 A was possible in each filament.

a   What is the expected resonance frequency v_0 in this NMR device of a ^1H atomic nucleus?
A test sample of 300 μL of 1 mM ethanol dissolved in TMS was used in the device.

b   If a general magnetic sample consisting of identical atoms of nuclear spin quantum number of I has a bulk magnetization M_0 given by the sum of all magnetic moments per unit volume, show that M_0 is proportional to $I(I + 1)B$ in an external magnetic field of magnitude B, stating any assumptions. Assuming that all single proton atomic nuclei in the ethanol sample have the same resonance frequency v_0, estimate its bulk magnetization.

c   The average measured resonance frequency v of ^1H in the sample was slightly different to v_0 by an amount of Δv. Explain why this is, and estimate Δv.

[You can assume that the vacuum permeability $\approx 1.3 \times 10^{-6}$ H m^{-1}; hint: the sum of n natural squares is $n(n + 1)(2n + 1)/6$].

Answers

a   The number of wire turns n' in the solenoid for tightly packed wires is given roughly by

$$n' = (6.0 - 5.0)/0.85 \approx 11 \; complete \; turns$$

However, each wire contains ~500 filaments, so the number of total turns n in solenoid $n = 11 \times 500 = 5500$ turns. Assuming the long solenoid approximation, the B-field is given by

$$B = (1.3 \times 10^{-6}) \times 5500 \times 100 / (0.07) = 10.2 \, T$$

^1H resonance frequency is 400 MHz for a 9.4 T field; thus, here the resonance frequency will be

$$v = 400 \times (10.2 / 9.4) = 434 \, MHz$$

b   If the magnetization is given by the sum of the magnetic moments per unit volume, this is the same as the total number of all magnetic moments per unit volume multiplied by the *expected* magnetic moment. The expected value of the magnet moment is given by $\langle \mu \rangle$ the probability-weighted sum over all possible μ values. The probability p_m of a general spin quantum number m is given by the Boltzmann factor for that energy state normalized by the sum of all possible Boltzmann factors of all energy states. Therefore,

$$
\begin{aligned}
\langle \mu \rangle &= \sum_{m=-I}^{I} p_m \mu(m) = \frac{\sum_{m=-I}^{m=I} (\gamma mh/2\pi) \exp[-E_m/k_BT]}{\sum_{m=-I}^{m=I} \exp[-E_m/k_BT]} \\
&= \frac{\sum_{m=-I}^{m=I} (\gamma mh/2\pi) \exp[\gamma mhB/2\pi k_BT]}{\sum_{m=-I}^{m=I} \exp[\gamma mhB/2\pi k_BT]} \\
&= \frac{\gamma h \sum_{m=-I}^{m=I} m \exp[-m]}{2\pi \sum_{m=-I}^{m=I} \exp[-m]}
\end{aligned}
$$

The magnitude of αm is small at $\sim(1/180) \times \frac{1}{2} \approx 0.01$, so use a Taylor expansion for the exponential (and using hint for sum of natural squares):

$$\therefore \langle \mu \rangle \approx \frac{\gamma h}{2\pi} \frac{\sum_{m=-1}^{m=1} m(1+\alpha m+\cdots)}{\sum_{m=-1}^{m=1} m(1+\alpha m+\cdots)} = \frac{\gamma h}{2\pi} \frac{\sum_{m=-1}^{m=1} m(1+\alpha m^2+\cdots)}{\sum_{m=-1}^{m=1} m(1+\alpha m+\cdots)}$$

$$= \frac{\gamma h}{2\pi} \frac{\sum_{m=-1}^{m=1}(m)+\alpha \sum_{m=-1}^{m=1}(m^2)+\cdots}{\sum_{m=-1}^{m=1}(1)+\alpha \sum_{m=-1}^{m=1}(m)+\cdots}$$

$$= \frac{\alpha \gamma h}{2\pi} \frac{0 + I(I+1)(2I+1)/3}{(2I+1)+0} = \frac{\gamma^2 h^2}{6\pi k_B T}I(I+1)B$$

Thus, if the number of magnetic moments per unit volume (i.e., per m³ in SI units) is N

$$M_0 = N \frac{\gamma^2 h^2}{6\pi k_B T}I(I+1)B$$

For each ethanol molecule, there are five ^{1}H atoms, and a 1 mM concentration (i.e., 1 mmol L^{-1}) will give

$$N = 5 \times (1 \times 10^{-3}) \times 6.02 \times 10^{23} \times 10^3 \approx 3 \times 10^{22} \text{ atoms m}^{-3}$$

Using the value for γ for ^{1}H indicated by Table 5.1 is $\sim42.6 \times 2\pi \approx 268$ MHz T^{-1}. Therefore,

$$M_0 \approx 3 \times 10^{22} \times (268 \times 10^6)^2 \times (6.6 \times 10 \times 10^{-34})^2 \times 0.5 \times (0.5+1)/$$
$$(6\pi \times 1.38 \times 10^{-23} \times 4) \approx 7 \times 10^{-7} \text{Am}^{-1}$$

The magnetic susceptibility χ of this sample is thus $\sim(7 \times 10^{-7})/10.2 \approx 7 \times 10^{-8}$.

c   Nuclear shielding of the ^{1}H atoms in ethanol results in a chemical shift of the resonance frequency, which is an *average* per ^{1}H atom of $\sim$2.2 ppm in TMS solvent. Therefore, the B-field required to excite these atoms will be *greater* by this factor, as will the resonance frequency, which scales with B. Therefore,

$$\Delta v \approx 434 \times 10^6 (2.2 \times 10^{-6}) = 955 \text{ Hz}$$

5.5 TOOLS THAT USE GAMMA RAYS, RADIOISOTOPE DECAYS, AND NEUTRONS

Several other high-energy particles can be used as biophysical probes. Alpha and beta particles and gamma rays are relevant to the radioactive decay of isotopes, which can be used as reporter tags in biomolecules, especially useful in investigating the kinetics of biochemical processes. Gamma rays are also relevant to *Mössbauer spectroscopy*. Also, structural details of biomolecules can be investigated using the scattering/diffraction of *thermal neutrons*.

5.5.1 MÖSSBAUER SPECTROSCOPY

The Mössbauer effect consists of recoilless emission and absorption of gamma rays by/from an atomic nucleus in a solid or crystal lattice. When an excited nucleus emits a gamma ray, it must recoil to conserve momentum since the gamma ray photon has momentum. This implies that the emitted gamma ray photon has an energy, which is slightly too small to excite an equivalent atomic nucleus transition due to absorption of another identical atomic nucleus in the vicinity. However, if the gamma ray–emitting atomic nuclei are located inside a solid lattice, then, under sufficiently low temperatures, the atomic nucleus emitting the gamma ray photon cannot recoil individually but instead the effective recoil is that of the *whole* large lattice mass.

Under these conditions, the energy of a gamma ray photon may not be high enough to excite phonon energy loss through the whole lattice and therefore these results in negligible recoil energy loss of the emitted gamma ray photon. Thus, this photon can be absorbed by another identical atomic nucleus to excite an atomic nuclear transition, with consequent emission of a gamma ray photon, which therefore results in absorption resonance within the sample. However, in a similar way to the fine structure of NMR resonance peaks discussed previously in this chapter, the local chemical and physical environment can result in *hyperfine splitting* of the atomic nucleus energy transition levels in atomic nuclear energy levels (due to magnetic Zeeman splitting, quadrupole interactions, or isomer shifts, which are relevant to nonidentical atomic radii between absorber and emitter), but which can shift the resonance frequency by a much smaller amount than that observed in NMR, here by typically just one part in $\sim 10^{12}$.

An important consequence of these small energy shifts, however, is that any relative motion between the source and absorber of speed around a few millimeters per second can result in comparable small shifts in the energy of the absorption lines; this can therefore result in absorption resonance in a manner that depends on the relative velocity between the gamma ray emission source and absorber. A typical *Mössbauer spectrometer* has a gamma ray source mounted on a drive, which can move at different velocities up to several millimeters per second, relative to a fixed absorber. A radiation *Geiger counter* is placed behind the absorber. When the source moves and Doppler shifting of the radiated energy occurs, resonance absorption in the fixed absorber decreases the measure transmission on the Geiger counter since excited nuclei reradiate over a time scale of $\sim 10^{-7}$ s but isotropically.

Several candidate atomic isotopes are suitable for Mössbauer spectroscopy; however, the iron isotope ^{57}Fe is ideal in having both a relatively low-energy gamma ray, which is a prerequisite for the Mössbauer effect, and relatively long-lived excited state, thus manifesting as a high-resonance signal-to-noise ratio. The cobalt isotope ^{57}Co decays radioactively to ^{57}Fe with emission of a 14.4 keV gamma ray photon and is thus typically used as the moving gamma ray source for performing ^{57}Fe Mössbauer spectroscopy in the fixed absorber sample.

Iron is the most abundant transition metal in biological molecules and ^{57}Fe Mössbauer spectroscopy has several biophysical applications, for example, biomolecules such as the oxygen carrier hemoglobin inside red blood cells, various essential enzymes in bacteria and plants, and also multicellular tissues that have high iron content, such as the liver and spleen. In essence, the information obtained from such experiments are very sensitive estimates for the number of distinct iron atom sites in the sample, along with their oxidation and spin states. Importantly, a Mössbauer spectrum is still observed regardless of the actual oxidation or spin state of the iron atoms, which differentiates from the EPR technique. These output parameters then allow predictions of molecular structure and function in the vicinity of the detected iron atoms to be made.

KEY BIOLOGICAL APPLICATIONS: NMR

Determining atomic-level precise molecular structures without the need for crystals; Identifying specific chemical bonds.

5.5.2 RADIOISOTOPE DECAY

An example of a radioactive isotope (or *radioisotope*) in ^{57}Co was discussed earlier in the context of being a gamma ray emitter in decaying to the more stable ^{57}Fe isotope. But there are a range of different radioisotopes that have direct biophysical application in acting as a source of detectable radioactivity, which can be tagged onto a specific region of a biomolecule. This radioactive tag therefore acts as a biochemical reporter or *tracer* probe.

The kinetics of radioactivity can be modeled as a simple first-order process:

$$\frac{dN}{dt} = -\lambda N$$

(5.27)

where N is the number of radioisotopes of a specific type, with a decay constant λ. This results in a simple exponential decay for N. The half-life $t_{1/2}$ is the time taken to reduce N by 50%, which is simple to demonstrate as $\ln 2/\lambda$, while the mean lifetime of a given radioisotope is given by $1/\lambda$.

The radioisotope is introduced in place of a normal relatively nonradioactive isotope, typically to detect components or metabolites in a biological system in time-resolved investigations. Radioisotopes have relatively unstable atomic nuclei and their presence can be detected from their emission of different specific types of radiation generated during the radioactive decay process in which an energetic lower energy (i.e., more stable) atomic nucleus is formed. The type of radiation produced depends on the isotope but can typically be detected by a Geiger counter or scintillation phosphor screen, often in combination with a CCD or PMT detector. In combination with *stopped-flow* techniques, biochemical reactions can be quenched at intermediate stages and the presence of radioisotopes measured in the detected metabolites, which thus allows a picture of the extent of different biochemical processes to be built up.

Common types of radiation emitted in radioisotope decay are gamma rays, beta particles (high-energy electrons), and alpha particles ($^4He^2+$, in other words helium nuclei with no atomic electrons). Alpha particles have a small depth of penetration (e.g., they are stopped by just a few centimeters of air) and are thus not useful as tracers but find application in radiotherapy. Common radioisotope tracers used in the life sciences include: ^3H, ^{14}C, ^{32}P and ^{33}P, ^{35}S, ^{45}Ca, and ^{125}I. But ^{99m}Tc has a more focused application as a biomedical tracer. In addition, a number of radioisotopes decay with output of a *positron*, which are relevant as biomedical tracers in *positron emission tomography*, or *PET* (biomedical applications are discussed more generally in Chapter 7).

5.5.3 NEUTRON DIFFRACTION AND SMALL-ANGLE SCATTERING

Neutron diffraction works on similar principles to that of x-ray diffraction but utilizing an incident beam composed of *thermal neutrons*. Thermal neutrons can be generated by two principal methods. One is to use a *thermal nuclear reactor*. These utilize the fission of the ^{235}U isotope, which releases an average of ~2.4 extra neutrons for every fission event. An example of ^{235}U fission following neutron absorption is

$$n + U_{92}^{235} \rightarrow U_{92}^{236}$$
$$U_{92}^{235} \rightarrow K_{36}^{89}r + B_{56}^{144}a + 3n + 177\,MeV$$

(5.28)

where just one of these released neutrons is required to sustain a *chain reaction*. Neutrons formed from uranium fission have an average energy of ~2 MeV. These neutrons are slowed down by a neutron *moderator* around the fission core (typically composed of water or graphite) so that emergent neutrons are in thermal equilibrium with their surroundings (hence, the term *thermal* neutrons), which have a mean energy of just ~0.025 eV with an equivalent de Broglie wavelength of ~0.2 nm.

KEY POINT 5.3

Put very simply, there are just three types of high-energy particles whose effective wavelength is comparable to interatomic spacings in biological matter, which are electrons, x-ray photons, and neutrons and which thus can all be used in diffraction experiments to generate information about the positions of atoms in biomolecules.

The other approach to generating thermal neutrons is to use *spallation neutron sources*. These utilize particle accelerators and/or synchrotrons to generate intense, high-energy proton beams, which are directed at a heavy metal target (e.g., made from *tantalum*) whose impact can split the atomic nuclei to generate more neutrons. Proton synchrotron radiation impacted on such a metal target can generate >10 neutrons from a nuclear reactor, with an effective wavelength of ~10^{-10} m. Here, the scattering is due to interaction between the atomic nuclei as opposed to the electron cloud.

Neutron diffraction has a significant advantage over x-ray diffraction in that hydrogen atomic nuclei (i.e., single protons) will measurably scatter a neutron beam, and this scatter signal can be further enhanced by chemically replacing any solvent-accessible labile hydrogen atoms with deuterium, D, typically by solvating the target molecule in heavy water (D_2O) rather than normal water (H_2O) prior to crystallization. This allows the position of the hydrogen atoms to be measured directly, resulting in more accurate bond length predictions, but with disadvantages of requiring larger crystals (length ~1 mm) and a nearby nuclear reactor.

Small-angle neutron scattering (*SANS*) uses elastic scattering of thermal neutrons by a sample to generate structural information over a length scale of ~1–100 nm. The principles of operation are similar to SAXS performed with an incident x-ray beam. However, since neutrons scatter from atomic nuclei, unlike x-rays, which are scattered from atomic electron orbitals, the signal-to-noise ratio of diffraction intensity peaks is greater than SAXS for lower-atomic-number elements. SANS has been applied to determine the structural details of several macromolecular complexes, for example, including ribosomes and various biopolymer architectures such as the *dendritic fibers* of nerves in solution.

KEY BIOLOGICAL APPLICATIONS: RADIOISOTOPES AND NEUTRONS

Tracking metabolic processes; Determining macromolecular structures.

Worked Case Example 5.3: Radioisotope Decay

A radioisotope A contains a nucleus which decays with a constant λ_A, into another radioisotope B whose nucleus also decays but with a smaller constant λ_B into a stable isotope C.

a   *If there are $N_A(0)$ initial atoms of A and none of B, determine a formula for the number of atoms of B, $N_B(t)$, after time t.*

b   *A controlled experiment was performed to simulate the effects of radiation damage to biological tissue during a nuclear reactor leak in which the ultimate product, isotope Z with a half-life ~20,000 years, is produced from a chain reaction involving the radioactive decay of isotope X, which decays with a half-life of 2.4 days to isotope Y via beta decay, which in turn decays to Z also by beta decay but with a half-life of 23.5 minutes. What percentage of the number of X atoms initially present will be Y atoms after 1 hour?*

Answers

a   The rate of change in number of B atoms is the rate of formation of B from A *minus* the rate of decay of B into C. Using the general radiation decay Equation 5.27:

$$\frac{dN_A}{dt} = -\lambda_A N_A \therefore \int \frac{dN_A}{N_A} = -\int \lambda_A dt \therefore N_A(t) = N_A(0)\exp(-\lambda_A t)$$

The rate of formation of B from A is negative, that is, the rate of decay of A from B (since one atom "lost" from A is "gained" by B), or $\lambda_A N_A$ or $N_A(0)\lambda_A \exp(-\lambda_A t)$. Rate of decay of B to C is $-\lambda_B N_B$ so rate of change of number of B atoms is:

$$\frac{dN_B}{dt} = \lambda_A N_A(0)\exp(-\lambda_A t) - \lambda_B N_B$$

This type of slightly more complicated rate equation can be solved using the *integrating factor method* such that in the general case $dy/dx + f(x)y = g(x)$ the integrating factor $I(x)$ is $\exp(\int f(x)dx)$, and by multiplying the original differential equation by $I(x)$ and integrating, gives a solution $y = \int g(x)I(x)dx/I(x)$, which can then be solved given appropriate boundary conditions. After a bit of mental gymnastics and using the given initial conditions, this therefore simplifies to:

$$N_B = \frac{\lambda_A N_A(0)}{\lambda_B - \lambda_A}\left(\exp(-\lambda_A t) - \exp(-t\lambda_B)\right)$$

b   The half-life $t_{1/2}$ is the decay time taken to reduce the number of atoms twofold, so from the first part of the answer to (a) above $\lambda = \ln 2/t_{1/2}$. The half-life of X is several orders of magnitude higher than the other two isotopes so can be considered "stable" here, so the formula you derived from part (a) applies, with:

$\lambda_X = \ln 2/(2.4 \text{ days} \times 24 \times 60 \times 60) = 8.0 \times 10^{-5}$ counts/s
$\lambda_Y = \ln 2/(23.5 \text{ min} \times 60) = 4.9 \times 10^{-4}$ counts/s

Substituting in these values for 1 hour or $t = 60 \times 60 = 3600$ s indicates $N_Y/N_X(0)$ ≈ 0.11 or 11%.

5.6 SUMMARY POINTS

- EM is a powerful structural biology tool, but care must be taken to avoid overinterpretation from sample preparation artifacts.
- High-energy electrons, x-rays, and neutrons can be used in diffraction experiments to reveal atomic locations in biomolecules.
- X-ray crystallography is particularly useful for determining structures of biomolecules where large crystals with few imperfections can be grown, including many proteins and nucleic acid complexes.
- NMR spectroscopy results in a unique molecular signature and is particularly useful in determining structures of membrane-integrated proteins, which are difficult to crystallize.
- Diffraction techniques may also be extended to longer length scale investigations than single atoms, such as biological fibers.

QUESTIONS

5.1   A sample of a protein was prepared for TEM on a 120 keV machine using evaporative platinum rotary shadowing. A platinum atom is ~0.5 nm in diameter; however, the smallest observable metal particles were composed of five atoms.
   a   What is the practical spatial resolution in this experiment?

The protein was known to exist in two structural states, A for 25% of the time and B for the rest of 75%. Each single TEM image of a protein was assigned as being in one of the two states so as to generate an average image from each subclass. Sequence data suggested the A–B transition results in a component of the molecule displacing by 5 nm.

b   If each TEM images contains a mean of ~12 protein molecules, estimate how many images one would need to analyze to be able to measure this 5 nm translation change to at least a 5% accuracy.

5.2   A dehydrated cell membrane was placed in a 200 kV electron microscope normal to the beam of electrons, resulting in some of the transmitted electrons being diffracted as they passed through the membrane. A first-order diffraction peak was measured at a deflection angle of 3.5°. Estimate the lipid molecule separation in the cell membrane, stating any assumptions you make.

5.3   Over what range of accelerating voltage do classical and relativistic predictions for the matter wavelength of an electron agree to better than 1%?

5.4   What is an x-ray free-electron laser? Discuss the advantages that this tool has for determining structures of biological molecules over more traditional x-ray methods.

5.5   A single cell contains ^{1}H atomic nuclei, which are mainly associated with water molecules, but with a significant number associated the $-CH_2-$ chemical motif found in lipids. A population of cells was homogenized and probed using ^{1}H NMR spectroscopy in a "400 MHz" machine, which results in a 1.4 kHz difference in the resonance frequency between H_2O and $-CH_2-$ protons. Estimate the equivalent chemical shift in units of ppm.

5.6   Why is NMR spectroscopy often described as the biophysical tool of choice for determining structures of membrane-integrated proteins?

5.7   A bespoke solenoid for an NMR spectroscopy device was constructed with a mean diameter of 50 mm by tightly winding an electrical wire of diameter 0.1 mm with a total length of 10 m. If a coil current of 5 A is run through the solenoid wire, estimate the average chemical shift of ^{1}H atom in ethanol dissolved in TMS.

5.8   With reference to drawing a representative NMR spectrum, explain why the protons in an ethanol molecule are often categorized as being in one of three types, in terms of their response to nuclear magnetic resonance.

5.9   What is the "phase problem" in x-ray crystallography? How can it be overcome?

5.10 If the probability of a radioisotope decaying in small time Δt is equal to the number N_1 of radioisotopes present multiplied by some constant λ_1, develop an expression for the number of radioisotopes present after a general time t.

a   What do we normally call λ_1 for a general radioisotope?

b   Derive a relation for the half-life and mean lifetime for these radioisotopes.

c   This radioisotope was found to decay into a second radioisotope, of number in the sample denoted by N_2, that decayed with a constant λ_2 into a nonradioactive atom, of number N_3 in the sample. Derive an expression for N_3 as a function of time.

d   Another radioactive decay chain was found to consist of several more (n – 2) radioactive intermediates from an initial radioisotope, before decaying to an nth that was not radioactive, with N_n atoms in the sample. Derive a general expression for N_n as a function of time, and comment on the relevance of one isotope in the chain having a significantly longer mean lifetime than the other in the series.

5.11 Electrons can be accelerated to far higher speeds than are currently used in modern electron microscopes, which therefore would have a smaller Bragg wavelength and better spatial resolution. Why then not use these to look at biological samples?

REFERENCES

KEY REFERENCE

Thibault, P. et al. (2008). High-resolution scanning x-ray diffraction microscopy. *Science 321*:379–382.

MORE NICHE REFERENCES

Fröhlich, H. (1968). Long-range coherence and energy storage in biological systems. *Int. J. Quant. Chem.* 2:641–649.

Gabor, D. (1948). A new microscopic principle. *Nature 161*:777–778.

Humphry, M.J. et al. (2012). Ptychographic electron microscopy using high-angle dark-field scattering for sub-nanometre resolution imaging. *Nat. Commun.* 3:730.

Leake, M.C. (2001). *Investigation of the extensile properties of the giant sarcomeric protein titin by single-molecule manipulation using a laser-tweezers technique.* PhD dissertation, London University, London.

Lundholm, I.V. et al. (2015). Terahertz radiation induces non-thermal structural changes associated with Fröhlich condensation in a protein crystal. *Struct. Dyn.* 2:054702.

Ortega, R. et al. (2012). X-ray absorption spectroscopy of biological samples. A tutorial. *J. Anal. At. Spectrom.* 27:2054–2065.

Schneider, T.R. (2008). Synchrotron radiation: Micrometer-sized x-ray beams as fine tools for macromolecular crystallography. *HFSP J.* 2:302–306.

Weightman, P. (2012). Prospects for the study of biological systems with high power sources of terahertz radiation. *Phys. Biol.* 9:053001.

Forces

Methods that Measure and/or Manipulate Biological Forces or Use Forces in Their Principal Mode of Operation on Biological Matter

> What would happen if we could arrange the atoms one by one the way we want them?
>
> —**Richard Feynman, Physicist (1959)**

General Idea: Several biophysical methods can both measure and manipulate biological forces across a range of length and time scales. These include methods that characterize forces in whole tissues, down through to single cells and to structures inside cells, right down to the single-molecule level. The force fields that are generated to probe the biological forces originate from various sources including hydrodynamic drag effects, solution pressure gradients, electrical attraction and repulsion, molecular forces, magnetism, optical forces, and mechanical forces. All of which are explored in this chapter.

6.1 INTRODUCTION

All experimental biophysical techniques clearly involve measurement and application of forces in some form or another. However, there is a subset of methods that are designed specifically to either measure the forces generated in biological systems, or to control and manipulate them. Similarly, there are tools that do not characterize biological forces directly, but which primarily utilize force methods in their mode of operation, for example, in using pressure gradients to purify biomolecular components.

There now exist several methods that permit both highly controlled measurement and manipulation of the forces experienced by single biomolecules. These various tools all come under the banner of *force transduction devices;* they convert the mechanical molecular forces into some form of amplified, measurable signal. Many of these single-molecule force techniques share several common features, for example, single molecules are not in general manipulated directly but are in effect physically conjugated, usually via one or more chemical links, to some form of adapter that is the real force transduction element in the system. The principal forces that are used to manipulate the relevant adapter include optical, magnetic, electrical, and mechanical. These are all coupled into an environment of complex feedback electronics and stable, noise-minimizing microscope stages, both for purposes of *measurement* and for *manipulation.*

Single-molecule biophysics methods extend beyond just force tools, which we explore here, encompassing also a range of advanced imaging techniques that we explored previously in Chapters 3 and 4. However, an important point to note here about single-molecule methods concerns the *ergodic hypothesis* of statistical thermodynamics. The ergodic hypothesis maintains that there is an equivalence between ensemble and single-molecule properties. In essence, over long periods of time, all accessible microstates are equally probable. This means that an ensemble average measurement (e.g., obtained from the mean average from many thousands of molecules) will be the same as the time-averaged measurement taken from one single molecule over a long period of time. The key difference with a single-molecule experiment is that one can sample the whole probability distribution of all microstates as opposed to just determining the mean value from all microstates as is the case from a bulk ensemble average experiment, though the caveat is that in practice this often involves generating significant amounts of data from single-molecule experiments to properly sample the underlying probability distribution.

KEY POINT 6.1

The ergodic hypothesis, that all accessible microstates with the same energy are equally probable over a long time, is relevant to single-molecule methods since it implies that the population mean measurement from a bulk ensemble experiment, involving typically several thousand molecules or more, will be the same as the mean of several measurements made on a single molecule sampled over a long period of time.

Statistical thermodynamics implicitly assumes ensemble average parameters. That is, a system with many, many particles. For example, a single microliter of water contains $\sim 10^{19}$ molecules. To apply the same concepts to a single molecule requires the ergodic hypothesis.

Intuitively, one might think that the mean average property of thousands upon thousands of molecules is an adequate description for any given single molecule. In some very simple, or exceptional, molecular systems, this is, in fact, the case. However, in general, this is not strictly true. The reason is that single biomolecules often exist in multiple microstates, which is in general intrinsically related to their biological function. A microstate here is essentially a measure of the free energy locked into that molecule, which is a combination of mainly chemical binding energy, the so-called enthalpy, and energy associated with how disordered the molecule is, or *entropy*. There are many molecules that, for example, exist in several different spatial conformations; a good illustration of which are molecular machines, whose theory of translocation is discussed later in Chapter 8. In other words, the prime reason for studying biology at the level of single molecules is the prevalence of *molecular heterogeneity*.

In the case of molecular machines, although there may be one single conformation that has a lower free energy microstate than the others, and thus is the most stable, several other shorter-lived conformations exist that are utilized in different stages of force and motion generation. The mean ensemble average usually looks similar to the most stable of these different conformations, but this single average parameter tells us very little of the behavior of the other shorter lived but functionally essential conformational states. What cannot be done with bulk ensemble average analysis is to probe such multistate molecular systems. The power of single-molecule experiments is that these subpopulations of molecular microstates can be explored directly and individually. Such subpopulations of states are a vital feature of the proper functioning of natural molecular machines.

As discussed in Chapter 2, there is a fundamental energetic instability in molecular machines, which allows them to switch between multiple states as part of their underlying physiological function. There are, however, many experimental biophysical methods that can be employed in bulk ensemble investigations to synchronize a molecular population. For example, these include thermal and chemical jumps such as stopped-flow reactions, electric and optical methods to align molecules, as well as freezing and/or crystallizing a population. A risk with such approaches is that the normal physiological functioning may be different. Some biological tissues, for example, muscles and cell membranes, are naturally ordered on a bulk scale. It is thus no mystery why these have historically generated the most physiologic-ally relevant ensemble data.

The lack of temporal and/or spatial synchronicity in ensemble average experiments is the biggest challenge in obtaining molecular level information. Different molecules in a large population may be doing different things at different times. For example, molecules may be in different conformational states at any given time, so the mean ensemble average snapshot encapsulates all *temporal fluctuations*, resulting in a broadening of the distribution of what-ever statistical parameter is being measured. A key problem of molecular asynchrony is that a typical ensemble experiment is in a steady state. That is, the rate of change between forward and reverse molecular states is the same. If the system is momentarily taken out of steady state, then transient molecular synchrony can be obtained, for example, by forcing all molecules into just one state; however, this, by definition, is a short-lived effect, so practical measurements are likely to be very transient.

Some ensemble average techniques overcome this problem by forcing the majority of the molecules in a system a single microstate, for example, with crystallography. But, in general, this widening of the measurement distribution presents challenges of result interpretation since there is no easy way to discriminate between anticipated widening of an experimental measurement due to, for example, finite detector sensitivity, and the more biologically rele-vant widening of the distribution due to underlying molecular asynchrony.

Thermal fluctuations in the surrounding solvent water molecules often act as the driving force for molecular machines switching between different states. This is because the typical energy difference between different molecular microstates is very similar to the thermal scale of $\sim k_B T$ energy associated with any molecule coupled to the thermal reservoir at a given tem-perature. However, it is not so much the heat energy of the biomolecule itself, which drives change into a different state, but rather that associated with each surrounding water molecule. The density of water molecules is significantly higher in general than that of the biomolecules themselves, so each biomolecule is bombarded by frequent collisions with water molecules ($\sim 10^9$ per second), and this change of momentum can be transformed to mechanical energy of the biomolecule. This may be sufficient to drive a change of molecular state. Biomolecules are thus often described as existing in a *thermal bath*.

There is a broad range in concentration of biomolecules inside living cells, though the actual number directly involved in any given biological process at any one time is generally low. Biological processes occur under typically minimal stoichiometry conditions in which stochastic molecular events become important. Paradoxically, it can often be these rarer, single-molecule events that are the most significant to the functioning of cellular processes. It becomes all the more important to strive to monitor biological systems at the level of single molecules.

KEY POINT 6.2

Temporal fluctuations in biomolecules from a population result in broadening the distribution of a measured parameter from an ensemble average experiment, which can be difficult to interpret physiologically. Thermal fluctuations are driven pri-marily by collisions from surrounding water molecules, which can drive biomolecules into different microstates. In an ensemble average experiment, this can broaden the measured value, which makes reliable inference difficult.

Single-molecule force methods include variants on *optical tweezer* and *magnetic tweezer* designs. They also include *scanning probe microscopy* (*SPM*) methods, the most important of which in a biophysical context is *atomic force microscopy* (*AFM*), which can be utilized both for imaging and in *force spectroscopy*. Electrical forces in manipulating biological objects, from molecules through to cells, are also relevant, such as for electric current measurements across membranes, for example, in *patch clamping*. On a larger length scale, rheological and hydrodynamic forces form the basis of several biophysical methods. Similarly, elastic forces are important components of techniques that permit whole cells and tissues to be mechanically probed.

6.2 RHEOLOGY AND HYDRODYNAMICS TOOLS

Rheology is the study of matter *flow*, principally in a liquid state. For the investigation of living matter, the liquid state is primarily concerned with water, namely, hydrodynamics forces, especially those that operate primarily through viscous drag forces on biological material, but is also concerned with the force response of fluid states of cellular structures. For example, how lipid membranes in cells, which have many properties consistent with those of liquid crystals, respond to external force and also how components inside the cell membrane impart rheological forces on neighboring components. In this section, we discuss a range of hydrodynamics force techniques used to study biological matter, as well as rheological force methods for probing cellular liquid/soft-solid states. These include a range of standard but invaluable tools (e.g., chromatography can arguably be considered a rheological force method) and also methods that utilize centrifugation and osmosis to characterize and/or isolate biological components. We also discussed techniques that result in plastic/viscoelastic rheological deformation of biological soft matter in response to applied forces.

6.2.1 CHROMATOGRAPHY TECHNIQUES

Chromatography is a standard biophysical tool used to separate components in an *in vitro* biological sample on the basis of different molecular properties such as mass and charge. In many biochemistry textbooks, this might not be considered in the context of being a "force method"; however, it does rely on a range of cohesive forces in water in particular, and so we discussed it here. Related methods include *high-performance liquid chromatography, gas chromatography, gel filtration, thin-layer chromatography*, and even standard *paper chromatography*. Molecular components bind to an immobile substrate to form a *stationary phase* for a characteristic *dwell time*, dependent on the physical and chemical features nature of the substrate. The *mobile phase* moves through the chromatography device via diffusion often facilitated by a driving pressure gradient.

Sepharose beads (sepharose is the trade name of a type of polysaccharide sugar generically called agarose, which is purified from seaweed and used for several purposes in experimental biology) of diameter ~40–400 μm are often used as the immobile substrate, tightly packed into a glass column for gel filtration chromatography and related *affinity chromatography* that uses specific *antibodies* bound to the beads, or different bead surface charges in *ion-exchange* chromatography. These factors, in addition to chemical parameters of the media in the mobile phase such as pH and ionic strength, determine the dwell time in the stationary phase, and thus the mean *drift speed* of each molecular component through the device. The end result is to separate out different molecular components on the basis of their relative binding strengths to the immobile substrate and of their mean speed of translocation through the chromatography device, with emerging components often detected using an optical absorption technique at a specific wavelength.

Size exclusion chromatography (SEC) is a chromatography method in which molecules in solution are separated on the basis of their size, and/or molecular weight, usually applied to large molecules such as proteins and nucleic acids. In SEC, a small molecule can penetrate more regions of the stationary phase pore system compared to larger molecules and so will have a slower drift speed, thus enabling larger and smaller molecules to emerge as different fractions at the bottom of a gel filtration column.

Reversed phase chromatography uses an electrically polar aqueous mobile phase and a hydrophobic stationary phase. Hydrophobic molecules preferentially adsorb to this stationary phase, and thus hydrophilic molecules have a faster drift speed in the mobile phase and will elute first from the bottom of the column. This enables separation of hydrophobic and hydrophilic biomolecules.

6.2.2 CENTRIFUGATION TOOLS

Sedimentation methods can be used to purify and characterize different components in *in vitro* biological samples. They rely on the formation of a sedimented pellet when it is spun in a centrifuge, depending on the frictional viscous drag of the sample and its mass. Quantitative measurements may be made using *analytical ultracentrifugation*, which generates centripetal forces ~300,000 times that of gravity and also have controlled cooling to avoid localized heating in the sample, which may be damaging in the case of biological material. By estimating the sedimentation speed, we can infer details of the size and shape of biological molecules and large complexes of molecules, as well as their molecular mass. Balancing the centripetal force on a particle of mass m being spun at angular velocity ω at a radius r from the axis of rotation with the buoyancy force from the displacement of the solvent by the particle and the viscous drag force due to moving through the solution with sedimentation speed v leads to a relation for the *sedimentation coefficient s*:

$$(6.1) \qquad s = \frac{v}{\omega^2 r} = \frac{m\left(1 - \rho_{solvent}/\rho_{particle}\right)}{\gamma}$$

where
  ρ is the density
  γ is the frictional drag coefficient

Diffusion causes the shape of the sedimenting boundary of the spun solution to spread with time. This can be monitored using either optical absorption or interference techniques, allowing both the sedimentation coefficient and the translation diffusion coefficient D to be determined. The Stokes–Einstein relationship (see Chapter 2) is then used to determine γ from D, which can be used to estimate the molecular mass.

A mix of different biological molecules (e.g., several different enzymes) may sometimes be separated on the basis of sedimentation rates in a standard centrifugation device, and a *density gradient* of suitable material (sucrose and cesium chloride are two commonly used agents) is created, such that there is a higher density of that substance toward the bottom of a centrifuge tube. By centrifuging the mix into such a gradient, the different chemicals may separate out as bands at different heights in the tube and subsequently be extracted as appropriate.

Field flow fractionation is a hydrodynamic separation technique that involves forward flow of a suspension of particles in a sample flow cell plus an additional hydrodynamic force applied normal to the direction of this flow. This perpendicular force is typically provided by centrifugation of the whole sample flow cell. Particles with a higher sedimentation coefficient will drift toward the edge of the flow cell due to this perpendicular force more rapidly than particles with a lower sedimentation coefficient. Under nonturbulent *laminar flow* conditions, known as "Poiseuille flow" (see Chapter 7) in a typical cylindrical-shaped pipe containing the sample, the speed profile of the fluid normal to the pipe long axis is *parabolic* (i.e., maximum in the center of the pipe, zero at the edges); thus, the particles with higher sedimentation coefficients are shifted more from the fastest on-axis flow lines on the pipe and will have a smaller drift speed through the flow cell. This therefore enables particles to be separated on the basis of sedimentation coefficient—put simply, to separate larger from smaller particles.

Microfluidics can use these principles to separate out different biological components on the basis of flow properties. Here, flow channels are engineered to have typical widths on the length scales of a few tens of microns, and different channels can be connected to generate a complex flow-based device. These systems are discussed fully in Chapter 7.

6.2.3 TOOLS THAT UTILIZE OSMOTIC FORCES

Dialysis, or *ultrafiltration*, has similar operating principles to chromatography in that the sample mobility is characterized by similar factors, but the solvated sample is on one side of a dialysis membrane that has a predefined pore size. This sets an upper molecular weight limit for whether molecules can diffuse across the membrane.

This *selectively permeable membrane* (also referred to as a *semipermeable membrane*) results in an osmotic force driven by entropy. On either side of the membrane, there is a concentration gradient, that is, the concentration of solvated molecules on one side of the membrane is different from that on the other side. The water molecules in the solution that has a *higher* concentration have *more* overall order since there are a greater relative number of available solute molecules to which they bond, usually via electrostatic and/or hydrogen bonding. This entropy difference between the two solutions either side of the membrane is manifested as a statistical/entropic driving force when averaged over time scales are much larger than the individual water molecule collision time, which acts in a direction to force a net flux of water molecules from the low-to high-concentration solutions (note that this is also the physical basis of *Raoult's law*, which states that the partial vapor pressure of each component of an ideal mixture of liquids is equal to the vapor pressure of the pure component multiplied by its mole fraction in that mixture; in other words, the components act independently of each other). This process can be used to separate different populations of biomolecules of the basis of molecular weight, often facilitated by a pressure gradient. The use of multiple dialysis stages using membranes with different pore sizes can be used to purify a complex mix of different molecules.

Osmotic pressure can also be used in the study of live cells. Lipid membranes of cells and subcellular cell organelles are selectively permeable. Although some ions undergo passage diffusion through pores in the membrane, in general the passage of water, ions, and various biomolecules is highly selective and often tightly regulated. Enclosure of solutes inside a cell membrane, therefore, results in a strong osmotic pressure on cells, exerted from the inside of the cell onto the membrane toward the outside.

As discussed previously (see Chapter 2), there are various mechanisms to prevent cells from exploding due to this osmotic pressure depending on the cell type, for example, cell walls in bacteria and plant cells, and/or regulation of ion and water pumps that are especially important in eukaryotic cells that in general have no stiff cell wall barrier. These mechanisms can be explored in an *osmotic chamber*. This is a device that allows cells to be visualized using light microscopy in their normal aqueous environment but allowing the external pressure exerted through the liquid environment to be carefully controlled, up to pressures or several tens of atmospheres. Combining cellular pressure control with fluorescence microscopy to probe cell wall proteins and ion channel components has proved informative in our understanding of cellular osmoregulatory mechanisms.

6.2.4 DEFORMING BIOLOGICAL MATTER WITH FLOW

The scientific study of flow and deformation of matter is known as *rheology*. It addresses the interesting observation that all ultimately matter "flows" and deforms under mechanical stress, but the time scales over which this occurs, and how the matter responds after the driving force is removed, varies widely across different materials, and this can tell us a great deal about the underpinning physical properties of that material which are often more complex than just considering how single molecules behave, but rather how molecules behave when they cooperate together in a so-called *emergent* way. This is particularly true for the soft matter that makes up biological systems, which can have highly diverse rheological properties, such as *viscoelasticity* (i.e., a complex combination of viscous and elastic components) over a range of different length and time scales.

The traditional tool which enables rheology measurements is a *rheometer*. These benchtop devices are broadly divided into two types, extensional and shear rheometers, depending up whether strain or stress deformation is induced. Shear rheometers have a largely consensus design of an internal often cylindrical metal plate, which can be rotated concentric to an outer fixed cylindrical plate over a range of speeds enabling the response of the fluid-like sample inside, deformed by the shearing force between the plates, to be measured as a function of distance between from plates typically using optical sensing methods. Extensional rheometers have more varied designs, for example including piezoelectric and acoustic variants, and are most useful for high-viscosity systems in the range 1–1000 cP in which biological properties are related to tensile deformation, such as in filamentous biopolymers.

Microrheology methods can be used to measure rheological properties over a scale of microns down to a few tens of nanometers, much smaller than the capability of traditional rheometers. *Passive microrheology* uses intrinsic thermal fluctuations of a tracer probe (e.g. a micron-sized bead) diffusing through the sample and tracked using optical localization microscopy methods either in bright-field, fluorescence, or laser darkfield (see Chapter 4). Autocorrelation and Fourier analysis of the tracked position of the probe as a function of time can reveal the distinct viscous and elastic components of the material, which act 90° out of phase with each other such that the smaller spatial frequency data are more associated with viscous components whereas higher spatial frequencies are more sensitive to elastic components.

The primary limitation is the size of the probe, in that larger probes experience greater friction drag and so limit the time scale over which measurements can be made and limit the spatial resolution. In this regard, the most ideal probes currently are gold beads of a few tens of nanometers in diameter, which can be tracked with sub-millisecond time resolution and nanoscale spatial resolution using laser darkfield microscopy over long durations since, unlike fluorescent probes, they do not photobleach, and experiments can be high throughput since multiple beads can be tracked simultaneously across relatively large fields of view of up to a few tens of microns in width. Another bead tracking approach is *tethered particle motion* (TPM, see section 6.6.8 for a fuller descrption). Here, typically a biopolymer under investigation is tethered to a tracker bead while the other end of the molecule is tethered to a coverslip surface. By tracking the motion of the tethered bead using similar localization microscopy methods, some of the biomolecular rheological properties can be measured.

Active microrheology uses similar physics concepts but also applies a driving force to induce deformation over this microscopic length scale, for example using optical or magnetic tweezers to controllably deforming single biopolymer molecules (see sections 6.3 and 6.4), which provides high-precision force and extension measurements but is intrinsically low throughput. Higher throughput trapping methods involving *acoustic trapping* of beads, which are tethered via a biomolecule under investigation to a coverslip surface similar to a TPM arrangement, are a useful compromise since these allow several beads in a field of view to be monitored while sacrificing some precision in terms of force and extension (see section 6.6.8).

Aqueous flow can be used to straighten relatively long, filamentous biomolecules in a process called *molecular combing*. For example, by attaching one end of the molecule to a microscope coverslip using specific antibody binding or a specific chemical conjugation group on the end of the molecule, very gentle fluid flow is sufficient to impart enough viscous drag on the molecule to extend it parallel to the direction of flow (for a theoretical discussion of the mechanical responses of biopolymers to external forces, see Chapter 8).

This technique has been applied to filamentous protein molecules such as titin, a large muscle protein discussed later in this chapter in the context of single-molecule force transduction techniques, which can then facilitate imaging of the full extent of the molecule, for example, using fluorescence imaging if fluorophore probes can be bound to specific regions of the molecule, or using transmission electron microscopy. Binding a micron-sized bead to the other end of the molecule increases the viscous drag in the fluid flow and allows higher forces to be exerted on the molecule, and thus larger molecular extensions. This has been used on single DNA molecules *in vitro*, for example, in the study of DNA replication. This technique of extending a biomolecule-tethered bead by flow can also be used in conjunction with optical and magnetic tweezers to facilitate the initial stable trapping of the bead.

The molecular combing technique can be adapted to significantly improve throughput. This is seen most dramatically in the *DNA curtains* technique (Finkelstein et al., 2010). Here, DNA molecules are tethered on a nanofabricated microscope coverslip containing etched platforms for tether attachment such that the tethered end of a molecule is clear from the coverslip surface, thus minimizing the effects of surface forces on the molecule, which are often difficult to quantify and can impair the biological function of DNA. Optimization of the tethering incubation conditions allows several individual DNA molecules to be tethered in line, spaced apart on the coverslip surface by only a few hundred nanometers.

The molecules can be visualized by labeling the DNA using a range of DNA-binding dyes and imaging the molecules in real time using fluorescence microscopy. This can be used to investigate the topological, polymer physics properties of single DNA molecules, but can also be used in investigating a variety of different molecular machines that operate by binding to DNA by labeling a component of the machine with a different color fluorophore and then utilizing dual-color fluorescence imaging to monitor the DNA molecules and molecular machines bound to them simultaneously. The key importance of this technique is that it allows several tens of individual DNA molecules to be investigated simultaneously under the same flow and imaging conditions, improving statistical sampling, and subsequent biological interpretation of the data, enormously.

KEY BIOLOGICAL APPLICATIONS: RHEOLOGYTOOLS

Molecular separation and identification.

6.3 OPTICAL FORCE TOOLS

There are several biophysical techniques that utilize the linear momentum associated with a single photon of light to generate forces that then can be used to probe and manipulate single biomolecules and even whole cells. *Optical tweezers* utilize this approach, as do the related *optical stretcher* technology. Optical tweezers are an exceptionally powerful tool for manipulating single biomolecules and characterizing many aspects of their force-dependent features, and for this reason we explore the theory of their operation in detail here. Although single biomolecules themselves cannot be optically trapped with any great efficiency (some early optical tweezers experiments toyed with rather imprecise manipulation of chromosomes), they can be manipulated via a micron-sized optically trapped bead. But there are also methods that can utilize the angular momentum of photons to probe the rotary motion of the biological material. Other applications of optics, which allow monitoring of biological forces, include Brillouin scattering, polarization microscopy, and Förster resonance energy transfer (FRET).

6.3.1 BASIC PRINCIPLES OF OPTICAL TWEEZERS

The ability to trap particles using laser radiation pressure was reported first by Arthur Ashkin, the pioneer of optical tweezers (also known as *laser tweezers*) (Ashkin, 1970). This was a relatively unstable 1D trap consisting of two juxtaposed laser beams whose photon flux resulted in equal and opposite forces on a micron-sized glass bead. The modern form of the standard 3D *optical trap* (specifically described as a *single-beam gradient force trap*), developed in the 1980s by Ashkin et al. (1986), results in a net optical force on a refractile, dielectric particle, which has a higher refractive index than the surrounding medium, roughly toward the intensity maximum of a focused laser. These optical force transduction devices have since been put to very diverse applications for the study of single-molecule biology (for older but still rewarding reviews, see Svoboda and Block, 1994 for an accessible explanation of the physics, and Moffitt et al., 2008 for a compilation of some of the applications).

Photons of light carry linear momentum p given by the de Broglie relation $p = E/c = h\nu/c = h/\lambda$, for a wave of energy E, frequency ν, and wavelength λ where c is the speed of light and h Plank's constant. Photon momentum results in radiation pressure if photons are scattered from an object. Also, if refraction occurs at the point of a photon emerging from an optically transparent particle, there is a change in beam direction and intensity, and thus a change in momentum, which results in an equal and opposite force on the particle. Standard optical tweezers utilize this effect as a gradient force in the focal plan of a light microscope.

KEY POINT 6.3

Basics of optical tweezers:

1 If a refractile particle changes the direction of a photon, then a force acts on it according to Newton's third law, since photons have momentum.
2 The intensity to generate optical forces large enough to overcome thermal forces at room temperature is high and so requires a laser.
3 If a laser beam is brought to a steep focus, the combination of scatter and refractive force results in a net force roughly toward the laser focus.

To understand the principles of optical trapping, we can consider the material in the particle through which photons propagate to be composed of multiple electric dipoles on the same length scale as individual atoms. A propagating electromagnetic light wave through the particle imparts a small force on each electric dipole, which time averages to point in the direction of the intensity gradient of the photon beam. A full derivation of the forces involved requires a solution of *Maxwell's electromagnetic equations* (see Rohrbach, 2005); however, we can gain qualitative insight by considering a *ray-optic* depiction of the passage of light through a particle (Figure 6.1a).

The photon energy flux dE in a small time dt of a laser beam parallel to the optic (z) axis of total power P propagating through the particle is given by

$$(6.2) \qquad\qquad dE = Pdt = cdp$$

where
  c is the speed of light
  p is the total momentum of the beam of photons, such that dp is the small associated change in momentum in time dt

If we assume that the lateral optical trapping force F arises mainly from photons traveling close to the optic axis, which is exerted as photons exit the particle at a slightly deviated direction from the incident beam by a small angle θ, then F is given by the rate of change of photon momentum projected onto the x-axis:

$$(6.3) \qquad\qquad F = \sin\theta\frac{dp}{dt} \approx \theta\frac{dp}{dt} = \frac{P\theta}{c}$$

where we assume the small-angle approximation if θ is measured in radians. Typically, for optical tweezers, θ will be a few tens of milliradians, equivalent to a few degrees (see Worked Case Example 6.1).

Typically, a single-mode laser beam (the so-called TEM00 that is the lowest-order fundamental transverse mode of a laser resonator head and has a Gaussian intensity profile across the beam) is focused using a high NA objective lens onto a refractile, dielectric particle (typically a bead of diameter $\sim 10^{-6}$ m composed of latex or glass) whose refractive index ($\sim 1.4–1.6$) is higher than that of the surrounding water solution (~ 1.33) to form a confocal intensity volume (see Chapter 4). Optical trapping does not require symmetrical particles though most often the particles used are spheres. A stably trapped particle is located slightly displaced axially by the forward scatter momentum from the laser focus, which is the point

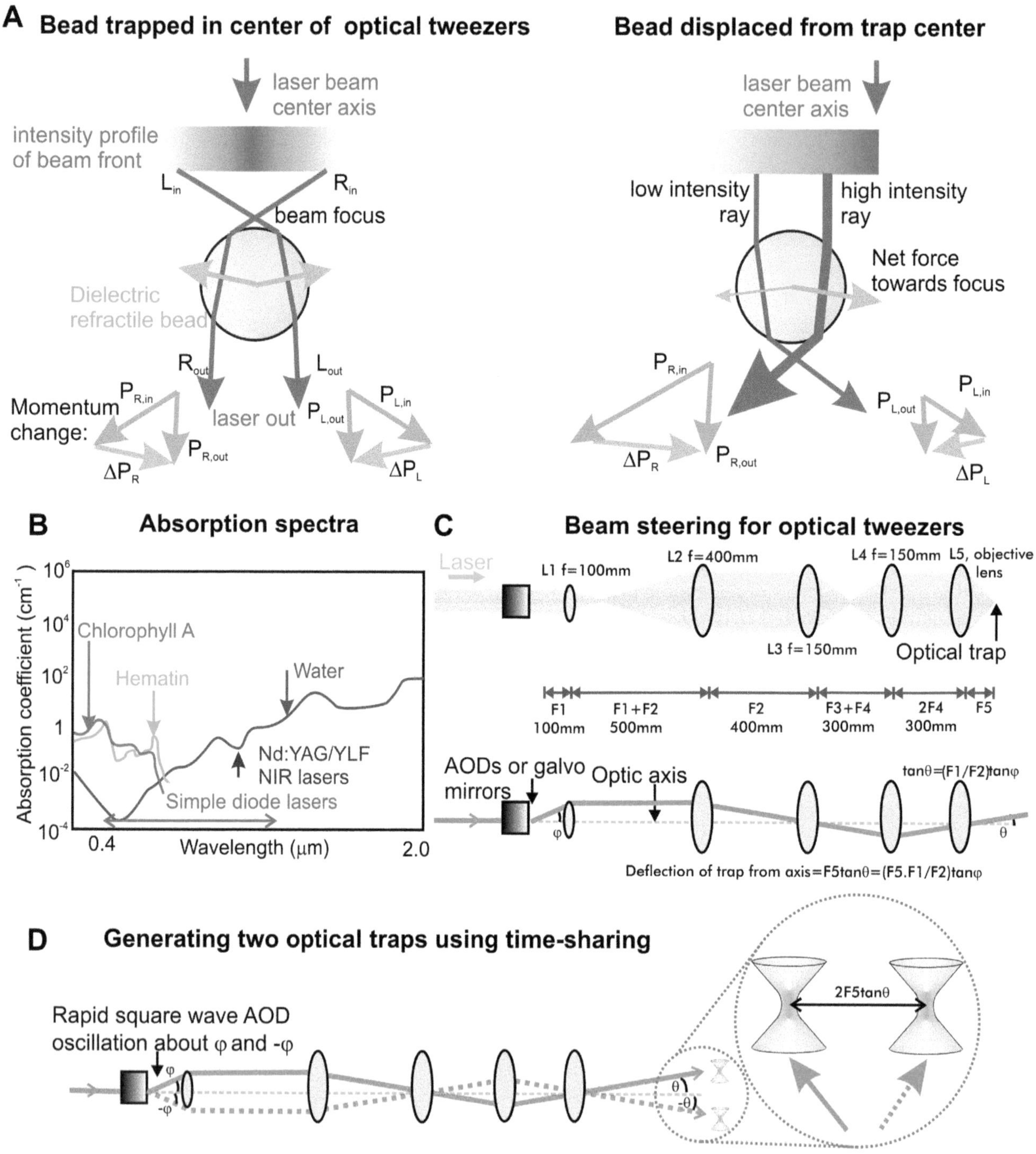

A Bead trapped in center of optical tweezers

Bead displaced from trap center

laser beam
center axis

intensity profile
of beam front

L_{in} R_{in}

beam focus

Dielectric
refractile bead

R_{out} L_{out}

Momentum $P_{R,in}$ $P_{L,in}$
change: laser out $P_{L,out}$
      $P_{R,out}$
ΔP_R ΔP_L

laser beam
center axis

low intensity
ray

high intensity
ray

Net force
towards focus

$P_{R,in}$ $P_{L,in}$
                $P_{L,out}$
ΔP_R $P_{R,out}$ ΔP_L

B   Absorption spectra

Chlorophyll A

Hematin

Water

Nd:YAG/YLF
NIR lasers

Simple diode lasers

Absorption coefficient (cm^{-1})

10^6
10^4
10^2
1
10^{-2}
10^{-4}

0.4   Wavelength (μm)   2.0

C   Beam steering for optical tweezers

Laser

L1 f=100mm   L2 f=400mm   L4 f=150mm L5, objective
                                                lens

L3 f=150mm   Optical trap

F1   F1+F2   F2   F3+F4   2F4   F5
100mm   500mm   400mm   300mm   300mm

AODs or galvo
mirrors   Optic axis

$\tan\theta = (F1/F2)\tan\varphi$

φ θ

Deflection of trap from axis $= F5\tan\theta = (F5.F1/F2)\tan\varphi$

D   Generating two optical traps using time-sharing

Rapid square wave AOD
oscillation about φ and -φ

$2F5\tan\theta$

φ
$-\varphi$ θ
                                        $-\theta$

FIGURE 6.1 (See color insert.) Optical tweezers. (a) The sum of refractive forces through a bead in optical tweezers results in a net force roughly toward the laser focus. (b) Many biomolecules have a strong absorption in the visible light range, illustrated here with chlorophyll in plants and hematin in the blood; water absorbs weakly in the visible region, increasing absorption in the near infrared, but with a local dip in absorption at ~1 μm wavelength. (c) Typical arrangement of beam steering (and expansion) for optical tweezers. (d) Two (or more) optical traps can be generated by time-sharing of the laser beam using rapid deflection by an AOD.

at which the gradient of the intensity of the focused laser light in the lateral xy focal plane of the microscope is zero.

If the particle is displaced laterally from the focus, then the refraction of the higher-intensity light fraction through the particle close to the focus causes an equal and opposite force on the particle, which is greater than that experienced in the opposite direction due to refraction of the lower-intensity portion of the laser beam. The particle therefore experiences a net restoring force back to the laser focus and hence is "trapped," provided any external force perturbations on the particle do not displace it beyond the physical extent of the optical tweezers.

In practice, stable optical tweezers require a diffraction-limited focus; photons entering the focal waist of the confocal volume at a steep angle relative to the optical axis result in high-intensity gradients across the trap profile and so contribute the most to the optical restoring force. To achieve this steepness of angle requires a high NA objectives lens in the range ~1.2–1.5 often combined with marginally overfilling the back aperture of the objective lens with collimated incident laser light. The actual size of the optical tweezer trapping volume is determined by the spatial extent of the diffraction-limited interference pattern in the vicinity of the laser focus, which laterally (xy) has a width of ~λ, whereas axially (z) this is more like two to three times times λ (see Chapter 4). This implies that the intensity gradient is reduced by the same factor. Combining this reduction in axial gradient stiffness with a weakness of the axial trapping force due to forward scatter radiation pressure results in axial trap stiffness values (i.e., a measure of the restoring force for a given small displacement of the particle) that are smaller than the lateral stiffness by a factor of ~3–8, depending on the particle size and specific wavelength used.

6.3.2 OPTICAL TWEEZER DESIGNS IN PRACTICE

Typical bead diameters are ~0.2–2 μm, though optical trapping has been demonstrated on gold-coated particles with a diameter as small as 18 nm (Hansen et al., 2005). The wavelength used is normally near infrared (NIR) of ~1 μm, the choice being made on the basis of optimization of trap stiffness and size while minimizing sample photodamage. Some damage is due to a localized heating effect from laser absorption either by the water solvent or chromophores in the biological sample, at a level of ~1–2 K for every 100 mW of NIR laser power. However, the most likely cause of biological damage is due to the generation of free radicals in water through single- and multiphoton absorption effects found at high local intensities at the focus of a trap, which can bind indiscriminately to biological structures.

The choice of wavelength used is a compromise between two competing absorption factors. One is that absorption of electromagnetic radiation by water itself increases sharply from visible into the infrared, peaking at a wavelength of ~3 μm. However, natural biological chromophores can absorb strongly at visible light wavelengths, as well as increasing the likelihood for generating free radicals; therefore, a wavelength of ~1 μm is a good compromise. At wavelengths between 1 and 1.2 μm, there is also a small local dip in the water absorption spectrum, which makes Nd:YAG (λ = 1.064 μm) and Nd:YLF (λ = 1.047 μm) crystal lasers attractive choices (Figure 6.1b).

In most applications, optical tweezers are coupled to a light microscope. An NIR laser beam is expanded usually to marginally overfill the back aperture of a high NA objective lens, which is steered by upstream optics to rotate the beam through the back aperture, resulting in lateral displacement of the optical trap at the focal plane in a microscope flow cell (Figure 6.1c). Steering of the optical trap can be done using mirrors positioned in a conjugate plane to the objective lens back aperture. However, it is common in many applications to use higher bandwidth steering with *acousto-optic deflectors* (*AODs*), discussed in the following text. The laser beam for generating a conventional gradient force optical trap can be split before reaching the sample, either using a space-dividing optical component such as a glass splitter cube or by time-sharing the beam along different optical paths in the microscope setup to generate more than one optical tweezers (Figure 6.1d). Time-sharing is most popularly obtained by passing the initial beam through AODs.

An AOD is composed of an optical crystal, typically of *tellurium dioxide* (TeO$_2$) in a synthetic tetragonal structure (also known as the crystal *paratellurite*). In this form, TeO$_2$ is a nonlinear optical crystal that is transparent through the visible and into the mid-infrared range of the electromagnetic spectrum, with a high refractive index of ~2.2, exhibiting a relatively slow shear-wave propagation along the [110] crystal plane. These crystals exhibit *photoelasticity*, in that mechanical strain in the crystal results in a local change in optical permittivity, manifest as there being a spatial dependence on refractive index. These factors facilitate standing wave formation in the crystal parallel to the [110] plane from acoustic vibration if a radio-frequency forcing function is applied from a piezoelectric transducer from one end of the crystal, with the other end of the crystal at the far end of the [110] plane acting as a fixed point in being coupled to an acoustic absorber (Figure 6.2a). The variation in refractive index can be modeled as

$$(6.4) \qquad n(z,t) = n_0 + \Delta n \cos(\omega t - kz)$$

where
  n_0 is the unstrained refractive index
  ω is the angular frequency of the forcing function
  k is the wave vector of the sound wave parallel to the z-axis (taken as parallel to the [110] plane)

The factor Δn is given by the photoelastic tensor parameters. The result is a sinusoidally varying function of n with a typical spatial periodicity of around a few hundred nanometers, which thus has similar attributes to a diffraction grating for visible/infrared light. The diffracted light is a mixture of two types, which due to *Raman–Nath diffraction* can occur at an arbitrary angle of incidence at lower acoustic frequencies (most prevalent at ~10 MHz or less), and that due to Bragg diffraction (see Chapter 4) at higher acoustic frequencies more typically >100 MHz, which occurs at a specific angle of incident θ_B such that

$$(6.5) \qquad \sin\theta_B = \frac{-\lambda f}{2 n_i v}\left(1 + \frac{v^2}{\lambda^2 f^2}\left(n_i^2 - n_d^2\right)\right)$$

where
  λ is the free-space wavelength of the incident light
  f is the acoustic wave frequency
  n_i and n_d are the incident and diffracted wave refractive indices of the medium, respectively
  v is the acoustic wave speed

AODs are normally configured to use the first-order Bragg diffraction peak angle θ_d for beam steering, which satisfies $\sin(\theta_d) = \lambda/\Lambda$ where Λ is the acoustic wavelength. The maximum efficiency of an AOD is ~80% in terms of light intensity propagated into the first-order Bragg diffraction peak (the remainder composed of Raman–Nath diffraction and higher-order Bragg peaks), and for steering in the sample focal plane in both x and y requires two orthogonal AODs; thus, ~40% of incident light is not utilized, which can be disadvantageous if a very high stiffness trap is desired.

An AOD has a frequency response of >10^7 Hz, and so the angle of deflection can be rapidly alternated between ~5° and 10° on the submicrosecond time scale, resulting in two time-shared beams separated by a small angle, which can then each be manipulated to generate a separate optical trap. Often, two orthogonally crossed AODs are employed to allow not only time-sharing but also independent full 2D control of each trap in the lateral focal plane of the microscope, over a time scale that is three orders of magnitude faster than the relaxation time due to viscous drag on a micron-sized bead. This enables feedback type experiments to be applied. For example, if there are fluctuations to the molecular force of a tethered single molecule, then the position of the optical trap(s) can be rapidly adjusted to maintain a constant molecular force (i.e., generating a *force clamp*), which allows, for example, details of the kinetics of molecular unfolding and refolding to be explored in different precise force regimes.

An alternative method to generating multiple optical traps involves physically splitting the incident laser beam into separate paths using splitter cubes that are designed to transmit a certain proportion (normally 50%) of the beam and reflect the remainder from a dielectric interface angled at 45° to the incident beam so as to generate a reflected beam path at 90° to the original beam. Other similar optical components split the beam on the basis of its linear polarization, transmitting the parallel (p) component and reflecting the perpendicular (s) component, which has an advantage over using nonpolarizing splitter cubes in permitting more control over the independent laser powers in each path by rotating the incident *E*-field polarization vector using a half-wave plate (see Chapter 3). These methods can be used to generate independently steerable optical traps.

The same principle can be employed to generate more than two optical traps; however, in this case, it is often more efficient to use either a *digital micromirror array* or a spatial light modulator (SLM) component. Both optical components can be used to generate a phase modulation pattern in an image plane conjugate to the Fourier plane of the sample's focal plane, which results in controllable beam deflection into, potentially, several optical traps, which can be manipulated not only in x and y but also in z. Such approaches have been used to generate tens of relatively weak traps whose position can be programmed to create an *optical vortex* effect, which can be used to monitor fluid flow around biological structures. The primary disadvantage of digital micromirror array or SLM devices is that they have relatively low refresh bandwidths of a few tens of Hz, which limit their utility to monitoring only relatively slow biological processes, if mobile traps are required. But they have an advantage in being able to generate truly 3D optical tweezers, also known as *holographic optical traps* (Dufresne and Grier, 1998).

Splitting light into a number of N traps comes with an obvious caveat that the stiffness of each trap is reduced by the same factor N. However, there are many biological questions that can be addressed with low stiffness traps, but the spatial fluctuations on trapped beads can be >10 nm, which often swamps the molecular level signals under investigation. The theoretical upper limit to N is set by the lowest level of trap stiffness, which will just be sufficient to prevent random thermal fluctuations pushing a bead out of the physical extent of the trap. The most useful multiple trap arrangement for single biomolecule investigations involves two standard Gaussian-based force gradient traps, between which a single biomolecule is tethered.

6.3.3 CHARACTERIZING DISPLACEMENTS AND FORCES IN OPTICAL TWEEZERS

The position of an optically trapped bead can be determined using either the bright-field image of the bead onto a charge-coupled device (CCD) camera or *quadrant photodiode* (*QPD*) or, more commonly, to use a laser interferometry method called back focal plane (BFP) detection. The position of the center of a bead can be determined using similar centroid determination algorithms to those discussed previously for super-resolution localization microscopy (see Chapter 4). QPDs are cheaper than a CCD camera and have a significantly higher bandwidth, allowing determination of x and y from the difference in voltage signals between relevant halves of the quadrant (Figure 6.2b) such that

(6.6)
$$x = \alpha\left(\left(V_2 + V_3\right) - \left(V_1 - V_4\right)\right)$$
$$y = \alpha\left(\left(V_1 + V_2\right) - \left(V_3 - V_4\right)\right)$$

where α is a predetermined calibration factor. However, bright-field methods are *shot noise* limited—shot noise, also known as *Poisson noise*, results from the random fluctuations of the number of photons detected in a given sampling time window and of the electrons in the

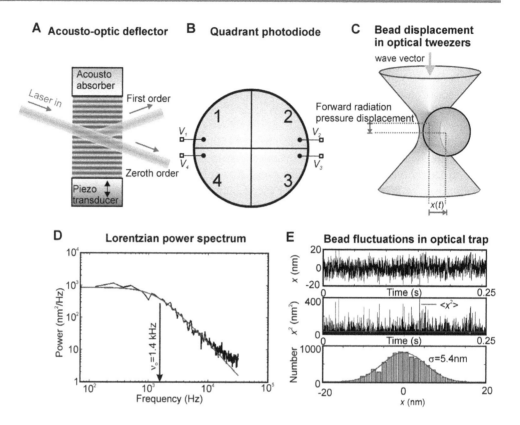

FIGURE 6.2   Controlling bead deflections in optical tweezers. (a) In an AOD, radio-frequency driving oscillations from a piezo transducer induce a standing wave in the crystal that acts as diffraction grating to deflect an incident laser beam. (b) Schematic of sectors of a quadrant photodiode. (c) Bead displacement in optical tweezers. (d) Bead displacements in an optical trap, here shown with a trap of stiffness 0.15 pN/nm, have a Lorentzian-shaped power, resulting in a characteristic corner frequency (here 1.4 kHz) that allows the trap stiffness to be determined, which can also be determined from (e) the root mean squared displacement, shown here for data of the same trapped bead (see Leake, 2001).

photon detector device, approximated by a Poisson distribution. The relatively small photon budget limits the speed of image sampling before shot noise in the detector swamps the photon signal in each sampling time window.

For BFP detection, the focused laser beam used to generate an optical tweezer trap propagates through a specimen flow cell and is typically recollimated by a condenser lens. The BFP of the condenser lens is then imaged onto a QPD. This BFP image represents the Fourier transform of the sample plane and is highly sensitive to phase changes of the trapping laser propagating through an optically trapped bead. Since the trapping laser is highly collimated, interference occurs between this refracted beam and the undeviated laser light propagating through the sample. The shift in the intensity centroid of this interference pattern on the QPD is a sensitive metric of the displacement between the bead center and the center of the optical trap.

In contrast to bright-field detection of the bead, BFP detection is not shot noise limited and so the effective photon budget for detection of bead position in the optical trap is large and can be carved into small submicrosecond sampling windows with sufficient intensity in each to generate sub-nanometer estimates on bead position, with the high sampling time resolution limited only by the ~MHz bandwidth of QPD detectors. Improvements in localization precision can be made using a separate BFP detector laser beam of smaller wavelength than the trapping laser beam, coaligned to the trapping beam.

The stiffness k of an optical trap can be estimated by measuring the small fluctuations of a particle in the trap and modeling this with the *Langevin equation*. This takes into account the restoring optical force on a trapped particle along with its viscous drag coefficient γ due to the viscosity of the surrounding water solvent, as well as random thermally driven fluctuations in force (the *Langevin force*, denoted as a random functional of time *F(t)*):

$$(6.7) \qquad kx + \gamma v = F(t)$$

where x is the lateral displacement of the optically trapped bead relative to the trap center and v is its speed (Figure 6.2c), and $\langle F(t) \rangle$ when averaged over large times is zero. The inertial term in the Langevin equation, which would normally feature, is substantially smaller than the other two drag and optical spring force terms due to the relatively small mass of the bead involved and can be neglected. The motion regime in which optically trapped particles operate can be characterized by a very small *Reynolds number*, with the solution to Equation 6.7 being under typical conditions equivalent to *over-damped simple harmonic motion*. The Reynolds number R_e is the measure of ratio of the inertial to drag forces:

$$(6.8) \qquad R_e = \frac{\rho v l}{\eta}$$

where
   ρ is the density of the fluid (this case water) of viscosity η (specifically termed the "dynamic viscosity" or "absolute viscosity" to distinguish it from the "kinematic viscosity," which is defined as η/ρ)
   l is a characteristic length scale of the particle (usually the diameter of a trapped bead)

The viscous drag coefficient on a bead of radius r can be approximated from *Stokes law* as $6\pi r \eta$, which indicates that its speed v, in the presence of no other external forces, is given by

$$(6.9) \qquad v = \frac{kv}{6\pi r \eta}$$

Note that this can still be applied to nonspherical particles in which r then becomes the effective *Stokes radius*. Thus, the maximum speed is given when the displacement between bead and trap centers is a maximum, and since the physical size of the trap in the lateral plane has a diameter of ~λ, this implies that the maximum x is ±λ/2. A reasonably stiff optical trap has a stiffness of ~10^{-4} N m^{-1} (or ~0.1 pN nm^{-1}, using the units that are commonly employed by users of optical tweezers). The speed v of a trapped bead is usually no more than a few times its own diameter per second, which indicates typical R_e values of ~10^{-8}. As a comparison, the values associated with the motility of small cells such as bacteria are ~10^{-5}. This means that there is no significant *gliding* motion as such (in either swimming cells or optically trapped beads). Instead, once an external force is no longer applied to the particle, barring random thermal fluctuations from the surrounding water, the particles come to a halt. To arrive at the same sort of Reynolds number for this non-gliding condition of cells for, for example, a human swimming, they would need to be swimming in a fluid that had a viscosity of *molasses* (or *treacle*, for readers in the United Kingdom).

Equation 6.7 describes motion in a parabolic-shaped energy potential function (if k is independent of x, the integral of the trapping force kx implies trapping potential of $kx^2/2$). The position of the trapped bead in this potential can be characterized by the power spectral density $P(v)$ as a function of frequency v of a *Lorentzian* shape (see Wang, 1945) given by

$$(6.10) \qquad P(v) = \frac{k_B T}{2\pi^3 \left(v^2 + v_0^2\right)}$$

The power spectral density emerges from the Fourier transform solution to the bead's equation of motion (Equation 6.7) in the presence of the random, stochastic Langevin force. Here, ν_0 is the *corner frequency* given by $k/2\pi\gamma$. The corner frequency is usually ~1 kHz, and so provided the x position is sampled at a frequency, which is an order of magnitude or more greater than the corner frequency, that is, >10 kHz, a reasonable fit of P to the experimental power spectral data can be obtained (Figure 6.2d), allowing the stiffness to be estimated. Alternatively, one can use the *equipartition theorem* of thermal physics, such that the mean squared displacement of an optically trapped bead's motion should satisfy

$$(6.11) \qquad \frac{k\langle x^2 \rangle}{2} = \frac{k_B T}{2} \therefore k = \frac{k_B T}{\langle x^2 \rangle}$$

Therefore, by estimating the mean squared displacement of the trapped bead, the trap stiffness may be estimated (Figure 6.2e). The Lorentzian method has an advantage in that it does not require a specific knowledge of a bead displacement calibration factor for the positional detector used for the bead, simply a reasonable assumption that the response of the detector is linear with small bead displacements.

Both methods only generate estimates for the trap stiffness at the center of the optical trap. For low-force applications, this is acceptable since the trap stiffness is constant. However, some single-molecule stretch experiments require access to relatively high forces of >100 pN, requiring the bead to be close to the physical edge of the trap, and in this regime there can be significant deviations from a linear dependence of trapping force with displacement. To characterize, the position of an optical trap can be oscillated using a square wave at ~100 Hz of amplitude ~1 μm; the effect at each square wave alternation is to rapidly (depending on the signal generator, in <10^{-7} s) displace the trap focus such that the bead is then at the very edge of the trap almost instantaneously. Then, the speed v of movement of the bead back toward the trap center can be used to calculate the drag force; using Equation 6.7 and averaging over many cycles such that the mean of the Langevin force is zero imply that the average drag force should equal the trap restoring force at each different value of x, and therefore the trap stiffness can be characterized for the full lateral extent of the trap. Similarly, optical tweezers can be scanned across a surface-immobilized bead in order to determine the precise response of the BFP detector at different relative separations between a bead center and optical trap center.

6.3.4 APPLICATIONS OF OPTICAL TWEEZERS

Appropriate latex or silica-based microspheres suitable for optical trapping can be commercially engineered to include a chemical coating of a variety of different compounds, most importantly carboxyl, amino, and aldehyde groups that can be used as adapter molecules to conjugate to biomolecules. Using standard bulk conjugation chemistry, these chemical groups on the bead surface can be bound either directly to biomolecules or more commonly linked to an adapter molecule such as a specific antibody or a biotin group that will then bind to a specific region of a biomolecule of interest (see Chapter 7). Chemically functionalizing microspheres in this way allows single biomolecules to be attached to the surface of an optically trapped bead and tethered to a fixed surface such as a microscope coverslip (Figure 6.3a).

Several of the first optical tweezers experiments involved the large muscle protein titin (Tskhovrebova et al., 1997), which enabled the mechanical elasticity of single titin molecules to be probed as a function of its molecular extension by laterally displacing the microscope stage to stretch the molecule relative to the trapped bead. This technique was further modified to tether a single titin molecule between an optically trapped bead and a micropipette, which secured to a second bead attached to the other end of the molecule by suction forces (Figure 6.3b), which offered some improvement in fixing the tether axis to be parallel to the lateral plane of movement of the trap thus making the most out of the lateral trapping force available (Kellermayer et al., 1997).

This method was also employed to measure the mechanical properties of single DNA molecules (Smith, 1996), which enabled estimation of the *persistence length* of DNA of ~50 nm based on worm-like chain modeling (see Chapter 8) as well as enabling observations of phenomena such as the *overstretch transition* in which the stiffness of DNA suddenly drops at forces in the range 60–70 pN due to structural changes to the DNA helix. Similarly, optical tweezers have been used to measure the force dependence of folding and unfolding of model structural motifs, such as the RNA hairpin (see Chapter 2, and Liphardt et al., 2001). These techniques quantify the refolding of a molecule, indicating that they are far from a simple reversal of the unfolding mechanism (see Sali et al., 1994).

Tethering a single biomolecule between two independent optically trapped beads (Figure 6.3c) offers further advantages of fast feedback experiments to clamp both the molecular force and position while monitoring the displacements of two separate beads at the same time (Leake et al., 2004). Typically, a single-molecule tether is formed by tapping two optically trapped beads together, one chemically conjugated to one end of the molecule, while the other is coated with chemical groups that will bind to the other end. The two optically trapped beads are tapped together and then pulled apart over several cycles at a frequency of a few hertz. There is, however, a probability that the number of molecules tethered between the two beads is >1. If the probability of a given tether forming is independent of the time, then this process can be modeled as a *Poisson distribution*, such that probability $P_{teth}(n)$ for forming n tethers is given by $\langle n \rangle n \exp[-\langle n \rangle]/n!$, with $\langle n \rangle$ the average number of observed tethers formed between two beads (see Worked Case Example 6.1).

The measurement of the displacement of a trapped bead relative to the center of the optical trap allows the axial force experienced by a tethered molecule to be determined from knowledge of the optical tweezer stiffness. The relationship between the force and the end-to-end extension of the molecule can then be experimentally investigated. In general, the main contribution to this force is entropic in origin, which can be modeled using a variety of polymer physics formulations to determine parameters such as equivalent chain segment lengths in the molecule, discussed in Chapter 8.

Several single-molecule optical tweezers experiments are performed at relatively low forces of just a few piconewtons, which is relevant to the physiological forces experienced in living cells for a variety of different *motor proteins* (see Chapter 2). These studies famously have included those of the muscle protein myosin interacting with actin (Finer et al., 1994), the

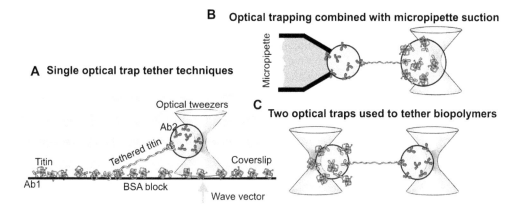

FIGURE 6.3 Tethering single biopolymers using optical tweezers. (a) A biopolymer, exemplified here by the giant molecule title found in muscle tissue, can be tethered between a microscope coverslip surface and an optically trapped bead using specific antibodies (Ab1 and Ab2). (b) A biopolymer tethered may also be formed between an optically trapped bead and another bead secured by suction from a micropipette. (c) Two optically trapped beads can also be used to generate a single-molecule biopolymer tether, enabling precise mechanical stretch experiments.

kinesin protein involved in cell division (Svoboda et al., 1993), as well as a variety of proteins that use DNA as a track. The state of the art in optical tweezers involves replacing the air between the optics of a bespoke optical tweezers setup with helium to minimize noise effects due to the temperature-dependent refraction of lasers through gases, which has enabled the transcription of single-nucleotide base pairs on a single-molecule DNA template by a single molecule of the ribonucleic acid polymerase motor protein enzyme to be monitored directly (Abbondanzieri et al., 2005).

6.3.5 NON-GAUSSIAN BEAM OPTICAL TWEEZERS

"Standard" optical tweezers are generated from focusing a Gaussian profile laser beam into a sample. However, optical trapping can also be enabled using non-Gaussian profile beams. For example, a Bessel beam may be used. A Bessel beam, in principle, is diffraction free (Durnin et al., 1987). They have a Gaussian-like central peak intensity of width roughly one wavelength, as with a single-beam gradient force optical trap; however, they have in theory zero divergence parallel to the optic axis. In practice, due to finite sizes of optical components used, there is some remaining small divergence at the ~mrad scale, but this still results in minimal spreading of the intensity pattern over length scales of 1 m or more.

The main advantage of optical trapping with a Bessel beam, a *Bessel trap*, is that since there is minimal divergence of the intensity profile of the trap with depth into the sample, which is ideal for generating optical traps far deeper into a sample than permitted with conventional Gaussian profile traps. The Bessel trap profile is also relatively unaffected by small obstacles in the beam path, which would cause a significant distortion for standard Gaussian profile traps; a Bessel beam can reconstruct itself around an object provided a proportion of the light waves is able to move past the obstacle. Bessel beams can generate multiple optical traps that are separated by up to several millimeters.

Optical tweezers can also be generated using optical fibers. The numerical aperture of a single-mode fiber is relatively low (~0.1) generating a divergent beam from its tip. Optical trapping can be achieved using a pair of juxtaposed fibers separated by a gap of a few tens of microns (Figure 6.4a). A refractile particle placed in the gap experiences a combination of forward scattering forces and lateral forces from refraction of the two beams. This results in an optical trap, though 10–100 times less stiffness compared to conventional single-beam gradient force traps for a comparable input laser power. Such an arrangement is used to trap relatively large single cells, in a device called the "optical stretcher."

The refractive index of the inside of a cell is in general heterogeneous, with a mean marginally higher than the water-based solution of the external environment (see Chapter 3). This combined with the fact that cells have a defined compliance results in an optical stretching effect in these optical fiber traps, which has been used to investigate the mechanical differences between normal human cells and those that have a marginally different stiffness due to being cancerous (Gück et al., 2005). The main disadvantage with the method is that the laser power required to produce measurable probing of cell stiffness also results in large rises in local temperature at the NIR wavelengths nominally employed—a few tens of degree centigrade above room temperature is not atypical—which can result in significant thermal damage to the cell.

It is also possible to generate 2D optical forces using an evanescent field, similar to that discussed for TIRF microscopy (see Chapter 3); however, to trap a particle stably in such a geometry requires an opposing, fixed structure oriented against the direction of the force vector, which is typically a solid surface opposite the surface from which the evanescent field emanates (the light intensity is greater toward the surface generating the evanescent field and so the net radiation pressure is normal to that away from the surface). This has been utilized in the cases of nanofabricated photonic waveguides and at the surface of optical fibers. There is scope to develop these techniques into high-throughput assays, for example, applied in a multiple array format of many optical traps, which could find application in new biosensing assays.

6.3.6 CONTROLLING ROTATION USING "OPTICAL SPANNERS"

Cells have, in general, rotational asymmetry and so their angular momentum can be manipulated in the optical stretcher device. However, the trapped particles used in conventional Gaussian profile optical tweezers are usually microspheres with a high degree of symmetry and so experience close to zero net angular momentum, about the optic axis. It is, in general, technically nontrivial to controllably impart a nonzero mean torque using this approach.

Cells have, in general, rotational asymmetry and so their angular momentum can be manipulated in an optical stretcher device. However, the trapped particles used in conventional Gaussian profile optical tweezers are usually symmetrical microspheres and so experience zero net angular momentum about the optic axis. Therefore, it is not possible to controllably impart a nonzero mean torque.

There are two practical ways that achieve this using optical tweezers; however, both can lay claims to being in effect *optical spanners*. The first method requires introducing an asymmetry into the trapped particle system to generate a lever system. For example, one can controllably fuse two microspheres such that one of the beads is chemically bound to a biomolecule of interest to be manipulated with torque, while the other is trapped using standard Gaussian profile optical tweezers whose position is controllably rotated in a circle centered on the first bead (Figure 6.4b). This provides a wrench-like effect, which has been used for studying the F1-ATPase enzyme (Pilizota et al., 2007). F1 is one of the rotary molecular motors, which, when coupled to the other rotary machine enzyme Fo, generates molecules of the universal biological fuel ATP (see Chapter 2). Fused beads can be generated with reasonable efficiency by increasing the ionic strength (usually by adding more sodium chloride to the solution) of the aqueous bead media to reduce to the Debye length for electrostatic screening (see Chapter 8) with the effect of reducing surface electrostatic repulsion and facilitating hydrophobic forces to stick beads together. This generates a mixed population of bead *multimers* that can be separated into bead pairs by centrifugation in a viscous solution composed of sucrose such that the bead pairs are manifested as a distinct band where the hydrodynamic, buoyancy, and centripetal forces balance.

The second method utilizes the angular momentum properties of light itself. *Laguerre–Gaussian* beams are generated from higher-order laser modes above the normal TEM00 Gaussian profile used in conventional optical tweezers, by either optimizing for higher-order lasing oscillation modes from the laser head itself or by applying phase modulation optics in the beam path, typically via an SLM. Combining such asymmetrical laser profiles (Simpson et al., 1996) or Bessel beams with the use of helically polarized light on multiple particles on single birefringent particles that have differential optical polarizations relative to different spatial axes such as certain crystal structures (e.g., calcite particles, see La Porta and Wang, 2004) generates controllable torque that has been used to study interactions of proteins with DNA (Forth et al., 2011).

6.3.7 COMBINING OPTICAL TWEEZERS WITH OTHER BIOPHYSICAL TOOLS

Standard Gaussian profile dual-optical tweezers have been used to generate so-called negative supercoiling in single-molecule tethered DNA samples (King et al, 2019). Here, although the two optically trapped beads at either end of a single DNA molecule tether are spherical and ostensibly identical, some level of nonzero torsional control can be enabled through exploiting the fact that B-DNA at low tension is intrinsically twisted; if this molecule is then stretched by moving the two beads apart, then its natural twist will try to unravel. This imposes a small torque on the biotin–streptavidin covalent bond links between the beads and the end of the DNA, which can be sufficient to increase the likelihood for stochastically breaking these bonds during the time scale of a typical experiment. During this transient bond breakage before stochastically rebinding, the DNA can then unravel by a few turns, hence negative supercoiling is inserted into the molecule thereby facilitating a range of experiments to investigate the effect of negative supercoiling on DNA function and mechanics.

This technique does not use direct angular momentum control on the optically trapped beads, though small asymmetries in the beads may perhaps result in small nonzero torque which might work to slow down free rotation of the beads in the optical traps, but relies more on the fact that the frictional drag of the relatively large 4.5 μm diameter beads used is relatively high and so the time scale of free bead rotation of several milliseconds which results is higher than the DNA unraveling time scale when the biotin-avidin bonds are transiently broken.

Optical tweezers can be incorporated onto a standard light microscope system, which facilitates combining other single-molecule biophysics techniques that utilize nanoscale sample manipulation and stages and light microscopy–based imaging. The most practicable of these involves single-molecule fluorescence microscopy. To implement optical trapping with simultaneous fluorescence imaging is, in principle, relatively easy, in that a NIR laser–trapping beam can be combined along a visible light excitation optical path by using a suitable dichroic mirror (Chapter 3), which, for example, will transmit visible light excitation laser beams but reflect NIR, thus allowing the laser-trapping beam to be coupled into the main excitation path of a fluorescence microscope (Figure 6.4c).

This has been used to combine optical tweezers with TIRF to study the unzipping of DNA molecules (Lang et al., 2003) as well as imaging sections of DNA (Gross et al., 2010). A dual-optical trap arrangement can also be implemented to study motor proteins by stretching a molecular track between two optically trapped microspheres while simultaneously monitoring using fluorescence microscopy (Figure 6.4d), including DNA motor proteins. Such an arrangement is similar to the DNA curtains approach, but with the advantage that both motor protein motion and molecular force may be monitored simultaneously by monitoring the displacement fluctuations of the trapped microspheres. Lifting the molecular track from the microscope coverslip eradicates potential surface effects that could impede the motion of the motor protein.

A similar technique is the dumbbell assay (Figure 6.4e), originally designed to study motor protein interactions between the muscle proteins myosin and actin (Finer et al., 1994), but since utilized to study several different motor proteins including kinesin and DNA motor complexes. Here, the molecular track is again tethered between two optically trapped microspheres but is lowered onto a third surface-bound microsphere coated in motor protein molecules, which results in stochastic power Stoke interactions, which may be measured by monitoring the displacement fluctuations of the trapped microspheres. Combining this approach with fluorescence imaging such as TIRF generates data for the position of the molecular track at the same time, resulting in a very definitive assay.

Another less widely applied combinatorial technique approach has involved using optical tweezers to provide a restoring force to electro-translocation experiments of single biopolymers to controllably slow down the biopolymer as it translocates down an electric potential gradient through a nanopore, in order to improve the effective spatial resolution of ion-flux measurements, for example, to determine the base sequence in DNA molecule constructs (Schneider et al., 2010). There have been attempts at combining optical tweezers with AFM imaging discussed later in this chapter, for example, to attempt to stretch a single-molecule tether between two optically trapped beads while simultaneously imaging the tether using AFM; however, the vertical fluctuations in stretched molecules due to the relatively low vertical trap stiffness have to date been high enough to limit the practical application of such approaches.

Optical tweezer Raman spectroscopy, also known as *laser tweezer Raman spectroscopy*, integrates optical tweezers with confocal Raman spectroscopy. It facilitates manipulation of single biological particles in solution with their subsequent biochemical analysis. The technique is still emerging but been tested on the optical trapping of single living cells, including red and white blood cells. It shows diagnostic potential at discriminating between cancerous and noncancerous cells.

6.3.8 OPTICAL MICROSCOPY AND SCATTERING METHODS TO MEASURE BIOLOGICAL FORCES

Some light microscopy and scattering techniques have particular utility in investigating forces in cellular material. Polarization microscopy (see Chapter 3) has valuable applications

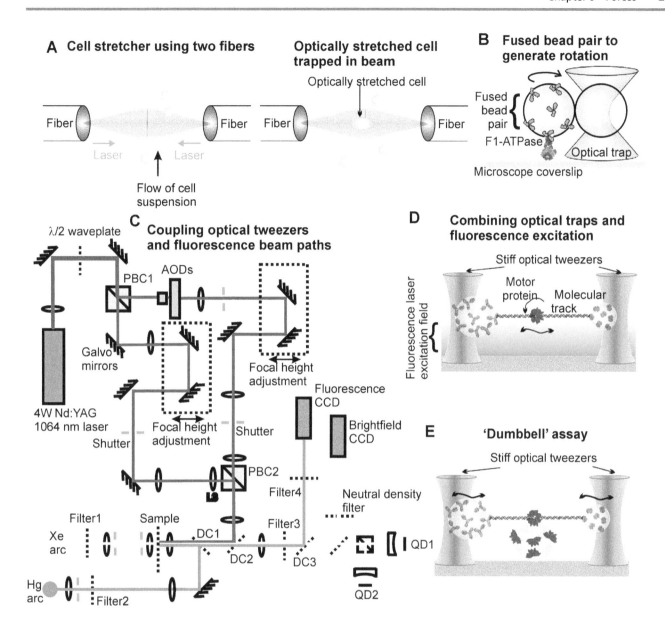

A Cell stretcher using two fibers

Optically stretched cell trapped in beam

Optically stretched cell

Fiber   Fiber   Fiber   Fiber

Laser   Laser

Flow of cell suspension

B Fused bead pair to generate rotation

Fused bead pair

F1-ATPase

Optical trap

Microscope coverslip

C Coupling optical tweezers and fluorescence beam paths

λ/2 waveplate

PBC1

AODs

Galvo mirrors

Focal height adjustment

4W Nd:YAG 1064 nm laser

Shutter

Focal height adjustment

Shutter

Fluorescence CCD

Brightfield CCD

PBC2

Filter4

Neutral density filter

Filter1

Xe arc

Sample

DC1

Filter3

DC2   DC3

QD1

Hg arc

Filter2

QD2

D Combining optical traps and fluorescence excitation

Stiff optical tweezers

Motor protein

Molecular track

Fluorescence laser excitation field

E 'Dumbbell' assay

Stiff optical tweezers

FIGURE 6.4 (See color insert.) More complex optical tweezers applications. (a) Cell stretcher, composed of two juxtaposed optical beams generating a stable optical trap that can optically stretch single cells in suspension. (b) Rotary molecular motors, here shown with the F1-ATPase component of the ATP synthase that is responsible for making ATP in cells (see Chapter 2), can be probed using optical trapping of fused bead pairs. (c) Trapping a fluorescence excitation beam paths can be combined (d) to generate optical traps combined with fluorescence imaging. (e) A three-bead "dumbbell" assay, consisting of two optically trapped beads and a fixed bead on a surface, can be used to probe the forces and displacements of "power strokes" due to molecular motors on their respective tracks.

for measuring the orientation and magnitude of forces experienced in tissues, and how these vary with mechanical strain. The usual reporters for these mechanical changes are birefringent protein fibers in either connective tissue or the cytoskeleton. In particular, collagen fibrils form an anisotropic network in cartilage and bone tissue, which has several important mechanical functions, largely responsible for tensile and shear stiffness. This method has the advantage of being label-free and thus having greater physiological relevance. The resulting anisotropy images represent a tissue force map and can be used to monitor damage and repair mechanisms of collagen during tissue stress resulting from disease.

FRET (see Chapter 4) can also be utilized to monitor mechanical forces in cells. Several synthetic molecular probes have been developed, which undergo a distinct bimodal conformational change in response to local changes in mechanical tension, making a transition from a compact, folded state at low force to an unfolded open conformation at high force (de Souza, 2014). This transition can be monitored using a pair of FRET dyes conjugated to the synthetic construct such that in the compact state, the donor and acceptor molecules are close (typically separated by ~1 nm or less) and so undergo measurable FRET, whereas in the open state the FRET, dyes are separated by a greater distance (typically >5 nm) and so exhibit limited FRET. Live-cell smFRET has been used in the form of *mechanical force detection* across the cell membrane (Stabley et al., 2011). Here, a specially designed probe can be placed in the cell membrane such that a red Alexa647 dye molecule and a FRET acceptor molecule, which acts as a quencher to the donor at short distances, are separated by a short extensible linker made from the polymer *polyethylene glycol* (PEG). Local mechanical deformation of the cell membrane results in extension of the PEG linker, which therefore has a *dequenching* effect. With suitable calibration, this phenomenon can be used to measure local mechanical forces across the cell membrane.

The forward and reverse transition probabilities between these states are dependent on rates of mechanical stretch (see Chapter 8). By generating images of FRET efficiency of a cell undergoing mechanical transitions, local cellular stresses can be mapped out with video-rate sampling resolution with a localization precision of a few tens of nanometers. The technique was first utilized for measurement of mechanical forces at cell membranes and the adhesion interfaces between cells; however, since FRET force sensors can be genetically encoded in much the same way as fluorescent proteins (for a fuller description of genetic encoding technology see Chapter 7), this is now being applied to monitoring internal *in vivo* forces inside cells. Variants of FRET force sensors have also been developed to measure the forces involved in molecular crowding in cells.

Finally, Brillouin light scattering in transparent biological tissue results from coupling between propagated light and acoustic phonons (see Chapter 4). The extent of this inelastic scattering relates to the biomechanical properties of the tissue. Propagating acoustic phonons in a sample result in expansion and contraction, generating periodic variation in density. For an optically transparent material, this may result in spatial variation of refractive index, allowing energetic coupling between the propagating light in the medium and the medium's acoustic vibration modes.

This is manifested as both an upshift (Stokes) and downshift (anti-Stokes) in photon frequency, as a function of frequency, similar to Raman spectroscopy (Chapter 4), resulting in a characteristic *Brillouin doublet* on the absorption spectrum whose separation is a metric of the sample's mechanical stiffness. The typical shift in photon energy is only a factor of $~10^{-5}$ due to the relatively low energy of acoustic vibration modes, resulting in GHz level frequency shifts for incident visible light photons. This technique has been combined with confocal scanning to generate spatially resolved data for the stiffness of extracted transparent components in the human eye, such as the cornea and the lens (Scarcelli and Yun, 2007), and to investigate biomechanical changes in the eye tissue as a function of tissue age (Bailey et al., 2010). It has advantages of conventional methods of probing sample stiffness in being minimally perturbative to the sample since it is a noncontact and nondestructive technique, without requiring special sample preparation such as labeling.

KEY BIOLOGICAL APPLICATIONS: OPTICAL FORCE TOOLS

Measuring molecular and cellular viscoelasticity; Quantifying biological torque; Cellular separations.

Worked Case Example 6.1: Optical Tweezers

Two identical single-beam gradient force optical tweezers were generated for use in a single-molecule mechanical stretch experiment on the muscle protein titin using an incident laser beam of 375 mW power and wavelength 1047 nm, which was time-shared equally to form two optical traps using an AOD of power efficiency 80%, prior to focusing each optical tweezers into the sample, with each trapping a latex bead of diameter 0.89 μm in water at room temperature. One of the optically trapped beads was found to exert a lateral force of 20 pN when the bead was displaced 200 nm from its trap center.

a   *Estimate the average angle of deviation of laser photons in that optical trap, assuming that the lateral force arises principally from photons traveling close to the optical axis. At what frequency for the position fluctuations in the focal plane is the power spectral density half of its maximum value? (Assume that the viscosity of water at room temperature is ~0.001 Pa·s.)*
The bead in this trap was coated in titin, bound at the C-terminus of the molecule, while the bead in the other optical trap was coated by an antibody that would bind to the molecule's N-terminus. The two beads were tapped together to try to generate a single-molecule titin tether between them.

b   *If one tether binding event was observed on average once in every n_{tap} tap cycles, what is the probability of not binding a tethered molecule between the beads?*

c   *By equating the answer to (b) to $P_{teth}(n = 0)$ where $P_{teth}(n)$ is the probability of forming n tethers between two beads, derive an expression for $\langle n \rangle$ in terms of n_{tap}.*
We can write the fraction a of "multiple tether" binding events out of all binding events as $P_{teth}(>1)/(P_{teth}(1) + P_{teth}(>1))$.

d   *Use this to derive an expression for a in terms of $\langle n \rangle$.*

e   *If the bead pair are tapped against each other at a frequency of 1 Hz and the incubation conditions have been adjusted to ensure a low molecular surface density for titin on the beads such that no more than 0.1% of binding events are due to multiple tethers, how long on average would you have to wait before observing the first tether formed between two tapping beads? (This question is good at illustrating how tedious some single-molecule experiments can sometimes be!)*

Answers

a   Since there is an 80% power loss propagating through the AOD and the laser beam is then time-shared equally between two optical traps, the power in each trap is

$$(0.375W) \times 0.8 / 2 = 0.15W$$

Using Equation 6.3, the angle of deviation can be estimated as

$$\theta = (20 \times 10^{-12}N) \times (3 \times 10^8 ms^{-1}) / (150 \times 10^{-3}W) = 40 \text{ mrad} = 2.3°$$

Modeling the power spectrum of the bead's lateral position as a Lorentzian function indicates that the power will be at its maximum at a frequency of zero, therefore at half its maximum $P(v)/p(0) = 1/2 = v_0^2 / (v^2 + v_0^2)$. Thus, $v = v_0$, the corner frequency, also given by $k/2\pi\gamma$. The optical trap stiffness k is given by

$$\frac{20 \times 10^{-12}N}{200 \times 10^{-9}m} = 1 \times 10^{-4} Nm^{-1}$$

The viscous drag γ on a bead of radius r in water of viscosity η is given by 6πrη; thus, the corner frequency of the optical trap is given by

$$\frac{1 \times 10^{-4}}{2\pi \times 6\pi \times 0.89 \times 10^{-6} \times 1 \times 10^{-3}} = 949 \text{ Hz}$$

b   The probability of not forming a tether is simply equal to $(1 - 1/n_{tap})$.

c   Using the Poisson model for tether formation between

$$<n>^n exp(-<n>)/n!$$

tapping beads, the probability $P_{teth}(n)$ for forming n tethers is given by $(\langle n \rangle n \exp[-\langle n \rangle]/n!$, thus at $n = 0$,

$$P_{tech}(0) = \exp[-\langle n \rangle] = \left(1 - 1/n_{top}\right)$$

Thus,

$$\langle n \rangle = \ln\left(n_{tap}/\left(n_{tap} - 1\right)\right)$$

d   Fraction of binding events in which >1 tether is formed:

$$\alpha = \frac{p_{tech}(>1)}{p_{tech}(1)p_{tech}(>1)} = \frac{1 - p_{tech}(0) - p_{tech}(1)}{1 - p_{tech}(0)}$$

$$= \frac{1 - \exp[-\langle n \rangle] - \langle n \rangle \exp[-\langle n \rangle]}{1 - \exp[-\langle n \rangle]}$$

e   No more than 0.1% tethers due to multiple tether events implies 1 in 10^3 or less multiple tethers. At this threshold value, $a = 0.001$, indicating (after, e.g., plotting the dependence of a on n_{tap} from [d] and interpolating) $n_{tap} \sim 600$ cycles. At 1 Hz, this is equivalent to ~600 s, or ~10 min for a tether to be formed on average.

6.4 MAGNETIC FORCE METHODS

Magnetism has already been discussed as a useful force in biophysical investigations in the context of structural biology determination in NMR spectroscopy as well as for the generation of x-rays in cyclotrons and synchrotrons for probing biological matter (Chapter 5). But magnetic forces can also be utilized to identify different biomolecules from a mixed sample and to isolate and purify them; for example, using *magnetic beads* bound to biological material to separate different molecular and cellular components, or using a magnetic field to deflect electrically charged fragments of biomolecules with the workhorse analytical technique of biophysics, which is *mass spectrometry*. Also, magnetic fields can be manipulated to generate exquisitely stable *magnetic tweezers*. Magnetic tweezers can trap a suitable magnetic particle, imposing both force and torque, which can be used to investigate the mechanical properties of single biomolecules if tethered to the magnetic particle.

6.4.1 MAGNETIC BEAD–MEDIATED PURIFICATION METHODS

Magnetic beads are typically manufactured using a latex matrix embedded with iron oxide nanoscale particulates, or other similar ferromagnetic materials such as chromium dioxide. If the concentration of ferromagnetic material in a bead is sufficiently small, then in the absence of an external B-field such beads possess no net magnetic moment. In the presence of an external B-field, the whole resultant bead is magnetized by induction of a magnetic moment aligned with the B-field, but which is lost once the external B-field is removed. This is a property of *paramagnetic* materials, distinct from *ferromagnetic* materials, which can retain a net magnetic moment after the external B-field is removed. This is a particularly useful feature of beads used for biological purification/isolation methods, since removal of an imposed B-field can then permit separation of components after being isolated from a mixture using a magnetic field. Beads can be chemically functionalized on their surface, using the same technology as for microspheres used for optical tweezers, to permit conjugation to a variety of biomolecules, such as antibodies and short specific sequences of nucleic acid (see Chapter 7). A suspension of magnetic beads is mixed with the target biological components.

Once bound to a specific biomolecule, then an applied magnetic field can pull the bead, and the biomolecule and any other biological structures attached to that biomolecule (e.g., a whole cell, in the case of the biomolecule attached to a cell membrane receptor protein) away from a mix of heterogeneous biological components/cells, and any unbound material can be removed by aspiration/washing, facilitating purification of that bound component. Magnetic bead–mediated separations result in comparatively low mechanical stress to the biological components being isolated and are rapid, cheap, and high throughput.

Applications include purification of nucleic acids by using short nucleotide sequences (or *oligonucleotides*, sometimes known simply as *oligos*) bound to the paramagnetic beads, which are complementary to sequences in the specific target nucleic acid molecules. When coated with specific antibodies, magnetic beads can also be used to purify proteins and to isolate various specific cell types including eukaryotes such as human cells and bacteria, and also smaller subcellular structures and viruses, subcellular organelles, and individual proteins.

6.4.2 MASS SPECTROMETRY

For a bulk ensemble average *in vitro* biophysical technique, mass spectrometry (often shorted to *mass spec*) is one of the most quantitatively robust. Here, $\sim 10^{-15}$ kg of sample (small, but still equivalent to several millions of molecules) is injected into an *ionizer.* Ionization generates fragments of molecules with different mass and charge. The simplest machine is the *sector mass spectrometer* that accelerates ion fragments in a vacuum using an electric field E and deflects them using a magnetic field sector oriented at right angles of magnitude B to this so that the ions follow a roughly circular path. The circle radius r is a function of the *mass-to-charge ratio* m_q of particles in the beam, which can easily be derived by equating the magnetic to the centripetal force:

(6.12)

$$r = \frac{\sqrt{2Em_q}}{B}$$

Different ionized molecule fragments are collected and analyzed depending upon the detector position (often termed a *velocity selector*) in the circular ion path. This generates a *mass spectrum* that can yield detailed data concerning the relative proportions of different ionic species in the sample.

Variants of the basic mass spectrometer include *ion trap* (injected ions trapped in a cavity using electric fields), *Fourier transform* (ions injected into a cyclotron cell and resonated into orbit using an oscillating electric field generating a radio-frequency signal that is detected and subsequently Fourier transformed to yield mass spectrum), *time of flight* (an ion vapor pulse is created using a high-energy laser and high-energy ions are accelerated using an electric field with time taken to travel a given distance measured), and *quadrupole* (accelerated ion beam passed between four metal rods to which direct current (DC) and alternating current (AC) potentials are applied causing resonance to the ion beam such that only ions with a narrow range of m_q will pass through the rod cavity into the detector unit) mass spectrometers.

The state of the art for quantitative mass spectrometry includes *matrix-assisted laser desorption ionization (MALDI,* also known as MALDI imaging spectroscopy) and *stable isotopic labeling by amino acids in cell culture (SILAC)* techniques. MALDI is an imaging technique in which the sample, typically a thin section of biological tissue (which can be applied both to animal and plant tissues), raster scans in 2D while the mass spectrum is recorded using focused laser ablation of the biological material to generate ion fragments. To enhance laser adsorption, a strongly laser-absorbing chemical reagent matrix is normally sprayed onto the sample surface. MALDI can enable the imaging of the localization of different biomolecule types including proteins/peptides, lipids, and several other small molecules such as synthetic drugs.

SILAC is a popular technique used in *quantitative proteomics*, in detecting differences in protein amounts from cell samples using nonradioactive isotope labeling. Typically, two populations of cells are grown in culture medium, one contains normal amino acids, while the other contains amino acids labeled with stable nonradioactive heavy isotopes, usually replacing the normal carbon ^{12}C isotope with the heavier ^{13}C with one more labeled amino acid type in the growth medium. For example, if arginine (which contains six carbon atoms per molecule) was used, then all peptides and proteins containing arginine in the cells would be ~6 Da heavier per molecule of arginine present, compared to the "normal" cell population grown in parallel. Another approach involves more uniform labeling with ^{13}C or the heavier ^{15}N isotope of nitrogen. Both cell populations are then analyzed using mass spectrometry and then compared in a pairwise fashion for chemically identical peptide ion fragments. The measured ratio of signal intensity of such paired fragments in the mass spectrum is an estimate for the relative abundance of a cellular protein that contains those specific heavier amino acids. It can thus be used as a tool to measure the different expression levels of different proteins from a live-cell population.

Biophysical applications of mass spectrometry are significant and include sensitive biological particle detection. The detection sensitivity is around one particle per liter, which compares favorably relative to other bulk ensemble average techniques. A particle can be detected with a sampling time resolution of a few minutes. The technique has been applied for investigations of sample purity quality control, detection of relatively subtle mutations in nucleic acids, protein conformation and folding studies, and proteomics experiments investigating protein–protein interactions. The *Simple Analysis at Mars (SAM)* instrument suite of NASA's rover that landed on Mars on August 5, 2012, included a portable mass spectrometer device for detection of putative biological material. Also, the spatial resolution of state-of-the-art mass spec devices now permit precision down to the level of just a few cells and, in some exceptional cases of relatively large cells, just a single cell.

6.4.3 MAGNETIC TWEEZERS

Magnetic particles that have a length scale range of hundreds to thousands of nanometers can be controlled directly and efficiently via the manipulation of the local external B-field using field strengths in the milli-Tesla (mT) range. This force transduction device is commonly referred to as *magnetic tweezers*. Both paramagnetic and ferromagnetic beads of around a micron diameter are typical probes used. This has been used to great success for investigating the mechanical properties of several types of biopolymer molecules, especially DNA (Manosas et al., 2010).

The external B-field in the magnetic tweezers setup is usually built as a module to an inverted optical microscope, with either two permanent bar magnets mounted to have juxtaposed poles or a combination of multiple electromagnetic coils to generate a suitable mT B-field placed around the magnetic probe (Figure 6.5a). By moving the microscope stage, a candidate bead can be captured in the locally generated B-field. Ferromagnetic beads contain a permanent magnetic dipole moment m, and the interaction between this and the local B-field, as indicated by the gradient of their dot product, results in a force F on the bead and a torque τ, which results from their cross product, rotating the bead in a direction so as to align the magnetic moment with the B-field:

$$\vec{F} = \nabla\left(\vec{m}.\vec{B}\right) \tag{6.13}$$

$$\vec{\tau} = \vec{m} \times \vec{B} \tag{6.14}$$

It is more common to use paramagnetic beads in magnetic tweezers. In this instance, a magnetic dipole moment is induced in the bead by the external B-field, and in typical milli-Tesla field strengths, the magnetization M (given by m/V where V is the bead volume) saturates at a value M_{max}. For the most common permanent magnet pair arrangement, the B-field is parallel to the focal plane of the microscope in between the opposite magnet poles, which means that the B-field gradient is zero everywhere apart from the vector normal to the focal plane. Thus, the magnetic force is parallel to optic axis (z) in a direction away from the microscope coverslip surface:

$$(6.15) \qquad F_z = M_{max} V \frac{dB}{dz}$$

Thus, a biopolymer tethered between the coverslip and a magnetic bead will be stretched vertically until balanced by the opposing molecular force that increases with molecular extension.

The trapped bead's position is still free to fluctuate in the lateral plane. Considering displacements parallel to the focal plane, the small angle $\delta\theta$ satisfies

$$(6.16) \qquad \tan\delta\theta = \frac{x}{z} = \frac{F_x}{F_z}$$

where
  z is the molecular extension of the biopolymer parallel to the optic axis
  x is the displacement from the equilibrium in the focal plane

The equipartition theorem can be applied similarly as for optical tweezers to estimate the stretching force parallel to the optic axis:

$$(6.17) \qquad F_z = \frac{k_B T_z}{\langle x^2 \rangle}$$

Measurement of x can be achieved using similar techniques to optical tweezers bead detection, including bright-field detection of the bead image onto a QPD or CCD, or to use BFP detection that is less common for magnetic tweezers systems since it requires an additional focused detection laser to be coaligned with the magnetic trap. As the bead moves above and below the focal plane, its image on a CCD camera contains multiple diffraction rings. The diameter of these rings is a metric for z, which can be determined by precalibration. Measuring the torque on a tethered molecule requires knowledge not only of the magnetic dipole moment and the local magnetic field strength but also the angle between their two vectors. However, since a magnetic bead is spherically symmetrical, this can be difficult to determine unless asymmetry is added, for example, in the form of a marker on the bead for angle of rotation, such as a fluorescent quantum dot (see Chapter 3) fused to the magnetic bead.

The B-field vector can be rotated either by differential phasing of the AC current input through each different electromagnetic coil or by mechanically rotating the two permanent magnets, which thus results in rotation of the magnetic bead. A paramagnetic bead may be similarly rotated by first inducing a magnetic moment in the bead by the presence of a separate nearby permanent magnet.

Usually, the magnetic bead is conjugated to a single biomolecule of interest, which in turn is tethered via its opposite end to a microscope slide or coverslip. By moving the stage vertically relative to the permanent magnets or coils, for example, by changing the focus, the molecule's end-to-end extension can be controllably adjusted. Therefore, the mechanical properties of individual molecules can be probed with this approach in much the same way as for optical tweezers. One advantage of magnetic tweezers over optical tweezers is that there is potentially less damage to the biological sample, since high stiffness optical tweezers at least require a few hundred milliwatts of NIR laser power, which is sufficient to raise the sample temperature and induce phototoxic effects.

However, a particular advantage of magnetic tweezers is that their relatively easy ability to *rotate* a particle, in the form of a magnetic bead, compared to technically less trivial optical rotation methods, which enables controllable torque to be applied to a single biomolecule tethered to bead provided appropriate *torsional constraints* that are inserted into the links between the tether and slide and tether and bead (in practice these are just multiple repeats of the chemical conjugation groups at either end of the biomolecule). This is a more direct and technically simpler method than can be achieved for optical tweezers, which would need either to utilize an extended optical handle or use the rotating polarization of a non-Gaussian mode laser.

Magnetic tweezers–mediated torque control has been used on DNA–protein complexes, for example, to study DNA replication. DNA in living cells is normally a negative supercoiled structure (see Chapter 2), with the supercoiling moderated by *topoisomerase* enzymes. However, to undergo replication or repair, or to express peptides and proteins from the genes, this supercoiled structure needs first to relax into an uncoiled conformation. To access the individual strands of the double helix then requires this helical structure itself to be unwound, which in turn is made possible by enzymes called helicases. It is likely that many of these torque-generating molecular machines work in a highly coordinated fashion.

A disadvantage of magnetic tweezers over optical tweezers is that they are slower by a factor of $\sim 10^3$ since they do not utilize fast AOD components as optical tweezers can and traditionally require using relatively large micron-sized beads to have a sufficiently large magnetic moment but with the caveat of a relatively large frictional drag, which ultimately limits how fast they can respond to changes in external B-field—a typical bandwidth for magnetic tweezers is ~ 1 kHz, so they are limited to detect changes over time scales >1 ms. Also, traditionally, it has not been possible to visualize a molecule that has been stretched through application of magnetic tweezers at the same time as monitoring its extension and force, for example, using fluorescence microscopy if the biomolecule in question can be tagged with a suitable dye. This is because the geometry of conventional magnetic tweezers is such that the stretched molecule is aligned parallel to the optic axis of the microscope and so cannot be visualized extended in the lateral focal plane. To solve this problem, some groups are developing *transverse magnetic tweezers* systems (Figure 6.5b). The main technical issue with doing so is that there is often very confined space in the microscope stage region around a sample to physically position magnets or coils in the same lateral plane as the microscope slide. One way around this problem is to use very small electromagnetic coils, potentially microfabricated, integrated into a bespoke flow cell.

Other recent improvements have involved using magnetic probes with a much higher magnetic moment, which may allow for reductions in the size of the probe, thus incurring less viscous drag, with consequent improvements to maximum sampling speeds. One such probe uses a *permalloy* of nickel and chromium manufactured into a disk as small as ~ 100 nm diameter (Kim et al., 2009) and still have a sufficiently high magnetic moment to

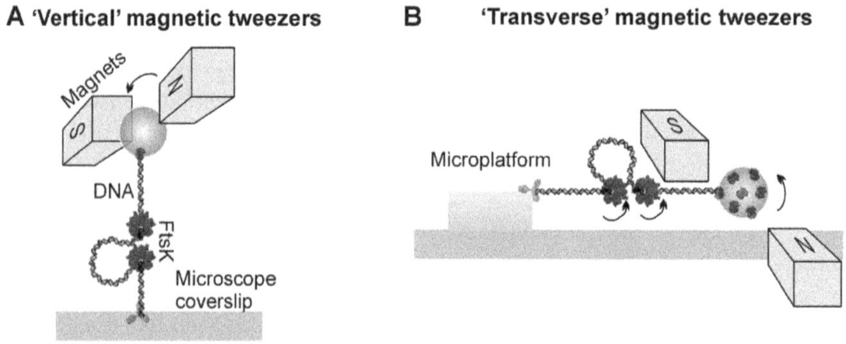

KEY BIOLOGICAL APPLICATIONS: MAGNETIC FORCE TOOLS

Quantifying biological torque; Molecular and cellular separation and identification; Measuring biomolecular mechanics.

A 'Vertical' magnetic tweezers

Magnets
N
S
DNA
FtsK
Microscope coverslip

B 'Transverse' magnetic tweezers

Microplatform
S
N

FIGURE 6.5 Magnetic tweezers. Single-molecule mechanical experiments can be performed in both (a) vertical and (b) transverse geometries, for example, to probe the mechanical properties of DNA molecules and of machines, such as FtsK shown here, which translocate on DNA.

permit torque experiments in principle on biomolecules such as DNA, which would thus enable faster molecular rotation experiments to be performed.

6.5 SCANNING PROBE MICROSCOPY AND FORCE SPECTROSCOPY

SPM includes several techniques that render topographic information of a sample's surface. There are in excess of 20 different types of SPM that have been developed and can measure a variety of different physical parameters through detection of forces as the probe is placed in proximity to a sample surface, and the variation of these physical parameters across the surface are measured by scanning the probe laterally across the sample. Scanning near-field optical microscopy is one such technique that was discussed previously in Chapter 4. However, the most useful SPM technique in terms of obtaining information on biological samples is *AFM*. AFM can be utilized both as an imaging tool, but also as a probe to measure mechanical properties of biological matter including cell walls and membranes and, especially, single biomolecules. But there are also a range of other SPM techniques such as *scanning tunneling microscopy* (*STM*) and *surface ion conductance microscopy* (*SICM*), which have biological applications.

6.5.1 PRINCIPLES OF AFM IMAGING

AFM imaging (Binnig et al., 1986), sometimes known as *scanning force microscopy*, is the most frequently used SPM technique. AFM had been applied to imaging-purified biomolecule samples conjugated to flat surfaces such as mica, as well as cells. It has also been used to image the topographic surface features of some native living tissues also, for example, blood vessels (see Mao, 2009). This technique shows some promise for *in vivo* imaging, though the challenge mainly lies in the relatively slow lateral scanning of the sample resulting in problems of sample drift due to the large areas of tissue, which might move before the scan probe can image their extent fully. This is why AFM imaging of biological matter has experienced more successful developments when applied to single cells and purified molecular components.

In AFM imaging, a small, solid-state probe tip is scanned across the surface of the sample (Figure 6.6a), using piezoelectric technology, to generate topographical information. The tip is usually manufactured from either *silicon* or more commonly the ceramic insulator *silicon nitride*, Si_3N_4, with a typical pyramidal or tetrahedral shape of a few microns to tens of microns edge length and height scale. However, a standard AFM tip has a *radius of curvature* of ~10 nm, which is primarily what determines the spatial resolution, though in some specially sharpened tips this can be an order of magnitude smaller.

The AFM tip is attached to a cantilever, which is normally manufactured from the same continuous piece of material (a Si-based wafer) via photolithography/masking/wet etching (see Chapter 7). The silicon or silicon nitride cantilever may be subsequently coated with a metal on the "topside," which can be used to enhance laser reflection for positional detection (see in the following text). The cantilever acts as a force actuator, with the tip detecting a superposition of different types of forces between it and the sample, both attractive and repulsive, which operate over long lengths scales in excess of 100 nm from the surface through to intermediate and much shorter length scale forces over distances of just a single atom of ~0.1 nm. The thin, flexible metallic cantilever strip is ~0.1 mm wide and a few tenths of a millimeters long, and as the tip approaches the surface repulsive forces dominate and cause the metallic strip to bend upward. In essence, this bending of cantilever gives a readout of distance between the tip and the sample, which allows the surface topography to be mapped out.

6.5.2 FORCES EXPERIENCED DURING AFM IMAGING

The tip–sample interaction can be described as the total potential energy U_{total}, which is the sum of three potentials U_{SD} (which is due to the sample deformation as the tip approaches),

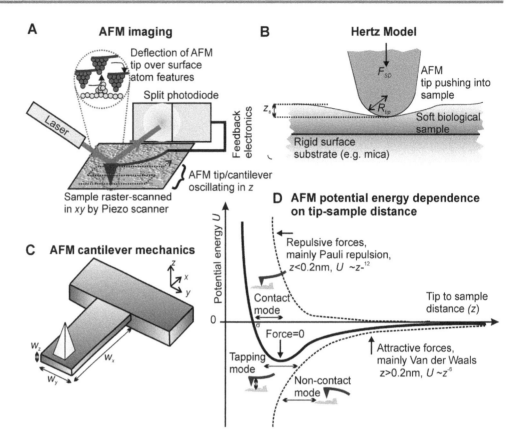

FIGURE 6.6   Atomic force microscopy imaging. (a) Samples are raster scanned laterally relative to an AFM tip, such that the cantilever on the tip will deflect due to topographical features, which can be measured by the deflection of a laser beam. (b) Schematic for the Hertz model of biological material deformation. (c) Standard AFM cantilever size and shape. (d) Shape of energy potential curve for a tip–cantilever system as a function of the tip–sample distance.

U_{ME} (which is due to the mechanical elastic potential of the cantilever bending), and U_{IP} (which is the interaction potential between the sample and the tip), resulting in the total force F on the AFM tip:

$$(6.18) \qquad F = -\nabla U_{total} = -\nabla U_{ME} = \nabla U_{SD} + \nabla U_{IP}) = \vec{F}_{ME} + \vec{F}_{SD} + \vec{F}_{IP}$$

Sample deformation forces are not trivial to characterize due to the often heterogeneous and nonlinear elastic response of soft matter. The *Hertz model* offers a reasonable approximation, however, for a small local sample indentation z_s such that the restoring force F_{SD} (see Figure 6.6b) is given by

$$(6.19) \qquad F_{SD} \approx \frac{4Y_{sample}}{3(1-\mu^2)} \sqrt{R_{tip} z_s^3}$$

where R_{tip} is the radius of curvature of the AFM tip being pushed toward a sample of Young's modulus Y_{sample}, with the Poisson ratio of the soft-matter material given by μ.

The restoring mechanical force resulting from deflections of the cantilever is easier to characterize since it can be modeled as a simple *Hookean spring*, such that $F_{ME} = -k_z z$ where k_z is the cantilever's vertical stiffness and z is the small vertical displacement of the cantilever from its equilibrium zero force position. Thus, this mechanical elastic potential has a simple parabolic shape. Whereas the local stiffness of the sample is often difficult to determine directly, the well-defined geometry and mechanical properties of the cantilever enable a more accurate estimation of its stiffness to be made.

For standard mode force sensing, the AFM cantilever oscillations can be approximated as those of a solid beam flexing about its point of attachment, namely, that of a beam-like structure that has one relatively compliant axis (normal to the cantilever flat surface) in terms of mechanical stiffness, with the two remaining orthogonal axes having comparatively high stiffness values. The beam bending equation of classical mechanics, for which one end of the beam is fixed while the other is allowed to undergo deflections, can be used to estimate the bending stiffness parallel to the z-axis, k_z as

(6.20)
$$k_z = \frac{Yw_y}{4}\left(\frac{w_y}{w_x}\right)^3$$

where
    w_i is the width of the cantilever beam parallel to the ith axis (Figure 6.6c)
    Y is the *Young's modulus*

This beam–cantilever system also has a resonance frequency v_0 given by

(6.21)
$$v_0 = \frac{1}{2\pi}\sqrt{\frac{k_z}{m_0}} = 0.162\frac{w_z}{w_x^2}\sqrt{\frac{Y}{\rho}}$$

where
    m_0 is the effective total mass of the cantilever and AFM tip
    ρ is the cantilever density

Measurement of the resonance frequency far from the surface (such that the potential energy function is dominated solely by the mechanical elasticity of the cantilever with negligible contributions from surface-related forces) can be used to estimate k_z. However, this relies on accurate knowledge of the material properties of the cantilever, which are often not easy to obtain due to the variability from cantilever to cantilever in a given manufactured batch.

An alternative method involves measuring the vertical mean squared displacement $\langle z^2 \rangle$ far away from the sample surface, which requires only an accurate method of measuring z. Deflections of the cantilever can be very accurately detected usually involving focusing a laser onto the reflecting back of the polished metal cantilever and imaging the reflected image onto a split photodiode detector. The voltage response is converted ultimately to a corresponding distance displacement of the tip, provided that the cantilever stiffness has been determined.

We can then use the equipartition theorem in a similar way as for quantifying the stiffness of optical and magnetic tweezers to estimate $k_z = k_B T/\langle z^2 \rangle$. At a room temperature of ~20°C, $k_B T$ is ~4.1 pN·nm (see Chapter 2). Thus, if the voltage output V from a photodiode results in a volts per nm of cantilever vertical deflection, the z stiffness is given roughly by $4.1a^2/\langle V^2 \rangle$ in units of pN/nm. Typical cantilever stiffness values used for probing biological material in AFM are ~0.1 pN/nm (see Worked Case Example 6.2).

The interaction forces experienced by the tip include electrostatic and chemical forces as well as van der Waals (vdW) forces. The *Morse potential* is a good qualitative model for the chemical potential, in characterizing the potential energy due to the separation, z, of two atoms that can form a chemical bond to create a diatomic molecule when they approach each other to within ~0.1 nm, as might occur between atoms of the sample and the approaching tip. The shape of the potential energy curve has a minimum corresponding to the equilibrium atom separation, σ, which is ~0.2 nm, so the term *potential energy well* is appropriate. The functional form of the Morse potential is

(6.22)
$$U_{\text{Morse}} = -E_{\text{bond}}(2\exp[-\kappa(z-\sigma) - \exp[-2\kappa(z-\sigma)]])$$

where E_{bond} is the *bonding energy*, with a *decay length* κ. Quantum mechanics can predict an exact potential energy curve for the H_2^+ diatomic molecule system, which the Morse potential fits very well, whose form exhibits the qualitative features at least of the real AFM chemical potential energy function. However, for more complex chemical interactions of higher atomic number atoms involving anisotropic molecular orbital effects, as occurred in practice with the tip–sample system for AFM, empirical models are used to approximate the chemical potential energy, including the *Stillinger–Weber potential* and *Tersoff potential*, with the Stillinger–Weber potential showing most promise from *ab initio* calculations for interactions involving silicon-based materials (as is the case for AFM tips). The functional form of this potential energy involves contributions from nearest-neighbor and next nearest-neighbor atomic interactions. The nearest-neighbor component U_{NN} is given by

$$(6.23) \qquad U_{NN} = E_{bond} A \left[B \left(\frac{z}{\sigma'} \right)^{-p} - \left(\frac{z}{\sigma'} \right)^{-q} \right] \exp\left[\frac{1}{z/\sigma' - a} \right]$$

where A, B, a, p, and q are all constants to be optimized in a heuristic fit. The next nearest-neighbor component, U_{NNN}, is a more complex formulation that embodies angular dependence of the atomic orbitals:

$$(6.24) \qquad U_{NN} = E_{bond} \left[h\left(x_{ij}, z_{ik}, \theta_{jik} \right) + h\left(x_{ji}, z_{jk}, \theta_{ijk} \right) + h\left(x_{ki}, z_{kj}, \theta_{ijk} \right) \right]$$

such that

$$(6.25) \qquad h\left(x_{ij}, z_{ik}, \theta_{jik} \right) = \lambda \exp\left[\frac{1}{\left(z_{ij}/\sigma' - a \right)} + \frac{1}{\left(z_{ik}/\sigma' - a \right)} \right] \left(\cos\theta_{ijk} + \frac{1}{3} \right)^2$$

where
  λ is a constant
  i, j, and k are indices for three interacting atoms

If an AFM tip is functionalized to include electrostatic components, these can interact with electrostatic components on the biological sample surface also. The functional form can be approximated as $U_{ES} \approx \pi \varepsilon R_{rip} V^2 / 2z^2$ where ε is electrical permittivity of the aqueous solvent surrounding the tip and sample, R_{tip} is again the AFM tip radius of curvature and V is the electrical potential voltage across a vertical distance z between tip and sample.

The most significant of the interaction forces for AFM are the vdW forces, modeled as the *Lennard–Jones potential* (also known as the *6–12 potential*, introduced in Chapter 2):

$$(6.26) \qquad U_{LJ} = -E_{bond} \left(2\frac{z^6}{\sigma^6} - \frac{z^{12}}{\sigma^{12}} \right)$$

The vdW forces arise from a combination of the fluctuations in the electric dipole moment, and the coupling between these fluctuating dipole moments, and the exclusion effect between paired electrons. The longer-range $\sim z^6$ dependence is the attractive component, while the shorter-range $\sim z^{12}$ component results from the *Pauli exclusion principle* between paired electrons that prohibits electrons with the same spin and energy state from occupying the same position in space and thus results in a repulsive force at very short tip–sample separations.

However, approximating the tip as a sphere of radius R_{tip} and assuming interactions with a planar sample surface, we can integrate the incremental contributions from the sphere using Equation 6.26 to give

$$(6.27) \qquad\qquad U_{LJ} \approx -\frac{zA_H R_{rip}}{6z}$$

where A_H is the *Hamaker constant* depending on the tip and sample electrical polarizability and density. As Figure 6.6d illustrates, the combination of multiple independent potential energy functions experienced by an AFM tip, which operate over different length scale regimes, results in a highly nonlinear force–distance response curve.

6.5.3 AFM IMAGING MODES

During AFM imaging, the tip–cantilever force actuator can be used either in *contact mode, noncontact mode, tapping mode*, or a relatively newly developed *torsional mode*. During contact mode imaging, the cantilever deflection is kept *constant* throughout as the tip is scanned across the sample surface using fast feedback electronics from the photodiode detector to a piezo actuator controlling the cantilever z position, to maintain a constant force on the tip (and hence constant height above the surface, assuming the material properties of the sample remain the same). Here, although the tip itself does not make direct contact as such with the sample, it is placed in relatively close contact to it (typically less than the equilibrium atom separation of ~0.2 nm) such that the overall force detected by the tip from the sample is in the short-range repulsive force regime.

As Figure 6.6d suggests, the force, as the gradient of the potential energy curves, varies dramatically with vertical displacement, with typical forces being in the range 10^{-6} to 10^{-9} N. This high sensitivity to vertical displacement allows potentially atomic-level resolution to be obtained in contact mode. However, shearing forces at short distances from the sample are high potentially resulting in sample distortion, in addition to sample damage from scraping of soft sample features by the AFM tip during lateral scanning.

Although atomic-level resolution in z can in principle be obtained in contact mode, the finite AFM tip radius of curvature results in a limit on the absolute maximum measurable z displacement (i.e., height) between neighboring surface features. If, for example, similar sharp surface features are separated by a characteristic displacement d in the lateral surface plane, then the maximum height Δz, which an AFM tip of radius of curvature R_{tip} could measure, is given from simple space constraints as

$$(6.28) \qquad\qquad \Delta_z \approx \frac{d^2}{8R_{tip}}$$

In contact mode imaging, the AFM tip can penetrate beyond water layers bound to the surface to image the sample molecules directly, manifested in a greater spatial resolution. However, the finite sharpness of the AFM tip itself means that some sample surface features will be inaccessible with a resultant tip broadening convolution artifact (see Worked Case Example 6.2). The AFM tip experiences a lateral force from a stiff object on the surface when the AFM tip is pushed down vertically during imaging. If the half angle of the tip's triangular cross-section is θ, then simple geometrical considerations indicate that the tip broadening coefficient κ, defined as the ratio of the apparent measured width r' of the stiff object (modeled as a sphere with a circular cross-section of radius r), satisfies

$$(6.29) \qquad\qquad \kappa = \frac{r'}{r} = \tan\theta + \sec\theta$$

In noncontact mode imaging, the AFM tip is kept far from the sample surface often much greater than the equilibrium atom separation of ~0.2 nm such that forces experienced by the tip are in the attractive force regime of the force–displacement response curve. In this regime, the tip can be several nanometers to tens of nanometers from the sample surface, resulting in much weaker forces compared to contact mode, of ~10^{-12} N (i.e., ~pN). Noncontact mode imaging therefore has an advantage in minimizing the vertical force on a sample, important in general cases of relatively compliant material, therefore minimizing sample deformation and also minimizing the risk of sample damage from lateral scanning. However, the spatial resolution is poorer compared to contact mode with slower scanning speeds permissible. Also noncontact mode ideally requires hydrophobic samples to minimize the thickness of adsorbed water solvent on the sample, which would otherwise impair the scanning by trapping the tip in the adsorbed layer. If a nonhydrophobic sample is in physiological water-based pH buffer environment (as opposed to a high vacuum as would be the case to generate the lowest measurement of noise), then noncontact mode will in general image not only the sample surface but also the first few shells of water molecules, which reduces the imaging spatial resolution.

In tapping mode imaging, the cantilever is driven to oscillate vertically at a distance of 100–200 nm from the sample surface with an amplitude of at least ~10 nm, with a frequency marginally above its resonance frequency. As the tip approaches the sample surface during its periodic oscillation, the increase in attractive forces results in a decrease of the amplitude of oscillation, with an associated change of phase. Depending upon the measurement system, either the change in frequency can be detected (*frequency modulation*) or the change in amplitude or phase (*amplitude modulation*), the latter of which is relatively sensitive to the type of sample material being imaged. These detected signal changes can all be converted after suitable calibration into a distance measurement from the sample. Such AFM imaging has been able to measure several types of single biomolecules (Arkawa et al., 1992), including snapshots of the motor protein myosin during its power stroke cycle and also to visualize artificial lipid bilayers containing integrated single protein complexes in a physiological aqueous environment.

A traditional weakness of AFM is the relatively slow imaging speeds due to slow scanning and feedback electronics. Recent improvements have been made in improving the imaging speed, the so-called high-speed AFM, using a combination of sharp AFM tips with a greater natural resonance frequency, or by using torsional mode imaging. Here, the AFM cantilever oscillates through a twisting vibrational mode as opposed to a vertical one. This has resulted in video-rate imaging speeds (Hobbs et al., 2006), which have been employed for measuring actual real-time stepping motions of the key molecular motor molecule of myosin, which is responsible for the contraction of muscle tissue, though the lateral spatial resolution is in principle slightly worse than nontorsional AFM for an equivalent tip due to the additional torsional sweep of the end of the probe.

AFM imaging and scanning electron microscopy (SEM) (see Chapter 5) can both generate topographical data from biological samples. AFM has some advantages to SEM: no special sample staining or elaborate preparation is required and the spatial resolution can be at the nanometer scale as opposed to tens of nanometers with SEM. However, there are also disadvantages: AFM has relatively slow scanning speeds for comparable levels of spatial resolution, which can lead to issues of drift in the sample (most usually thermal related drift effects), the scan area for AFM is much smaller than for SEM (usually a few tens of micron squared for AFM, as opposed to several hundred micron squared for SEM), differences between AFM tips and cantilever even in the same manufactured batch can lead to image artifacts as well as there being unavoidable tip broadening artifacts, and also piezoelectric material driving the lateral scanning of the tip and the vertical displacement of the cantilever can suffer cross talk between the x, y, and z axes. Also, AFM can image samples under liquid/physiological conditions with a suitable enclosed cell, whereas SEM cannot. However, the choice of which biophysics technique to choose to render topographic detail should clearly be made on a case-by-case basis.

6.5.4 SINGLE-MOLECULE AFM FORCE SPECTROSCOPY

AFM can also be used to investigate the mechanical elasticity of single biomolecules (Figure 6.7a) in a technique called AFM force spectroscopy. In the simplest form, AFM force spectroscopy experiments involve nonspecific binding of the biomolecule in question to a gold- or platinum-coated coverslip followed by dipping an AFM tip into the surface solution. Upon retracting the AFM tip back, there is a probability that a section of a molecule is nonspecifically tethered between the tip, for example, by hydrophobic binding, and the coverslip. In having tethered a section of a single molecule, the tip–cantilever system can then be used to investigate how the molecular restoring force varies with its end-to-end extension, similar to optical and magnetic tweezers discussed previously in this chapter.

Simple AFM force spectroscopy devices can be limited to just one axis of controllable movement for the vertical axis controlled by a piezo actuator to move the AFM tip relative to the sample (these one-axis instruments in effect relying on lateral sample drift to move to a different region of the sample, so there is a paradoxical benefit in having a marginally unstable system). Single-molecule AFM force spectroscopy experiments are often performed on modular proteins, either purified from the native source or using smaller synthetic molecules that allow shorter sections of the native molecules to be probed in a more controllable way than the whole native molecule. The large muscle protein titin, discussed previously in the context of optical tweezers, has proved to be an invaluable model system in AFM force spectroscopy studies. In one of the best examples of such pioneering experiments, single molecule constructs consisting of up to eight repeats of the same protein "Ig" domain (Rief et al., 1997).

The properties of the molecule titin are worth discussing in greater detail due to its importance in force spectroscopy experiments and our subsequent understanding of molecular mechanical properties. Titin is an enormous molecule whose molecular weight lies in the MDa range, consisting of ~30,000 individual amino acid residues and is part of a filamentous system in muscle, which act as springs to align the functional subunits of muscle tissue called sarcomeres. Most of the molecule is composed of repeating units of β-barrel modules of ~100 amino acid residues each, which either belong to a class called "fibronectin" (Fn) or "immunoglobulin" (Ig), with a combined total in excess of 370 combined Fn and Ig domains.

The increased likelihood of unfolding of the β-barrel structure of Fn or Ig domains as force is increased on the titin possibly confers a *shock-absorber* effect, which ensures that the *myofibril*, the smallest fully functional filamentous subunit of muscle tissue compared to multiple repeating sarcomeres, can maintain structural integrity even in the presence of anomalously high forces, which could damage the muscle. Titin is made in a variety of different forms with different molecular weights depending on the specific type of muscle tissue and its location in the body, and there is good evidence to indicate that this allows the titin molecular stiffness to be catered to the range of force experienced in a given muscle type.

In fishing for surface-bound titin constructs, a variable number of Ig modules in the range 1–8 can be tethered between the gold surface and the AFM tip depending on the essentially random position of the nonspecific binding to both. These domains unfold in the same manner as those described for mechanical stretch experiments on titin using optical tweezers, with a consequent sudden drop in entropic force from the molecule and increase in molecular extension of ~30 nm due to an Ig domain making a transition from a folded to an unfolded conformation. Thus, the resultant force-extension relation has a characteristic *sawtooth* pattern, with the number of "teeth" corresponding to the number of Ig domains unfolded in the stretch, and therefore varying in the range 1–8 in this case (Figure 6.7b). These sawtooth patterns are important since they indicate the presence of a single-molecule tether, as opposed to multiple tethers, which might be anticipated if the surface density of molecules is sufficiently high. The sawtooth pattern thus denotes a *molecular signature*.

AFM force spectroscopy can also be used with greater binding specificity by chemically functionalizing both the AFM tip and the gold or platinum surface. Many AFM force spectroscopy devices are also used in conjunction with an *xy* nanostage that allows lateral control of the sample to allow reproducible movements to different sample regions as well as force feedback to clamp the molecular force to a preset value to enable definitive observation of force-dependent kinetics of unfolding and subsequently refolding of molecular domains (details of analytical methods to model these processes are discussed in Chapter 8).

6.5.5 AFM "CUT AND PASTE"

The spatial reproducibility and resolution of many AFM systems is high enough such that single molecules may be pulled clear from a surface, moved laterally by anything from a few nanometers to several microns, and then controllably repositioned by pressing the tip back into the surface (Figure 6.7c).

This in effect is a molecular *cut-and-paste* device (Kufer et al., 2008). By combining AFM cut and paste with the specificity of DNA base pairing, it has been possible to use a complementary DNA strand conjugated to an AFM tip to specifically capture surface-bound DNA constructs from a "depot" region on the surface of the sample and repositioned elsewhere on the surface, offering future potential for smart designed *synthetic biology* applications at the nanometer length scale (discussed in Chapter 9).

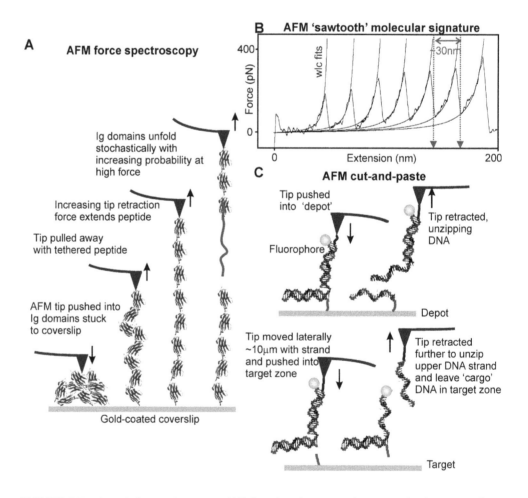

FIGURE 6.7 Atomic force microscopy (AFM) molecular manipulation methods. (a) AFM force spectroscopy using a "fishing" approach to tether a peptide construct, consisting of repeat "Ig domain" subunits, which results in (b), a characteristic sawtooth response of the force versus extension relation. (c) An AFM tip can be used to controllably relocate single DNA "cargo" molecules to different regions of a gold-coated coverslip.

6.5.6 AFM AND FLUORESCENCE MICROSCOPY

Single-molecule experiments are increasingly characterized by combinatorial approaches—combining simultaneous measurements on the same molecule but using different single-molecule methods. An example of this is the combination of AFM force spectroscopy with fluorescence microscopy (Sarkar et al., 2004). For example, it is possible to engineer constructs that have a single Ig domain bounded by fluorescent protein FRET dye pairs. In the unfolded Ig domain conformation, the separation of the FRET pairs is <5 nm and therefore results in a measurable FRET efficiency (see Chapter 4). If the molecule is stretched using AFM force spectroscopy, then the unfolding of the Ig domain results in an increase in the FRET dye pair separation by ~30 nm, thus resulting in no measurable FRET efficiency between the acceptor and donor pair. This therefore constitutes a "double" molecular signature that gives a significant increase in confidence that one is really observing a single-molecule event. Single-molecule fluorescence imaging has also been utilized in AFM cut-and-paste techniques to confirm the correct placement of repositioned DNA molecules.

6.5.7 AFM TO MEASURE CELLULAR FORCES

An AFM cantilever–tip force actuator can be used as a probe to measure cellular mechanical forces by probing regions of the cell that are accessible to the AFM tip. These include elastic and viscoelastic forces present in the cell membrane and cell wall, as well as the mechanical properties of the cytoskeleton just beneath the cell membrane and forces of adhesive, which are used to stick certain cells together in structural tissues. AFM can be used to generate maps of mechanical properties including the Young's modulus and stiffness values across the surface of a cell. Often, to avoid damage to the relatively soft cell membrane, an AFM tip can be modified by conjugating a larger "blunter" probe onto the end of the AFM tip, for example, a latex microsphere. Mechanical forces are particularly important in characterizing cancerous cells, since these often change in stiffness compared to noncancerous cell types as well as having reduced cell-to-cell adhesion forces (investigated, for example, in cells called fibroblasts present in connective tissue), which thus increases their chance of breaking free from a tumor or *metastasizing*, that is, spreading to other parts of the body. Measurement of mechanical properties at a single-cell level may therefore have future diagnostic potential in biomedicine.

AFM force spectroscopy (as well as optical and magnetic tweezers) suffers similar issues in regard to being applicable to measurements *inside* the living cells, as opposed to being applied to features that are accessible from the cell surface. Recent developments in synthetic transmembrane adaptor molecules, which can integrate into a cell membrane but bind both to internal cellular substructures and/or molecular complexes and to external molecules on the outside of the cell, may in the near future be functional as suitable "handles" for AFM tip probes (and indeed for trapped microspheres inside the cell) to allow access to monitoring internal cellular force-dependent processes.

A novel variant of AFM that has applications to cellular force measurements and probing is the *fluid force microscopy* (or FluidAFM). Here, a microfluidics channel (see Chapter 7) is engineered into the cantilever extending down through the AFM tip, which allows the tip end to function as a suction nanopipette. This can be used, for example, to manipulate single cells and/or probe their cell membranes.

6.5.8 SCANNING TUNNELING MICROSCOPY

STM was the seed of all SPM techniques. It was developed in the early 1980s (Binnig et al., 1982) and uses a solid-state scanning probe tip, similar to that used for AFM imaging experiments, but which can conduct electrons away from the surface of an electrically conducting sample in a vacuum environment. Tips are made from gold, platinum/iridium, or tungsten, manufactured usually by chemical etching (though historically a narrow wire could be cut to produce a sharp enough tip). Nonmetallic but electrically conducting carbon nanotubes have also been utilized, which have some advantages of better manufacturing reproducibility and mechanical stability. Electrical conduction at the tip is mediated through the atom of the tip closest to the sample surface, and so the effective radius of curvature is one to two orders of magnitude smaller than that for AFM tips.

No physical contact is made between tip and sample, and therefore there is a classically forbidden energy gap across which electrons must *quantum tunnel* across for electrical conduction to occur (Figure 6.8a). In the classical picture, if an electron particle of speed v and charge q has kinetic energy E of ~$mv^2/2$, then it will not be able to travel to a region of space, which involves crossing a potential energy barrier P of qV where V is the electrical potential voltage difference if $qV > mv^2/2$ since this implies an electron with negative kinetic energy in the barrier itself, and so instead the electron is reflected back from the boundary.

However, in a quantum mechanics model the electron particle is also an *electron wave*, which has a finite probability of tunneling through the potential barrier. This can be demonstrated by solving *Schrödinger's wave equation* in the barrier that results in a nonperiodic evanescent wave solution whose transmission coefficient, T, for a rectangular shaped barrier of width z, takes the form

$$(6.30) \qquad T(E) = \exp\left[-2z\sqrt{\frac{2m}{h}(V-E)}\right]$$

The tunneling electron current I_T depends on the tip–sample separation z and the effective width w_{xy} in the lateral xy plane of the sample over which tunneling can occur as

$$(6.31) \qquad I_T = I_0\exp\left[-Az^{1/2}w_{xy}\right]$$

where

I_0 is the equivalent current at zero gap between the tip and the sample
A is a constant equal to $(4\pi/h)(2m)^{1/2}$ where h is Plank's constant and m the free electron mass

The free electron depends upon the electrical potential energy gap between tip and sample, and so A is sometimes written in non-SI units as 1.025 Å $eV^{-1/2}$. The exponential dependence of I_T in terms of tip–sample distance makes it very difficult to measure the current experimentally if the distance is greater than a few tenths of a nanometers from a weakly conducting surface such as a biological sample.

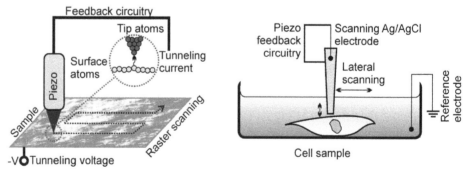

FIGURE 6.8 Other scanning probe microscopies, (a) Scanning tunneling microscopy and (b) scanning ion conductance microscopy.

STM is usually operated in a *constant current* imaging mode, analogous to the constant height or force mode of AFM, that is, I_T is kept constant using feedback electronics to vary the height of the tip from the sample, typically using a highly sensitive low-noise piezoelectric device, while the probe is laterally raster scanned. The variation in measured sample height is thus a measure of the sample topography, which can be converted into a 3D contour plot of the sample surface in much the same way as for AFM. A less common mode of operation is for the tip–sample distance z to be kept a constant such that the variation in tunneling current itself can be converted into topographical information. This has the advantage of not requiring electronic feedback, which ultimately can permit faster imaging, though it requires a sample to be, in effect, smooth at the atomic level, and so is of limited use for imaging biological material.

The spatial resolution of STM is less sensitive to the tip's size and shape as is the case for AFM and is ~0.1 nm laterally (i.e., the length scale of a single atom) and ~0.01 nm vertically. STM therefore provides lateral information at an atomic resolution but topographical data at a *subatomic* resolution. The main limitation for its use in biology is that most biological matter is only very weakly electrically conducting and so generates small values of I_T that are difficult to measure above experimental noise. However, STM has been used to image single DNA molecules, protein complexes made up of large *macroglobulin* molecules, and single virus particles (Arkawa et al., 1992).

AFM has also been combined with STM and *Kelvin probe microscopy*. Here, an ultra-cold probe tip with at a temperature of just ~5 K is used to measure the actual distribution of electronic charge in a single molecule, in this case an organic molecule called naphthalocyanine. This has been used in the context of developing *single-molecule logic switches* for bionanotechnological purposes (Mohn et al., 2012).

6.5.9 SCANNING ION CONDUCTANCE MICROSCOPY

For scanning ion conductance microscopy (*SICM*), the probe consists of a glass pipette drawn out such that its end diameter is only ~20–30 nm (Hansma et al., 1989). This technique combines the scanning probe methods of SPM with the ion-flux measurements methods of patch clamping (discussed later in this chapter). An electric potential is applied across the end of the tip, which results in a measureable ion current in physiological ionic solutions. However, as the tip is moved to within its own diameter from the biological sample being scanned, the ion flow is impeded.

Using fast feedback electronics similar to those described previously for AFM and STM, this drop in ion current can be used to maintain a constant distance between the nanopipette tip and the sample. This can generate topographical information as the tip is laterally scanned across the surface (Figure 6.8b). SICM has a poorer spatial resolution compared to STM or AFM of ~50 nm, but with an advantage of causing less sample damage. Recent improvements, primarily in narrowing the diameter of the pipette to ~10 nm, have enabled noncontact imaging of collections of single protein molecular complexes on the outer membrane surface of live cells. Also, SICM has been used in conjugation with single-molecule folding kinetics studies of fluorescent proteins by using the same nanopipette to deliver a denaturant to chemically unfold, and hence photobleach, single fluorescent protein molecules prior to their refolding and gaining photoactivity (Klenerman et al., 2011).

6.5.10 ULTRASONIC FORCE MICROSCOPY

Ultrasonic force microscopy (*UFM*) (Kolosov and Yamanaka, 1993), also referred to as *atomic force acoustic microscopy* and essentially the same as *scanning near-field ultrasound holography*, applies similar principles of AFM imaging in using a silicon or silicon nitride tip attached to a metallic cantilever, which is scanned laterally across the sample surface. However, in UFM, the sample is coupled to a piezoelectric transducer below the sample and a second transducer to the cantilever, which both emit longitudinal acoustic waves of slightly different frequencies close to the resonance frequency of the tip–cantilever system. These waves are propagated into the sample and then result in an acoustic interference pattern. The acoustic vibrations from the interference pattern are picked up by the tip and transmitted into the cantilever, either through the sample media of air or water or through direct contact with the sample in the case of vacuum imaging. Cantilever oscillations are then detected in the same manner as for AFM using laser beam reflected and imaged onto a split photodiode.

Greatest sensitivity is achieved using force acoustic frequencies slightly higher than the resonance frequency of the normal flexure mode of the tip–cantilever system, usually from ~10 kHz up to ~5 MHz. The perturbations to both phase and amplitude across the sample surface in the acoustic standing wave are locally monitored by the tip, which acts as an acoustic antenna via a *lock-in amplifier*. Note that there are many common standard forms of instrumentation used in biophysics, which we will not explore in depth of this book; however, the lock-in amplifier is of particular use and is singled out here. It is an electronic amplifier common to many other applications of biophysics that can pull out and amplify a signal from a very specific frequency carrier wave from an otherwise extremely noisy environment, often in cases where the signal amplitude is up to six orders of magnitude smaller than the typical noise amplitude. As such it has myriad uses in single-molecule biophysics signal detection and amplification in particular.

Monitoring perturbations of the AFM tip in this way not only generates topographical information from the sample but also is a direct measure of elastic properties. For example, it can be used to infer the Young's modulus deeper in the sample in the region directly below the scanned tip, with an effective spatial resolution of 10–100 nm (Shekhawat and Dravid, 2005). Biological applications have included the imaging of malarial parasites buried deep inside red blood cells and monitoring aggregation effects *in vitro* of amyloid peptides (important precursors of various forms of dementia when misfolded, see Chapter 2).

KEY BIOLOGICAL APPLICATIONS: AFM

Imaging biological surface topography; Measuring molecular viscoelasticity and mechanics.

Worked Case Example 6.2: AFM Imaging

In an AFM imaging experiment, the cantilever–tip system was composed of a tetrahedral-shaped silicon nitride tip of height 10 μm with tip end radius of curvature 15 nm, which was fixed to a cantilever of mass 25 pg, such that the tip was located at the end of the cantilever's 0.1 mm long axis. The cantilever width was equal to the tetrahedron tip edge length.

a *What is the value of the cantilever resonance frequency in kHz and its stiffness in units of pN/nm? Comment on how these compare with the stiffness of typical "stiff" optical tweezers.*

b *How does the effective mass of the AFM tip–cantilever system compare with the mass of the cantilever and the mass of the tip?*

  AFM imaging was performed using this tip–cantilever system in contact mode to investigate an array of spike structures called "pili" (singular = "pilus") covering the surface of a spherical bacterial cell of ~2 μm diameter. Prior SEM imaging suggested that pili have a mean ~1 μm length and are composed of ~1000 copies of a sub-unit protein called "PapA," with pili expressed over the surface uniformly with a surface density that did not depend on the cell size. When the tip was scanned between consecutive pilus spikes, the estimated vertical force on the cantilever dropped by ~20 nN.

c *If the range of cell diameters in a population is ~1–3 μm, estimate the range in PapA copy numbers per cell, stating assumptions you make.*

d *The estimated mean width of a single pilus from earlier SEM imaging was 2.5 nm; however, AFM imaging suggested a width of almost 5 nm. Explain this discrepancy.*

(You may assume that the density and Young's modules of steel are 8.1×10^3 kg m^{-3} and 210 GPa, and the density of silicon nitride is 3.4×10^3 kg m^{-3}.)

Answers

a   We can first deduce the dimensions of the cantilever: $w_x = 0.1$ mm, w_y is equal to the edge length a of a tetrahedron of height h of 10 μm—tetrahedral trigonometry indicates that $w_y = a = h(3/2)^{1/2} = 12.2$ μm. The volume V of the cantilever is calculated from mass/density:

$$V = (25 \times 10^{-12})/(8.1 \times 10^3) = 3.1 \times 10^{-15} \text{ m}^3$$

Thus,

$$w_z = V - (w_x w_y) = (3.1 \times 10.15)/(1 \times 10^{-4} \times 1.22 \times 10^{-5}) = 2.5 \text{ μm}$$

The resonance frequency equation for a deflecting beam thus indicates

$$\begin{aligned} v_0 &= 0.162 \times (2.5 \times 10^{-6})/(1 \times 10^{-4})^2 \times ((2.1 \times 10^8)/(8.1 \times 10^3))^{1/2} \\ &= 6521 \text{ Hz or } \sim 6.5 \text{ kHz} \end{aligned}$$

$$\begin{aligned} k_z &= (1/4) \times (2.1 \times 10^8) \times (12.2 \times 10^{-5}) \times ((2.5 \times 10^{-6})/(1 \times 10^{-4}))^3 \\ &= 0.1 \text{ Nm}^{-1} \text{ or } \sim 100 \text{ pN/nm} \end{aligned}$$

"Stiff" optical tweezers have a typical stiffness of ~0.1 pN/nm, so this AFM cantilever is ~4 orders of magnitude stiffer.

b   Rearranging the beam equation to give the total effective mass of the cantilever–tip system indicates
$m_0 = 0.1/(6.1 \times 10^3)2 \times 4\pi^2 = 6.8 \times 10^{-11} \text{kg} = 68\text{pg}$,   which   is   three   times   the mass of the cantilever. The volume of the tip is given by the tetrahedron formula of $(1/3) \times$ base area $\times$ height $= 2.1 \times 10^{-16}$ m^3

Thus, the tip mass $= (3.4 \times 10^3) \times (2.1 \times 10^{-16}) = 0.7$ pg, which is two orders of magnitude smaller than the effective mass of the cantilever–tip system and thus negligible.

c   From Hooke's law, the change in vertical displacement of the cantilever corresponding to a 20 nN drop in force is

$$(20 \times 10^{-9})/0.1 = 200 \text{ nm}$$

Assuming sample deformation effects are negligible, this indicates that the average lateral spacing between pili on the surface is

$$(8 \times 15 \times 10^{-9} \times 200 \times 10^{-9})^{1/2} = 155 \text{ nm}$$

Model the occupancy area around each pilus as a circle of radius $155/2 = 78$ nm. Thus, the mean surface density of pili is

$$1/(\pi \times (78 \times 10^{-9})^2) = 5.2 \times 10^{13} \text{ pili per m}^2$$

The range of cell diameters in a population is 1–3 μm, equating to an area range of 3.1–28.3 μm². Thus, the range of number of pili in the population is 161–1450 per cell, 1000 PapA molecules per pilus, and therefore the copy number, assuming all cellular PapA expressed in pili, is 1.6×10^5 to 1.5×10^6 PapA molecules per cell.

d   The tip broadening artifact results in an apparently larger pilus width. The half angle of tip is ~(180 – 109.5) = 35.3° where 109.5° is the tetrahedral angle, so tip broadening is tan(35.3°) + sec(35.3°) = 1.9, so the apparent width measured by AFM imaging of pilus will be ~2.5 × 1.9 = 4.8 nm, as observed.

6.6 ELECTRICAL FORCE TOOLS

Electrical forces can be used to control the mobility of molecules (e.g., in *gel electrophoresis*) as well as generating stable traps for biological particles (*anti-Brownian electrophoretic/electrokinetic traps* [*ABEL traps*]) and controlling their rotation (*electrorotation*). One of the most important applications of electrical forces is in the measurements of ion fluxes through *nanopores*. If a small hole is made in a sheet of electrical insulator surrounded by an ionic solution and an electrostatic potential difference applied across the sheet, then ions will flow through the hole. If the hole itself has a length scale of nanometers (i.e., a "nanopore"), then this simple principle can form the basis of several biophysical detection techniques, most especially *patch clamping*.

6.6.1 GEL ELECTROPHORESIS

Gel electrophoresis is one the of most ubiquitous and routine biophysical techniques in modern biochemical research laboratories, but is still in many ways one of the most useful for its ability to separate the components of a complex *in vitro* sample composed of a mixture of several different biomolecules in a simple, relatively fast and cheap way, and to characterize these molecules, on their basis of their size, charge, and shape. Usually, a sample of a few tens of microliters in volume is injected into a semiporous gel and exposed to an electric field gradient. The gel is composed of either *polyacrylamide* (for protein samples) or *agarose* (for samples containing nucleic acids), with space for ~10 parallel channels in each gel so that different samples can be run under the same conditions simultaneously. Gels are cast with a characteristic concentration that affects the distribution of pore sizes in the gel matrix. Smaller molecules will therefore diffuse faster through this mesh of pores than larger molecules.

Equating the electrical force to the drag force indicates that the *drift speed* v_d of a molecule of net charge q during gel electrophoresis across an electric field E is given by

(6.32)
$$v_d = \frac{Eq}{\gamma}$$

Protein samples may either be first denatured by heating and combined with an ionic *surfactant* molecule such as *sodium dodecyl sulfate* (*SDS*), or may be run in a nondenatured native state. SDS disrupts noncovalent bonds in the protein and so disrupts the normal molecular conformation to generate a less globular structure with significantly higher surface negative charge compared to the native state. Each SDS molecule binds nonspecifically to peptides with a ratio of roughly one molecule per two amino acid residues, resulting in a large net negative charge from the sulfate groups of the SDS, which mask smaller surface charges of the substituent group of each amino acid.

The net charge q is thus roughly proportional to the total number of amino acids in the peptide.

Proteins in this state electrophoretically migrate in a typically ellipsoidal conformation with their major axis significantly extended parallel to the electric field, normally oriented vertically. This elongated conformation bears no necessary resemblance to the original molecular structure and so the molecular mobility is relatively insensitive to the original molecular shape, but is sensitive to the molecular weight, that is, the elongation of the charged molecule during electrophoresis is also roughly proportional to its molecular weight. Thus, Equation 6.32 might appear to suggest that the effects of frictional drag and charge cancel out. This is largely true when we consider frictional drag in a homogeneous medium, and the fact that larger molecules electrophoretically migrate more slowly than smaller ones, due primarily to the complex interaction between the molecule and the pores of the gel matrix (the complex physics are explored comprehensively in Viovy, 2000). The end result is that a protein molecule will have traveled a characteristic distance in a given time dependent on its molecular

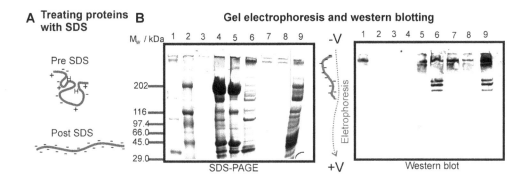

FIGURE 6.9 Gel electrophoresis methods. (a) The detergent sodium dodecyl sulfate (SDS) is used to denature and linearize proteins—native proteins are coated in positive and negative charge, as well as having regions of the protein stabilized by nonelectrostatic forces, such as hydrophobic bonding (marked here with "H"). SDS eradicates hydrophobic binding and coats all the peptide chain uniformly with them with negatively charged sulfate groups. (b) These linearized and negatively charged proteins then migrate down an *E*-field gradient, such as in SDS polyacrylamide gel electrophoresis (left panel) here shown with molecular weight lines on the left on the scanner gel, with proteins stained to reveal their location on the gel at a specific time point after starting the electrophoresis in distinct bands (the different channels here show various different mixtures of muscle proteins). The proteins in such gels may also be blotted onto a separate substrate and probed with a specific antibody, which only binds to a specific region of a particular protein, called a "western blot" (right panel).

weight, such that higher-mass molecules appear as a band toward the top of the gel, whereas lower-mass molecules appear as a band toward the bottom end of the gel (Figure 6.9). Note that enhanced separation of proteins (e.g., with a higher molecular weight cutoff) is also possible using gel electrophoresis in rationally designed microfabricated arrays, instead of the standard slab-cast gel systems.

Double-stranded DNA samples are often run on agarose gels in a native state. Single-stranded nucleic acids, either single-stranded DNA or RNA, have a propensity to form a range of secondary structures *in vitro* due to transient Watson–Crick base pairing interactions, resulting in a range of mobility during gel electrophoresis. This presents potential problems in interpretation, and so single-stranded nucleic acid samples are often denatured first using *urea* or *formamide* treatment.

In *native gel electrophoresis*, the proteins in the sample are not denatured. Such gels can discriminate molecular components on the basis of shape, net electrical charge, and molecular weight and are often run using *2D gel electrophoresis* (*2-DE*). In 2-DE, a population of native molecules (usually proteins) in an *in vitro* sample are first separated electrophetically by their mobility on a gel that contains a *pH gradient*. A molecule will therefore migrate down the electric field to the position on the gel equivalent to an overall zero charge of that molecule at that particular pH in the gel, that is, their isoelectric point (see Chapter 2). The sample is then treated with SDS to denature the proteins and an electric field generated at 90° to the original field to induce electrophoretic movement horizontally as opposed to vertically. Thus, instead of a 1D ladder of bands separating different molecules with different molecular weights, there are 2D blobs whose position on the gel, after given electrophoresis times in both dimensions, can be related to their molecular weight and native electrical charge properties, with the caveat that it is a technically more challenging and time-consuming technique.

Molecules can be visualized on a gel using chemical staining, either directly in visible light (e.g., *Coomassie Blue* is a standard stain of choice for proteins, though *silver staining* may also be applied if greater sensitivity is required) or using fluorescence emission via excitation of a stain from ultraviolet light (e.g., *ethidium bromide stain*, used for nucleic acids). Each band/blob may also be carefully extracted to reconstitute the original, now purified, sample. Thus, this technique may be used both for purification and for characterization, for example, to estimate the molecular weight of a sample by interpolation of the positions on a gel against a reference calibration sample. Isolation of a protein using 2D-E is often a precursor to analysis using mass spectrometry.

6.6.2 ELECTROPHYSIOLOGY

The lipid bilayer architecture of cell membranes is disrupted by natural nanopores of ion channels. These are protein structures that enable controllable ion flow into and out of cells. They generally involve high specificity in terms of the ions allowed to translocate through the pore and often use sensitive voltage gating and other molecular mechanisms to achieve this. The presence of these nanopore molecular complexes can be investigated using *patch clamping*.

The resistance of an open ion channel in a cell membrane is a few GΩ; therefore, any probe measuring electric current through the channel must have a resistance seal with the membrane of at least a GΩ, hence the term *gigaseal*. For a nanopore of cross-sectional area A through which ions in a solution of electrical resistivity ρ translocate a total axial distance length l, then the nominal resistance is given by $\rho l/A$ as expected from *Ohm's law*, plus an additional *access resistance* (see Hall, 1975) due to either ion entry or ion exit to/from a circular aperture radius a of $\rho/4a$. Thus, the total electrical resistance $R_{channel}$ of an ion channel is approximated by

$$(6.33) \qquad R_{channel} = \frac{\rho}{a}\left(\frac{l}{\pi a} + \frac{1}{4}\right)$$

Usually a glass micropipette tipped with a silver electrode is pressed into suction contact to make a seal with very high electrical resistance greater than the GΩ level (Figure 6.10a). Time-resolved ion-flux measurements are performed with the micropipette in contact either with the whole intact cell or with the attached patch of membrane excised from the cell either by keeping the current clamped using feedback circuitry and measuring changes in voltage across the membrane patch or, more commonly, by clamping the voltage to a set value and measuring changes in current. Current measurements are often made in conjunction with physical or chemical interventions that are likely to affect whether the ion channel is opened or closed and to probe a channel's mode of operation, for example, by adding a ligand or drug inhibitor or by changing the fixed voltage level, typically set at ~100 mV (Figure 6.10b).

The physical basis of the equilibrium level of voltage across a selectively permeable barrier, such as the cell membrane with pores through which ions can selectively diffuse, is established when the osmotic force due to differences in ion concentration either side of the membrane is balanced by the net electrostatic force due to the electrochemical potential on the charged ion in the presence of the membrane voltage potential. As discussed previously in this chapter, the osmotic force is entropic in origin; however, the electrostatic force is enthalpic. The combination of both forces gives rise to another depiction of the Nernst equation (see Chapter 2):

$$(6.34) \qquad V_{mem} = \frac{RT}{nF}\ln\frac{[A]_{out}}{[A]_{in}}$$

where V_{mem} is the equilibrium voltage across the cell membrane with charged ion A with n its ionic charge in equivalent number of electrons per ion, having concentrations (A) inside and outside the cell (R is the molar gas constant, T the absolute temperature, and F the Faraday constant). With several ions, the equilibrium potential can be calculated from the fractional contribution of each using the more general *Goldman equation*. For many cell types, V_{mem}

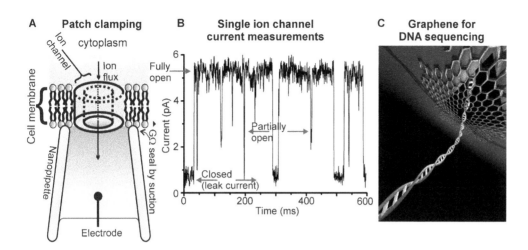

FIGURE 6.10   Electric current flow through nanopores. (a) Patch clamping to capture one or a few single-ion channels in a patch of membrane at the end of an electrical "gigaseal" nanopipette tip, which can generate (b) time-resolved current measurements at the ion channel opens and closes. (c) Synthetic nanopores can also be manufactured about solid substrate, for example, graphene is then but mechanically very stiff and stable and so offers potential for high-resolution characterization of biopolymers, for example, sequencing of DNA as each nucleotide base pair on translating through the nanopore results in different characteristic drop in electric current. (Courtesy of Cees Dekker, TU Delft, the Netherlands.)

is in the range −50 to −200 mV, with the negative sign due to energized net pumping out of sodium ions compared against a smaller influx of potassium ions.

In many cells, the size of the equilibrium transmembrane voltage potential is finely controlled, for example, in bacteria, V_{mem} is very closely regulated to −150 mV. In some cells, for example, during nerve impulse conduction, V_{mem} can vary due to a wave of depolarization of voltage. In the resting state sodium ions are actively pumped out in exchange for potassium ions that are pumped into the cell, energized by the hydrolysis of ATP. These sodium–potassium ion-exchange pumps are essentially selective ion channels that exchange with the ratio of three sodium to every two potassium ions, hence resulting in a net negative V_{mem}, with the so-called resting potential of ca. −70 mV. During nerve impulse conduction the ion pumps transiently open to both sodium and potassium causing depolarization of V_{mem}, rising over a period of ~1 ms to between +40 and +100 mV depending on nerve cell type, with the resting potential reestablished a few milliseconds later. The recovery time required before another action potential can be reached is typically ~10 ms, so the maximum nerve *firing rate* is ~100 Hz.

An open ion channel current is open in around one to a few tens of picoamperes. This is roughly a millionth-millionth the level of electric current that a TV or a kettle uses and is equivalent to ~10^6–10^8 ions per second, which even when sampled with fast GHz detector bandwidth struggles to be *single ion* detection. Rather, the ion flux is the detection signature for the presence of a single ion channel and of its state of opening or closure, or indeed somewhere in between as appears to be the case for some channels.

The area of membrane encapsulated by the patch clamp may contain more than one ion channel, which impairs the ability to measure single ion channel properties, for example, investigating whether there are heterogeneous short-lived states between a channel being fully open or closed. The measured current for a membrane patch enclosing multiple ion channels will be the sum of the currents through each channel, and since each may be open or closed in a stochastic (i.e., asynchronous) manner, this leads to difficulties in interpretation of the experimental ion-flux data.

Genetically modifying the cell to generate a lower surface density of ion channels reduces this risk, as does inhibiting ion channel protein levels of expression from their genetic source. Similarly, growing larger cells reduces the effective ion channel surface density.

However, none of these modifications is ideal as they all affect the native cell physiology. The use of smaller diameter pipettes is a less perturbative improvement—as for SICM, glass micropipettes may be heated and controllably stretched to generate inner diameters down to a few tens of nanometers. Ion channel current measurements may also be performed in combination with fluorescence imaging—if a fluorescence maker can be placed on a component of the nanopore, then it may be possible to count how many ion channels are present in the patch clamp region directly, through controllably placing a fluorescent tag on a nanopore but avoiding impairment of the ion channel function is nontrivial.

Many researchers also utilize similar electrophysiology techniques on larger tissue samples. The most popular biological systems to study involve muscle and nerve tissue. Much of the early historical research involving biophysical techniques used electrophysiology approaches, but many of these methods are still relevant today. In essence, they involve either excised tissue or, as is sometimes the case for cardiac muscle studies, experiments using whole living animal models. Electrodes are relatively large in length scale compared to the thinned micropipettes used for patch clamp methods, for example, consisting of metal needles or micron length scale diameter micropipettes filled with electrolyte solution.

Although lacking some of the finesse of patch clamping, traditional electrophysiology methods have a distinct advantage in generating experimental data in a physiologically relevant tissue level environment. The importance of this is that single cells respond electrically to both chemical and mechanical triggers of their neighbors in addition to their intrinsic electrical properties at the single-cell level. These effects are very important in the emergence of larger length scale properties of whole tissues, for example, in determining the complicated beating rhythms of a whole heart. There is also significant scope for valuable biophysical modeling of these complex whole tissue electrical events, and the cross length scale features are often best encapsulated in *systems biophysics* approaches (i.e., systems biology in the context of biophysical methodology), which are discussed in Chapter 9.

6.6.3 SOLID-STATE NANOPORES

Modern nanofabrication methods now make it possible to reproducibly manufacture nanopores using synthetic silicon-based solid-state substrate. One popular method to manufacture these involves *focused ion beam* (*FIB*) technology. FIB devices share many similarities to TEMs in generating a high-intensity beam of electrons on the sample. The beam is focused onto a thin sheet consisting of silicon nitride, which generates a hole. By varying the power of the beam the size of the nanopore can be tuned, resulting in reproducible pore diameters as low as ~5 nm (van den Hout et al., 2010). Such nanopores have been applied successfully in the detection of single molecules of a variety of biopolymers including nucleic acids (Rhee and Burns, 2006) and also have been used to measure the unfolding macromolecules.

Molecular detection using ion flux through solid-state nanopores involves first applying a voltage across either side of the nanopore, which causes ion flow through the pore in the case of a typical physiological solution. However, any biopolymer molecules in the solution will in general possess a nonzero net charge due to the presence of charges on the molecular surface, resulting in the whole molecule migrating down the voltage gradient. Due to the large size of biopolymer molecules, their drift speed down the voltage gradient will be much slower than that of the ion flow through the nanopore. When a biopolymer molecule approaches the nanopore, the flow of ions is impeded, maximally as the molecule passes through the nanopore. The drop in ion current is experimentally measurable if the translocation speed through the nanopore is sufficiently slow. The specific shape of the drop in current with time during this translocation is a signature for that specific type of molecule, and so can be used as a method of single-molecule detection.

With greater spatial precision than is currently possible, a hope is to consistently measure different nucleotide bases of nucleic acids as a single molecule of DNA migrates through the nanopore, hence sequencing a single DNA molecule rapidly. The main problem with this scenario is the speed of migration: even for the lowest controllable voltage gradient the translocation speeds are high for unconstrained DNA molecules leading to the unreliability in experimental measurements of the ion-flux signature. One method to slow down a DNA molecule as it translocates through a nanopore is by controllably pulling on the molecule from the opposite direction to the electrostatic force using optical tweezers (Keyser et al., 2006).

An additional issue with DNA sequencing through a solid-state nanopore is the finite translocation length. The minimum width of a structurally stable silicon nitride sheet is ~20 nm, equivalent to ~50 nucleotide base pairs of DNA (see Chapter 2) assuming the double-helical axis of the molecule is stretched parallel to the central nanopore axis. Attempts to circumvent this problem have involved reducing the substrate thickness by using a monolayer of graphene (Schneider et al., 2010). Graphene is a 2D single atomic layer of carbon atoms packed into a honeycomb shape with a thickness of only ~0.3 nm but which is structurally stable. This is comparable to just a single-nucleotide base pair (Figure 6.10c).

Graphene is not an easy substrate to work with, however, being mechanically quite brittle, and also graphene is only as strong as its weakest link, such that imperfections in its manufacture can seed extensive cracks in its structure. Also, graphene nominally has a high hydrophobicity that can causes problems when working with physiological solutions. An alternative compromise being developed is to use a molybdenum disulfide three-atom layer substrate. This has an inferior larger thickness of ~0.8 nm, but fewer of the problems are described earlier.

Simulation studies for the translocation of single biopolymers through a nanopore that incorporate some degree of realistic flexibility of the nanopore wall actually suggest that allowing the pore, some level of compliant wiggle can increase the speed of biopolymer translocation (see Cohen et al., 2011). In this case, nanopores composed of a less stiff material than graphene, molybdenum disulfide, or silicon nitride might be an advantage, such as those composed of soft matter, discussed in the following text.

6.6.4 SYNTHETIC SOFT-MATTER NANOPORES

A number of natural pore-forming proteins exist, which can self-assemble within a phospholipid bilayer, and are much more compliant than the synthetic silicon-based nanopores discussed earlier. The best characterized of these is a protein called α-hemolysin. This is a poison secreted by the *Staphylococcus aureus* bacterium to kill other species of competing bacteria (a version of *S. aureus* that is resistant to certain antibiotics has been much in the news due to its increasing prevalence in hospitals, called *methicillin-resistant S. aureus*). α-Hemolysin binds to cell membranes of these nearby competing bacteria and spontaneously punches a hole in the phospholipid bilayer significantly impairing these cells' viability by disrupting the proton motive force across the membrane, which thus allows protons to leak uncontrollably through the hole and destroy their ability to manufacture ATP from the oxidative phosphorylation process (see Chapter 2).

An α-hemolysin pore is formed by self-assembly from seven monomer subunits (Figure 6.11a). These nanopores can be used in a controlled environment in an artificial phospholipid bilayer and utilized in a similar manner to solid-state nanopores to study the translocation of various biomolecules through the nanopore by measuring the molecular signature of the ion current as the molecule translocates through the nanopore (see Bayley, 2009). These naturally derived protein nanopores have advantages over solid-state nanopores. First, their size is consistent and not prone to manufacturing artifacts. Second, they can be engineered to operate both with additional adapter molecules such as *cyclodextrin*, which allows greater ion current measuring sensitivity for translocating molecules such as DNA, and in addition the amino acid residues that make up the inside surface of the pore can be modified, for example, to alter their electrostatic charge, which can be used to provide additional selectivity on which biomolecules are permitted to translocate through the pore. This nanopore technology is a prime candidate to first achieve the goal of reliable, consistent, rapid single-molecule sequencing of important biopolymers such as DNA in the near future.

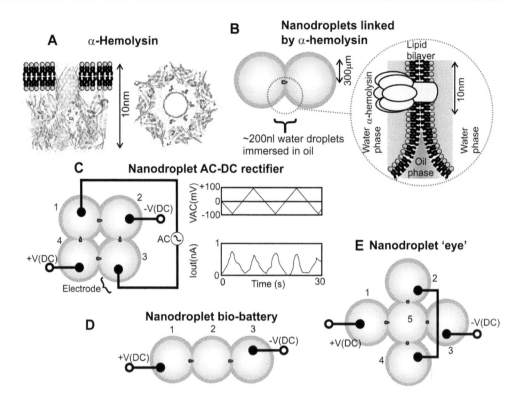

FIGURE 6.11 Synthetic *soft* nanopores using protein adapters in lipid bilayers. (a) Structure of protein complex α-hemolysin shown in a side view (left panel, integrated in a lipid bilayer) and plan view (right panel), which can (b) form a link between two adjacent aqueous droplets with a lipid monolayer border that becomes a bilayer where the droplets touch, surrounded by oil that stabilizes the lipid monolayer. The α-hemolysin protein complex allows flux of ions, water, and narrow molecules between the two droplets. These droplets can form complex, functional devices by linking together multiples of droplets, such as (c) an AC–DC voltage rectifier, (d) a *biobattery*, and (e) a nanodroplet photoreceptor (or *nanoeye*).

These soft-matter nanopores can also be used in constructing complex nanodroplet systems. Here, 200 nL droplets have an internal aqueous phase separated by an artificial phospholipid monolayer that remains structurally stable due to centrally acting hydrophobic forces imposed from an external oil phase (Figure 6.11b). These droplets can be positioned directly by capturing onto the tip of an agarose-coated Ag/AgCl 100 μm diameter electrode using surface tension from the aqueous phase, which in turn is connected to a micromanipulator. Multiple droplets may be positioned adjacent to each other relatively easily in a 2D array, with droplets sharing common phospholipid bilayer interfaces and joined by one or more α-hemolysin nanopores integrated in the bilayer.

By modifying the amino acid residues in the pore lumen to give all positive charges, it was found that these nanopores would be open in the presence of a positive voltage potential, but closed in the presence of a negative potential, presumably due to some induced conformational change blocking the pore lumen (Maglia et al., 2009). This modified nanopore is therefore voltage gated and acts as an electrical diode. As a proof of principle, it was possible to join four such nanodroplets to form a full-wave AC–DC rectification system (Figure 6.11c).

Other complex arrangements of nanodroplets have led to a tiny nanodroplet *biobattery* (Figure 6.11d), in its simplest form made from a linear arrangement of three nanodroplets in which the central droplet is connected to the others via either a positive or a negative ion selective nanopore, resulting in a small current flow between the electrode termini of ~50 pA located at the outer two nanodroplets. There is also a *nanoeye* in which photons of light can be detected (Holden et al., 2007). This biomimetic system consists of five nanodroplets

(Figure 6.11e) with nanopores consisting either of α-hemolysin or the photosensitive protein *bacteriorhodopsin*. Bacteriorhodopsin is a cell membrane nanopore in bacteria, which
utilizes the absorbed energy from a single green wavelength photon of light to pump a single
proton across a phospholipid bilayer. This constitutes a small current, which the nanodroplet
arrangement can detect. Although it is possible to controllably implement such a system with
only one to two α-hemolysin complexes in each common phospholipid bilayer interface, the
number of bacteriorhodopsin molecules required to generate a measurable current is of the
order of thousands, but as a proof of principle, this shows great promise. Currently, most
nanodroplet arrangements are 2D, but there are recent developments toward implementing
more complex nanoscale biosynthetic systems in 3D. Although nanodroplet systems are
clearly not natively cellular, they represent a synthetic biological system that is moving in the
direction of an exceptionally cell-like physiological behavior.

6.6.5 ELECTROROTATION

An electric dipole can be induced on an electrically polarizable particle between
microelectrodes generating an electric field in a bespoke microscope flow cell (Figure 6.12a).
A suitable particle could include a latex microsphere that has been functionalized on its
surface with electrically charged chemical groups, such as negatively charged carboxyl or
amino groups that become protonated and hence positively charged in aqueous solution (see
Chapter 2). In a nonuniform electric field, there is a nonzero E-field gradient that imparts
a force on the particle in a direction parallel to the E-field gradient, termed "dielectrophoresis" (Figure 6.12b), the same driving force of electrical molecular mobility used in gel
electrophoresis.

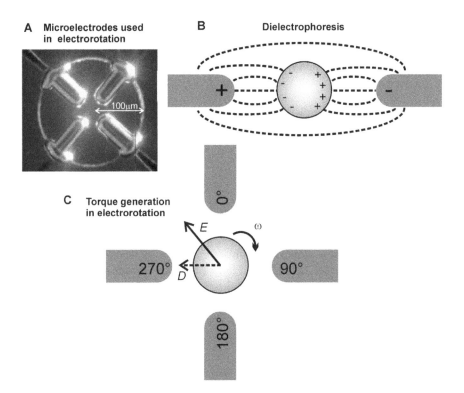

FIGURE 6.12 Electrorotation. (a) Microfabricated electrodes in quadrature using for
electrorotation. (Courtesy of Hywel Morgan, University of Southampton, Southampton, U.K.)
(b) Induced electrical dipole by dielectrophoresis. (c) Generation of torque during electrorotation
due to phase lag between direction of driving *E*-field and induced dipole in bead.

However, if the E-field is uniform, the particle's dipole moment aligns parallel to the field lines. If this E-field vector rotated around the particle, then the finite time taken for the induced dipole to form (the dipole *relaxation time*), resulting from charge redistribution in/on the particle, lags behind the phase of the E-field, a phenomenon that becomes increasingly more significant with increasing rotational frequency of the E-field. This results in a nonzero angle between the E-field vector and the dipole at any given time point, and therefore there is a force on the induced electrical dipole in a direction of realignment with the rotating E-field. In other words, the particle experiences a torque (Figure 6.12c) that causes the particle to rotate out of phase with the field, either with or against the direction E-field rotation depending on whether the phase lag is less or more than half an E-field period. This effect is called electrorotation.

The speed of electrorotation depends on the particle's surface density of electrical charge, its radius a (assuming a spherical particle) and the magnitude E, and frequency v of the electric field. The torque G experienced by the bead is given by

(6.35)
$$G = -4\pi\varepsilon_w a^3 E^2 \mathrm{Im}\left[K(v)\right]$$

where

  ε_w is the permittivity of the surrounding water-based pH buffer that embodies the charge qualities of the solution
  $\mathrm{Im}[K(v)]$ is the imaginary component of K, which is the *Clausius–Mossotti factor*, which embodies the charge properties of the bead

The full form of the Clausius–Mossotti factor is given by $\left(\varepsilon_b^* - \varepsilon_w^*\right)/\left(\varepsilon_b^* + 2\varepsilon_w^*\right)$ where ε_w^* is the complex permittivity of the surrounding water solution and ε_b^* is the complex permittivity of the bead. A general complex electrical permittivity $\varepsilon^* = \varepsilon - ik/2\pi v$ where ε is the real part of the complex permittivity and k is the electrical conductivity of the water solution.

Typically, there are four microelectrodes whose driving electric currents are phased in quadrature, which produces a uniform AC E-field over an area of a few square microns in between the microelectrodes (Figure 6.12a). For micron-sized charged beads, an E-field rotational frequency of 1–10 MHz with ~50 V amplitude voltage drop across a microelectrode gap of a few tens of microns will produce a bead rotation frequency in the range 0.1–1 kHz. Electrorotation experiments have been applied to studies of single rotary molecular machines, such as the bacterial flagellar motor (Rowe et al., 2003), to characterize the relation between the machine's rotational speed and the level of torque it generates, which is indicative of its mechanism of operation.

6.6.6 ABEL TRAPPING

An arrangement of four microelectrodes, similar to that used for electrorotation described earlier, can be also used in a DC mode. In doing so, an electrically charged particle in the center of the electrodes can be controllably moved using dielectrophoresis to compensate for any random fluctuations in its position due to Brownian diffusion. The potential electrical energy on a particle of net charge q in an electric field of magnitude E moving through a distance d parallel to the field is qEd, and by the equipartition theorem this indicates that the mean distance fluctuations at the center of the trap will be ~$2k_B T/qE$. Also, the size of the dielectrophoretic force F is given by

(6.36)
$$F = 2\pi a^3 \varepsilon_w \nabla\left(E^2\right)\mathrm{Re}\left[K(v)\right]$$

where $\mathrm{Re}\left[K(v)\right]$ is the real component of the Clausius–Mossotti factor. If the electric field can be adjusted using rapid feedback electronics faster than the time scale of diffusion of the particle inside a suitable microscope sample flow cell, then the particle's position in the focal plane of the microscope can be confined to a region of space covering an area of just a few square microns. This arrangement is called an anti-Brownian electrophoretic/electrokinetic (ABEL) trap . It can operate on any object that can be imaged optically, which can acquire an electric charge in water, and was first demonstrated on fluorescently labeled microspheres using a device whose effective trap stiffness was four orders of magnitude smaller than that of a typical single-beam gradient force optical trap (Cohen and Moerner, 2005). Further refinements include real-time positional feedback parallel to the z-axis (e.g., automated refocusing by fast feedback of the bead's detected position to a nanostage) to ensure that the particle lies in the same lateral plane as the four microelectrodes.

The application of ABEL traps permits longer continuous observation of, for example, molecular machines in solution that otherwise may diffuse away from their point of action relatively quickly over a time scale of milliseconds away from the detector field of view. Earlier, similar approaches for confining a single biomolecule's Brownian motion directly (i.e., without using a relatively large adapter particle such as a micron-sized bead) used surface binding either via surface tethering of molecules or surface binding of lipid vesicles containing a small number of molecules for use in smFRET investigations (see Chapter 4); however, the advantage of the ABEL trap is that there are no unpredictable surface forces present that could interfere with molecular properties.

ABEL trapping has been applied at the single-molecule level to provide ~1 nm precise trapping. This level of spatial resolution opens the possibility for measuring molecular conformational transitions in single biomolecules in solution in real time. For example, this approach has been used to monitor differences in electrokinetic mobility of single fluorescently labeled DNA molecules in the presence or absence of a DNA-binding protein called "RecA" (which is involved in repairing damaged DNA in the living cell) over periods of several seconds (Fields and Cohen, 2011).

6.6.7 PIEZOELECTRIC TECHNOLOGIES

The *piezoelectric effect* is a consequence of electrical charge redistribution in certain solid materials dependent on mechanical stress, typically a ~0.1% change in mechanical strain resulting in a measurable piezoelectric current. Such *piezoelectric materials* have reversibility in that they also exhibit a *converse piezoelectric effect*, such that the application of an electrical field creates mechanical deformation in the solid. Piezoelectric materials include various crystals (quartz being the most common) and synthetic ceramics and semiconductors but also include natural biological material including bone, certain proteins and nucleic acids, and even some viruses (see Chapter 9), with a role being potentially one of a *natural force sensor*.

The piezoelectric effect involves a linear electromechanical interaction between the mechanical and the electrical state in crystalline materials, which possess no *inversion symmetry*. Crystals that have inversion symmetry contain a structure comprising a repeating unit cell (there is a point in each known as the *inversion center* that is indistinguishable from that point in any other unit cell), whereas piezoelectric material has no equivalent inversion center. The piezoelectric effect results from a change of bulk electric polarization of the material with mechanical stress caused either by a redistribution of electric dipoles in the sample or their reorientation. The change in electrical polarization results in a variation of electrical charge density on the surface of the material. The strength of the piezoelectric effect is characterized by its *dielectric constant*, which for the most common piezoelectric synthetic ceramic of *lead zirconate titanate* (also known as "PZT") is in the range ~300–3850 depending on specific doping levels in the crystal, with an equivalent *dielectric strength* (the ratio of measured voltage change across faces of the crystal to the change in separation of the faces) of ~8–25 MV m^{-1} (equivalent to ~1 mV for a single atomic diameter separation change).

Primary uses of piezoelectric material in biophysical techniques are either as sensitive *force actuators* or *force sensors*. Actuators utilize the converse piezoelectric effect and can involve relatively simple devices such as mechanical valves in microfluidics devices and for the fine control of the steering of optical components as well for scanning probe microscopes discussed previously in this chapter. They reach a state of the art in controlling the 3D deflection of "smart" microscope nanostages to sub-nanometer precision (see Chapter 7).

The biophysical application of piezo sensor is best exemplified in the *quartz crystal microbalance (QCM)*, especially the *QCM with dissipation monitoring (QCM-D)* that uses very sensitive acoustic detection technology to determine the thickness of an absorbed layer of biomolecules in a liquid environment. A QCM-D measures the variation in mass per unit area from the change in the natural resonance frequency of the quartz crystal. As we have seen, mechanical stress on a piezoelectric material induces a small voltage change across faces of the material, but this in turn generates an electrical force that acts to push the material in the opposite direction, making such a material to naturally oscillate as a *crystal resonator*. The resonance frequency of the manufactured quartz crystal resonator being in the range of a few kHz up to hundreds of MHz depends on the size of the crystal. This is the basis of the timing signature of cell phones, computers, and digital watches, with the standard normally set for a wristwatch being 32.768 kHz.

In a QCM-D, the resonance frequency is changed by the addition or removal of very small masses on one of the quartz surface faces, with the unbound state resulting in a typical resonance frequency of ~10 MHz. For example, a QCM-D can be used to determine the binding affinity of biomolecules to chemically functionalized surfaces, with an equivalent monolayer of bound biomolecules reducing the resonance frequency of the quartz crystal resonator of the QCM-D by a few MHz. A typical application here is that of an antibody binding to its recognition sites that might be expressed controllably on the surface. Similarly, to monitor the formation of artificial phospholipid bilayers on a surface, since the QCM-D is sufficiently sensitive to discriminate between a lipid monolayer and a bilayer bound to the surface.

KEY BIOLOGICAL APPLICATIONS: ELECTRICAL FORCE TOOLS

Molecular separation and identification; Quantifying biological torque; Measuring ionic currents.

6.6.8 TETHERED PARTICLE MOTION AND ACOUSTIC TRAPPING

Tethered particle motion (TPM) involves tracking the position of a tracer bead, typically of a micron diameter or less, which is tethered to one end of a filamentous biopolymer, the other end of which is immobilized onto a coverslip surface. The forces involved are the thermally driven *Langevin force* of the surrounding solvent molecules on the bead and the biopolymer, which results in random thermal fluctuations in the bead position, and a trapping force derived from the elasticity of the tether, which, as we will see later in Chapter 8, is primarily entropic in origin. The extent of displacement of the bead, and how this is correlated in time, is a measure of the mechanical properties of the tether; TPM is often used to quantify the biomolecule's persistence length by modeling the positional data using typically the *Kratky–Porod model*, which derives from a worm-like chain approximation to the molecule's elasticity (fast forward to section 8.3.3 for details).

As discussed previously for microrheology investigations that use tracer tracking (section 6.2.4), bead frictional drag imposes a limit on the time resolution of TPM, as does drag from the tethered biomolecule itself. As with tracer tracking in microrheology, laser darkfield using gold nanobeads is currently the best compromise approach, which maximizes time and space resolution while also allowing extended duration observations due to the absence of photobleaching effects. Simple stochastic binding to the coverslip is relatively easy to configure, but for better throughput and consistency, it is also possible to print conjugation chemicals using a compliant substrate such as polydimethylsiloxane (PDMS) (see section 7.6.2) into well-defined grid patterns on a coverslip surface, enabling up to several hundred tethered beads to be monitored in a single camera detector field of view simultaneously; however, the bottleneck then becomes the video-rate speeds of the camera sampling available to such a wide pixel area (up to ~1 kHz), whereas lower throughput detection methods such as laser interferometry back focal plane detection (see section 6.3.3) can yield sampling rates two orders of magnitude faster.

Acoustic trapping or *acoustic tweezers* can trap microscale particles in the standing wave nodes created from the interference of ultrasonic waves. It its most useful form it can be seen as a variant to TPM in that beads tethered to surface-bound single biopolymer molecules can be stably trapped and so by then varying the height of the coverslip by manipulating the microscope stage, different levels of force can be applied to extend the molecules. It has much of the capabilities of vertical magnetic tweezers though cannot apply torsion but is arguably easier to configure in high-throughput modes.

6.7 TOOLS TO MECHANICALLY PROBE CELLS AND TISSUES

Several tissue types exhibit a range of important mechanical properties. These can be investigated using a range of biophysical biomechanical tools. Much research in this field has involved study of muscle tissue in particular, but several types of connective and bone tissues in animals have also been studied, as have mechanical forces relevant to plant tissues.

6.7.1 MECHANICAL STRETCH TECHNIQUES ON MUSCLE FIBERS AND MYOFIBRILS

A variety of mechanical stretching apparatus has been developed for various tissue samples, most especially exemplified by bundles of *muscle fibers* to subject them to mechanical stretching and subsequent relaxation. For example, by attaching controllable electrical motors conjugated to the ends of muscle fiber bundles while subjecting the muscle fibers to different biochemical stimuli to explore the onset of active muscle contraction. Active contraction requires the hydrolysis of ATP through the interaction of myosin and actin protein filament systems, as well as the maintenance of passive elasticity through other muscle filaments such as *titin* already discussed in this chapter.

Stretched muscle fiber bundles can also be monitored using various optical diffraction techniques. Muscle fibers have several structural features that are spatially highly periodic, which therefore can act as diffraction gratings for appropriate incident wavelengths of electromagnetic radiation. Visible light *laser diffraction* through fiber bundles can be used to estimate the dynamic change in length of the sarcomere, the repeating structural subunit of myofibrils from which muscle fibers are assembled. Fluorescence microscopy can also be combined with myofibril stretching to indicate the change in position to specific parts of filamentous molecules, for example, using fluorescently labeled antibodies that target specific locations in the giant muscle molecule titin, to explore the relative elasticity of different regions of the titin molecule.

X-ray diffraction (see Chapter 5) can also be used on muscle fiber bundles to investigate smaller molecular length scale changes to the protein architecture during muscle contraction. For example, using both small-angle x-ray scattering to explore large length scale changes to the sarcomere unit and higher-angle diffraction investigates more subtle changes to the binding of myosin to action. This has contributed to a very detailed knowledge of the operation of molecular motors, which is now being complemented by a range of cutting-edge single-molecule methods such as optical tweezers.

6.7.2 MECHANICAL STRESS TECHNIQUES ON NONMUSCLE TISSUES

Developing bone tissue has also been investigated using similar mechanical stretch apparatus, as has *connective tissue* (the tissue that connects/separates different types of tissues/organs in the body), and epithelial tissue (the tissue that typically lines surface structures in the body), including *skin*. Stretch-release experiments on such tissues can also generate bulk tissue mechanical parameters such as the Young's modulus, which can be linked back to biological structural details mathematical modeling such as discretized *finite element analysis* and biopolymer physics *mesoscale modeling* approaches (see Chapter 8). Other forms of continuum mathematical modeling of elasticity, also discussed in Chapter 8, include *entropic spring* approaches such as characterizing the elasticity by *a freely jointed chain* or *worm-like chain* in addition to modeling the *viscous relaxation* effects of tissues manifest as energy losses in tissue stretch-relaxation cycles in characteristic *hysteresis loops*, which again can be linked back to specific biological structures in the tissues.

Other more diverse tissues have been investigated using modified AFM probes discussed previously in this chapter to press into the tissue to measure the highly localized spatial dependence of tissue compliance. These include studying the mechanics of microbial biofilms, as well as developing root tissues of plants, in order to understand how these structures are assembled and maintained.

As discussed previously in this chapter, single cells may also be mechanically probed directly using light by using a cell stretcher optical tweezer device. Cultured cells can also be mechanically stressed and investigated using simple techniques that involve mechanically stressing the solid substrate on which the cells are grown. An example of this involves bespoke compliant cell growth chambers made from an optically transparent form of silicone rubber called "polydimethylsiloxane" (*PDMS*), a versatile solid substrate that can be controllably cast from a liquid state using a combination of chemical and UV curing, which is used widely now for the manufacture of microfluidics flow-cell devices (see Chapter 7).

In a range of related methods known as traction force microscopy (TFM), PDMS cell chambers in conjunction with a solid cell substrate (e.g., the polysaccharide sugar agarose, which is mechanically stable, optically transparent as well as possessing pores that are large enough to permit nutrients and gases to diffuse to and from cells, and is also comparatively non-insert in terms of its chemical interactions with cells), a suitable cell growth surface medium can be cast onto the PDMS and the cells grown in a physiologically relevant environment. However, since PDMA is compliant, it can be stretched and subsequently relaxed by external force control, for example, something as simple as a pair of fine-pitch screws located either side of the cell growth chamber. This propagates mechanical forces to the walls or membranes of the growing cells and, if combined with light microscopy, can be used to investigate the cellular responses to these mechanical interventions.

Of recent interest is how various diseases can impair biomechanically important tissues. This has led to developing methods of *tissue engineering* and *regenerative medicine* to either replace damaged structures with *biomimetic materials,* or to encourage the regeneration of native structures, for example, by using *stem cell therapy* (see Chapter 9). Techniques that can accurately measure the biomechanical properties of such synthetic materials are therefore particularly useful.

KEY BIOLOGICAL APPLICATIONS: CELL AND TISSUE MECHANICS TOOLS

Cell and tissue stretch experiments.

Worked Case Example 6.3: Tethered Particle Motion

A   In TPM, the usual model used to account for the apparent end-to-end length of the tethered molecule is the Kratky–Porod model (see section 8.3.3, Equation 8.46). Show that if the molecule is relatively floppy, then the model reduces to a Gaussian chain (see section 8.3.2) whereas if it is stiff it reaches the rodlike limit.

b   A TPM experiment was performed at room temperature using a B-DNA construct of ~15 kbp (i.e., 15,000 base pairs) in length using a 1 μm diameter latex bead imaged using bright-field microscopy with video-rate sampling of 33 ms per frame. The center bead height was set at 1 μm throughout using feedback on the microscope stage based on the diffraction rings of the bead like that used for vertical magnetic tweezers (see section 6.4.3). The maximum lateral deflection of the bead from its equilibrium position over time was 38% larger than its own radius. Assuming the B-DNA is relatively floppy, estimate its persistence length.

c   A nucleoid associated protein, or NAP (NAPs constitute a range of bacterial proteins which bind to DNA) was added to the sample flow cell, resulting in the maximum diameter of the center of the fluctuating bead changing to almost 10 times the bead's radius. Explain this observation.

d   If the equivalent Stokes radius of the DNA in the absence of any NAP is given approximately by its persistence length, estimate how long it would take the tethered bead to travel diametrically across the circle, which encloses the measured positions of the bead center assuming the viscosity of the surrounding pH buffer is ~1 cP and comment of whether video-rate sampling is adequate for these types of measurements.

Answers

a   The Kratsky–Porod model states that the end-to-end separation is:

$$\left(R^2_{WLC}\right) = 2l_p R_{max} - 2l^2_p \left(1 - \exp\left(-R_{max}/l_p\right)\right)$$

The local pointing direction of a filamentous "floppy" molecule changes often over distances that are relatively short compared to its maximum end-to-end length R_{max}, also known as the molecule's contour length, indicating that its persistence length $l_p \ll R_{max}$ where, and so:

$$\left(R^2_{WLC}\right) = 2l_p R_{max} - 2l^2_p \left(1 - 1/\left(1 + \alpha + \alpha^2/2! + \dots\right)\right) \approx 2l_p R_{max}$$

if $R_{max}/l_p = \alpha$ and $\alpha \gg 1$. In the Gaussian chain approximation (Equation 8.42), the mean squared distance of one end of a polymer to the other for n random "polymer walk steps" of size b, if we take other end as the origin, is:

$$\left(R^2_{GC}\right) = \int_{-\infty}^{+\infty} (bx)^2 \, pdx = b^2 n$$

where $p = \dfrac{1}{\sqrt{2\pi n}} \exp\left(\dfrac{-x^2}{2n^2}\right)$ (from Equation 8.42). The length b is the Kuhn length, which is roughly $2l_p$ for a floppy polymer (see Equation 8.47) thus:

$$\left(R^2_{WLC}\right) \approx \left(R^2_{GC}\right)$$

In the case a stiff molecule, $l_p \gg R_{max}$ so:

$$\left(R^2_{WLC}\right) = 2l_p R_{max} - 2l^2_p \left(1 - \left(1 + (-\alpha) + \dfrac{(-\alpha)^2}{2!} + \dots\right)\right) \approx R^2_{max}$$

In other words, the end-to-end distance is the same as the contour length of the molecule, with is the definition of the rodlike limit.

b   Here, R_{max} is $15,000 \times 0.34$ nm (see section 2.3.5) or ~5,100 nm or 5.1 μm. The bead is a vertical distance 1 μm from the point of attachment of the DNA on the coverslip, so the distance between the center of the bead at this maximum deflection of 38% of its own radius (i.e., 0.688 μm) and its point of attachment to the surface of the coverslip with some simple trigonometry is $\sqrt{(1.0^2 + 0.688^2)} = 1.214$ μm (note the point of attachment of the DNA on the bead at maximum deflection will not be vertically directly below the center of the bead since the direction of force from the tether will act through the bead center by conservation of angular momentum—that is, the bead must rotate as it is laterally deflected). So, the mean end-to-end distance of the DNA tether is 1.214 minus the bead radius, that is, 1.214 – 0.5 or 0.714 μm. So, assume the floppy molecule result of part (a), this implies:

$$l_p \approx 0.714/(2 \times 5.1) = 0.050 \text{ μm or 50 nm.}$$

c   10 times the bead radius is 5.0 μm so, from the same trigonometry analysis as for part (b), the end-to-end distance of the DNA is $\sqrt{(1.0^2 + 5.0^2)} - 0.5 = 4.6$ μm. This is 4.6/5.1 or ~90% of R_{max}, so the binding of the NAP has shifted DNA toward rodlike limit, that is, a stiff filament.

d   If we model the diffusion coefficient by the Stokes–Einstein relation (see Equation 2.11) and crudely approximate the local diffusion of the DNA as that of a sphere of radius l_p, then the diffusion coefficient of the bead in this case is greater by an order of magnitude. If we thus neglect the DNA drag entirely and assume the drag will be mainly due to the bead itself, then its diffusion coefficient is:

~$1.38 \times 10^{-21} \times 300 /(6 \times \pi \times 0.001 \times 0.5 \times 10^{-6}) = 4.4 \times 10^{-11}$ m²/s. The time t taken to diffuse the 2D distance of 0.688 μm is from Equation 2.12 will be given roughly by:

$t = (0.688 \times 10^{-6})^2/(4 \times 4.4 \times 10^{-11}) = 2.7 \times 10^{-3}$ s or ~3 ms. This is an order of magnitude smaller than the sampling time. The answer to this question is therefore *maybe*. Although the video-rate imaging is not fast enough to adequately track the bead from frame-to-frame, in a sense for this simple analysis it does not matter since we can still measure what the maximum bead displacement is by sampling enough frames, albeit relatively slowly though in reality the bead might move too quickly to be trackable *per se*, so we would need to rely on the blur image to deduce the bead's maximum extent. However, if we want to measure more nuanced properties like dynamic affects, for example, how to monitor the real-time changes to the bead deflection as a NAP is added to the flow cell, then we would need to definitively track the bead from frame-to-frame, so sampling at least an order of magnitude faster.

6.8 SUMMARY POINTS

- Optical tweezers, magnetic tweezers, and AFM can all probe single biomolecule mechanics.
- Magnetic tweezers and modified optical tweezers can probe single-biomolecule torque.
- AFM imaging can generate sub-nanometer precise topological information of biological samples.
- Electrical forces can be used to monitor currents through natural ion channels in cells and through artificial nanopores for biosensing.
- Electric field can control the rotation and displacements of particles attached to biological structures or of biomolecules directly.
- Whole tissues can be mechanically studied using relatively simple stretch devices.

QUESTIONS

6.1   Ficoll is a synthetic polymer of sucrose used to change osmotic pressure and/or viscosity in biophysical experiments. A version of Ficoll with molecular weight 40 kDa had viscosities relative to water at room temperature of [1, 5, 20, 60, 180, 600] corresponding to (w/v) % in water of [0, 10, 20, 30, 40, 50], respectively. If a fluorophore-labeled antibody has a Stokes radius of 8 nm and the viscosity of water at room temperature is 0.001 Pa·s, estimate the molarity of the Ficoll needed to be present to observe labeled antibodies *in vitro* unblurred in solution using a wide-field epifluorescence microscope of 1.45 NA capable of sampling at 40 ms per image frame.

6.2   In a two-bead optical tweezers tapping style mechanical stretch experiment on a single molecule of linear DNA, 1600 separate bead pairs were generated over the course of a week by a diligent student. They used a constant bead tapping frequency with a triangle wave profile. The student thought there were three different populations of molecules characterized by different estimated values of persistence length based on worm-like chain model fits applied to the force-extension data (see Chapter 8), which indicated 1497 molecules having a persistence length of ~50 nm. There are 39 molecules that had a persistence length of close to ~20–30 nm, and the remainder had a persistence length of ~10–15 nm. For the group of molecules with persistence length close to 50 nm, the DNA molecular stiffness was observed to decrease at values of molecular force above ~65 pN. Explain these observations.

6.3   A "vertical" magnetic tweezers experiment was performed on a single molecule of DNA tethered between a microscope coverslip and a magnetic bead, when the molecular motor FtsK, which uses DNA as a track on which to translocate, was added to the microscope sample chamber; the length of the distance between the coverslip and the magnetic bead was observed to decrease.

   a   Some researchers have used this as evidence that there might be two FtsK molecular motors acting together on the DNA—explain why this makes sense.

   b   What other explanations could there be?

6.4   Optical tweezers using a focused laser of wavelength 1047 nm exerted a lateral force of 80 pN on a latex bead of diameter 1000 nm suspended in water at room temperature when the bead is displaced 500 nm from the trap center.

   a   If the trapping laser power passing through the bead is 220 mW, estimate the average angle of deviation of laser photons, assuming the lateral force arises principally from photons traveling close to the optical axis.

   b   Make an annotated sketch of the frequency power spectral density of the microsphere's positional fluctuations.

   c   At what frequency is the power spectral density half of its maximum value for these optical tweezers? The incident laser beam is then divided up using a time-share approach with an AOD of efficiency 75% into several beams of equal power to generate several independent optical tweezers.

   d   If each optical trap must exert a continuous high force of 20 pN, estimate the maximum number of traps that can be used.

   e   For experiments not requiring continuous high force, estimate the maximum theoretical number of optical traps that can be generated by this method if you assume that a *stable* trap is such that mean displacement fluctuations of a trapped particle position do not extend beyond the physical dimensions of the trap.

   f   How many such trapped beads would be required to push on a single *molecular motor* molecule to prevent it from undergoing a force-generating molecular conformational change known as a *power stroke* of average magnitude 5 pN?

6.5   Most bacteria have an outer rigid cell wall composed of proteins and sugars (peptidoglycan, see Chapter 2), which allows them to withstand osmotic pressures of 15 bar or more, but semipermeable so allows a variety of small molecules, including water, to diffuse through.

   a   A virus known to infect bacteria has a diameter of 50 nm—is it likely that the reason why each bacterial cell ultimately splits after a virus infects the cell and multiplies inside the cell is due to the buildup of pressure due to the large number of virus particles?

   b   If each virus consists of a maximum of 2000 protein molecules in its "capsid" coat (see Chapter 2), which were to spontaneously split apart from each other, would this make any difference to your answer? (*Hint:* treat a population of viruses in a cell as an ideal gas whose concentration is limited by tight packing.)

6.6   An ABEL trap was used to constrain a 20 nm latex bead in water whose surface contained the equivalent of ~3000 delocalized electrons. A mean *E*-field strength of 9000 V m^{-1} was applied to the trap's electrodes. How many frames per second must a camera sample the bead's position to ensure that the expected distance diffused by

Brownian motion each image frame is less than the displacement fluctuations in the ABEL trap due to its finite stiffness? (Assume the room temperature viscosity of water is 0.001 Pa·s.)

6.7 Assume solutions of the form $A(\omega) \exp(i\omega t)$ for the Langevin force at angular frequency ω.

    a   Derive an expression for the displacement $x(t)$ at time t in an optical trap.

    b   If the power spectral density $G(\omega)d\omega$ is defined as $|A(\omega)^2|$, derive an expression for the mean squared displacement in terms of G.

    c   Show that the power spectral density should be a Lorentzian function. (Assume for "white noise" that G is a constant and that the equipartition theorem predicts that the mean squared displacement at each separate frequency is associated with a mean energy of $k_B T/2$.)

6.8 An experimental protocol was devised using BFP detection to monitor the lateral displacement of a 200 nm diameter latex bead attached to a rotary molecular motor of the bacterial flagellar motor (which enables bacteria to swim) via a stiff filament stub to a live bacterium, which was free to rotate in a circle a short distance above the cell, which itself is stuck firmly to a microscope coverslip. The motor is expected to rotate at speeds of ~100 Hz and is made up of around ~20 individual subunits in a circle that each are thought to generate torque independently to push the filament around.

    a   What is the minimum sampling bandwidth of the QPD in order to see all of the torque-generating units?

        In practice, it is difficult to make a completely stiff filament; in a separate experiment using a completely unstiffened filament attached to a 500 nm diameter latex bead, it was found that the filament compliance resulted in a relaxation drag delay to bead movement following each ratchet of a few tenths of a microseconds, whereas a 1000 nm diameter bead had an equivalent response time ~10 times slower.

    b   Explain these observations and discuss which bead is the best choice to try to monitor rotation mechanism of the flagellar motor.

        It is possible to make some of the 20 ratchet subunits nonfunctional without affecting the others.

    c   How many subunits need to be made nonfunctional to detect individual activity of each torque-generating subunit?

    d   New evidence suggests that there may be cooperativity between the subunits—how does this affect your previous answers?

6.9 AFM force spectroscopy and optical tweezers are both used to investigate single-molecule mechanics stretching single biomolecules, as well as observing domain unfolding and refolding of modules inside the molecules. Explain with reasoning if one technique is better.

6.10 What is a "molecular signature," and why are they needed? The "sawtooth" pattern of a force-extension trace as obtained from AFM force spectroscopy on certain molecules is an example of a molecular signature. Can you think of other molecular signatures?

6.11 Single-molecule force spectroscopy is normally performed on purified molecules or on the surface of cells. Why? Under what circumstances experiments might be performed inside living cells?

6.12 At a prestigious biophysics tools and techniques awards dinner, a helium balloon escaped and got loosely trapped just under the ceiling. Assuming no lateral friction, how long would it take a red laser pointer of 1 mW power output to push the balloon 10 m across the length of the dinner hall ceiling using forward photon pressure alone? How would this change using a fancier 5 mW green laser pointer? Would it make significant difference to encourage all the other ~500 people attending the awards dinner to assist in getting out their laser pointers and performing this in parallel? (This is what the author attempted in the not too distant past. It demonstrates the great merit in doing theoretical calculations in advance of experiments.)

6.13 A membrane protein was imaged using AFM in contact mode. In one experiment, the protein was purified and inserted into an artificial lipid bilayer on a flat surface. This indicated protein topographic features 0.5 nm pointing above the bilayer. When the AFM tip was pushed into these features and then retracted, it was found that the tip experienced an attractive force toward the membrane. When the same experiment was performed using a living cell, similar topographic features could be imaged, but when the tip was pushed into the sample with the same force limit set as before and then retracted, no such pulling force was experienced. Explain these observations.

6.14 An AFM image was obtained for a hard spherical nanoparticle surface marker between live cells stuck to a mica surface. The image obtained for the nanoparticle did not indicate a sphere but a hump shape whose width was ~150 nm larger than its estimates obtained from transmission electron microscopy.

    a   Explain this.
         The AFM tip was a tetrahedron with a base edge length of 900 nm and a base tip height of 10,000 nm.

    b   What is the diameter of the nanoparticle?

6.15 Physiological "Ringer" solution has a resistivity of 80 $\Omega \cdot$cm. What is the total electrical resistance measured across a typical open sodium ion channel of length 5 nm and pore diameter 0.6 nm?

6.16 A silicon-substrate nanopore of 5 nm diameter was used to detect the translocation of a polymeric protein in the pH buffer "PBS" using a 120 mV voltage across the nanopore. The protein consists of 5 α-helices containing 10–20 amino acids each connected by a random coil of 5–10 amino acids. The protein had a small net positive charge and it was found that there were just two cysteine residues separated by 20 amino acids. When the electric current through the nanopore was measured, it indicated that for most of the time the current had reasonably stable value of 50 pA, but also had much shorter-lived 40, 42, 44, and 46 pA. However, when 5 mM DTT (see Chapter 2) was added to the solution the short-lived current values were measured at 40, 42, 44, 46, and 48 pA. Explain these results.

6.17 Graphene is a very thin yet strong structure and also electrically conducting. Is this an advantage or disadvantage to using it as the nanopore substrate for sequencing single DNA molecules?

6.18 Fick's first law of diffusion (see Chapter 8) states that the vector particle flux $J = -D \cdot$ grad(n) where D is the diffusion coefficient and n is the number of particles per unit volume.

    a   Modeling an ion channel as a 1D cylinder of radius a, derive an expression for the channel current due solely to diffusion of univalent ions of molar concentration C, stating any assumptions you make.

    b   In a patch clamp experiment, an extracted region of cell membrane contained ~10 Na^+ ion channels each of diameter 1 nm. When a voltage of –150 mV was applied across the membrane patch in a solution of 175 mM NaCl, the measured current was found to fluctuate with time from a range of zero up to a maximum at which the observed resistance of the patch was measured as 2.5×10^9 Ω.

    c   Estimate the current through a single Na^+ channel and the minimum sampling frequency required to monitor the passage of a single ion. How significant is diffusion to ion flux through a single channel?

6.19 A cell was placed in a physiological solution consisting of 100 mM NaCl, 20 mM KCl at room temperature. The cell membrane had several open Cl^- channels; using single-molecule fluorescence imaging, their internal concentration of Cl^- ions was measured at 20 mM, while that of K^+ was 30 mM.

    a   What is the transmembrane voltage on the basis of the Cl^- concentration? Why is it sensible to use Cl^- concentrations for this calculation and not K^+?

    b   It was found that K^+ would on average not spontaneously translocate out of the cell, but rather that this required energy to pump K^+ out. Why is this?

    c   A chemical *decoupler* was applied that forced all Na^+ and K^+ ion channels to open, and the ions then moved across the membrane to reach electrochemical equilibrium. Would you expect the K^+ ion concentration inside and outside the cell to be equal?

6.20 Using scanning conductance microscopy, images of a purified enzyme of ATP synthase (see Chapter 2) could just be discerned when the enzymes were stuck to a flat microscope coverslip. But, when potassium channels were overexpressed in a cell membrane and a patch excised and imaged on a flat surface, no clear images of the channels could be obtained. Why?

6.21 In an AFM imaging experiment, the maximum vertical displacement of the cantilever was limited by the height of the silicon nitride tip of 10 µm, giving a full-scale deflection of the photodiode output of 5 V. At room temperature with the tip far away from the sample the rms photodiode output was 2.6 mV with the laser reflecting onto the back of the cantilever switch on, and 0.9 mV when the laser was switched off. The machine was used to image single "myosin" molecules on a flat surface (molecular motors found in muscle tissue), whose head regions generated forces of 5 pN each when performing a "power stroke." When the tip is just in contact with a head region, what offset voltage should be applied to just cause a power stroke to stall?

6.22 In an ABEL trap experiment, a fluorescently labeled protein could be recentered when experiments were performed at "low" salt in NaCl (equivalent to an ionic strength of ~10 mM) but was not able to recenter the protein at "high" salt (ionic strength >0.2M). What could be the explanation?

REFERENCES

KEY REFERENCE

Rief, M., Gautel, M., Oesterhelt, F., Fernandez, J.M., and Gaub, H.E. (1997). Reversible unfolding of individual titin immunoglobulin domains by AFM. *Science 276*:1109–1112.

MORE NICHE REFERENCES

Abbondanzieri, E.A. et al. (2005). Direct observation of base-pair stepping by RNA polymerase. *Nature 438*:460–465.

Arkawa, H., Umemura, K., and Ikai, A. (1992). Protein images obtained by STM, AFM and TEM. *Nature 358*:171–173.

Ashkin, A. (1970). Acceleration and trapping of particles by radiation pressure. *Phys. Rev. Lett. 24*:156–159.

Ashkin, A., Dziedzic, J.M., Bjorkholm, J.E., and Chu, S. (1986). Observation of a single-beam gradient force optical trap for dielectric particles. *Opt. Lett. 11*:288–290.

Bailey, S.T. et al. (2010). Light-scattering study of the normal human eye lens: Elastic properties and age dependence. *IEEE Trans. Biomed. Eng. 57*:2910–2917.

Bayley, H. (2009). Piercing insights. *Nature 459*:651–652.

Binnig, G. et al. (1982). Tunneling through a controllable vacuum gap. *Appl. Phys. Lett. 40*:178–180.

Binnig, G., Quate, C.F., and Gerber, C. (1986). Atomic force microscope. *Phys. Rev. Lett. 56*:930–933.

Cohen, A.E. and Moerner, W.E. (2005). Method for trapping and manipulating nanoscale objects in solution. *Appl. Phys. Lett. 86*:093109.

Cohen, J.A., Chaudhuri, A., and Golestanian, R. (2011). Active polymer translocation through flickering pores. *Phys. Rev. Lett. 107*:238102.

De Souza, N. (2014). Tiny tools to measure force. *Nat. Methods 11*:29.

Dufresne, E.R. and Grier, D.G. (1998). Optical tweezer arrays and optical substrates created with diffractive optical elements. *Rev. Sci. Instrum. 69*:1974–1977.

Durnin, J., Miceli, J.J., and Erberly, J.H. (1987). Diffraction-free beams. *Phys. Rev. Lett. 58*:1499–1501.

Fields, A.P. and Cohen, A.E. (2011). Electrokinetic trapping at the one nanometer limit. *Proc. Natl. Acad. Sci. USA 108*:8937–8942.

Finer, J.T., Simmons, R.M., and Spudich, J.A. (1994). Single myosin molecule mechanics: Piconewton forces and nanometre steps. *Nature 368*:113–119.

Finkelstein, I.J., Visnapuu, M.L., and Greene, E.C. (2010). Single-molecule imaging reveals mechanisms of protein disruption by a DNA translocase. *Nature 468*:983–987.

Forth, S., Deufel, C., Patel, S.S., and Wang, M.D. (2011). A biological nano-torque wrench: Torque-induced migration of a Holliday junction. *Biophys. J. 101*:L05–L07.

Gross, P., Farge, G., Peterman, E.J.G., and Wuite, G.J.L. (2010). Combining optical tweezers, single-molecule fluorescence microscopy, and microfluidics for studies of DNA–protein interactions. *Methods Enzymol. 475*:427–453.

Gück, J. et al. (2005). Optical deformability as an inherent cell marker for testing malignant transformation and metastatic competence. *Biophys. J. 88*:3689–3698.

Hall, J.E. (1975). Access resistance of a small circular pore. *J. Gen. Physiol. 66*:531–532.

Hansen, P.M., Bhatia, V.K., Harrit, N., and Oddershede, L. (2005). Expanding the optical trapping range of gold nanoparticles. *Nano Letters 5*:1937–1942.

Hansma, P.K. et al. (1989). The scanning ion-conductance microscope. *Science 243*:641–643.

Hobbs, J.K., Vasilev, C., and Humphris, A.D.L. (2006). VideoAFM—A new tool for high speed surface analysis. *Analyst 131*:251–256.

Holden, M.A., Needham, D., and Bayley, H. (2007). Functional bionetworks from nanoliter water droplets. *J. Am. Chem. Soc. 129*:8650–8655.

Kellermayer, M.S., Smith, S.B., Granzier, H.L., and Bustamante, C. (1997). Folding-unfolding transitions in single titin molecules characterized with laser-tweezers. *Science 276*:1112–1116.

Keyser, U.F., van der Does, J., Dekker, C., and Dekker, N.H. (2006). Optical tweezers for force measurements on DNA in nanopores. *Rev. Sci. Instrum. 77*:105105.

Kim, D.-H. et al. (2009). Biofunctionalized magnetic-vortex microdiscs for targeted cancer-cell destruction. *Nat. Methods 9*:165–171.

King, G.A. et al (2019). Supercoiling DNA optically. *Proc Natl Acad Sci U S A* 116:26534–26539.

Klenerman, D., Korchev, Y.E., and Davis, S.J. (2011). Imaging and characterisation of the surface of live cells. *Curr. Opin. Chem. Biol. 15*:696–703.

Kolosov, O.V. and Yamanaka, K. (1993). Nonlinear detection of ultrasonic vibrations in an atomic force microscope. *Jpn. J. Appl. Phys. Part 2-Lett. 32*:L1095–L1098.

Kufer, S.K., Puchner, E.M., Gumpp, H., Liedl, T., and Gaub, H.E. (2008). Single-molecule cut-and-paste surface assembly. *Science 319*:594–596.

La Porta, A. and Wang, M.D. (2004). Optical torque wrench: Angular trapping, rotation, and torque detection of quartz microparticles. *Phys. Rev. Lett. 92*:190801.

Lang, M.J., Fordyce, P.M., and Block, S.M. (2003). Combined optical trapping and single-molecule fluorescence *J. Biol. 2*:6.

Leake, M.C. (2001). *Investigation of the extensile properties of the giant sarcomeric protein titin by single-molecule manipulation using a laser-tweezers technique.* PhD dissertation, London University, London.

Leake, M.C., Wilson, D., Gautel, M., and Simmons, R.M. (2004). The elasticity of single titin molecules using a two-bead optical tweezers assay. *Biophys. J. 87*:1112–1135.

Liphardt, J. et al. (2001). Reversible unfolding of single RNA molecules by mechanical force. *Science 292*:733–737.

Maglia, G. et al. (2009). Droplet networks with incorporated protein diodes show collective properties. *Nat. Nanotechnol. 4*:437–440.

Manosas, M. et al. (2010). Magnetic tweezers for the study of DNA tracking motors. *Methods Enzymol. 475*:297–320.

Mao, Y. et al. (2009). In vivo nanomechanical imaging of blood-vessel tissues directly in living mammals using atomic force microscopy. *Appl. Phys. Lett. 95*:013704

Moffitt, J.R., Chemla, Y.R., Smith, S.B., and Bustamante, C. (2008). Recent advances in optical tweezers. *Annu. Rev. Biochem. 77*:205–228.

Mohn, F., Gross, L., Moll, N., and Meyer, G. (2012). Imaging the charge distribution within a single molecule. *Nat. Nanotechnol. 7*:227–231.

Pilizota, T., Bilyard, T., Bai, F., Futai, M., Hosokawa, H., and Berry, R.M. (2007). A programmable optical angle clamp for rotary molecular motors. *Biophys. J. 93*:264–275.

Rhee, M. and Burns, M.A. (2006). Nanopore sequencing technology: Research trends and applications. *Trends Biotechnol. 24*:580–586.

Rohrbach, A. (2005). Stiffness of optical traps: Quantitative agreement between experiment and electromagnetic theory. *Phys. Rev. Lett. 95*:168102.

Rowe, A., Leake, M.C., Morgan, H., and Berry, R.M. (2003). Rapid rotation of micron and sub-micron dielectric particles measured using optical tweezers. *J. Mod. Opt. 50*:1539–1555.

Sali, A., Shakhnovich, E., and Karplus, M. (1994). How does a protein fold? *Nature 369*:248–251.

Sarkar, A., Robertson, R.B., and Fernandez, J.M. (2004). Simultaneous atomic force microscope and fluorescence measurements of protein unfolding using a calibrated evanescent wave. *Proc. Natl. Acad. Sci. USA 101*:12882–12886.

Scarcelli, G. and Yun, S.H. (2007). Confocal Brillouin microscopy for three-dimensional mechanical imaging. *Nat. Photonics 2*:39–43.

Schneider, G.F. et al. (2010). DNA translocation through graphene nanopores. *Nano Letters 10*:3163–3167.

Shekhawat, G.S. and Dravid, V.P. (2005). Nanoscale imaging of buried structures via scanning near-field ultrasound holography. *Science 310*:89–92.

Simpson, N.B., Allen, L., and Padgett, M.J. (1996). Optical tweezers and optical spanners with Laguerre-Gaussian modes. *J. Mod. Opt. 43*:2485–2491.

Smith, S.B., Cui, Y., and Bustamante, C. (1996). Overstretching B-DNA: The elastic response of individual double-stranded and single-stranded DNA molecules. *Science 271*:795–799.

Stabley, D.R., Jurchenko, C., Marshall, S.S., and Salaita, K.S. (2011). Visualizing mechanical tension across membrane receptors with a fluorescent sensor. *Nat. Methods 9*:64–67.

Svoboda, K. and Block, S.M. (1994). Biological applications of optical forces. *Annu. Rev. Biophys. Biomol. Struct. 23*:247–285.

Svoboda, K., Schmidt, C.F., Schnapp, B.J., and Block, S.M. (1993). Direct observation of kinesin stepping by optical trapping interferometry. *Nature 365*:721–727.

Tskhovrebova, L., Trinick, J., Sleep, J.A., and Simmons, R.M. (1997). Elasticity and unfolding of single molecules of the giant muscle protein titin. *Nature 387*:308–312.

van den Hout, M. et al. (2010). Controlling nanopore size, shape, and stability. *Nanotechnology 21*:115304.

Viovy, J.-L. (2000). Electrophoresis of DNA and other polyelectrolytes: Physical mechanisms *Rev. Mod. Phys. 72*:813–872.

Wang, M.C. and Uhlenbeck, G.E. (1945). On the theory of the Brownian motion II. *Rev. Mod. Phys. 17*:323–341.

Complementary Experimental Tools

Valuable Experimental Methods that Complement Mainstream Research Biophysics Techniques

Anything found to be true of *E. coli* must also be true of elephants.

—**Jacques Monod, 1954** (from Friedmann, 2004)

General Idea: There are several important accessory experimental methods that complement techniques of biophysics, many of which are invaluable to the efficient functioning of biophysical methods. They include controllable chemical techniques for gluing biological matter to substrates, the use of "model" organisms, genetic engineering tools, crystal preparation for structural biology studies, and a range of bulk sample methods, including some of relevance to biomedicine.

7.1 INTRODUCTION

The key importance for a student of physics with regard to learning aspects of biophysical tools and technique is to understand the physics involved. However, the devil is often in the detail, and the details of many biophysical methods include the application of techniques that are not directly biophysical as such, but which are still invaluable, and sometimes essential, to the optimal functioning of the biophysical tool. In this chapter, we discuss the key details of these important, complementary approaches. We also include a discussion of the applications of biophysics in biomedical techniques. There are several textbooks dedicated to expert-level medical physics technologies; however, what we do here is highlight the important biophysical features of these to give the reader a basic all-round knowledge of how biophysics tools are applied to clinically relevant questions.

7.2 BIOCONJUGATION

Bioconjugation is an important emerging field of research in its own right. New methods for chemical derivatization of all the major classes of biomolecules have been developed, many with a significant level of specificity. As we have seen from the earlier chapters in this book that outline experimental biophysics tools, bioconjugation finds several applications in biophysical techniques, especially those requiring molecular level precision, for example, labeling biomolecules with a specific fluorophore tag or EM marker, conjugating a molecule to a bead for optical and magnetic tweezers experiments, chemically modifying surfaces in order to purify a mixture of molecules.

7.2.1 BIOTIN

Biotin is a natural molecule of the B-group of vitamins, relatively small with a molecular weight roughly twice that of a typical amino acid residue (see Chapter 2). It binds with high affinity to two structurally similar proteins called "avidin" (found in egg white of animals) and "streptavidin" (found in bacteria of the genus *Streptomyces;* these bacteria have proved highly beneficial to humans since they produce >100,000 different types of natural antibiotics, several of which are used in clinical practice). Chemical binding affinity in general can be characterized in terms of a *dissociation constant* (K_d). This is defined as the product of all the two concentrations of the separate components in solution that bind together divided by the concentration of the bound complex itself and thus has the same units as concentration (e.g., molarity, or M). The biotin–avidin or biotin–streptavidin interaction has a K_d of 10^{-14} to 10^{-15} M. Thus, the concentration of "free" biotin in solution in the presence of avidin or streptavidin is exceptionally low, equivalent to just a single molecule inside a volume of a very large cell of ~100 μm diameter.

KEY POINT 7.1

"Affinity" describes the strength of a single interaction between two molecules. However, if multiple interactions are involved, for example, due not only to a strong covalent interaction but also to multiple noncovalent interactions, then this accumulated binding strength is referred to as the "avidity."

These strong interactions are very commonly used by biochemists in conjugation chemistry. Biotin and streptavidin–avidin pairs can be chemically bound to a biomolecule using accessible reactive groups on the biomolecules, for example, the use of carboxyl, amine, or sulfhydryl groups in protein labeling (see in the following text). Separately, streptavidin–avidin can also be chemically labeled with, for example, a fluorescent tag and used to probe for the "biotinylated" sites on the protein following incubation with the sample.

7.2.2 CARBOXYL, AMINE, AND SULFHYDRYL CONJUGATION

Carboxyl (–COOH), amine (–NH$_2$), and sulfhydryl (–SH) groups are present in many biomolecules and can all form covalent bonds to bridge to another chemical group through the loss of a hydrogen atom. For example, conjugation to a protein can be achieved via certain amino acids that contain reactive amine groups—these are called "primary" (free) amine groups that are present in the side "substituent" group of amino acids and do not partake in peptide bond formation. For example, the amine acid lysine contains one such primary amine (see Chapter 2), which under normal cellular pH levels is bound to a proton to form the *ammonium ion* of –NH$_3^+$. Primary amines can undergo several types of chemical conjugation reactions, for example, *acylation, isocyanate formation*, and *reduction*.

Similarly, some amino acids (e.g., *aspartic acid* and *glutamic acid*) contain one or more reactive carboxyl groups that do not participate in peptide bond formation. These can be coupled to primary amine groups using a cross-linker chemical such as *carbodiimide* (*EDC* or *CDI*). The stability of the cross-link is often increased using an additional coupler called "sulfo-*N*-hydroxysuccinimide" (sulfo-NHS).

Chemically reactive sulfhydryl groups can also be used for conjugation to proteins. For example, the amino acid cysteine contains a free sulfhydryl group. A common cross-linker chemical is *maleimide*, with others including *alkylation reagents* and *pyridyl disulfide*.

Normally, however, cysteine residues would be buried deep in the inaccessible hydrophobic core of a protein often present in the form of two nearby cysteine molecules bound together via their respective sulfur atoms to form a disulfide bridge –S–S– (the subsequent cysteine dimer is called *cystine*), which stabilize a folded protein structure. Chemically interfering with the sulfhydryl group's native cysteine amino acid residues can, therefore, change the structure and function of the protein.

However, there are many proteins that contain no native cysteine residues. This is possibly due to the function of these proteins requiring significant dynamic molecular conformational changes that may be inhibited by the presence of –S–S– bonds in the structure. For these, it is possible to introduce one or more foreign cysteine residues by modification of the DNA encoding the protein using genetic engineering at specific sequence DNA locations. This technique is an example of *site-directed mutagenesis* (*SDM*), here specifically *site-directed cysteine mutagenesis* discussed later in this chapter. By introducing nonnative cysteines in this way, they can be free to be used for chemical conjugation reactions while minimizing impairment to the protein's original biological function (though note that, in practice, significant optimization is often still involved in finding the best candidate locations in a protein sequence for a nonnative cysteine residue so as not to affect its biological function).

Binding to cysteine residues is also the most common method used in attaching spin labels for ESR (see Chapter 5), especially through the cross-linker chemical *methanethiosulfonate* that contains an –NO group with a strong ESR signal response.

7.2.3 ANTIBODIES

An *antibody*, or *immunoglobulin* (*Ig*), is a complex protein with bound sugar groups produced by cells of the *immune system* in animals to bind to specific harmful infecting agents in the body, such as bacteria and viruses. The basic structure of the most common class of antibody is Y-shaped (Figure 7.1), with a high molecular weight of ~150 kDa. The types (or *isotypes*) of antibodies of this class are mostly found in mammals and are *IgD* (found in milk/saliva), *IgE* (commonly produced in allergic responses), and *IgG*, which are produced in several immune responses and are the most widely used in biophysical techniques. Larger variants consisting of multiple Y-shaped subunits include *IgA* (a Y-subunit dimer) and *IgM* (a Y-subunit pentamer). Other antibodies include *IgW* (found in sharks and skates, structurally similar to IgD) and IgY (found in birds and reptiles).

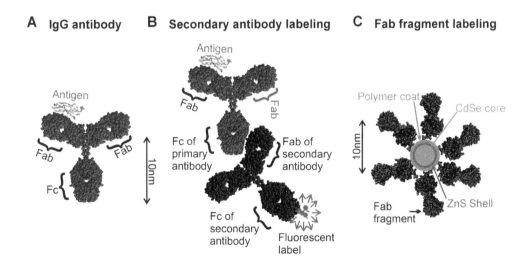

A IgG antibody   **B** Secondary antibody labeling   **C** Fab fragment labeling

FIGURE 7.1 Antibody labeling. Use of (a) immunoglobulin IgG antibody directly and (b) IgG as a primary and a secondary IgG antibody, which is labeled with a biophysical tag that binds to the Fc region. (c) Fab fragments can also be used directly.

The stalk of the Y structure is called the *Fc region* whose sequence and structure are reasonably constant across a given species of animal. The tips of the Y comprise two *Fab regions* whose sequence and structure are highly variable and act as a unique binding site for a specific region of a target biomolecule (known as an *antigen*), with the specific binding site of the antigen called the "epitope." This makes antibodies particularly useful for specific biomolecule conjugation. Antibodies can also be classed as *monoclonal* (derived from identical immune cells and therefore binding to a single epitope of a given antigen) or *polyclonal* (derived from multiple immune cells against one antigen, therefore containing a mixture of antibodies that will potentially target different epitopes of the same antigen).

The antibody–antigen interaction is primarily due to significantly high van der Waals forces due to the tight-fitting surface interfaces between the Fab binding pocket and the antigen. Typical affinity values are not as high as strong covalent interactions with K_d values of $\sim 10^{-7}$ M being at the high end of the affinity range.

Fluorophores or EM gold labels, for example, can be attached to the Fc region of IgG molecules and to isolated Fab regions that have been truncated from the native IgG structure to enable specific labeling of biological structures. Secondary labeling can also be employed (see Chapter 3); here a primary antibody binds to its antigen (e.g., a protein on the cell membrane surface of a specific cell type) while a secondary antibody, whose Fc region has a bound label, specifically binds to the Fc region of the primary antibody. The advantage of this method is primarily one of cost since a secondary antibody will bind the Fc region of all primary antibodies from the same species and so circumvents the need to generate multiple different labeled primary antibodies.

Antibodies are also used significantly in single-molecule manipulation experiments. For example, single-molecule magnetic and optical tweezer experiments on DNA often utilize a label called "digoxigenin" (DIG). DIG is a steroid found exclusively in the flowers and leaves of the plants of the *Digitalis* genus, highly toxic to animals and perhaps as a result through evolution has highly *immunogenic* properties (meaning it has a high ability to provoke an immune response, thus provoking the production of several specific antibodies to bind to DIG), and antibodies with specificity against DIG (called generally "anti-DIG") have very high affinity. DIG is often added to one end of a DNA molecule, while a trapped bead that has been coated in anti-DIG molecule can then bind to it to enable single-molecule manipulation of the DNA.

DIG is an example of a class of chemical called "haptans." These are the most common secondary labeling molecule for *immuno-hybridization* chemistry due to their highly immunogenic properties (e.g., biotin is a haptan). DIG is also commonly used in *fluorescence in situ hybridization* (*FISH*) assays. In FISH, DIG is normally covalently bound to a specific nucleotide triphosphate probe, and the fluorescently labeled IgG secondary antibody anti-DIG is subsequently used to probe for its location on the chromosome, thus allowing specific DNA sequences, and genes, to be identified following fluorescence microscopy.

A powerful application of FISH is for RNA imaging. *RNA FISH* can be used to visualize specific mRNA transcripts in living cells and in tissue sections. The state-of-the-art is *smFISH*, which can enable single-molecule detection on RNA transcripts of chemically fixed cell samples.

7.2.4 "CLICK" CHEMISTRY

Click chemistry is the general term that describes chemical synthesis by joining small-molecule units together both quickly and reliably, which is ideally modular and has high yield. It is not a single specific chemical reaction. However, one of the most popular examples of click chemistry is the *azide–alkyne Huisgen cycloaddition*. This chemical reaction uses copper as a catalyst and results in a highly selective and strong covalent bond formed between *azide* (triple bonded N–N atoms) and alkyne (triple bonded C–C bonds) chemical groups to form stable *1,2,3-triazoles*. This method of chemical conjugation is rapidly becoming popular in part due to its specific use in conjunction with increased development of *oligonucleotide labeling*.

7.2.5 NUCLEIC ACID OLIGO INSERTS

Short sequences of nucleotide bases (~10 base pairs), known as oligonucleotides (or just oligos), can be used to label specific sites on a DNA molecule. A DNA sequence can be cut at specific locations by enzymes called "restriction endonucleases," which enables short sequences of DNA complementary to a specific oligo sequence to be inserted at that location. Incubation with the oligo will then result in binding to the complementary sequence. This is useful since oligos can be modified to be bound to a variety of chemical groups, including biotin, azide, and alkynes, to facilitate conjugation to another biomolecule or structure. Also, oligos can be derivatized with a fluorescent dye label either directly or via, for example, a bound biotin molecule, to enable fluorescence imaging visualization of specific DNA sequence locations.

7.2.6 APTAMERS

Aptamers are short sequences of either nucleotides or amino acids that bind to a specific region of a target biomolecule. These peptides and RNA- or DNA-based oligonucleotides have a molecular weight that is relatively low at ~8–25 kDa compared to antibodies that are an order of magnitude greater. Most aptamers are unnatural in being chemically synthesized structures, though some natural aptamers do exist, for example, a class of RNA structures known as *riboswitches* (a riboswitch is an interesting component of some mRNA molecules that can alter the activity of proteins that are involved in manufacturing the mRNA and so regulate their own activity).

Aptamers fold into specific 3D shapes to fit tightly to specific structural motifs for a range of different biomolecules with a very low unbinding rate measured as an equivalent dissociation constant in the pico- to nanomolar range. They operate solely via a structural recognition process, that is, no chemical bonding is involved. This is a similar process to that of an antigen–antibody reaction, and thus aptamers are also referred to as *chemical antibodies*.

Due to their relatively small size, aptamers offer some advantages over protein-based antibodies. For example, they can penetrate tissues faster. Also, aptamers in general do not evoke a significant immune response in the human body (they are described as *nonimmunogenic*). They are also relatively stable to heat, in that their tertiary and secondary structures can be denatured at temperatures as high as 95°C but will then reversibly fold back into their original 3D conformation once the temperature is lowered to ~50°C or less, compared to antibodies that would irreversibly denature. This enables faster chemical reaction rates during incubation stages, for example, when labeling aptamers with fluorophore dye tags.

Aptamers can recognize a wide range of targets including small biomolecules such as ATP, ions, proteins, and sugars, but will also bind specifically to larger length scale biological matter, such as cells and viruses. The standard method of aptamer manufacture is known as *systematic evolution of ligands by exponential enrichment*. It involves repeated binding, selection, and then amplification of aptamers from an initial library of as many as $\sim 10^{18}$ random sequences that, perhaps surprisingly, can home in on an ideal aptamer sequence in a relatively cost-effective manner.

Aptamers have significant potential for use as drugs, for example, to block the activity of a range of biomolecules. Also, they have been used in biophysical applications as markers of a range of biomolecules. For example, although protein metabolites can be labeled using fluorescent proteins, this is not true for nonprotein biomolecules. However, aptamers can enable such biomolecules to be labeled, for example, if chemically tagged with a fluorophore they can report on the spatial localization of ATP accurately in live cells using fluorescence microscopy techniques, which is difficult to quantify using other methods.

A promising recent application of aptamers is in the fluorescent labeling of RNA in living cells. To date, the best labeling technology for RNA *in situ* has been antibodies, using RNA FISH and smFISH (see section 7.2.3). However, a drawback with both is the size of the antibodies (Stokes radius ~10 nm) impairing functional activity of the RNA, also, for smFISH that the experiments need to be performed on chemically fixed (i.e., dead) cells.

KEY BIOLOGICAL APPLICATIONS: BIOCONJUGATION TECHNIQUES

Attaching biophysical probes; Molecular separation; Molecular manipulation.

New variants of aptamers have now been developed which can directly label RNA in live cells using a bright organic dye called (4-((2-hydroxyethyl)(methyl)amino)-benzylidene)-cyanophenylacetonitrile (*HBC*) bound to the aptamer. The size of these aptamers is around the same as that of fluorescent proteins, hence resulting in far less steric impairment of biological functional compared to antibody labeling. Aptamers can be genetically encoded into RNA transcripts, though a weakness with the approach still is that the HBC dye must be introduced into the cell by permeabilizing its membrane, which has some functional impairment and imparts some limitation on the extent of maximum labeling possible for cellular RNA.

7.3 MODEL ORGANISMS

Technical advances of light microscopy have now enabled the capability to monitor whole, functional organisms (see Chapter 3). Biophysics here has gone full circle in this sense, from its earlier historical conception in, in essence, physiological dissection of relatively large masses of biological tissue. A key difference now, however, is one of enormously enhanced spatial and temporal resolution. Also, researchers now benefit greatly from a significant knowledge of underlying molecular biochemistry and genetics. Much progress has been made in biophysics through the experimental use of carefully selected *model organisms* that have ideal properties for light microscopy in particular; namely, they are thin and reasonably optically transparent. However, model organisms are also invaluable in offering the researcher a tractable biological system that is already well understood at a level of biochemistry and genetics.

7.3.1 MODEL BACTERIA AND BACTERIOPHAGES

There are a few select model bacteria species that have emerged as model organisms. *Escherichia coli* (*E. coli*) is the best known. *E. coli* is a model Gram-negative organism (see Chapter 2) whose genome (i.e., total collection of genes in each cell) comprises only ~4000 genes. There are several genetic variants of *E. coli*, noting that the spontaneous mutation rate of a nucleotide base pair in *E. coli* is ~10^{-9} per base pair per cell generation, some of which may generate a selective advantage for that individual cell and so be propagated to subsequent generations through natural selection (see Chapter 2). However, there are in fact only four key cell sources from which almost all of the variants are in use in modern microbiology research, which are called K-12, B, C, and W. Of these, K-12 is mostly used, which was originally isolated from the feces of a patient recovering from *diphtheria* in Stanford University Hospital in 1922.

Gram-positive bacteria lack a second outer cell membrane that Gram-negative bacteria possess. As a result, many exhibit different forms of biophysical and biochemical interactions with the outside world, necessitating a model Gram-positive bacterium for their study. The most popular model Gram-positive bacterium is currently *Bacillus subtilis*, which is a soil-dwelling bacterium. It undergoes an asymmetrical spore-forming process as part of its normal cell cycle, and this has been used as a mimic for biochemically triggered cell shape changes such as those that occur in higher organisms during the *development* of complex tissues.

There are many viruses known to infect bacteria, known as *bacteriophages*. Although, by the definition used in this book, viruses are not living as such, they are excellent model systems for studying genes. This is because they do not possess many genes (typically only a few tens of native genes), but rather hijack the genetic machinery of their host cell; if this host cell itself is a model organism such as *E. coli*, then this can offer significant insights into methods of gene operation/regulation and repair, for example. The most common model bacterium-infecting virus is called "bacteriophage lambda" (or just *lambda phage*) that infects *E. coli*. This has been used for many genetics investigations, and in fact since its DNA genetic code of almost ~49,000 nucleotide base pairs is so well characterized, methods for its reliable purification have been developed, and so there exists a readily available source of this

DNA (called λ *DNA*), which is used in many *in vitro* investigations, including single-molecule experiments of optical and magnetic tweezers (see Chapter 6). Another model of bacterium-infecting virus includes *bacteriophage Mu* (also called *Mu phage*), which has generated significant insight into relatively large transposable sections of DNA called "transposons" that undergo a natural splicing out from their original location in the genetic code and relocated *en masse* in a different location.

KEY POINT 7.2

"Microbiology" is the study of living organisms whose length scale is around ~10^{-6} m, which includes mainly not only bacteria but also viruses that infect bacteria as well as eukaryotic cells such as yeast. These cells are normally classed as being "unicellular," though in fact for much of their lifetime, they exist in colonies with either cells of their own type or with different species. However, since microbiology research can perform experiments on single cells in a highly controlled way without the added complication of a multicellular heterogeneous tissue environment, this has significantly increased our knowledge of biochemistry, genetics, cell biology, and even developmental biology in the life sciences in general.

7.3.2 MODEL UNICELLULAR EUKARYOTES OR "SIMPLE" MULTICELLULAR EUKARYOTES

Unlike prokaryotes, eukaryotes possess a distinct nucleus, as well as other subcellular organelles. This added compartmentalization of biological function can complicate experimental investigations (though note that even prokaryotes have distinct areas of local architecture in their cells so should not be perceived as a simple "living test tube"). Model eukaryotes for the study of cellular effects possess relatively few genes and also are ideally easy to cultivate in the laboratory with a reasonably short cell division time, allowing cell cultures to be prepared quickly. In this regard, three organisms have emerged as model organisms. One includes the single-celled eukaryotic *protozoan* parasite of the *Trypanosoma* genus that causes African *sleeping sickness*, specifically a species called *Trypanosoma brucei*, which has emerged as a model cell to study the synthesis of lipids. A more widely used eukaryote model cell organism is yeast, especially the species called *Saccharomyces cerevisiae* also known as *budding yeast* or *baker's yeast*. This has been used in multiple light microscopy investigations, for example, involving placing a fluorescent tag on specific proteins in the cell to perform superresolution microscopy (see Chapter 4). The third very popular model eukaryote unicellular organism is *Chlamydomonas reinhardtii* (often shortened to "*C. reinhardtii*"). This is a green alga and has been used extensively to study photosynthesis and cell motility.

Dictyostelium discoideum is a more complex multicellular eukaryote, also known as slime mold. It has been used as a model organism in studies involving *cell-to-cell communication* and *cell differentiation* (i.e., how eukaryote cells in multicellular organisms commit to being different specific cell types). It has also been used to investigate the effects of *programmed cell death* or *apoptosis* (see the following text).

More complex eukaryotic cells are those that would normally reside in tissues, and many biomedical investigations benefit from model human cells to perform investigations into human disease. The main problem with using more complex cells from animals is that they normally undergo the natural process of programmed cell death, called apoptosis, as part of their cell cycle. This means that it is impossible to study such cells over multiple generations and also technically challenging to grow a cell culture sample. To overcome this, *immortalized cells* are used, which have been modified to overcome apoptosis.

An *immortal cell* derived from a multicellular organism is one that under normal circumstances would not proliferate indefinitely but, due to being genetically modified, is no longer limited by the *Hayflick limit*. This is a limit to future cell division set either by

DNA damage or by shortening of cellular structures called "telomeres," which are repeating DNA sequences that cap the end of chromosomes (see Chapter 2). Telomeres normally get shorter with each subsequent cell division such that at a critical telomere length cell death is triggered by the complex biochemical and cellular process of apoptosis. However, immortal cells can continue undergoing cell division and be grown under cultured *in vitro* conditions for prolonged periods. This makes them invaluable for studying a variety of cell processes in complex animal cells, especially human cells.

Cancer cells are natural examples of immortal cells but can also be prepared using biochemical methods. Common immortalized cell lines include the *Chinese hamster ovary, human embryonic kidney, Jurkat* (*T lymphocyte*, a cell type used in the immune response), and *3T3* (mouse *fibroblasts* from connective tissue) cells. However, the oldest and most commonly utilized human cell strain is the *HeLa* cell. These are *epithelial cervical cells* that were originally cultured from a cancerous cervical tumor of a patient named *Henrietta Lacks* in 1951. She ultimately died as a result of this cancer but left a substantial scientific research legacy in these cells. Although there are potential limitations to their use in having undergone potentially several mutations from the original normal cell source, they are still invaluable to biomedical research utilizing biophysical techniques, especially those that use fluorescence microscopy.

7.3.3 MODEL PLANTS

Traditionally, plants have received less historical interest as the focus of biophysical investigations compared to animal studies, due in part to the lower relevance to human biomedicine. However, global issues relating to food and energy (see Chapter 9) have focused recent research efforts in this direction in particular. Many biophysical techniques have been applied to monitoring the development of complex plant tissues, especially involving advanced light microscopy techniques such as light sheet microscopy (see Chapter 4), which has been used to study the development of plant roots from the level of a few cells up to complex multicellular tissue.

The most popular model plant organism is *Arabidopsis thaliana*, also known commonly as *mouse ear cress*. It is a relatively small plant with a short generation time and thus easy to cultivate and has been characterized extensively genetically and biochemically. It was the first plant to have its full genome sequenced.

7.3.4 MODEL ANIMALS

Two key model animal organisms for biophysics techniques are those that optimized for *in vivo* light microscopy investigations, including the zebrafish *Danio rerio* and the nematode flatworm *Caenorhabditis elegans*. The *C. elegans* flatworm is ~1 mm in length and ~80 μm in diameter, which lives naturally in soil. It is the simplest eukaryotic multicellular organism known to possess only ~1000 cells in its adult form. It also breeds relatively easily and fast taking three days to reach maturation, which allows experiments to be performed reasonably quickly, is genetically very well characterized and has many tissue systems that have generic similarities to those of other more complex organisms, including a complex network of nerves, blood vessels and heart, and a gut. *D. rerio* is more complex in having ~10^6 cells in total in the adult form, and a length of a few centimeters and several hundred microns thick and takes more like ~3 months to reach maturation.

These characteristics set more technical challenges on the use of *D. rerio* compared to *C. elegans*; however, it has a significant advantage in possessing a spinal cord in which *C. elegans* does not, making it the model organism of choice for investigating specifically *vertebrate* features, though *C. elegans* has been used in particular for studies of the nervous system. These investigations were first pioneered by the Nobel Laureate Sydney Brenner in the 1960s, but later involved the use of advanced biophysics optical imaging and stimulation methods using an invaluable technique of *optogenetics*, which can use light to control the expression of genes (see later in this chapter). At the time of writing, *C. elegans* is the only organism for which the *connectome* (the *wiring diagram* of all nerve cells in an organism) has been determined.

The relative optical transparency of these organisms allows standard *bright-field* light microscopy to be performed, a caveat being that adult zebrafish grow pigmented stripes on their skin, hence their name, which can impair the passage of visible light photons. However, mutated variants of zebrafish have now been produced in which the adult is colorless.

Among invertebrate organisms, that is, those lacking an *internal skeleton, Drosophila melanogaster* (the common *fruit fly*) is the best studied. Fruit flies are relatively easy to cultivate in the laboratory and breed rapidly with relatively short life cycles. They also possess relatively few chromosomes and so have formed the basis of several genetics studies, with light microscopy techniques used to identify positions of specifically labeled genes on isolated chromosomes.

For studying more complex biological processes in animals, rodents, in particular mice, have been an invaluable model organism. Mice have been used in several biophysical investigations involving deep tissue imaging in particular. Biological questions involving practical human biomedicine issues, for example, the development of new drugs and/or investigating specific effects of human disease that affects multiple cell types and/or multiply connected cells in tissues, ultimately involve larger animals of greater similarity to humans, culminating in the use of primates. The use of primates in scientific research is clearly a challenging issue for many, though such investigations require significant oversight before being granted approval from ethical review committees that are independent from the researchers performing the investigations.

KEY BIOLOGICAL APPLICATIONS: MODEL ORGANISMS

Multiple biophysical investigations requiring tractable, well-characterized organism systems to study a range of biological processes.

KEY POINT 7.3

A "model organism," in terms of the requirements for biologists, is selected on it being genetically and phenotypically/behaviorally very well characterized from previous experimental studies and also possesses biological features that at some level are "generic" in allowing us to gain insight into a biological process common to many organisms (especially true for biological processes in humans, since these give us potential biomedical insight). For the biophysicist, these organisms must also satisfy an essential condition of being experimentally very tractable. For animal tissue research, this includes the use of thin, optically transparent organisms for light microscopy. One must always bear in mind that some of the results from model organisms may differ in important ways from other specific organisms that possess equivalent biological processes under study.

7.4 MOLECULAR CLONING

The ability to sequence and then controllably modify the DNA genetic code of cells has complemented experimental biophysical techniques enormously. These genetic technologies enable the controlled expression of specific proteins for purification and subsequent *in vitro* experimentation as well as enable the study of the function of specific genes by modifying them through controlled mutation or deleting them entirely, such that the biological function might be characterized using a range of biophysical tools discussed in the previous experimental chapters of this book. One of the most beneficial aspects of this modern *molecular biology* technology has been the ability to engineer specific biophysical labels at the level of the genetic code, through incorporation either of label binding sites or of fluorescent protein sequences directly.

7.4.1 CLONING BASICS

Molecular cloning describes a suite of tools using a combination of genetic engineering, cell and molecular biology, and biochemistry to generate modified DNA to enable it to be replicated within a host organism ("cloning" simply refers to generating a population of cells all containing the same DNA genetic code). The modified DNA may be derived from the same or different species as the host organism.

In essence, for cloning of genomic DNA (i.e., DNA obtained from a cell's nucleus), the source DNA, which is to be modified and ultimately cloned, is first isolated and purified from its originator species. Any tissue/cell source can, in principle, be used for this provided the DNA is mostly intact. This DNA is purified (using a *phenol extraction*), and the number of purified DNA molecules present is amplified using polymerase chain reaction (PCR) (see Chapter 2). To ensure efficient PCR, *primers* need to be added to the DNA sequence (short sequences of 10–20 nucleotide base pairs that act as binding sites for initiating DNA replication by the enzyme DNA polymerase). PCR can also be used on RNA sample sources, but using a modified PCR technique of the *reverse transcription polymerase chain reaction* that first converts RNA back into *complementary DNA (cDNA)*, which is then amplified using conventional PCR. A similar process can also be used on synthetic DNA, that is, artificial DNA sequences not from a native cell or tissue source.

The amplified, purified DNA is then chemically broken up into fragments by restriction endonuclease enzymes, which cut the DNA at specific sequence locations. At this stage, additional small segments of DNA from other sources may be added that are designed to bind to specific cut ends of the DNA fragments. These modified fragments are then combined with *vector DNA*. In molecular biology, a *vector* is a DNA molecule that is used to carry modified (often foreign) DNA into a host cell, where it will ultimately be replicated and the genes in that recombinant DNA expressed can be replicated and/or expressed. Vectors are generally variants of either bacterial plasmids or viruses (see Chapter 2). Such a vector that contains the modified DNA is known as *recombinant DNA*. Vectors in general are designed to have multiple specific sequence *restriction sites* that recognize the corresponding fragment ends (called "sticky ends") of the DNA generated by the cutting action of the restriction endonucleases. Another enzyme called "DNA ligase" catalyzes the binding of the sticky ends into the vector DNA at the appropriate restriction site in the vector, in a process called ligation. It is possible for other ligation products to form at this stage in addition to the desired recombinant DNA, but these can be isolated at a later stage after the recombinant DNA has been inserted in the host cell.

KEY POINT 7.4

The major types of vectors are viruses and plasmids, of which the latter is the most common. Also, hybrid vectors exist such as a "cosmid" constructed from a lambda phage and a plasmid, and artificial chromosomes that are relatively large modified chromosome segments of DNA inserted into a plasmid. All vectors possess an origin of replication, multiple restriction sites (also known as multiple cloning sites), and one or more selectable marker genes.

Insertion of the recombinant DNA into the target host cell is done through a process called either "transformation" for bacterial cells, "transfection" for eukaryotic cells, or, if a virus is used as a vector, "transduction" (the term "transformation" in the context of animal cells actually refers to changing to a cancerous state, so is avoided here). The recombinant DNA needs to pass through the cell membrane barrier, and this can be achieved using both natural and artificial means. For natural transformation to occur, the cell must be in a specific physiological state, termed *competent*, which requires the expression of typically tens of different proteins in bacteria to allow the cell to take up and incorporate external DNA from solution (e.g., filamentous *pili* structures of the outer member, as well as protein complexes in the cell membrane to pump DNA from the outside to the inside). This natural phenomenon in bacteria occurs in a process called "horizontal gene transfer," which results in genetic diversity through the transfer of plasmid DNA between different cells, and is, for example, a mechanism for propagating antibiotic resistance in a cell population. It may also have evolved as a mechanism to assist in the repair of damaged DNA, that is, to enable the internalization of nondamaged DNA that can then be used as a template from which to repair native damaged DNA.

Artificial methods can improve the rate of transformation. These can include treating cells first with enzymes to strip away outer cells walls, adding divalent metal ions such as magnesium or calcium to increase binding of DNA (which has a net negative charge in solution due to the presence of the backbone of negatively charged phosphate groups), or increasing cell membrane fluidity. These also include methods that involve a combination of cold and heat shocking cells to increase internalization of recombinant by undetermined mechanisms as well as using ultrasound (*sonication*) to increase the collision frequency of recombinant DNA with host cells. The most effective method, however, is *electroporation*. This involves placing the aqueous suspension of host cells and recombinant DNA into an electrostatic field of strength 10–20 kV cm^{-1} for a few milliseconds that increases the cell membrane permeability dramatically through creating transient holes in the membrane through which plasmid DNA may enter.

Transfection can be accomplished using an extensive range of techniques, some of which are similar to those used for transformation, for example, the use of electroporation. Other more involved methods have been optimized specifically for host animal cell transfection, however. These include biochemical-based methods such as packaging recombinant DNA into modified liposomes that then empty their contents into a cell upon impact on, and merging with, the cell membrane. A related method is *protoplast fusion*, which involves chemically or enzymatically stripping away the cell wall from a bacterial cell to enable it to fuse in suspension with a host animal cell. This delivers the vector that may be inside the bacterial cell, but with the disadvantage of delivery of the entire bacterial cell contents, which may potentially be detrimental to the host cell.

But there are also several biophysical techniques for transfection. These include *sonoporation* (using ultrasound to generate transient pores in cell membranes), *cell squeezing* (gently massaging cells through narrow flow channels to increase the membrane permeability), *impalefection* (introducing DNA bound to a surface of a nanofiber by stabbing the cell), gene guns (similar to impalefection but using DNA bound to nanoparticles that are fired into the host cell), and *magnet-assisted transfection* or *magnetofection* (similar to the gene gun approach, though here DNA is bound to a magnetic nanoparticle with an external *B*-field used to force the particles into the host cells).

The biophysical transfection tool with the most finesse involves *optical transfection*, also known as *photoporation*. Here, a laser beam is controllably focused onto the cell membrane generating localized heating sufficient to form a pore in the cell membrane and allow recombinant DNA outside the cell to enter by diffusion. Single-photon absorption processes in the lipid bilayer can be used here, centered on short wavelength visible light lasers; however, better spatial precision is enabled by using a high-power near-infrared (IR) femtosecond pulsed laser that relies on two-photon absorption in the cell membrane, resulting in smaller pores and less potential cell damage.

Viruses undergoing transfection (i.e., viral transduction) are valuable because they can transfer genes into a wide variety of human cells in particular with very high transfer rates. However, this method can also be used for other cell types, including bacteria. Here, the recombinant DNA is packaged into an empty virus capsid protein coat (see Chapter 2). The virus then performs its normal roles of attaching to host cell and then injecting the DNA into the cell very efficiently, compared to the other transfection/transformation methods.

The process of inserting recombinant DNA into a host cell has normally low efficiency, with only a small proportion of host cells successfully taking up the external DNA. This presents a technical challenge in knowing which cells have done so, since these are the ones that need to be selectively cultivated from a population. This selection is achieved by engineering one or more *selectable markers* into the vector. A selectable marker is usually a gene conferring resistance against a specific antibiotic that would otherwise be lethal to the cell. For example, in bacteria, there are several resistance genes available that are effective against broad-spectrum antibiotics such as ampicillin, chloramphenicol, and kanamycin. Those host cells that have successfully taken up a plasmid vector during transformation will survive culturing conditions that include the appropriate antibiotic, whereas those that have not taken up the plasmid vector will die. Using host animal cells, such as human cells, involves a similar strategy to engineer a stable transfection such that the recombinant DNA is incorporated ultimately into the genomic DNA using a marker gene that is encoded into the genomic DNA conferring resistance against the antibiotic *Geneticin*. Unstable or transient transfection does not utilize marker genes on the host cell genome but instead retains the recombinant DNA as plasmids. These ultimately become diluted after multiple cell generations and so the recombinant DNA is lost.

7.4.2 SITE-DIRECTED MUTAGENESIS

SDM is a molecular biology tool that uses the techniques of molecular cloning described earlier to make controlled, spatially localized mutations to a DNA sequence, at the level of just a few, or sometimes one, nucleotide base pairs. The types of mutations include a single base change *(point mutation), deletion* or *insertion*, as well as multiple base pair changes. The basic method of SDM uses a short DNA primer sequence that contains the desired mutations and is complementary to the template DNA around the mutation site and can therefore displace the native DNA by hybridizing with the DNA in forming stable Watson–Crick base pairs. This recombinant DNA is then cloned using the same procedure as described in section 7.4.

SDM has been used in particular to generate specific cysteine point mutations. These have been applied for bioconjugation of proteins as already discussed in the chapter and also for a technique called *cysteine scanning* (or *cys-scanning) mutagenesis*. In cys-canning mutagenesis, multiple point mutations are made to generate several foreign cysteine sites, typically in pairs. The purpose here is that if a pair of such nonnative cysteine amino acids is biochemically detected as forming a disulfide bond in the resultant protein, then this indicates that these native residue sites that were mutated must be within ~0.2 nm distance. In other words, it enables 3D mapping of the location of different key residues in a protein. This was used, for example, in determining key residues used in the rotation of the F1Fo-ATP synthase that generates the universal cellular fuel of ATP (see Chapter 2).

A similar SDM technique is that of alanine scanning. Here, the DNA sequence is point mutated to replace specific amino acid residues in a protein with the amino acid alanine. Alanine consists of just a methyl ($-CH_3$) substituent group and so exhibits relatively little steric hindrance effects, as well as minimal chemical reactivity. Substituting individual native amino acid residues with alanine, and then performing a function test on that protein, can generate insight into the importance of specific amino acid side groups on the protein's biological function.

7.4.3 CONTROLLING GENE EXPRESSION

There are several molecular biology tools that allow control of the level of protein expression from a gene. The ultimate control is to delete the entire gene from the genome of a specific population of cells under investigation. These *deletion mutants*, also known as *gene knockouts*, are often invaluable in determining the biological function of a given gene, since the mutated cells can be subjected to a range of functionality tests and compares against the native cell (referred to as the *wild type*).

A more finely tuned, reversible method to modify gene expression is to use *RNA silencing*. RNA silencing is a natural and ubiquitous phenomenon in all eukaryote cells in which the expression of one or more genes is *downregulated* (which in molecular biology speaks for "lowered") or turned off entirely by the action of a small RNA molecule whose sequence is complementary to a region of an mRNA molecule (which would ultimately be translated to a specific peptide or protein). RNA silencing can be adapted by generating synthetic small RNA sequences to specially and controllably regulate gene expression. Most known RNA silencing effects operate through such *RNA interference*, using either *microRNA* or similar *small interfering RNA* molecules, which operate via subtly different mechanisms but which both ultimately result in the degradation of a targeted mRNA molecule.

Gene expression in prokaryotes can also be silenced using a recently developed technique that utilizes *clustered regularly interspaced short palindromic repeats* (*CRISPR*, pronounced "crisper," Jinek et al., 2012). *CRISPR-associated genes* naturally express proteins whose biological role is to catalyze the fragmentation of external foreign DNA and insert them into these repeating CRISPR sequences on the host cell genome. When these small CRISPR DNA inserts are transcribed into mRNA, they silence the expression of external DNA—it is a remarkable bacterial immune response against invading pathogens such as viruses. However, the CRISPR are also found in several species that are used as model organisms including *C. elegans* and zebrafish and can also be effective in human cells as a gene-silencing tool. CRISPR has enormous potential for revolutionizing the process of *gene editing*.

Transcription activator-like effector nucleases (*TALENs*) can also be used to suppress expression from specific to genes. TALENs are enzymes that could be encoded onto a plasmid vector in a host cell. These can bind to a specific sequence of DNA and catalyze cutting of the DNA at that point. The cell has complex enzyme systems to repair such a cut DNA molecule; however, the repaired DNA is often not a perfect replica of the original that can result in a nonfunctional protein expressed from this repaired DNA. Thus, although gene expression remains, no functional protein results from it.

RNA silencing can also use *upregulated* (i.e., "increased") gene expression, for example, by silencing a gene that expresses a transcription factor (see Chapter 2) that would normally represses the expression of another gene. Another method to increase gene expression includes *concatemerization of* genes, that is, generating multiple sequential copies under control of the same promoter (see Chapter 2).

Expression of genes in plasmids, especially those in bacteria, can be controlled through inducer chemicals. These chemicals affect the ability of a transcription factor to bind to a specific promoter of an operon. The operon is a cluster of genes on the same section of the chromosome that are all under control of the same promoter, all of which get transcribed and translated in the same continuous gene expression burst (see Chapter 2). The short nucleotide base pair sequence of the promoter on the DNA acts as an initial binding site for RNA polymerase and determines where transcription of an mRNA sequence translated from the DNA begins. Insight into the operation of this system was made originally using studies of the bacterial *lac operon*, and this system is also used today to control the gene expression of recombinant DNA in plasmids.

Although some transcription factors act to recruit the RNA polymerase, and so result in upregulation, most act as repressors through binding to the promoter that inhibits binding of RNA polymerase, as is the case in the lac operon. The lac operon consists of three genes that express enzymes involved in the internalization into the cell and metabolism of the disaccharide lactose into the monosaccharides glucose and galactose. Decreases in the cell's concentration of lactose result in reduced affinity of the repressor protein to the *lacI* gene that, in turn, is responsible for generating the LacI protein repressor molecule that inhibits expression of the operon genes and is by default normally switched "on" (note that the names of genes are conventionally written in italics starting with a lowercase letter, while the corresponding protein, which is ultimately generated from that gene following transcription and translation, is written in non-italics using the same word but with the first letter in uppercase). This prevents operon gene expression. This system is also regulated in the opposite direction by a protein called CAP whose binding in the promoter region is inversely proportional to cellular glucose concentration. Thus, there is negative feedback between gene expression and the products of gene expression.

The nonnative chemical *isopropyl-β-ᴅ-thio-galactoside* (*IPTG*) binds to LacI and in doing so reduces the LacI affinity to the promoter, thus causing the operon genes to be expressed. This effect is used in genetic studies involving controllable gene expression in bacteria. Here, a gene under investigation desired to be expressed is fused upstream of the lac promoter region in the lac operon and into a plasmid vector using the molecular cloning methods described earlier in this chapter. These plasmids are also replicated during normal cell growth and division and so get passed on to subsequent cell generations.

If IPTG is added to the growth media, it will be ingested by the cells, and the repressing effects of LacI will be inactivated; thus, the protein of interest will start to be made by the cells, often at levels far above normal wild type levels, as it is difficult to prevent a large number of plasmids from being present in each cell. Since IPTG does not have an infinite binding affinity to LacI, there is still some degree of suppression of protein production, but also the LacI repressor similarly is not permanently bound to the operator region, and so if even in the absence of IPTG, a small amount of protein is often produced (this effect is commonly described as being due to a *leaky plasmid*).

In theory, it is possible to cater the IPTG concentration to a desired cellular concentration of expressed protein. In practice though, the response curve for changes in IPTG concentration is steeply sigmoidal, the effect is largely all or nothing in response to changes in IPTG concentration. However, another operon system used for genetics research in *E. coli* and other bacteria is the *arabinose* operon that uses the monosaccharide arabinose as the equivalent repressor binder; here the steepness of the sigmoidal response is less than the IPTG operon system, which makes it feasible to control the protein output by varying the external concentration of arabinose.

A valuable technique for degrading the activity of specific expressed proteins from genes in prokaryotes is *degron-targeted proteolysis*. Prokaryotes have a native system for reducing the concentration level of specific proteins in live cells, which involves their controlled degradation by *proteolysis*. In the native cell, proteins are first marked for degradation by tagging them with a short amino acid *degradation sequence*, or *degron*. In *E. coli*, an adaptor protein called SspB facilitates binding of protein substrates tagged with the SsrA peptide to a *protease* called "ClpXP" (pronounced "Clip X P"). ClpXP is an enzyme that specifically leads to proteolytic degradation of proteins that possess the degron tag.

This system can be utilized synthetically by using molecular cloning techniques to engineer a foreign *ssrA* tag onto a specific protein that one wishes to target for degradation. This modification is then transformed into a modified *E. coli* cell strain in which the native gene *sspB* that encodes for the protein SspB has been deleted. Then, a plasmid that contains the *sspB* gene is transformed into this strain such that expression of this gene is under control on an inducible promoter. For example, this gene might then be switched "on" by the addition of extracellular arabinose to an arabinose-inducible promoter, in which case the SsrB protein is manufactured that then results in proteolysis of the SsrA-tagged protein.

This is a particularly powerful approach in the case of studying *essential proteins*. An essential protein is required for the cell to function, and so deleting the protein would normally be lethal and no cell population could be grown. However, by using this degron-tagging strategy, a cell population can first be grown in the absence of SspB expression, and these cells are then observed following controlled degradation of the essential protein after arabinose (or equivalent) induction.

KEY POINT 7.5

Proteolysis is the process of breaking down proteins into shorter peptides. Although this can be achieved using heat and the application of nonbiological chemical reagents such as acids and bases, the majority of proteolysis occurs by the chemical catalysis due to enzymes called proteases, which target specific amino acid sequences for their point of cleavage of a specific protein.

7.4.4 DNA-ENCODED REPORTER TAGS

As outlined previously, several options exist for fluorescent tags to be encoded into the DNA genetic code of an organism, either directly, in the case of fluorescent proteins, or indirectly, in the case of SNAP/CLIP-tags. Similarly, different segment halves of a fluorescent protein can be separately encoded next to the gene that expresses proteins that are thought to interact, in the BiFC technique, which generates a functional fluorescent protein molecule when two such proteins are within a few nanometers' distance (see Chapter 3).

Most genetically encoded tags are engineered to be at one of the ends of the protein under investigation, to minimize structural disruption of the protein molecule. Normally, a linker sequence is used of ~10 amino acids to increase the flexibility with the protein tag and reduce steric hindrance effects. A common linker sequence involves repeats of the amino acid sequence "EAAAK" bounded by alanine residues, which is known to form stable but flexible helical structures (whose structure resembles a conventional mechanical spring). The choice of whether to use the C- or N-terminus of a protein is often based on the need for binding at or near to either terminus as part of the protein's biological function, that is, a terminus is selected for tagging so as to minimize any disruption to the normal binding activities of the protein molecule. Often, there may be binding sequences at both termini, in which case the binding ability can still be retained in the tagged sequence by copying the end DNA sequence of the tagged terminus onto the very end of the tag itself.

Ideally, the native gene for a protein under investigation is entirely deleted and replaced at the same location in the DNA sequence by the tagged gene. However, sometimes this results in too significant an impairment of the biological function of the tagged protein, due to a combination of the tag's size and interference of native binding surfaces of the protein. A compromise in this circumstance is to retain the native untagged gene on the cell's genome but then create an additional tagged copy of the gene on a separate plasmid, resulting in a *merodiploid* strain (a cell strain that contains a partial copy of its genome). The disadvantage with such techniques is that there is a mixed population of tagged and untagged protein in the cell, whose relative proportion is often difficult to quantify accurately using biochemical methods such as western blots (see Chapter 6).

A useful tool for researchers utilizing fluorescent proteins in live cells is the *ASKA library* (Kitagawa et al., 2005), which stands for "A complete Set of *E. coli* K-12 ORF Archive." It is a collection (or "library") of genes fused to genetically encoded fluorescent protein tags. Here, each open reading frame (or "ORF"), that is, the region of DNA between adjacent start and stop codons that contains one or more genes (see Chapter 2), in the model bacterium *E. coli*, has been fused with the DNA sequence for the yellow variant of GFP, YFP. The library is stored in the form of DNA plasmid vectors under IPTG inducer control of the lac operon.

In principle, each protein product from all coding bacterial genes is available to study using fluorescence microscopy. The principal weakness with the ASKA library is that the resultant protein fusions are all expressed at cellular levels that are far more concentrated than those found for the native nonfusion protein due to the nature of the IPTC expression system employed, which may result in nonphysiological behavior. However, plasmid construct sequences can be spliced out from the ASKA library and used for developing genomically tagged variants.

Optogenetics (see Pastrana, 2010; Yizhar et al., 2011) specifically describes a set of techniques that utilize light-sensitive proteins that are synthetically genetically coded into nerve cells. These foreign proteins are introduced into nerve cells using the transfection delivery methods of molecular cloning described earlier in this chapter. These optogenetics techniques enable investigation into the behavior of nerves and nerve tissue by controlling the ion flux into and out of a nerve cell by using localized exposure to specific wavelengths of visible light. Optogenetics can thus be used with several advanced light microscopy techniques, especially those of relevance to deep tissue imaging such as multiphoton excitation methods (see Chapter 4). These light-sensitive proteins include a range of *opsin* proteins (referred to as *luminopsins*) that are prevalent in the cell membranes of single-celled organisms as channel protein complexes. These can pump protons, or a variety of other ions, across the membrane

Optogenetics

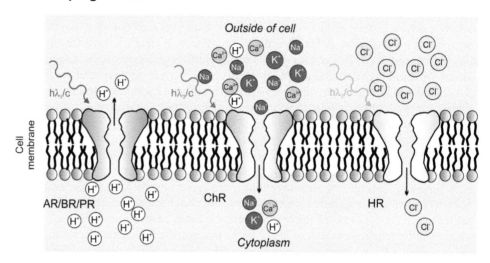

FIGURE 7.2   Optogenetics techniques. Schematic of different classes of light-sensitive opsin proteins, or luminopsins, made naturally by various single-celled organisms, which can be introduced into the nerve cells of animals using molecule cloning techniques. These luminopsins include proton pumps called archeorhodopsins, bacteriorhopsins, and proteorhodopsins that pump protons across the cell membrane out of the cell due to absorption of typically blue light (activation wavelength $\lambda_1 \sim 390–540$ nm), chloride negative ion (anion) pumps called halorhodopsins that pump chloride ions out of the cell (green/yellow activation wavelength $\lambda_2 \sim 540–590$ nm), and nonspecific positive ion (cation) pumps called channelrhodopsins that pump cations into the cell (red activation wavelength $\lambda_3 > 590$ nm).

using the energy from the absorption of photons of visible light, as well as other membrane proteins that act as ion and voltage sensors (Figure 7.2).

For example, *bacteriorhodopsin, proteorhodopsin*, and *archaerhodopsin* are all proton pumps integrated in the cell membranes of either bacteria or archaea. Upon absorption of blue-green light (the activation wavelengths (λ) span the range ~390–540 nm), they will pump protons from the cytoplasm to the outside of the cell. Their biological role is to establish a proton motive force across the cell membrane, which is then used to energize the production of ATP (see Chapter 2).

Similarly, *halorhodopsin* is a chloride ion pump found in a type of archaea known as *halobacteria* that thrive in very salty conditions, whose biological role is to maintain the osmotic balance of a cell by pumping chloride into their cytoplasm from the outside, energized by absorption of yellow/green light (typically 540 nm < λ < 590 nm). *Channelrhodopsin* (*ChR*), which is found in the single-celled model alga *C. reinhardtii*, acts as a pump for a range of nonspecific positive ions including protons, Na$^+$ and K$^+$ as well as the divalent Ca^{2+} ion. However, here longer wavelength red light ($\lambda > 590$ nm) fuels a pumping action from the outside of the cell to the cytoplasm inside.

In addition, light-sensitive protein sensors are used, for example, chloride and calcium ion sensors, as well as membrane voltage sensor protein complexes. Finally, another class of light-sensitive membrane integrated proteins are used, the most commonly used being the *optoXR* type. These undergo conformational changes upon the absorption of light, which triggers intracellular chemical signaling reactions.

The light-sensitive pumps used in optogenetics have a typical "on time" constant of a few milliseconds, though this is dependent on the local excitation of laser illumination. The importance of this is that it is comparable to the electrical conduction time from one end of a single nerve cell to the other and so in principle allows individual action potential pulses to be probed. The nervous conduction speed varies with nerve cell type but is roughly in the range

$1-100$ ms^{-1}, and so a signal propagation time in a long nerve cell that is a few millimeters in length can be as slow as a few milliseconds.

The "off time" constant, that is, a measure of either the time taken to switch from "on" to "off" following removal of the light stimulation, varies usually from a few milliseconds up to several hundred milliseconds. Some ChR complexes have a bistable modulation capability, in that they can be activated with one wavelength of light and deactivated with another. For example, *ChR2-step function opsins (SFOs)* are activated by blue light of peak $\lambda = 470$ nm and deactivated with orange/red light of peak $\lambda = 590$ nm, while a different version of this bistable ChR called *VChR1-SFOs* has the opposite dependence with wavelength such that it is activated by yellow light of peak $\lambda = 560$ nm, but deactivated with a violet light of peak $\lambda = 390$ nm. The off times for these bistable complexes are typically a few seconds to tens of seconds. Light-sensitive biochemical modulation complexes such as the optoXRs have off times of typically a few seconds to tens of seconds.

Genetic mutation of all light-sensitive protein complexes can generate much longer off times of several minutes if required. This can result in a far more stable on state. The rapid on times of these complexes enable fast activation to be performed either to stimulate nervous signal conduction in a single nerve cell or to inhibit it. Expanding the off-time scale using genetic mutants of these light-sensitive proteins enables experiments using a far wider measurement sample window. Note also that since different classes of light-sensitive proteins operate over different regions of the visible light spectrum, this offers the possibility for combining multiple different light-sensitive proteins in the same cell. Multicolor activation/deactivation of optogenetics constructs in this way result in a valuable *neuroengineering toolbox*.

Optogenetics is very useful when used in conjunction with the advanced optical techniques discussed previously (Chapter 4), in enabling control of the sensory state of single nerve cells. The real potency of this method is that it spans multiple length scales of the nervous sensory system of animal biology. For example, it can be applied to individual nerve cells cultured from samples of live nerve tissue (i.e., *ex vivo*) to probe the effects of sensory communication between individual nerve cells. With advanced fluorescence microscopy methods, these can be combined with the detection of single-molecule chemical transmitters at the *synapse* junctions between nerve cells to explore the molecular scale mechanisms of sensory nervous conduction and regulation. But larger length scale experiments can also be applied using intact living animals to explore the ways in which neural processing between multiple nerve cells occurs. For example, using light stimulation of optogenetically engineered parts of the nerve tissue in *C. elegans* can result in control of the swimming behavior of the whole organism. Similar approaches have been applied to monitor neural processing in fruit flies and also experiments on live rodents and primates using optical fiber activation of optogenetics constructs in the brain have been performed to monitor the effect on whole organism movement and other aspects of animal behavior relating to complex neural processing. In other words, optogenetics enables insight into the operation of nerves from the length scale of single molecules through to cells and tissues up to the level of whole organisms. Such techniques also have a direct biomedical relevance in offering insights into various neurological diseases and psychiatric disorders.

KEY BIOLOGICAL APPLICATIONS: MOLECULAR CLONING

Controlled gene expression investigations; Protein purification; Genetics studies.

7.5 MAKING CRYSTALS

Enormous advances have been made in the life sciences due to structural information of biomolecules, which is precise within the diameter of single constituent atoms (see Chapter 5). The most successful biophysical technique to achieve this, as measured by the number of different uploaded PDB files of atomic spatial coordinates of various biomolecule structures into the primary international PDB data repository of the Protein Data Bank (www.pdb.org, see Chapter 2), has been x-ray crystallography. We explored aspects of the physics of this technique previously in Chapter 5. At present, a technical hurdle in x-ray crystallography is the preparation of crystals that are large enough to generate a strong signal in the diffraction pattern while being of sufficient quality to achieve this diffraction to a high measurable spatial resolution. There are therefore important aspects to the practical methods for generating biomolecular crystals, which we discuss here.

7.5.1 BIOMOLECULE PURIFICATION

The first step in making biomolecule crystals is to purify the biomolecule in question. Crystal manufacture ultimately requires a *supersaturated* solution of the biomolecule (meaning a solution whose effective concentration is above the saturation concentration level equivalent to the concentration in which any further increase in concentration results in biomolecules precipitating out of solution). This implies generating high concentrations equivalent in practice to several mg mL^{-1}.

Although crystals can be formed from a range of biomolecule types, including sugars and nucleic acids, the majority of biomolecule crystal structures that have been determined relate to proteins, or proteins interacting with another biomolecule type. The high purity and concentration required ideally utilizes molecular cloning of the gene coding for the protein in a plasmid to overexpress the protein. However, often a suitable recombinant DNA expression system is technically too difficult to achieve, requiring less ideal purification of the protein from its native cell/tissue source. This requires a careful selection of the best model organism system to use to maximize the yield of protein purified. Often bacterial or yeast systems are used since they are easy to grow in liquid cultures; however, the quantities of protein required often necessitate the growth of several hundred liters of cells in culture.

The methods used for the extraction of biomolecules from the native source are classical biochemical purification techniques, for example, *tissue homogenization*, followed by a series of *fractionation precipitation* stages. Fractionation precipitation involves altering the solubility of the biomolecule, most usually a protein, by changing the pH and ionic strength of the buffer solution. The ionic strength is often adjusted by addition of *ammonium sulfate* at high concentrations of ~2.0 M, such that above certain threshold levels of ammonium sulfate, a given protein at a certain pH will precipitate out of a solution, and so this procedure is also referred to as *ammonium sulfate precipitation*.

At low concentrations of ammonium sulfate, the solubility of a protein actually increases with increasing ammonium sulfate, a process called "salting in" involving an increase in the number of electrostatic bonds formed between surface electrostatic amino acid groups and water molecules mediated through ionic salt bridges. At high levels of ammonium sulfate, the electrostatic amino acid surface residues will all eventually be fully occupied with salt bridges and any extra added ammonium sulfate results in the attraction of water molecules away from the protein, thus reducing its solubility, known as *salting out*. Different proteins salt in and out at different threshold concentrations of ammonium sulfate; thus, a mixture of different proteins can be separated by centrifuging the sample to generate a pellet of the precipitated protein(s) and then subject either the pellet or the suspension to further biochemical processing—for example, to use gel filtration chromatography to further separate any remaining mixtures of biomolecules on the basis of size, shape, and charge, etc., in addition to methods of dialysis (see Chapter 6). Ammonium sulfate can also be used in the final stage of this procedure to generate a high concentration of the purified protein, for example, to salt out, then resuspend the protein, and dissolve fully in the final desired pH buffer for the purified protein to be crystallized. Other *precipitants* aside from ammonium sulfate can be used depending on the pH buffer and protein, including *formate, ammonium phosphate*, the alcohol *2-propanol*, and the polymer *polyethylene glycol (PEG)* in a range of molecular weight value from 400 to 8000 Da.

7.5.2 CRYSTALLIZATION

Biomolecule crystallization, most typically involving proteins, is a special case of a thermodynamic phase separation in a nonideal mixture. Protein molecules separate from water in solution to form a distinct, ordered crystalline phase. The nonideal properties can be modeled

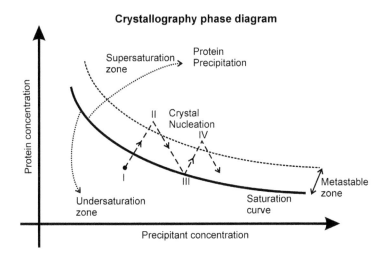

Crystallography phase diagram

FIGURE 7.3 Generating protein crystals. Schematic of 2D phase diagram showing the dependence of protein concentration on precipitant concentration. Water vapor loss from a concentrated solution (point I) results in supersaturated concentrations, which can result in crystal nucleation (point II). Crystals can grow until the point III is reached, and further vapor loss can result in more crystal nucleation (point IV).

as a virial expansion (see Equation 4.25) in which the parameter B is the second virial coefficient and is negative, indicative of net attractive forces between the protein molecules.

Once a highly purified biomolecule solution has been prepared, then, in principle, the attractive forces between the biomolecules can result in crystal formation if a supersaturated solution in generated. It is valuable to depict the dependence of protein solubility on precipitant concentration as a 2D phase diagram (Figure 7.3). The *undersaturation zone* indicates solubilized protein, whereas regions to the upper right of the *saturation curve* indicate the *supersaturation zone* in which there is more protein present than can be dissolved in the water available.

The crystallization process involves local decrease in entropy S due to an increase in the order of the constituent molecules of the crystals, which is offset by a greater increase in local disorder of all of the surrounding water molecules due to breakage of solvation bonds with the molecules that undergo crystallization. Dissolving a crystal breaks strong molecular bonds and so releases enthalpy H as heat (i.e., an exothermic process), and similarly crystal formation is an endothermic process. Thus, we can say that

$$\Delta G_{crystallization} = G_{crystal} - G_{solution} = \left(H_{crystal} - H_{solution} \right) - T\left(S_{crystal} - S_{solution} \right)$$

The change in enthalpy for the local system composed of all molecules in a given crystal is positive for the transition of disordered solution to ordered crystal (i.e., crystallization), as is the change in entropy in this local system. Therefore, the likelihood that the crystallization process occurs spontaneously, which requires $\Delta G_{crystallization} < 0$, increases at lower temperatures T. This is the same basic argument for a change of state from liquid to solid.

Optimal conditions of precipitant concentration can be determined in advance to find a *crystallization window* that maximizes the likely number of crystals formed. For example, static light scattering experiments (see Chapter 4) can be performed on protein solutions containing different concentrations of the precipitant. Using the Zimm model as embodied by Equation 4.30 allows the second virial coefficient to be estimated. Estimated preliminary values of B can then be plotted against the empirical crystallization success rate (e.g., number of small crystals observed forming in a given period of time) to determine an empirical crystallization window by extrapolating back to the ideal associated range in precipitant concentration, focusing on efforts for longer time scale in crystal growth experiments.

Indeed, the trick for obtaining homogeneous crystals as opposed to amorphous precipitated protein is to span the metastable zone between supersaturation and undersaturation by making *gradual changes* to the effective precipitant concentration. For example, crystals can be simply formed using a *solvent evaporation method*, which results in very gradual increases in precipitant concentration due to the evaporation of solvent (usually water) from the solution. Other popular methods include *slow cooling* of the saturated solution, *convection* heat flow in the sample, and *sublimation* methods under vacuum. The most common techniques, however, are *vapor diffusion methods*.

Two popular types of vapor diffusion techniques used are *sitting drop* and *hanging drop* methods. In both methods, a solution of precipitant and concentrated but undersaturated protein is present in a droplet inside a closed microwell chamber. The chamber also contains a larger reservoir consisting of a precipitant at higher concentration than the droplet but no protein, and the two methods only essentially differ in the orientation of the protein droplet relative to the reservoir (in the hanging drop method the droplet is directly above the reservoir, in the sitting drop method it is shifted to the side). Water evaporated from the droplet is absorbed into the reservoir, resulting in a gradual increase in the protein concentration of the droplet, ultimately to supersaturation levels.

The physical principles of these crystallization methods are all similar; in terms of the phase diagram, a typical initial position in the crystallization process is indicated by point I on the phase diagram. Then, due to water evaporation from the solution, the position of the phase diagram will translate gradually to point II in the *supersaturation zone* just above the *metastable zone*. If the temperature and pH conditions are optimal, then a crystal may *nucleate* at this point. Further evaporation causes crystal growth and translation on the phase diagram to point III on the *saturation curve*. At this point, any further water evaporation then potentially results in translation back up the phase transition boundary curve to the supersaturation point IV, which again may result in further crystal nucleation and additional crystal growth. Nucleation may also be seeded by particulate contaminants in the solution, which ultimately results in multiple nucleation sites with each resultant crystal being smaller than were a single nucleation site present. Thus, solution and sample vessel cleanliness are also essential for generating large crystals. Similarly, mechanical vibration and air disturbances can result in detrimental multiple nucleation sites. But the rule of thumb with crystal formation is that any changes to physical and chemical conditions in seeding and growing crystals should be made slowly—although some proteins can crystallize after only a few minutes, most research grade protein crystals require several months to grow, sometimes over a year.

Nucleation can be modeled as two processes of primary nucleation and secondary nucleation. Primary nucleation is the initial formation such that no other crystals influence the process (because either they are not present or they are too far away). The rate *B* of primary nucleation can be modeled empirically as

$$(7.2) \qquad\qquad B_1 = \frac{dN}{dt} = k_n \left(C - C_{sat} \right)^n$$

where

B_1 is the number of crystal nuclei formed per unit volume per unit time
N is the number of crystal nuclei per unit volume
k_n is a rate constant (an on-rate)
C is the solute concentration
C_{sat} is the solute concentration at saturation
n is an empirically determined exponent typically in the range 3–4, though it can be as high as ~10

The secondary nucleation process is more complex and is dependent on the presence of other nearby crystal nuclei whose separation is small enough to influence the kinetics of further crystal growth. Effects such as fluid shear are important here, as are collisions between preexisting crystals. This process can be modeled empirically as

$$(7.3) \qquad B_2 = k_1 M_T^j \left(C - C_{sat} \right)^b$$

where
  k_1 is the secondary nucleation rate constant
  M_T is the density of the crystal suspension
  j and b are empirical exponents of ~1 (as high as ~1.5) and ~2 (as high as ~5), respectively

Analytical modeling of the nucleation process can be done by considering the typical free energy change per molecule ΔG_n associated with nucleation. This is given by the sum of the bulk solution (ΔG_b) and crystal surface (ΔG_s) terms:

$$(7.4) \qquad \Delta G_n = \Delta G_b + \Delta G_s = \frac{-4\pi r^3 \Delta\mu}{\Omega} + 4\pi r^2 \alpha$$

where
  α is the *interfacial free energy* per unit area of a crystal of effective radius r
  Ω is the volume per molecule
  $\Delta\mu$ is the change in *chemical potential* of the crystallizing molecules, which measures the mean free energy change in a molecule transferring from the solution phase to the crystal phase

Standard thermodynamic theory for the chemical potential indicates

$$(7.5) \qquad \Delta\mu = k_B T \ln\left(\frac{C - C_{sat}}{C_{sat}} \right) = k_B T \ln\sigma$$

where σ is often called the "saturation." Inspection of Equation 7.4 indicates that there is a local maximum in ΔG_n equivalent to the free energy barrier ΔG^* for nucleation at a particular threshold value of r known as the critical radius r_c:

$$(7.6) \qquad r_c = \frac{2\Omega\alpha}{k_B T \sigma}$$

In practice, two interfacial energies often need to be considered, one between the crystal and the solid substrate and the other between the crystal and the solid surface substrate in which crystals typically form on. Either way, substituting r_c into Equation 7.5 indicates

$$(7.7) \qquad \Delta G^* = \frac{16\pi\alpha^3}{3} \left(\frac{\Omega}{3k_B T \sigma} \right)^2$$

Thus, the rate of nucleation J_n can be estimated from the equivalent Boltzmann factor:

$$(7.8) \qquad J_n = A\exp\left(\frac{-\Delta G_n}{k_B T} \right) = A\exp\left(\frac{-B\alpha^3}{\sigma^2} \right)$$

where A and B are constants. Thus, there is a very sensitive dependence on nucleation rate with both the interfacial energy and the supersaturation. This is a key thermodynamic explanation for why the process of crystal formation is so sensitive to environmental conditions and is considered by some to be tantamount to a *black art!* Similar thermodynamic arguments can be applied to model the actual geometrical shape of the crystals formed.

Many functional biomolecular complexes may be formed from multiple separate components. Obtaining crystals from these is harder since it requires not only a mixture of highly pure separate components but also one in which the relative stoichiometry of the components to each other is tightly constrained. Finding optimum temperature and pH conditions that avoid premature precipitation in the separate components is a key challenge often requiring significant experimental optimization. The use of microorganisms such as bacteria and unicellular eukaryotes to grow such crystals has shown recent promise, since the small volume of these cells can result in very concentrated intracellular protein concentration. Crystals for viral capsids (see Chapter 2) have been generated in this way, with the caveat that the crystal size will be limited to just a few microns length due to the small size of the cells used.

7.5.3 TREATMENT AFTER CRYSTALLIZATION

The precision of an x-ray crystal diffraction pattern is affected significantly by the homogeneity of the crystal, and its size. Controlled, gradual dehydration of crystals can result in an ultimate increase in crystal size, for example, using elevated concentration levels of PEG to draw out the water content, which in some cases can alter the shape of the crystal unit cell, resulting in more efficient packing in a larger crystal structure. Also, small *seed crystals* placed in the undersaturated solution can be efficient sites for nucleation of larger growing larger crystals.

The use of *crystallization robots* has significantly improved the high-throughput nature of crystallization. These devices utilize vapor diffusion methods to automate the process of generating multiple crystals. They include multiple arrays of microwell plates, resulting in several tens of promising crystals grown in each batch under identical physical and chemical conditions using microfluidics (see later in this chapter). These methods also utilize batch screening methods to indicate the presence of promising small crystals that can be used as seed crystals. The detection of such small crystals, which may have a length scale of less than a micron, by light microscopy is hard but may be improved by UV excitation and detection of fluorescence emission from the crystals or by using polarization microscopy. Second harmonic imaging (see Chapter 4) can also be used in small crystal identification, for example, in a technique called *second-order nonlinear optical imaging of chiral crystals*. The use of such high-throughput technologies in crystallization with robotized screening has enabled the selection of more homogeneous crystals from a population.

7.5.4 PHOTONIC CRYSTALS

A special type of crystal, which can occur naturally both in living and nonliving matter and also which can be engineered synthetically for biophysical applications, is *photonic crystals*. Photonic crystals are spatially periodic optical nanostructures that perturb the propagation of transmitted photons. This is analogous to the perturbation of electrons in ionic crystal structures and semiconductors, for example, there are certain energy levels that are forbidden in terms of propagation of photons, in the same manner that there are forbidden energy levels for electron propagations in certain spatially periodic solids.

Photonic crystals are spatially periodic in terms of dielectric constant, with the periodicity being comparable to the wavelength of visible or near visible light. This results in diffractive effects only for specific wavelengths of light. An allowed wavelength of propagation is a mode, with the summation of several modes comprising a band. Disallowed energy bands imply that photons of certain wavelengths will not propagate through the crystal, barring small quantum tunneling effects, and are called "photonic bandgaps."

Natural nonbiological photonic crystals include various gemstones, whereas biological photonic crystals include *butterfly wings*. Butterfly wings are composed of periodic scales made from fibrils of a protein called chitin combined with various sugar molecules in a matrix of other proteins and lipids that, like all crystals, appear to have many *self-assembly* steps involved in their formation (see Chapter 9). The chitin fibrils form periodic ridge structures, with the spacing between ridges being typically a few hundred nanometers, dependent on the butterfly species, resulting in photonic bandgaps, and the colorful, metallike appearance of many butterfly wings that remains constant whatever the relative angle of incident light and observation direction.

Synthetic photonic crystals come under the description of *advanced materials* or *metamaterials*. Metamaterials are those that are not found in nature; however, many of these have gained inspiration from existing biological structures, and in fact several can be described as *biomimetic* (see Chapter 9). Artificial photonic crystals utilize multilayered thin metallic films using microfabrication techniques (see the following section in this chapter), described as *thin-film optics*, and such technologies extend to generating photonic crystal fibers. For example, these have biophysical applications for lab-on-a-chip devices for propagating specific wavelengths of excitation light from a broadband white-light source, while another photonic crystal fiber propagates fluorescence emissions from a fluorescently labeled biological sample for detection that disallow propagation of the original excitation wavelength of light used, thus acting as a wavelength filter in a similar way to conventional fluorescence microscopy, but without the need for any additional large length scale traditional dichroic mirror or emission filter.

Structured light is used as the general term for light whose properties have been controllably engineered, many examples involving the application of photonic features, and since there are multiple examples of this in nature, it has led to several examples of bio-inspired technologies, so-called *biomicry*, utilizing structured light (see Chapter 9).

KEY BIOLOGICAL APPLICATIONS: CRYSTAL MAKING

Molecular structure determination through x-ray crystallography.

7.6 HIGH-THROUGHPUT TECHNIQUES

Coupled to many of these, more advanced biophysical characterization tools are a new wave of *high-throughput techniques*. These are technologies that facilitate the rapid acquisition and quantification of data and are often used in conjunction with several core biophysical methods, but we describe here in a devoted section due to their importance in modern biophysics research. These include the use of *microfluidics, smart microscope stage* designs and robotized sample control, the increasing prevalence of "omics" methods, and the development of smart fabrication methods including *microfabrication, nanofabrication*, and *3D printing* technologies leading to promising new methods of *bioelectronics* and *nanophotonics*.

7.6.1 SMART FABRICATION TECHNIQUES

Microfabrication covers a range of techniques that enable micron scale solid-state structures to be controllably manufactured, with nanofabrication being the shorter length scale precise end of these methods that permit details down to a few nanometer precision to be fabricated. They incorporate essentially the technology used in manufacturing integrated circuits and in devices that interface electronics and small mechanical components, or *microelectromechanical systems* (*MEMS*). The methods comprise *photolithography* (also known as *optical lithography*), chemical and *focused ion beam* (*FIB*) etching (also known as *electron beam lithography*), substrate doping, thin-layer deposition, and substrate polishing, but also incorporate less common methods of substrate etching including *x-ray lithography, plasma etching, ion beam etching*, and *vapor etching*.

The state-of-the-art microfabrication is typified with the publishing of the world's smallest book in 2007 entitled *Teeny Ted from Turnip Town*, which is made using several of these techniques from a single polished wafer of silicon generating 30 *micro pages* of size 70×100 μm, with the FIB generating letters with a line width of just ~40 nm. The book even has its own International Standard Book Number reference of ISBN-978-1-894897-17-4. However, it requires a suitably nanoscale precise imaging technology such as a scanning electron microscope to read this book (see Chapter 5).

Microfabrication consists of multiple sequential stages (sometimes several tens of individual steps) of manufacture involving treatment of the surface of a solid substrate through either controllably removing specific parts of the surface or adding to it. The substrate in question is often silicon based, stemming from the original application for integrated circuits, such as pure silicon and doped variants that include electron (n-type, using typical dopants of antimony, arsenic, and phosphorus) and electron hole (p-type, using typical dopants of aluminum, born, and gallium) donor atoms. Compounds of silicon such as silicon nitride and silicon dioxide are also commonly used. The latter (glass) also has valuable optical transmittance properties at visible light wavelengths. To generate micropatterned surfaces, a *lift-off* process is often used that, unlike surface removal methods, is additive with respect to the substrate surface. Lift-off is a method that uses a *sacrificial material* to creating topological surface patterns on a target material.

Surface removal techniques include chemical etching, which uses a strong acid or base that dissolves solvent accessible surface features, and focused ablation of the substrate using a FIB (see Chapter 5). Chemical etching is often used as part of *photolithography*. In photolithography, the substrate is first *spin-coated* with a *photoresist.* A photoresist is a light-sensitive material bound to a substrate surface, which can generate surface patterns by controllable exposure of light and chemical etching using an appropriate *photoresist developer.* They are typically viscous liquids prior to setting; a small amount of liquid photoresist is applied on the center of the substrate that is then centrifuged by spin-coating the surface controllably in a thin layer of photoresist. We can estimate the height $h(t)$ of the photoresist after a time t from spinning by using the *Navier–Stokes equation* assuming laminar flow during the spinning, resulting in equating frictional drag and centripetal forces on an incremental segment of photoresist at a distance r from the spinning axis with radial speed component v_r and height z above the wafer surface:

$$(7.9) \qquad \eta \frac{\partial^2 v_r}{\partial z^2} = \rho \omega^2 r$$

where the wafer is spun at angular frequency ω, η is the viscosity, and ρ is the density. Assuming no photoresist is created or destroyed indicates

$$\frac{\partial h}{\partial t} = \frac{-1}{r} \frac{\partial (rQ)}{\partial r}$$

where the flow rate by volume, Q, is given by

$$(7.10) \qquad Q = \int_0^{h(t)} v_r d_z$$

Assuming zero slip and zero sheer boundary conditions results in

$$(7.11) \qquad h(t) = \frac{h(0)}{\sqrt{\left\{ 2 + h(0)^2 \, t \left(\dfrac{4\rho \omega^2}{3\eta} \right) \right\}}}$$

Here, we assume an initial uniform thickness of $h(0)$. At long spin times, this approximates to

$$(7.12) \qquad h(t) \approx \sqrt{\frac{3\eta}{4\rho \omega^2 t}}$$

which is thus independent of $h(0)$. The thickness used depends on the photoresist and varies in the range ~0.5 µm up to 100 µm or more.

There are two types of photoresists. A *positive resist* becomes soluble to the photoresist developer on exposure to light, whereas a *negative resist* hardens and becomes insoluble upon exposure to light. A surface pattern, or *nanoimprimpt*, is placed on top of the photoresist. This consists of a dark printed pattern that acts as a mask normally of glass covered with chromium put on top of the thin layer of photoresist, which block outs the light in areas where chromium is present, and so areas of photoresist directly beneath are not exposed to light and become insoluble to the photoresist developer. The unexposed portion of the photoresist is dissolved by the photoresist developer (Figure 7.4a).

A popular photoresist in biophysical applications is the negative photoresist *SU-8*—an epoxy resin so called because of the presence of eight epoxy groups in its molecular structure. SU-8 has ideal adhesive properties to silicon-based substrates and can be spun out to form a range of thicknesses at the micron and submicron scale, which makes it ideal for forming a high-resolution mask on a substrate to facilitate further etching and deposition stages in the microfabrication process. It is for these reasons that SU-8 is a popular choice for biological hybrid MEMS devices, or *Bio-MEMS*, for example, for use as miniaturized *biosensors* on lab-on-a-chip devices (see Chapter 9).

A typical protocol for generating a surface pattern from SU-8 involves a sequence of spin-coating at a few thousand rpm for ca. 1 min, followed by lower temperature *soft baking* at 65°C–95°C prior to exposure of the nanoimprint-masked SU-8 to UV radiation, and then a post bake prior to incubation with the photoresist developer, rinsing, drying, and sometimes further *hard baking* at higher temperatures of ~180°C.

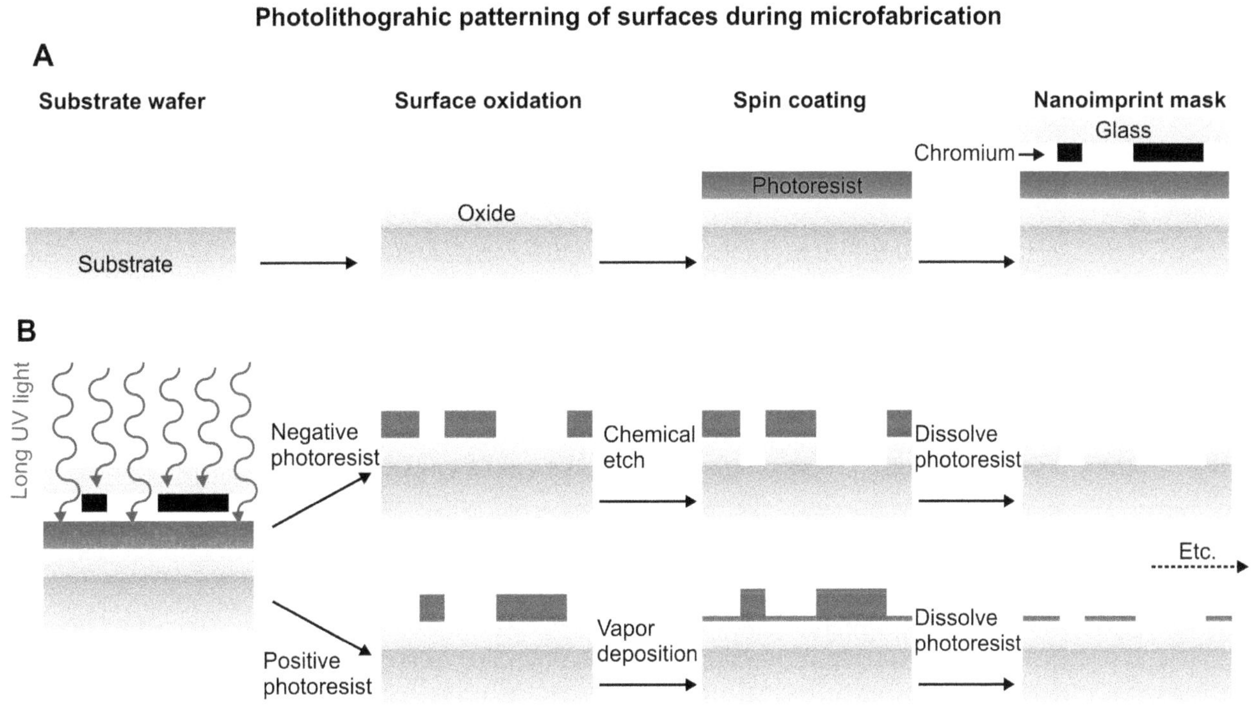

Photolithograhic patterning of surfaces during microfabrication

FIGURE 7.4 Microfabrication using photolithography. Schematic example of photolithography to engineer patterned surfaces. (a) A wafer of a substrate, typically silicon based can, for example, be oxidized if required, prior to spin-coating in a photoresist, which acts as a sacrificial material during this "lift-off" process. (b) Exposure to typically long UV light (wavelength ~400 nm) results in either hardening (positive photoresist) or softening (negative photoresist), such that the softened/unhardened photoresist can be removed by the solvent. At this stage, there are several possible options in the microfabrication process, either involving removal of material (such as the chemical etching indicated) or addition of material (such as due to vapor deposition, such as here with a surface layer of gold). The final removal of the photoresist using specific organic results in complexly patterned microfabricated surfaces.

Following the treatment with an appropriate photoresist developer leaves a surface pattern consisting of some regions of exposed substrate, where, in some of the regions, the photoresist still remains. The exposed regions are accessible to further chemical etching treatment, but regions masked by the remaining photoresist are not. Chemical etching treatment thus results in etched pattern onto the substrate itself. Also, at this stage, deposition or growth onto the patterned substrate can be performed, of one or more thin layers or additional material, for example, to generate electrically conducting, or insulating, regions of the patterned surface.

This can be achieved using a range of techniques including *thermal oxidation* and *chemical vapor deposition, physical vapor deposition* methods such as *sputtering* and *evaporative deposition*, and *epitaxy* methods (which deposit crystalline layers onto the substrate surface) (Figure 7.4b). Evaporative deposition is commonly used for controlled coating of a substrate in one or more thin metallic layers. This is typically achieved by placing the substrate in a high vacuum chamber (a common vacuum chamber used is a *Knudsen cell*) and then by winding solid metallic wire (e.g., gold, nickel, chromium are the common metals used) around a tungsten filament. The tungsten filament is then electrically heated to vaporize the metal wire, which solidifies on contact with the substrate surface. The method is essentially the same as that used for positive shadowing in electron microscopy (see Chapter 5). Following any additional deposition, any remaining photoresist can be removed using specific organic solvent treatment to leave a complex patterned surface consisting of etches and deposition areas.

Sputtering is an alternative to vapor deposition for coating a substrate in a thin layer of metal. Sputter deposition involves ejecting material from a metal target that is a source onto the surface of the substrate to be coated. Typically, this involves gas plasma of an inert gas such as argon. Positive argon ions, Ar+, are confined and accelerated onto the target using magnetic fields in a *magnetron* device to bombard the metal sample to generate ejected metal atoms of several tens of keV of energy. These can then impact and bind to the substrate surface as well as cause some resputtering of metal atoms previously bound to the surface.

Sputter deposition is largely complementary to evaporative deposition. One important advantage of sputtering is that it can be applied to metals with very high vaporization temperatures that may not be easy to achieve with typical evaporative deposition devices. Also, the greater speed of ejected metal atoms compared to the more passive diffusive speed from evaporative deposition results in greater adhesion to substrate surfaces in general. The principal disadvantage of sputtering over evaporative deposition is that sputtering does not generate a distinct metallic shadow around topographic features in the same way that evaporative deposition does because of the extra energy of the ejected metal atoms, resulting in a diffusive motion around the edges of these surface features; this can make the process of lift-off more difficult.

Microfabrication methods have been used in conjunction with biological conjugation tools (see the previous section of this chapter) to biochemically functionalize surfaces, for example, to generate platforms for adhesion of single DNA molecules to form DNA curtains (see Chapter 6). Also, by combining controlled metallic deposition on a microfabricated surface with specific biochemical functionalization, it is possible to generate smart *bioelectronics* circuitry. Smart surface structures can also utilize molecular *self-assembly techniques*, such as *DNA origami* (discussed in Chapter 9).

Important recent advances have been made in the area of nanophotonics using microfabrication and nanofabrication technologies. Many silicon-based substrates, such as silicon dioxide "glass," have low optical absorption and a reasonably high refractive index of ~1.5 in the visible light spectrum range, implying that they are optically transparent and can also act as photonic waveguides for visible light. A key benefit here in terms of biophysical applications is that laser excitation light for fluorescence microscopy can be guided through a silicon-based microfabricated device across significant distances; for example, if coupled to an optic fiber delivery system to extend to waveguide distance, this is potentially only limited by the optical fiber repeater distance of several tens of kilometers.

This potentially circumvents the need to have objective lens standard optical microscopy-based delivery and capture methods for light and so facilitates miniaturization of devices that can fluorescently excite fluorophore-tagged biomolecules and captures their fluorescence emissions. In other words, this is an ideal technology for developing miniaturized biosensors.

A promising range of nanophotonics biosensor devices use either evanescent field excitation or plasmon excitation or a combination of both. For example, a flow cell can be microfabricated to engineer channels for flowing through a solution of fluorescently labeled biomolecules from a sample. The waveguiding properties of the silicon-based channel can result in total internal reflection of a laser source at the channel floor and side walls, thus generating a 3D evanescent excitation field that can generate TIRF in a similar way to that discussed previously for light microscopy (see Chapter 3). Precoating the channel surfaces with a layer of metal ~10 nm thick allows surface plasmons to generate, in the same manner as for conventional SPR devices (see Chapter 3), thus presenting a method to generate kinetics of binding data for label-free non-fluorescent biomolecules if the channel surfaces are chemically functionalized with molecules that have high specific binding affinities to key biomolecules that are to be detected (e.g., specific antibodies). These technologies also can be applied to live-cell data.

The advantages of nanophotonics for such biosensing applications include not only miniaturization but also improvements in high-throughput sensing. For example, multiple parallel smart flow-cell channels can be constructed to direct biological samples into different detection areas. These improve the speed of biosensing by not only parallelizing the detection but also enabling multiple different biomolecules to be detected, for example, by using different specific antibodies in each different detection area. This ultimately facilitates the development of lab-on-a-chip devices (see Chapter 9).

Three-dimensional printing has emerged recently as a valuable, robust tool. For example, many components that are used in complex biophysical apparatus, such as those used in bespoke optical imaging techniques, consist of multiple components of nonstandard sizes and shapes, often with very intricate interfaces between the separate components. These can be nontrivial to fashion out conventional materials that are mechanically stable but light, such as aluminum, using traditional machining workshop tools, in a process of *subtractive manufacturing*. However, 3D printing technology has emerged as a cost-effective tool to generate such bespoke components, typically reducing the manufacturing time of traditional machining methods by factor of two or more orders of magnitude.

KEY POINT 7.6

Traditional machining methods utilize subtractive manufacturing—material is removed to produce the final product, for example, a hole is drilled into a metal plate, and a lathe is used to generate a sharpened tip. Conversely, 3D printing is an example of additive manufacturing, in which material is added together from smaller components to generate the final product.

A 3D printer operates on the principle of *additive manufacturing*, in which successful 2D layers of material are laid down to assemble the final 3D product. Most commonly, the method involves *fused deposition modeling*. Three-dimensional objects can be first designed computationally using a range of accepted file formats. A 3D printer will then lay down successive layers of material—liquid, powder, and paper can be used, but more common are *thermoplastics* that can be extruded as a liquid from a heated printer nozzle and then fused/solidified on contact with the material layer beneath. These layers correspond to a cross-section of the 3D model, with a typical manufacturing time ranging from minutes up to a few days, depending on the complexity of the model.

The spatial resolution of a typical 3D printer is ~25–100 μm. However, some high-resolution systems can print down to ~10 μm resolution. Several cost-effective desktop 3D printers cost, at the time of writing, less than $1000, which can generate objects of several tens of centimeter length scale. More expensive printers exist that can generate single printed objects of a few meters in length scale. Cheaper potential solutions exist generating large objects, for example, from attaching smaller objects together in a modular fashion and utilizing origami methods to fold several smaller printed sheetlike structures together to generate complex 3D shapes.

Worked Case Example 7.1: Microfabrication

A silicon substrate was spin-coated with an SU-8 photoresist by spinning at 3000 rpm in order to ultimately generate a layer of silicon oxide of the same thickness as the sacrificial photoresist material.

a   *To generate a 0.5 µm thick layer of silicon oxide, how many minutes must the spin-coating of the SU-8 proceed for?*
     In a subsequent step after the removal of the photoresist and deposition of the silicon oxide, silicon oxide was coated with a 10 nm thick layer of gold for a surface plasmon resonance application, employing evaporation deposition using a length of gold wire evaporated at a distance of 5 cm away from the silicon substrate.
b   *If the gold wire has a diameter of 50 µm and is wound tightly onto an electric heating filament under a high vacuum, which melts and vaporizes the gold completely, explain with reasoning how many centimeters of wire are needed to be used, stating any assumptions you make.*

(Assume that the density and dynamic viscosity of the SU-8 used are 1.219 g cm^{-3} and 0.0045 Pa · s.)

Answers

a   Assuming a high time approximation, we can rearrange Equation 7.7 to generate the time t required for a given photoresist thickness h of such that

$$t = \frac{3\eta}{4\rho\omega^2 h^2}$$

Thus,

$$t = (3 \times 0.0045) / (4 \times (1.219 \times 10^3) \text{ kg m}^{-3} \times (3000/60 \times 2\pi)^2$$
$$\text{rad}^2\text{s}^{-2} \times (20 \times 10^{-6})^2\text{m}) = 112 \text{ s} = 1.9 \text{ min}$$

b   If a mass m of gold vaporizes isotropically, then the mass flux per unit area at a distance d from the point of vaporization (here 5 cm) will be $m/4\pi d^2$. Thus, over a small area of the substrate δA, the mass of gold vapor deposited assuming it solidifies soon after contact will be

$$\delta m = \delta A \cdot m/4\pi d^2 = \rho_{Au}\delta A \cdot \delta Z$$

where the density of gold is ρ_{Au}, and the thickness of the deposited gold on the silicon oxide substrate is δz, which is 10 nm here. Thus,

$$\delta z = m/4\pi d^2 \rho_{Au}$$

But the mass of the gold is given by

$$m = l 4\pi r_{Au}^2 \rho_{Au}$$

where l is the length of gold wire used of radius r_{Au}.

Thus,

$$I = m/\pi r_{AU}^2 \rho_{Au} = \delta z \cdot 4\pi d^2 \rho_{Au}/\pi r_{AU}^2 \rho_{Au} = 4 \cdot \delta z \left(d/r_{AU} \right)^2$$

$$= 4 \times (10 \times 10^{-9} \text{ m}) \times ((5 \times 10^{-2} \text{ m})/(0.5 \times 50 \times 10^{-6} \text{ m}))^2$$

$$= 1.6 \times 10^{-1} \text{ m} = 16 \text{ cm}$$

7.6.2 MICROFLUIDICS

Microfluidics (for a good overview, see Whitesides, 2006) deals with systems that control the flow of small volumes of liquid, anything from microliters (i.e., volumes of 10^{-9} m³) down to femtoliters (10^{-18} m³), involving equivalent pipes or fluid channels of cross-sectional diameters of ~1 µm up to a few hundred microns. Pipes with smaller effective diameters down to ~100 nm can also be used, whose systems are often referred to as *nanofluidics*, which deal with smaller volumes still down to ~10^{-21} m³, but our discussion here is relevant to both techniques.

Under normal operation conditions, the flow through a microfluidics channel will be laminar. *Laminar flow* implies a Reynolds number (R_e) ca. < 2100 compared to *turbulent flow* that has an R_e ca. > 2100 (see Equation 6.8). Most microfluidics channels have a diameter in the range of ~10–100 µm and a wide range of mean flow speeds from ~0.1 up to ~10 m s⁻¹. This indicates a range of R_e of ~10^{-2} to 10^3 (see Worked Case Example 7.2).

The fluid for biological applications is normally water-based and thus can be approximated as *incompressible* and *Newtonian*. A *Newtonian fluid* is one in which viscous flow stresses are linearly proportional to the strain rate at all points in the fluid. In other words, its viscosity is independent of the rate of deformation of the fluid. Under these conditions, flow in a microfluidics channel can be approximated as *Hagen–Poiseuille flow*, also known as *Poiseuille flow* (for non-French speakers, Poiseuille is pronounced, roughly, "pwar-zay"), which was discussed briefly in Chapter 6. A channel of circular cross-section implies a parabolic flow profile, such that

$$(7.13) \qquad v_x(z) = -\frac{\partial p}{\partial x}\frac{1}{4\eta}(a^2 - z^2)$$

where
  η is the dynamic (or absolute) viscosity
  p is the fluid pressure along an axial length of channel x
  a is the channel radius
  $v_x(z)$ is the speed of flow of a streamline of fluid at a distance z perpendicular to x from the central channel axis

For a fully developed flow (i.e., far away from exit and entry points of the channel), the pressure gradient drop is constant, and so equals $\Delta p/l$ where Δp is the total pressure drop across the channel of length l. It is easy to demonstrate a dependence between Δp and the volume flow rate Q given by *Poiseuille's law*:

$$(7.14) \qquad \Delta p = \frac{8\eta l}{\pi a^4}Q = R_H Q$$

R_H is known as the hydraulic resistance, and the relation $\Delta p = R_H Q$ applies generally to noncircular cross-sectional channels. In the case of noncircular cross-sections, a reasonable approximation can be made by using the *hydraulic radius* parameter in place of *a*, which is defined as *2A/s* where *A* is the cross-sectional area of the channel and *s* is its contour length perimeter. For well-defined noncircular cross-sections, there are more accurate formulations, for example, for a rectangular cross-sectional channel of height *h* that is greater than the width *w*, an approximation for R_H is

(7.15)
$$R_H \approx \frac{12\iota l}{wh^2\left(1-0.63\tanh\left(1.57w/h\right)\right)}$$

The hydraulic resistance is a useful physical concept since it can be used in *fluidic circuits* in the same way that electrical resistance can be applied in electrical circuits, with the pressure drop in a channel being analogous to the voltage drop across an electrical resistor. Thus, for *n* multiple microfluidic channels joined in series,

(7.16)
$$R_{series,total} = \sum_{i-1}^{n} R_{H,i}$$

while, for *n* channels joined in parallel,

(7.17)
$$\frac{1}{R_{parallel,total}} = \sum_{i-1}^{n} \frac{1}{R_{H,i}}$$

Microfluidics devices consist not just of channels but also of several other components to control the fluid flow. These include *fluid reservoirs* and pressure sources such as *syringe pumps*, but often *gravity-feed systems*, for example, a simple open-ended syringe, which is placed at a greater height than the microfluidics channels themselves, connected via low-resistance Teflon tubing, which generates a small fluid flow but often works very well as it benefits from lower vibrational noise compared to automated syringe pumps. Other mechanisms to generate controllable flow include *capillary action, electroosmotic methods, centrifugal systems*, and *electrowetting* technologies. Electrowetting involves the controllable change in contact angle made by the fluid on the surface of a flow cell due to an applied voltage between the surface substrate and the fluid.

Other components include *valves*, fluid/particle *filters*, and various channel *mixers*. Mixing is a particular issue with laminar flow, since streamlines in a flow can only mix due to diffusion perpendicular to the flow, and thus mixing is in effect dependent on the axial length of the pipe (i.e., significant mixing across streamlines will not occur over relatively short channel lengths; see Worked Case Example 7.2). Often, such diffusive mixing can be facilitated by clever designs in channel geometries, for example, to introduce sharp corners to encourage transient turbulence that enables greater mixing between streamline components and similarly engineer herringbone chevron-shaped structures into the channels that have similar effects. However, differences in diffusion coefficients of particles in the fluid can also be utilized to facilitate filtering (e.g., one type of slow diffusing particle can be shunted into a left channel, while a rapid diffusing particle type can be shunted into a right channel). Several types of microvalve designs exist, including piezoelectric actuators, magnetic and thermal systems, and pneumatic designs.

Microfluidics often utilizes flow cells made from the silicone compound PDMS (discussed previously in the context of cell stretching devices in Chapter 6). The combination of mechanical stability, chemical inertness, and optical transparency makes PDMS an ideal choice for manufacturing flow cells in microfluidics devices, which involve some form of optical detection technique inside the flow cell, for example, detection of fluorescence emissions from living cells. Microfabrication methods can be used to generate a solid substrate mold into which liquid, degassed PDMS can be poured. Curing the PDMS is usually done either with UV light exposure and/or through baking in an oven. The cured PDMS can then be peeled

Microfluidics designs in practice

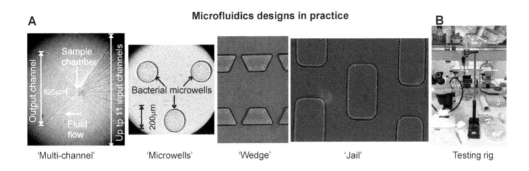

'Multi-channel'   'Microwells'   'Wedge'   'Jail'   Testing rig

FIGURE 7.5 Microfluidics. PDMS can be cast into a variety of microfluidics flow-cell designs using a solid substrate silicon-based mask manufactured using microfabrication techniques. (a) A number of designs used currently in the research lab of the author are shown here, including multichannel input designs (which enable the fluid environment of a biological sample in the central sample chamber to be exchanged rapidly in less than 1 s), microwells (which have no fluid flow, but consists of a simple PDMS mask placed over living cells on a microscope coverslip, here shown with bacteria, which can be used to monitor the growth of separate cell "microecologies"), a wedge design that uses fluid flow to push single yeast cells into the gaps between wedges in the PDMS design and in doing so immobilize them and thus enable them to be monitored continuously using light microscopy with the advantage of not requiring potentially toxic chemical conjugation methods, and a jail-type design that consists of chambers of yeast cells with a PDMS "lid" (which can be opened and closed by changing the fluid pressure in the flow cell, which enables the same group of dividing cells to be observed by up to eight different generations and thus facilitates investigation of memory effects across cell generations). (b) A simple testing rig for bespoke microfluidics designs, as illustrated from one used in the author's lab, can consist of a simple gravity-feed system using mounted syringes, combined with a standard "dissection" light microscope that allows low magnification of a factor of ca. 10–100 to be used on the flow cell to monitor the flow of dyes or large bead markers.

away from the mold, trimmed, and, if appropriate, bonded to a glass coverslip by drying the PDMS and subjecting both the PDMS and coverslip to plasma cleaning and then simply pressing the two surfaces together.

In this way, several complex, bespoke flow-cell designs can be generated (Figure 7.5). These enable biological samples to be immobilized in the sample chamber and observed continuously over long time scales (from minutes to several days if required) using light microscopy techniques. An important application uses multichannel inputs, which enables the fluid environment of the same biological sample (e.g., a collection of immobilized living cells on the microscope coverslip surface) to be exchanged rapidly typically in less than ~1 s. This has significant advantages in enabling observation of the effects of changing the extracellular environment on the exact same cells and in doing so circumvents many issues of cell-to-cell variability in a cell population that often makes definitive inference more challenging otherwise.

Microfluidics is also used in several high-throughput detection techniques, including FACS (discussed in Chapter 3). A more recent application has been adapted to traditional PCR methods (see the previous section of this chapter). Several commercial microfluidics PCR devices can now utilize microliter volumes in parallel incubation chambers. This can result in significant improvements in throughput. This general microfluidics-driven approach of reducing sample incubation volumes and parallelizing/multiplexing these volumes shows promise in the development of *next-generation sequencing* techniques, for example, in developing methods to rapidly sequence the DNA from individual patients in clinics and all parts of important progress toward greater *personalized medicine* (discussed in Chapter 9). Using microfluidics, it is now possible to isolate individual cells from a population, using similar fluorescence labeling approaches as discussed previously for FACS (see Chapter 3) and then sequence the DNA from that one single cell. This emerging technique of *single-cell sequencing* has been applied to multiple different cell types and has an enormous advantage of enabling correlation of phenotype of a cell, as exemplified by some biophysical metric such as the copy number of a particular protein expressed in that cell measured using some fluorescence technique (see Chapter 8) with the specific genotype of that one specific cell.

7.6.3 OPTICAL "OMICS" METHODS

As introduced in Chapter 2, there are several "omics" methods in the biosciences. Many of these share common features in the high-throughput technologies used to detect and quantify biomolecules. Typically, samples are prepared using a *cell lysate*. This comprises either growing an appropriate cell culture or preparing first cells of the desired type from a native tissue sample using standard purification methods (see Section 7.4) and then treating the cells with a cell bursting/permeabilizing reagent. An example of this is using an osmotically *hypotonic* solution, resulting in the high internal pressure of the cell bursting the cell membrane, which can be used with other treatments such as the enzyme *lysozyme* and/or various detergents to weaken the walls of cells from bacteria and plants that would normally be resistant to hypertonic extracellular environments.

The cell lysate can then be injected into a microfluidics device and flowed through parallel detection chambers, typically involving a microplate (an array of ca. microliter volume incubation wells, a standard design having 96 wells). A good example of this method is FISH (see Chapter 3). Here, the biomolecules under detection are nucleic acids, typically DNA. The microplate wells in this case are first chemically treated to immobilize DNA molecules, and a series of flow cycles and incubation steps then occurs in these microplate wells to incubate with fluorescently labeled oligonucleotide probes that bind to specific sequence regions of the DNA. After washing, each microplate can then be read out in a *microplate reader* that, for example, will indicate different colors of fluorescence emissions in each well due to the presence or not of bound probe molecule to the DNA. This technique is compatible with different probes simultaneously that are labeled with different colored fluorescent dyes.

FISH is a particularly power genomics tool. Using appropriate probes, it can be used diagnostically in clinical studies, for example, in the detection of different specific types of infectious bacteria in a diseased patient. Similar FISH techniques have also been used to study the species makeup of biofilms (see Chapter 2), also known as the *microbial flora*, for example, to use probes that are specific to different species of bacteria followed by multicolor fluorescence detection to monitor how multiple species in a biofilm evolve together.

Similar high-throughput binding-based assays can be used to identify biomolecules across the range of omics disciplines. However, proteomics in particular use several complementary techniques to determine the range of proteins in a cell lysate sample and the extent of the interactions between these proteins. For example, mass spectrometry methods have been developed for use in high-throughput proteomics (see Chapter 6). These can identify a wide range of protein and peptide fragment signatures and generate useful insight into the relative expression levels of the dominant proteins in a cell lysate sample.

To determine whether a given protein interacts with one or more other protein, the simplest approach is to use a biochemical bulk-ensemble-based *pull-down assay*. Traditional pull-down assays are a form of affinity chromatography in which a chromatography column is preloaded with a target protein (often referred to as *bait protein*) and the appropriate cell lysate flowed through the column. Any physical binding interactions with the target protein will be captured in the column. These are likely to be interactions with one or more other proteins (described as *prey protein* or sometimes *fish protein*), but may also involve other biomolecules, for example, nucleic acids. These captured binding complexes can then be released by changing either the ionic strength or pH of the eluting buffer in the column, and their presence determined using optical density measurements (see Chapter 3) on the eluted solution from the column.

A   **Single-molecule pull-down**

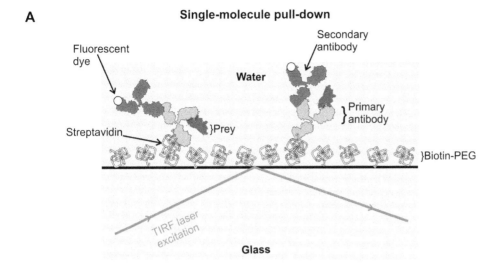

B   **Yeast two-hybrid assay**

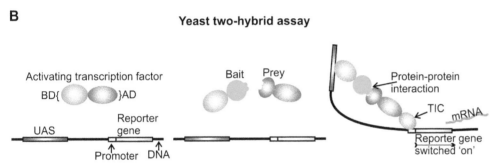

FIGURE 7.6 High-throughput protein detection. (a) Single-molecule pull-down, which uses typically immunofluorescence detection combined with TIRF excitation. (b) Yeast two-hybrid assay; an activation transcription factor is typically composed of binding domain (BD) and activation domain (AD) subunits (left panel). BD is fused to a bait protein and AD to a prey protein (middle panel). If bait and prey interact, then the reporter gene is expressed (right panel).

A smaller length scale version of this approach has been recently developed called "single-molecule pull-down" or "SimPull" (Jain, 2011), which is a memorable acronym to describe a range of surface-immobilization assays developed by several research groups using fluorescence microscopy to identify different proteins present in a solution from their capacity to bind to target molecules conjugated to a microscope slide or coverslip (Figure 7.6a). For example, the coverslip surface is conjugated first with a reagent such a PEG–biotin, which both serve to block the surface against nonspecific interactions with the glass from subsequent reagents used, and the flow cell is then subjected to a series of washes and incubation steps, first to flow in streptavidin/NeutrAvidin that will then bind to the biotin (see the previous section of this chapter). Biotinylated antibody is then flowed in, which can bind to the free sites on the streptavidin/NeutrAvidin (as discussed earlier, there are four available sites per streptavidin/NeutrAvidin molecule), which are not bound to the biotin attached to the PEG molecules.

This antibody has been designed to have binding specificity to a particular biomolecule to be identified from the cell lysate extract, which is then flowed in. This single-molecule prey protein can then be identified using immunofluorescence using a fluorescently labeled secondary antibody that binds to the Fc region of biotinylated primary antibody (see the previous section of this chapter) or directly if the prey protein has been tagged previously using a fluorescent protein marker. TIRF microscopy (see Chapter 3) can then be used to identify the positions and surface density of bound prey protein. Additional methods involving stepwise photobleaching of fluorescent dyes can also be used subsequently to determine the stoichi-ometry of subunits within a specific prey protein, that is, how many repeating subunits there are in that single molecule (see Chapter 8).

The most popular current technique for determining putative protein–protein is the *yeast two-hybrid assay* (also known as *two-hybrid screening*, the *yeast two-hybrid system*, and Y2H), which can also be adapted to probe for protein–DNA and DNA–DNA interactions. This uses a specific yeast gene to act as a reporter for the interaction between a pair of specific biomolecules, most usually proteins. The expression of a gene, that is, whether it is switched "off" or "on" such that the "on" state results in its DNA sequence being transcribed into an mRNA molecule that in turn is translated into a peptide or protein (see Chapter 2), is nor-mally under the control of transcription factors, which bind to the promoter region of the gene to either suppress transcription or activate it.

In the case of Y2H, an activating transcription factor is used. Activating transcription factors typically consist of two subunits. One subunit is called the "DNA-binding domain (BD)" that binds to a region of the DNA, which is upstream from the promoter itself, called the "upstream activating sequence." Another is called the "activation domain (AD)," which activates a *transcription initiation complex (TIC)* (also known as the *preinitiation complex*), which is a complex of proteins bound to the promoter region, which results in gene expres-sion being switched "on." In YTH, two fusion proteins are first constructed using two separate plasmids, one in which BD is fused to a specific bait protein and another in which AD is fused with a candidate prey protein, such that the BD and AD subunits will only be correctly bound to produce activation of the TIC if the bait and prey proteins themselves are bound together (Figure 7.6b). In other words, the gene in question is only switched "on" if bait and prey bind together.

In practice, two separate plasmids are transformed into yeast cells using different select-able markers (see Section 7.4). Different candidate prey proteins may be tried from a library of possible proteins, generating a different yeast strain for each. The reporter gene is typ-ically selected to encode for an essential amino acid. Therefore, cells that are grown onto agar plates that do not contain that specific essential amino acid will not survive. Colonies that do survive are thus indicative of the bait and prey combination used being interaction partners.

Y2H has implicitly high throughput since it utilizes cell colonies, each ultimately containing thousands of cells. An improvement to the speed of throughput may come from using fluor-escent protein tags on the separate AD and BD fusions. Work in this area is still in the early stages of development but may ultimately enable fluorescence microscopy screening to probe for potential colocalization of the AD and BD proteins, which to some extent competes with the lower-throughput BiFC method (see Chapter 4). There are a number of variants of Y2H, including a *one-hybrid assay* designed to probe protein–DNA interactions, which uses a single fusion protein in which the AD subunit is linked directly to the BD subunit, and a *three-hybrid assay* to probe RNA–protein interactions in which an RNA prey molecule links together the AD and BD subunits. Although optimized in yeast, and thus ideal to probe interactions of eukaryotic proteins, similar systems have now been designed to operate in model bacterial systems.

Methods that use cell lysates for probing the interactions of biomolecules are fast but run the risk that the spatial and temporal context of the native biomolecules is lost. The kin-etics of binding *in vivo* can be influenced by several other factors that may not be present in an *in vitro* assay and so can differ in some cases by several orders of magnitude. Y2H has advantages in that the protein interactions occur inside the live cell, though these may be affected by steric effects of the fusions used but also differences in local concentrations of the putatively interacting proteins being different to the specific region of the cell in which the interaction would naturally occur (as opposed to a specific region of the cell nucleus in the case of Y2H). These issues present problems for systems biology analysis that rely sig-nificantly on the integrity of molecular binding parameters (see Chapter 9).

7.6.4 "SMART" SAMPLE MANIPULATION

Several biophysical techniques are facilitated significantly by a variety of automated sample manipulation tools, which not only increase the throughput of sample analysis but can also enable high-precision measurements, which would be challenging using other more manual methods.

Several systems enable robotized manipulation of samples. At the small length scale, these include automated microplate readers. These are designed to measure typically optical absorption and/or fluorescence emissions over a range of different wavelengths centered on the visible light range, but extending into the UV and IR for spectroscopic quantification similar to traditional methods (see Chapter 3), but here on microliter sample volumes in each specific microplate well. Several microplate well arrays can be loaded into a machine and analyzed. In addition, automation also includes incubation and washing steps for the microplates. At the higher end of the length scale, there are *robotic sample processors*. These cover a range of automated fluid pipetting tasks and manipulation of larger-scale sample vessels such as *microfuge tubes*, flasks, and agar plates for growing cells. They also include crystallization robots mentioned in the previous section.

Light microscopy techniques include several tiers of smart automation. These often comprise user-friendly software interfaces to control multiple hardware such as the power output of bright-field illumination and lasers for fluorescence excitation. These also include a range of optomechanical components including shutters, flipper mounts for mirrors and lenses, *stepper motors* for optical alignment, and various optical filters and dichroic mirrors.

At the high precision end of automation in light microscopy are automated methods for controlling sample flow though a microfluidics flow cell, for example, involving switching rapidly between different fluid environments. Similarly, light microscope stages can be controlled using software interfaces. At a coarse level, this can be achieved by attaching stepper motors to a mechanical stage unit to control lateral and axial (i.e., focusing) movement to micron precision. For ultrasensitive light microscope applications, nanostages are attached to the coarse stage. These are usually based on piezoelectric technology (see Chapter 6) and can offer sub-nanometer precision movements over full-scale deflections up to several hundred microns laterally and axially. Both coarse mechanical stages and piezoelectric nanostages can be utilized to feedback on imaging data in real time. For example, pattern recognition software (see Chapter 8) can be used to identify specific cell types from their morphology in a low-magnification field of view that can then move the stages automatically to align individual cells to the center of the field of view for subsequent higher-magnification investigation.

Long time series acquisitions (e.g., data acquired on cell samples over several minutes, hours, or even days) in light microscopy are often impaired by sample drift, due either to mechanical slippage in the stage due to its own weight or to small changes in external temperatures, resulting in differential thermal expansion/contraction of optomechanical components, and these benefit from stage automation. Pattern recognition software is suitable for correcting small changes due to lateral drift (e.g., to identify the same cell, or group of cells, which have been laterally translated in a large field of view). Axial drift, or *focal drift*, is easier to correct by using a method that relies on total internal reflection. Several commercial "perfect focusing" systems are available in this regard, but the physics of their application is relatively simple: if a laser beam is directed at a supercritical angle through the light microscope's objective lens, then total internal reflection will occur, as is the case for TIRF (see Chapter 3). However, instead of blocking the emergent reflected beam using an appropriate fluorescence emission filter, as is the case for TIRF, this can be directed onto a split photodiode (Figure 7.7). Changes in height of the sample relative to the focal plane are then manifested in a different voltage response from the split photodiode; these can feedback via software control into the nanostage to then move the sample back into the focal plane.

KEY BIOLOGICAL APPLICATIONS: HIGH-THROUGHPUT TOOLS

Biosensing; Molecular separation; High-throughput microscopy.

Automated drift correction in light microscopy

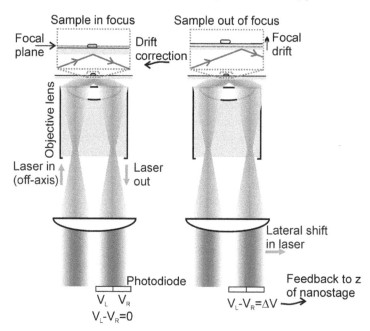

FIGURE 7.7   Automated drift correction. A total internal reflected laser beam can be directed on a split photodiode. When the sample is in focus the voltage from the left (V_L) and right (V_R) halves of the photodiode are equal (a). When the sample is out of focus, (b) the voltages from each half are not equal; this signal can be amplified and fed back into the **z**-axis controller of the nanostage to bring the sample back into focus.

Worked Case Example 7.2:   Using Microfluidics

A microfluidics channel was constructed consisting of a cylindrical pipe with a diameter of 20 μm using a water-based fluid of pH 7.5 with volume flow rate of 18.8 nL min^{-1}.

a State with reasoning whether the flow is laminar or turbulent.

b Derive Poiseuille's law starting only from the definition of viscosity and the assumption of laminar flow, incompressibility and that the fluid is Newtonian. In the case of the aforementioned channel, what is the maximum flow speed?

*   Somewhere along the channel's length, a second side channel joins this main channel from the bottom to continuously feed small volumes of a solution of the protein hemoglobin at pH 5.5 at low speed such that the protein is then swept forward into the main channel. After a given additional length L of the main channel, the mixed protein at pH 7.5 is injected into a microscope flow cell.*

c If the protein has a lateral diffusion coefficient of 7.0×10^{-7} cm^2 s^{-1} estimate, with reasoning, what the minimum value of L should be. Comment on this in light of lab-on-a-chip applications for analyzing a single drop of blood.

(Assume that the density and dynamic viscosity of water are $\sim 10^3$ kg m^{-3} and $\sim 10^{-3}$ Pa · s, respectively.)

Answers

a   The flow rate is given by $\pi a^2 <v>$ where a is the pipe radius and $<v>$ is mean speed of flow. Thus,

$$\langle v \rangle = \left(18.8 \times 10^{-9}\right) / 60 \times 10^{-3}\,s^{-1} / (\pi \times (10 \times 10^{-6})^2\,m^2)$$
$$= 1.0 \times 10^{-3}\,m\,s^{-1} = 1\,mm\,s^{-1}$$

The Reynolds number (R_e) is given by Equation 6.8, where we approximate the equivalent length parameter to the diameter of the pipe and the speed to the mean speed of flow; thus,

$$R_e = (1 \times 10^3) \times (1 \times 10^{-3}) \times (20 \times 10^{-6}) / (1 \times 10^{-3}) = 0.02$$

This is <2100, and so the flow is laminar.

b   Consider a cylindrical pipe shell of radius a with incremental axial length δx. The cylinder is full of fluid with pressure p at one end and $(p + \delta p)$ at the other end. Consider then a solid cylindrical shell of fluid of speed v, which is inside the pipe and has a radius z, and a shell thickness δz such that $z \le a$. The net force on the fluid shell is the sum of pressure and viscous drag components, which is zero:

$$\left((p + \delta p - p)\pi z^2 - \eta(2\pi z \delta x)\right)\frac{dv}{dr} = 0$$

$$\therefore \frac{dv}{dr} = \frac{r}{2\eta}\frac{\partial p}{\partial x}$$

Using boundary conditions of $v = 0$ at $z = a$ gives the required solution (i.e., a parabolic profile). The maximum speed occurs when $dv/dr = 0$, in which $z = 0$ (i.e., along the central axis of the pipe). The volume flow rate is given by

$$Q = \int_0^a v \cdot (2\pi r)\,dr$$

which gives the required solution. The mean speed is given by $Q/\pi a^2$, and it is easy to then show from the aforementioned that the maximum speed is twice the mean speed (which here is thus 2 mm s⁻¹).

c   Assume mixing is solely through diffusion across the streamlines. Thus, to be completely mixed with the fluid in the channel, a molecule needs to have diffused across the profile of streamlines, that is, to diffuse across the cross-section, a distance in one dimension equivalent to the diameter or 20 μm. In a time Δt, a protein molecule with diffusion coefficient D will diffuse a root mean square displacement of $\sqrt{(2D\Delta t)}$ in one dimension (see Equation 2.12). Equating this distance to 20 μm and rearranging indicates that

$$\Delta t = (20 \times 10^{-6}\,m)^2 / \left(2 \times 7 \times 10^{-7} \times 10^{-4}\,m^2 s^{-1}\right) = 20\,s.$$

The mean speed is 1 mm s⁻¹, therefore, the minimum pipe length should be ~20 mm or 2 cm. Comment: this is a surprisingly large length compared to that of the diameter of, say, a single drop of blood by ca. an order of magnitude. For proper mixing to occur in lab-on-a-chip devices, it requires either larger channel lengths (i.e., large chips) or, better, additional methods introduced into the mixing such as chevron structures to cause turbulence mixing.

7.7 CHARACTERIZING PHYSICAL PROPERTIES OF BIOLOGICAL SAMPLES IN BULK

There are several methods that enable experimental measurements on relatively macro-scopic volumes of biological material that use, at least in part, biophysical techniques but whose mainstream applications are in other areas of the biosciences, for example, test tube length scale level experiments to measure the temperature changes due to biochemical reactions. Also though, bulk samples of biological tissue can be probed to generate ensemble average data from hundreds or thousands of cells of the same type in that tissue, but also encapsulating the effect from potentially several other cell types as well as from extracellular material. This may therefore seem like a crude approach compared to the high spatial precision methods utilizing optical techniques discussed earlier in this chapter; however, what these methods lack in being able to dissect out some of the finer details of heterogeneous tissue features they make up for in generating often very stable signals with low levels of noise.

7.7.1 CALORIMETRY

One of the most basic biophysical techniques involves measuring heat transfer in biological processes *in vitro*, which ultimately may involve the absorption and/or emission of IR photons. The fact of calorimetry being a very established method takes nothing away from its scientific utility; in fact, it demonstrates a measure of its robustness. Changes in *thermodynamic potentials*, or *state variables*, such as *enthalpy* (*H*), may be measured directly experimentally. Other thermodynamic potentials that are more challenging to measure directly such as *entropy* (*S*), or the *Gibbs free energy* (*G*) that depends on entropy, need to be inferred indirectly from more easily measurable parameters, with subsequent analysis utilizing the first-order *Maxwell's relations* of thermal physics to relate the different thermodynamic potentials.

The most quantifiable parameter is sample temperature, which can be measured using specifically calibrated chambers of precise internal volumes, which typically include an integrated stirring device with chamber walls maximally insulated against heat flow to generate an *adiabatic* measuring system. Time-resolved temperature changes inside the chamber can easily be monitored with an electrical thermometer using a *thermistor* or *thermocouple*. Inside, a biological sample might undergo chemical and/or physical transitions of interest that may be *exothermic* or *endothermic*, depending on whether or not they generate or absorb heat, and enthalpic change can be very simply calculated from the change in temperature and knowledge of the specific heat capacity of the reactant mixture.

Isothermal titration calorimetry (*ITC*) is often used as an alternative. Here an adiabatic jacket made from a thermally highly conductive alloy is used to surround the sample cell, while an identical reference cell close enough to transfer heat very efficiently just to the sample cell contains a reference heater whose output is adjusted dynamically so as to maintain a constant measured temperature in the sample chamber. ITC has been used to study the kinetics and stoichiometry of reactants and products through monitoring estimated changes in thermodynamic potentials as a function of the titration of ligands injected into the sample cell that contains a suitable reactant, for example, of ligand molecules binding to proteins or DNA in the sample solution.

The heat transfer processes measured in biology are most usually due to a biochemical reaction, but potentially also involve phase transitions. For example, different mixtures of lipids may undergo temperature-dependent phase transition behavior that gives insight into the architecture of cell membranes. The general technique used to detect such phase transitions operates using similar isothermal conditions as for ITC and is referred to as *differential scanning calorimetry*.

7.7.2 ELECTRICAL AND THERMAL PROPERTIES OF TISSUES

Biological tissue contains both free and bound electrical charges and so has both electrically conductive and dielectric characteristics, which varies widely between different tissue types compared to other biophysical parameters. A comparison of, for example, the attenuation coefficients of clinical x-rays used in computer-assisted tomography (CAT)/computerized tomography (CT) scanning, a biophysical workhorse technology in modern hospitals (see the following section of this chapter), between the two most differing values from different tissues in the human body (fat and bone), indicates only a difference by a factor ~2. Blood and muscle tissue essentially have the same value, thus not permitting discrimination at all between these tissue types on x-ray images. The resistivity of different tissue types, however, varies by over two orders of magnitude and so offers the potential for much greater discrimination, in addition to a frequency dependence on the electrical impendence permitting even finer metrics of discrimination.

Electrical impedance spectroscopy (*EIS*), also known as *dielectric spectroscopy*, in its simplest form consists of electrodes attached across a tissue sample using sensitive amplification electronics to measure the impedance response of the tissue with respect to frequency of the applied AC voltage potential between the electrodes, which has been applied to a variety of different animal tissues primarily to explore the potential as a diagnostic tool to discriminate between normal and pathogenic (i.e., diseased) tissues. The cutting edge of this technology is the biomedical tool of *tissue impedance tomography*, discussed later in this chapter. A good historical example of EIS was in the original investigations of the generation of electrical potentials of nerve fibers utilizing the relatively large *squid giant axon*. The axon is the central tube of nerve fibers, and in squid these can reach huge diameters of up to 1 mm, making them relatively amenable for the attachment of electrodes, which enabled the electrical action potential of nervous stimuli to first be robustly quantified (Hodgkin and Huxley, 1952). But still these days similar EIS experiments are made on whole nerve fibers, albeit at a smaller length scale than for the original squid giant axon experiments, to probe the effect of disease and drugs on nervous conduction, with related techniques of *electrocardiography* and *electroencephalography* now accepted as clinical standards.

Different biological tissues also have a wide range of thermal conductivity properties. Biophysical applications of these have included the use of *radio frequency (RF) heating*, also known as *dielectric heating* in which a high-frequency alternating radio or microwave heats a dielectric material through an induced dipole resonance; this is essentially how microwave ovens work. This has been applied to the specific *ablation* of tissue, for example, to destroy diseased/dying tissue in the human body and to enable reshaping of damaged collagen tissue.

7.7.3 BULK MAGNETIC PROPERTIES OF TISSUES

Biological tissues have characteristic magnetic susceptibility properties, which is significantly influenced by the presence of blood in the tissue due to the iron component of hemoglobin in red blood cells (see Chapter 2), but can also be influenced by other factors such as the presence of myelin sheaths around nerve fibers and of variations in the local tissue biochemistry. The technique of choice for probing tissue magnetic properties involves using *magnetic resonance* to typically map out the variation of susceptibility coefficients χ_m across the extent of the tissue:

(7.18)
$$M = \chi_m H$$

where
   M is the magnetization of the tissue (i.e., the magnetic dipole moment per unit volume)
   H is the magnetic field strength

The technique of *magnetic resonance imaging* (*MRI*) is described in fuller detail later in this chapter.

7.7.4 TISSUE ACOUSTICS

The measure of resistance to acoustic propagation via *phonon* waves in biological tissue is the *acoustic impendence* parameter. This is the complex ratio of the *acoustic pressure* to the volume flow rate. The acoustic impendence in different animal tissues can vary by two orders of magnitude, from the lungs at the low end (which obviously contain significant quantities of air) and the bone at the high end, and is thus a useful physical metric for the discrimination of different tissue types, especially useful at the boundary interface of different tissue types since these often result in an *acoustic mismatch* that is manifested as a high *acoustic reflectance*, whose reflection signal (i.e., *echo*) can thus be detected. For example, a muscle–fat interface has a typical reflectance of only ~1%; however, a bone–fat interface is more like ~50%, and any soft water-based tissue with air has a reflectance of ~99.9%. This is utilized in various forms of *ultrasound imaging*.

KEY BIOLOGICAL APPLICATIONS: BULK SAMPLEBIOPHYSICSTOOLS

Multiple simple, coarse but robust mean ensemble average measurements on a range of different tissue samples.

KEY POINT 7.7

Bulk tissue measurements do not allow fine levels of tissue heterogeneity to be investigated, in that as an ensemble technique, their spatial precision is ultimately limited by the relatively macroscopic length scale of the tissue sample and any inference regarding heterogeneity in general is done indirectly through biophysical modeling; however, they are often very affordable techniques and relatively easy to configure experimentally and generate often very stable measurements for several different ensemble physical quantities, many of which have biomedical applications and can assist greatly in future experimental strategies of using more expensive and time-consuming techniques that are better optimized toward investigating heterogeneous sample features.

7.8 BIOMEDICAL PHYSICS TOOLS

Many bulk tissue techniques have also led to developments in biomedically relevant biophysical technologies. Whole textbooks are dedicated to specific tools of medical physics, and for expert insight of how to operate these technologies in a clinical context, I would encourage the reader to explore the *IPEM website* (www.ipem.ac.uk), which gives professional and up-to-date guidance of publications and developments in this fast-moving field. However, the interface between medical physics, that is, that performed in a clinical environment specifically for medical applications, and biophysics, for example for researching questions of relevance to biological matter using physics tools and techniques, is increasingly blurred in the present day due primarily to many biophysics techniques having a greater technical precision at longer length scales than previously, and similarly for medical physics technologies experiencing significant technical developments in the other direction of smaller-scale improvements in spatial resolution in particular, such that there is now noticeable overlap between the length and time scale regimes for these technologies. A summary of the principle of biophysical techniques relevant to biomedicine is therefore included here.

7.8.1 MAGNETIC RESONANCE IMAGING

MRI is an example of *radiology*, which is a form of imaging used medically to assist in diagnosis. MRI uses a large, cooled, electromagnetic coil of diameter up to ~70 cm, which can generate a high, stable magnetic field at the center of the coil in the range ~1–7 T (which compares with the Earth's magnitude field strength of typical magnitude ~50 μT). The physical principles are the same as those of NMR in which the nuclei of atoms in a sample absorb energy from the external magnetic field (see Chapter 5) and reemit electromagnetic radiation at an energy equal to the difference in nuclei spin energy states, which is dependent on the local physicochemical environment surrounding that atom and is thus a sensitive metric for probing tissue heterogeneity.

By moving the sample perpendicularly to the xy lateral sampling plane along the central z-axis of the scanner, full 3D xyz spatial maps can be reconstructed, with total scan times of a few tens of minutes. Diagnostic MRI can be used to discriminate relatively subtle differences in soft tissues that have similar x-ray attenuation coefficients and thus can reveal tissue heterogeneities not observed using CAT/CT scanning (see in the following text), for example, to diagnose deep tissue mechanical damage as well as small malignant tumors in a soft tissue environment.

MRI can also be used for *functional imaging*, defined as a method in biomedical imaging that can detect dynamic changes in metabolism. For MRI, this is often referred to as *functional MRI (fMRI)*. The best example of this is in monitoring of *blood flow*, for example, through the heart and major blood vessels, and to achieve this, a *contrast reagent* is normally applied to improve the discrimination of the fast-flowing blood against the soft tissues of the walls of the heart and the blood vessels, usually a paramagnetic compound such as a *gadolinium*-containing compound, which can be injected into the body via a suitable vein.

The spatial resolution of the best conventional MRI is limited to a few tens of microns and so in principle is capable of resolving many individual cell types. However, a new research technique called nitrogen vacancy MRI is showing potential for spatial resolution at the nanometer scale (see Grinolds et al., 2014), though it is at too early a stage in development to be clinically relevant.

7.8.2 X-RAYS AND COMPUTER-ASSISTED (OR COMPUTERIZED) TOMOGRAPHY

Ionizing radiation is so called because it carries sufficient energy to remove electrons from atomic and molecular orbitals in a sample. Well-known examples of ionizing radiation include alpha particles (i.e., helium nuclei) and beta radiation (high-energy electrons), but also x-rays (photons of typical wavelength $\sim 10^{-10}$ m generated from electronic orbital transitions) and gamma rays (higher energy photons of typical wavelengths $< 10^{-11}$ m generated from atomic nucleus energy state transitions), which are discussed in Chapter 5. All are harmful to biological tissue to some extent. X-rays were historically the most biomedically relevant, in that hard tissues such as the bone in particular have significantly larger attenuation coefficients for x-rays compared to that of soft tissues, and so the use of x-rays in forming relatively simple 2D images of the transmitted x-rays through a sample of tissue has grown to be very useful and is the standard technique used for clinical diagnosis.

Thus, T-rays (i.e., terahertz radiation) can be used in a similar way to x-rays for discriminating between soft and hard biological tissues (see Chapter 5). However, T-rays have a marginal advantage when specifically probing fine differences in water content between one tissue and another. These differences have been exploited for the detection of forms of epithelial cancer. But also, T-rays have been applied in generating images of the teeth. However, the widespread application of T-rays biomedically is more limited because of the lack of availability of commercial, portable T-ray sources and so is currently confined to research applications.

CAT or *CT*, also known as *computerized tomography*, involves scanners that utilize x-ray imaging but scan around a sample using a similar annulus scanner/emitter geometry to MRI scanners, resulting in a 2D x-ray *tomogram* of the sample in the lateral xy plane. As with MRI, the sample can be moved perpendicularly to the xy sampling plane along the central z-axis of the scanner, to generate different 2D tomograms at different incremental values of z, which can then be used to reconstruct full 3D xyz spatial maps of x-ray attenuation coefficients using offline interpolation software, representing a 3D map of different tissue features, with similar scan times. The best spatial resolution of commercial clinical systems is, in principle, a few hundred microns, in other words limited to a clump of a few cells. This is clinically very useful for diagnosing a variety of different disorders, for example, *cancer*, though in practice the smallest tumor that can be typically detected reliably in a soft tissue environment is ~ 2 cm in diameter.

Improvements to detection for use in dynamic functional imaging can be made with contrast rearrangements in a similar way to MRI. A good example of CT/CAT functioning imaging is in diagnosing gut disorders. These investigations involve the patient swallowing a suitable x-ray contrast reagent (e.g., a *barium meal*) prior to scanning.

7.8.3 SINGLE-PHOTON EMISSION CT AND POSITRON EMISSION TOMOGRAPHY

Nuclear imaging involves the use of gamma ray detection instead of x-rays, which are emitted following the radioactive decay of a radionuclide (also known as *tracer* or *radioisotope*), which can be introduced into the human body to bind to specific biomolecules. They are valuable functional imaging technologies. *Single-photon emission CT* (*SPECT*) works on similar 2D scanning and 3D reconstruction principles to CAT/CT and MRI scanning. Although there are several different radionuclides that can be used, including *iodine-123, iodine-131*, and *indium-111*, by far, the most commonly used is *technetium-99m*. This has a half-life of ~6 h and has been applied to various diagnostic investigations, including scanning of glands, the brain and general nerve tissue, white blood cell distributions, the heart and the bone, with a spatial resolution of ~1 cm.

There is an issue with the global availability of technetium-99m and, in fact, with a variety of other less commonly used radionuclides applied to biomedicine, referred to as the *technetium crisis;* in 2009 two key nuclear research reactors, in the Netherlands and Canada, were closed down, and these were responsible for generating ca. two-thirds of the global supply of *molybdenum-99*, which decays to form technetium-99m. There are other technologies being investigated to plug this enormous gap in supply, for example, using potentially cheaper linear particle accelerators, but at the time of writing, the sustainable and reliable supply of technetium-99m in particular seems uncertain.

Positron emission tomography (*PET*) works on similar gamma ray detection principles to SPECT, but instead utilizes *positron* radionuclide emitters to bring about gamma ray emission. Positrons are the *antimatter* equivalent of electrons that can be emitted from the radioactive decay of certain radionuclides, the most commonly used being carbon-11, nitrogen-13, oxygen-15, fluorine-18, and rubidium-82 (all of which decay with relatively short half-lives in the range ~1–100 min to emit positrons), which can be introduced into the human body to bind to specific biomolecules in a similar way to radionuclides used in SPECT. Emitted positrons, however, will annihilate rapidly upon interaction with an electron in the surrounding matter, resulting in the emission of two gamma ray photons whose directions of propagation are oriented at 180° to each other. This straight line of coincidence is particularly useful, since by detecting these two gamma rays simultaneously (in practice requiring a detector sampling time precision of $<10^{-9}$ s), it is possible to determine very accurately the *line of response* for the source of the positrons, since this line itself is oriented randomly, and so by intersecting several such lines, the source of the emission in 3D space can be determined, with a spatial resolution better than that of SPECT by a factor of ~2.

Time-of-flight PET (TOF PET) determines the difference δt in the arrival times of the two gamma ray photons generated from positron-electron annihilation produced at a 180° orientation. Multiple detectors are placed in a ring around the biological tissue sample (which can be a live human, since this is a very valuable new medical imaging technology), which has been doped with a positron emitter inside the scanner. To pinpoint the source of emission from the sample, with coincident signals detected at a rate of a few hundred MBq for typical doped samples (where 1 Bq is the SI unit of radioactivity corresponding to 1 disintegration per second) and sampled at GHz rates, the spatial displacement x is $c.\delta t/2$ where δt is the detection of coincidence timing resolution and c is the speed of light.

The rate of random coincidences k_2 from two identical gamma ray detectors oriented at 180° from each other, each with a random single detector rate k_1 during a sample time interval Δt is

(7.19)
$$k_2 = 2k_1^2 \Delta t$$

Thus, coincidence detection can result in a substantial reduction in random detection error. If the true signal rate from coincidence detection is k_S, then the effective single-to-noise ratio (SNR) is

(7.20)
$$\text{SNR} = \frac{k}{\sqrt{k_s + nk_2}}$$

Here, $n = 2$ for *delayed-coincidence methods*, which are the standard coincidence detection methods for PET involving one of the detector signals being held for several sampling time windows (up to $\sim 10^{-7}$ s in total), while the signal in the other detector is then checked. Recent improvements to this method involve parallel detector acquisition (i.e., no imposed delay) for which $n = 1$. For both methods, the k_S is much higher than k_2 and so the SNR scales roughly as $\sqrt{k_S}$, whereas for SPECT, this scales more as $k_S/\sqrt{k_1}$, which in general is $<\sqrt{k_S}$. Also, the signal rate for a single radionuclide atom is proportional to the reciprocal of its half-life, which is greater for PET than for SPECT radionuclides. These factors combined result in PET having a typical SNR that is greater than that of SPECT often by more than two orders of magnitude.

PET can also be combined with CAT/CT and MRI in some research development scanning systems, called *PET-CT* and *PET-MRI*, which have enormous future diagnostic potential in being able to overlay images from the same tissue obtained using the different techniques, but the cost of the equipment at present is prohibitive.

7.8.4 ULTRASOUND TECHNIQUES

The measurement of acoustic impedances using an ultrasound probe in direct acoustical contact with the skin is now commonplace as a diagnostic tool, for example, in monitoring the development of a fetus in the womb, detecting abnormalities in the heart (called an *echocardiogram*), diagnosing abnormal widening (*aneurysms*) of major blood vessels, and probing for tissue defects in various organs such as the liver, kidneys, testes, ovaries, pancreas, and breast. Deep tissue ultrasound scanning can also be facilitated by using an extension to enable the sound emitter/probe to get physically closer to the tissue under investigation.

A variant to this technique is *Doppler ultrasound*. This involves combined ultrasound acoustic impedance measurement with the *Doppler effect*. This results in the increase or decrease of the wavelength of the ultrasound depending on the relative movement of the propagation medium and so is an ideal biophysical tool for the investigation of the flow of blood through different chambers in the heart.

Photoacoustic imaging is another modification of standard ultrasound, using the *photoacoustic effect*. Here, absorbed light in a sample results in local heating that in turn can generate acoustical phonons through *thermal expansion*. The tissues of relevance absorb light strongly and have included investigations of skin disorders via probing the pigment *melanin*, as well as blood oxygenation monitoring since the oxygenated heme group in the hemoglobin molecule has a different absorption spectrum to the deoxygenated form. The technique can also be extended to RF electromagnetic wave absorption, referred to as *thermoacoustic imaging*.

7.8.5 ELECTRICAL SIGNAL DETECTION

The biophysical technique of using dynamic electrical signals from the electrical stimuli of heart muscle tissue, ideally from using up to 10 skin-contact electrodes both in the vicinity of the heart and at the peripheries of the body at the wrists and ankles, to generate an *electrocardiogram (EKG* or ECG) is a standard, cost-effective, and noninvasive clinical tool capable of assisting in the diagnosis of several heart disorders from the characteristic voltage–time data signatures. The *electroencephalogram* is an equivalent technique that uses multiple surface electrodes around the head to investigate disorders of the brain, most importantly for *epilepsy* diagnosis.

A less commonly applied technique is *electromyography*. This is essentially similar to ECG, but applied to *skeletal muscle* (also called "striated muscle"), which is *voluntarily controlled muscle*, mostly attached to the bones via collagen fibers called "tendons." Similarly *electronystagmography* is a less common tool, involving electrical measurements made in the vicinity of the nose, which is used in investigating the nerve links between the brain and the eyes.

7.8.6 INFRARED IMAGING AND THERMAL ABLATION

IR imaging (also known as *thermal imaging*, or *thermography*) utilizes the detection of IR electromagnetic radiation (over a wavelength range of ~9–14 μm) using thermal imaging camera detectors. An array of pixels composed of cooled *narrow gap semiconductors* are used in the most efficient detectors. Although applied to several biomedical investigations, the only clearly efficacious clinical application of IR imaging has been in *sports medicine* to explore irregular blood flow and *inflammation* around the muscle tissue.

Thermal ablation is a technique using either localized tissue heating via microwave or focused laser light absorption (the latter also called *laser ablation*), resulting in the removal of that tissue. It is often used in combination with *endoscopy* techniques, for example, in the removal of plaque blockages in major blood vessels used by the heart.

7.8.7 INTERNALIZED OPTICAL FIBER TECHNIQUES

Light can be propagated through *waveguides* in the form of narrow *optical fibers*. Cladded fibers can be used in standard endoscopy, for example, imaging the inside of the gut and large joints of the body to aid visual diagnosis as well as assisting in microsurgical procedures. Multimode fibers stripped of any cladding material can have a diameter as small as ~250 μm, small enough to allow them to be inserted into medical devices such as catheters and syringe needles and large enough to permit a sufficient flux of light photons to be propagated, for either detection or treatment.

Such thin fibers can be inserted into smaller apertures of the body (e.g., into various blood vessels) generating an internalized light source that can convey images from the scattered light of internal tissue features as well as allow propagating high intensity laser light for *laser microsurgery* (using localized *laser ablation of tissues*). The light propagation is not dependent upon external electromagnetic signals, and so optical fibers can be used in conjunction with several other biophysical techniques mentioned previously in this chapter, including MRI, CAT/CT scanning, and SPECT/PET.

7.8.8 RADIOTHERAPY METHODS

Radiation therapy (also known as *radiotherapy*) uses ionizing radiation to destroy *malignant cells* (i.e., cells of the body that divide and thrive uncontrollably that will give rise to cancerous tissue). The most common (but not exclusive) forms of ionizing radiation used are x-rays. Ionizing radiation results in damage to cellular DNA. The mechanism is thought to involve the initial formation of free radicals (see Chapter 2) generated in water from the absorption of the radiation, which then react with DNA to generate breaks, the most pernicious to the cell being *double-strand breaks* (*DSBs*), that is, localized breaks to both helical strands of the DNA.

DSBs are formed naturally in all cells during many essential processes that involve topological changing of the DNA, for example, in DNA replication, but these are normally very transient. Long-lived DSBs are highly reactive free ends of DNA, which have the potential for incorrectly *religating* to different parts of the DNA sequence through binding to DSBs in potentially a completely different region of DNA if it is accessible in the nucleus, which could have highly detrimental effects on the cell. Cellular mechanisms have unsurprisingly evolved to repair DSBs, but a competing cellular strategy, if repair is insufficient, is simply to destroy the cell by triggering *cell death* (in eukaryotes this is through a process of apoptosis, and prokaryotes have similar complex mechanisms such as the *SOS response*).

The main issue with radiotherapy is that similar doses of ionizing radiation affect normal and cancerous cells equally. The main task then in successful radiotherapy is to minimize the relative dose between normal and cancerous tissue. One way to achieve this is through specific internal localization of the ionizing radiation source. For example, iodine in the blood is taken up preferentially by the thyroid gland. Thus, the iodine-131 radionuclide, a positron emitter generating gamma rays used in PET scanning, can be used to treat thyroid cancer. *Brachytherapy*, also known as *internal radiotherapy* or *sealed source radiotherapy*, uses a sealed ionizing radiation source that is placed inside or next to a localized cancerous tissue (e.g., a tumor). *Intraoperative radiotherapy* uses specific surgical techniques to position an appropriate ionizing radiation source very close to the area requiring treatment, for example, in *intraoperative electron radiation therapy* used for a variety of different tissue tumors.

A more common approach, assuming the cancer itself is suitably localized in the body to a tumor, is to maximize the dose of ionizing radiation to the cancerous tissue relative to the surrounding normal tissue by using a narrow x-ray beam centered on the tumor and then at subsequent x-ray exposures to use a different relative orientation between the patient and the x-ray source such that the beam still passes through the tumor but propagates through a different region of normal tissue. Thus, this is a means of "focusing" the x-ray beam by time sharing its orientation but ensuring it always passes through the tumor. Such treatments are often carried out over a period of several months, to assist the regrowth of normal surrounding tissue damaged by the x-rays.

KEY BIOLOGICAL APPLICATIONS: BIOMEDICAL PHYSICS TOOLS

Multiple health-care diagnostic and treatment applications.

7.8.9 PLASMA PHYSICS IN BIOMEDICINE

Plasma medicine (not to be confused with *blood plasma*, which is the collection of essential electrolytes, proteins, and water in the blood) is the controlled application of physical plasmas (i.e., specific *ionized gases* induced by the absorption of strong electromagnetic radiation) to biomedicine. A standard clinical use of such plasmas is the rapid *sterilization* of medical implements without the need for bulky and expensive *autoclave* equipment that rely on superheated steam to destroy biological, especially microbial, contaminants. Plasmas are also used to modify the surfaces of artificial *biomedical implants* to facilitate their successful uptake in native tissues. In addition, therapeutic uses of plasmas have involved improving wound healing by localized destruction of *pathogenic microbes* (i.e., *nasty germs* that can cause wound *infections*).

Worked Case Example 7.3: PET Scanning

A time-of-flight positron emission tomography (TOF-PET) scan was performed on a biological tissue sample at 2.7 GHz sampling rate using a delayed-coincidence method, with the uncertainty in the difference of arrival times being ~10% of the sampling time.

a   Estimate the spatial resolution for localizing the position–electron annihilation events in the sample.
b   To explore the sensitivity of the TOF-PET instrument, the doping of the sample was lowered to produce a rate of coincident detection from two gamma ray detectors that was only 1 ± 0.1 MBq and the rate of random signal detection from and single gamma ray detector was 300,000 ± 80,000 counts per second. What is the signal-to-noise (SNR) in light of the precision of its measurement for true coincident signal detection?

Answers

a   Since the spatial displacement $x = c.\delta t/2$ the spatial precision $\Delta x = c.\Delta(\delta t)/2$. The error in the arrival time measurement $\Delta(\delta t)$ here is 10% of the sampling time $(1/f)$ where f is the sampling frequency so:

$$\Delta x = 3 \times 10^8 \times 0.1 \times (1/(2.7 \times 10^9)/2 = 5.6 \times 10^{-3} \text{ m or } 5.6 \text{ mm}$$

b   This question illustrates the importance of properly calculating errors from the known precision of multiple measurement parameters. The $SNR = k_s/\sqrt{(k_s+nk_2)}$ where k_s is the signal rate of the coincidence detection, $n = 2$ for a delayed confidence method and $k_2 = 2k_1$. Δt is the rate of random detected coincidences given a random detection rate k_1 from a single detector. Under normal applications, k_2 is much smaller than k_s and so SNR is $\sim \sqrt{k_s}$; however, here k_s and k_2 have similar orders of magnitude and so both need to be accounted for. So

$SNR = 1 \times 10^6/\sqrt{(1 \times 10^6 + (2 \times 3 \times 10^5))} = 790.568$ counts/s to three decimal places. But we need to quote this properly considering the precision by which we can measure the SNR,
$\Delta(SNR)$, which is given by the error on its measurement considering the error on both k_s and k_2. Using the general method of differentials to estimate errors, this indicates that:

$$\Delta(SNR) = \sqrt{((\Delta k_s.\partial(SNR)/\partial k_s)^2+(\Delta k_2.\partial(SNR)/\partial k_2)^2)}$$

The partial derivatives can be worked through as:

$$\partial(SNR)/\partial k_s = (1/\sqrt{(k_s+nk_2)}) - (1/2)k_s/(k_s+nk_2)^{3/2}$$
$$= (1/\sqrt{(1 \times 10^6 + (2 \times 3 \times 10^5))} -(0.5 \times 1 \times 10^6)/((1 \times 10^6 + (2 \times 3 \times 10^5))^{1.5}$$
$$= 5.4 \times 10^{-4} \text{ s per count}$$

$$\partial(SNR)/\partial k_2 = -(1/2)nk_s/(k_s+nk_2)^{3/2} = -(0.5 \times 2 \times 1 \times 10^6)/((1 \times 10^6 + (2 \times 3 \times 10^5))^{1.5}$$
$$= -4.9 \times 10^{-4} \text{ s per count. So:}$$

$$\Delta(SNR) = \sqrt{((0.1 \times 10^6 \times 5.4 \times 10^{-4})^2 + (80 \times 10^3 \times (-4.9 \times 10^{-4}))^2)}$$
$$= 67 \text{ counts/s}$$

So:

$$SNR = 791 \pm 67 \text{ counts/s}$$

7.9 SUMMARY POINTS

- There is a plethora of well-characterized chemical methods to specifically conjugate biomolecules to other biomolecules or to nonliving substrates with high affinity.
- Thin model organisms of nematode flatworms and zebrafish have proved particularly useful in generating *in vivo* biological insight from light microscopy techniques.
- Molecular cloning tools have developed mainly around model microbial organisms and can be used to genetically modify DNA and insert it into foreign cells.
- High-quality crystal formation is in general the bottleneck in crystallography.

- Microfluidics has transformed our ability to monitor the same biological sample under different fluid environments.
- Bulk tissue measurements can provide useful ensemble average information and have led to several developments in biomedical techniques.

QUESTIONS

7.1 A polyclonal IgG antibody that had binding specificity against a monomeric variant of GFP was bound to a glass coverslip surface of a microscope flow cell by incubation, and then GFP was flowed through the flow cell and incubated, and any unbound GFP was washed out. The coverslip was then imaged using TIRF, which results in bright distinct spots visible that were sparsely separated on the camera image much greater than their own point spread function width. Roughly 60% of these spots had a total brightness of ~5000 counts on the camera used, while the remaining 40% had a brightness of more like ~10,000 counts. Explain with reasoning what this could indicate in light of the antibody structure. (For more quantitative discussion of this type of *in vitro* surface-immobilization assay, see Chapter 8.)

7.2 What are the ideal properties of a model organism used for light microscopy investigations? Give examples. Why might a biofilm be a better model for investigating some bacteria than a single cell? What problems does this present for light microscopy, and how might these be overcome?

7.3 Why does it matter whether a genetically encoded tag is introduced on the C-terminus or N-terminus of a protein? Why are linkers important? What are the problems associated with nonterminus tagging?

7.4 Outline the genetic methods available for both increasing and decreasing the concentration of specific proteins in cells.

7.5 Image analysis was performed on distinct fluorescent spots observed in Slimfield images of 200 different cells in which DNA replication was studied. In bacteria, DNA replication is brought about by a structure of ~50 nm diameter called the replisome that consists of at least 11 different proteins, several of which are used in an enzyme called the DNA polymerase. One protein subunit of the DNA polymerase called ε was fused to the yellow fluorescent protein YPet. Stepwise photobleaching of the fluorescent spots (see Chapter 8) indicated three e-YPet molecules per replication fork. In this cell strain, the native gene that encoded for the ε protein was deleted and replaced entirely with ε fused to YPet. It was found that there was a 1/4 probability for any randomly sampled cell to contain ~80 e-YPet molecules not associated with a distinct replisome spot and the same probability that a cell contained ~400 e-YPet molecules per cell.

   a   Estimate the mean and standard error of the number of e-YPet molecules per cell. In another experiment, a modified cell strain was used in which the native gene was not deleted, but the e-YPet gene was instead placed on a plasmid under control of the lac operon. If no IPTG was added, the mean estimated number of e-YPet molecules per cell was ~50, and using stepwise photobleaching of the fluorescent replisome spots, this suggested only ~1–2 ε-YPet molecules per spot. When excess IPTG was added, the stepwise photobleaching indicated ~3 molecules per spot and the mean number of nonspot e-YPet molecules per cell was ~850.

   b   Suggest explanations for these observations.

7.6 Live bacteria were immobilized to a glass microscope coverslip in a water-based medium expressing a fluorescently labeled cytoplasmic protein at a low rate of gene expression resulted in a mean molar concentration in the cytoplasm of C. The protein was found to assemble into one or two distinct cytoplasmic complexes in the cell of mean number of monomer protein subunits per complex given by P.

   a   If complexes are half a cell's width of 0.5 μm from the coverslip surface and the depth of field of the objective lens used to image the fluorescence and generate a

TIRF evanescent excitation field is D nm, generate an approximate expression for the SNR of the spots using such TIRF excitation.

b   Explain under what conditions it might be suitable to use TIRF to monitor gene expression of these proteins.

7.7   If two crystal structures, A and B, for a given protein molecule M are possible that both have roughly the same interfacial energies and saturation values, but structure B has a more open conformation such that the mean volume per molecule in the crystal is greater than that in structure A by a factor of ~2, explain with reasoning what relative % proportion by number of crystals one might expect when crystals are grown spontaneously from a purified supersaturated solution of M.

7.8   Platinum wire of length 20 cm and diameter 75 μm was wound around a tungsten electrical resistance heating filament that was then heated to evaporate all of the platinum 3 cm away from the surface of a glass coverslip surface of size 22 × 22 mm under high vacuum to coat the surface in a thin layer of platinum for manufacturing an optical filter. The coated coverslip was used in an inverted fluorescence microscope flow cell for detecting a specific protein that was labeled with a single fluorescent dye molecule, conjugated to the coated coverslip surface. If the brightness of a single fluorescently labeled protein on an identical coverslip not coated in platinum was measured at ~7600 counts under the same imaging conditions, estimate with reasoning what range of spot brightness values you might observe for the case of using the platinum-coated coverslip. (Assume that the optical attenuation is roughly linear with thickness of the platinum layer, equivalent to ~15% at 50 nm thickness, and wavelength is independent across the visible light spectrum.)

7.9   A leak-free horizontal microfluidics device of length 15 mm was made consisting of three cylindrical pipes each of length 5 mm with increasing diameters of 10, 20, and 30 μm attached end to end. If the flow was gravity driven due to a reservoir of fluid placed 50 cm above the entrance of the flow cell connected via low-friction tubing, estimate the time it takes for a 1 μm bead to flow from one end of the flow cell to the other.

7.10   A competing flow-cell design to Question 7.9 was used, which consisted of three pipes of the same diameters as the previous but each of length 15 mm connected this time in parallel. How long will a similar bead take to flow from one end of the flow cell to the other?

7.11   High-throughput methods for measuring protein–protein interaction kinetics using cell lysate analysis *in vitro* can generate large errors compared to the *in vivo* kinetics. Give a specific example of a protein–protein interaction whose kinetics can be measured inside a living cell using a biophysical technique. What are the challenges to using such assays for systems biology? Suggest a design of apparatus that might provide high-throughput measurement and automated detection?

7.12   In internal radiotherapy for treating a thyroid gland of a total effective diameter of 4 cm using a small ionizing radiation source at its center to treat a central inner tumor with a diameter of 1 cm, estimate with reasoning and stating assumptions what proportion of healthy tissue will remain if 99% of the tumor tissue is destroyed by the treatment.

7.13   In developing a simple microfluidics-based pH sensor for biological samples, two parallel input channels, one containing a biological fluid sample of unknown pH, and the other containing a fluid pH indicator which changes color on mixing with the sample, were pumped under laminar flow into a single common detection chamber. In the first design, a color change in the detection chamber was observed only beyond a distance of ~3 mm from the point of entry of both input flow channels, whereas in a second design which had chevron shapes etched into the detection chamber surface the color changed after more like ~1 mm distance from the point of entry. Why is this?

REFERENCES

KEY REFERENCE

Kitagawa, M. et al. (2005). Complete set of ORF clones of *Escherichia coli* ASKA library (a complete set of *E. coli* K-12 ORF archive): Unique resources for biological research. *DNA Res. 12*:291–299.

MORE NICHE REFERENCES

Friedmann, H.C. (2004). From "butyribacterium" to *"E. coli"*: An essay on unity in biochemistry. *Perspect. Biol. Med. 47*:47–66.

Grinolds, M.S. et al. (2014). Subnanometre resolution in three-dimensional magnetic resonance imaging of individual dark spins. *Nat. Nanotechnol. 9*:279–284.

Hodgkin, A.L. and Huxley, A.F. (1952). A quantitative description of membrane current and its application to conduction and excitation in nerve. *J. Physiol. 177*:500–544.

Jain, A. et al. (2011). Probing cellular protein complexes using single-molecule pull-down. *Nature 473*:484–488.

Jinek, M. et al. (2012). A programmable dual-RNA-guided DNA endonuclease in adaptive bacterial immunity. *Science 337*:816–821.

Pastrana, E. (2010). Optogenetics: Controlling cell function with light. *Nat. Methods 8*:24.

Whitesides, G.M. (2006). The origins and the future of microfluidics. *Nature 442*:368–373.

Yizhar, O. et al. (2011). Optogenetics in neural systems. *Neuron 71*:9–34.

Theoretical Biophysics

Computational Biophysical Tools and Methods that Require a Pencil and Paper

8

It is nice to know that the computer understands the problem. But I would like to understand it too.

—Nobel laureate Eugene Wigner (1902–1995), one of the founders of modern quantum mechanics

General Idea: The increase in computational power and efficiency has transformed biophysics in permitting many biological questions to be tackled largely inside multicore computers. In this chapter, we discuss the development and application of several such *in silico* techniques, many of which benefit from computing that can be spatially delocalized both in terms of the servers running the algorithms and those manipulating and archiving significant amounts of data. However, there are also many challenging biological questions that can be tackled with theoretical biophysical tools consisting of a pencil and a piece of paper.

8.1 INTRODUCTION

Previously in this book we have discussed a range of valuable biophysical tools and techniques that can be used primarily in experimental investigations. However, biological insight from any experiment demands not only the right hardware but also a range of appropriate techniques that can, rather broadly, be described as *theoretical biophysics*. Ultimately, genuine insight into the operation of complex processes in biology is only gleaned by constructing a theoretical model of some sort. But this is a subtle point on which some life and physical scientists differ in their interpretation. To some biologists, a "model" is synonymous with speculation toward explaining experimental observations, embodied in a hypothesis. However, to many physical scientists, a "model" is a real structure built from sound physical and mathematical principles that can be tested robustly against the experimental data obtained, in our case from the vast armory of biophysical experimental techniques described in Chapters 3 through 7 in particular.

Theoretical biophysics methods are either directly coupled to the experiments through analyzing the results of specific experimental investigations or can be used to generate falsifiable predictions or simulate the physical processes of a particular biological system. This predictive capability is absolutely key to establishing a valuable theoretical biophysics framework, which is one that in essence contains more summed information than was used in its own construction, in other words, a model that goes beyond simply redefining/renaming physical parameters, but instead tells us something *useful*.

The challenges with theoretical biophysics techniques are coupled with the length and time scales of the biological processes under investigation. In this chapter, we consider regimes that extend from femtosecond fluctuations at the level of atomistic effects, up to the time scales of several seconds at the length scale of rigid body models of whole organisms.

DOI: 10.1201/9781003336433-8

Theoretical biophysics approaches can be extended much further than this, into smaller and larger length and time scales to those described earlier, and these are discussed in Chapter 9.

We can broadly divide theoretical biophysics into continuum level analysis and discrete level approaches. Continuum approaches are largely those of *pencil and paper* (or pen, or quill) that enable exact mathematical solutions to be derived often involving complex differential and integral calculus approaches and include analysis of systems using, for example, *reaction–diffusion kinetics, biopolymer physics modeling, fluid dynamics* methods, and also *classical mechanics*. Discrete approaches carve up the dimensions of space and time into incrementally small chunks, for example, to probe small increments in time to explore how a system evolves stochastically and/or to divide a complex structure up into small length scale units to make them tractable in terms of mathematical analysis. Following calculations performed on these incremental units of space or time, then each can be linked together using advanced *in silico* (i.e., computational) tools. Nontrivial challenges lie at the interface between continuum and discrete modeling, namely, how to link the two regimes. A related issue is how to model low copy number systems using continuum approaches, for example, at some threshold concentration level, there may simply not be any biomolecule in a given region of space in a cell at a given time.

The *in silico* tools include a valuable range of *simulation* techniques spanning length and time scales from *atomistic simulations* through to *molecular dynamics simulations* (*MDS*). Varying degrees of *coarse-graining* enable larger time and length scales to be explored. Computational discretization can also be applied to biomechanical systems, for example, to use *finite element analysis* (*FEA*). A significant number of computational techniques have also been developed for *image analysis*.

8.2 MOLECULAR SIMULATION METHODS

Theoretical biophysics tools that generate positional data of molecules in a biological system are broadly divided into *molecular statics* (*MS*) and *molecular dynamics* (*MD*) simulation methods. MS algorithms utilize *energy minimization* of the potential energy associated with forces of attraction and repulsion on each molecule in the system and estimate its local minimum to find the zero force equilibrium positions (note that the molecular simulation community use the phrase *force field* in reference to a specific type of potential energy function). MS simulations have applications in nonbiological analysis of nanomaterials; however, since they convey only static equilibrium positions of molecules, they offer no obvious advantage to high-precision structural biology tools and are likely to be less valuable due to approximations made to the actual potential energy experienced by each molecule. Also, this static equilibrium view of structure can be misleading since, in practice, there is variability around an average state due to thermal fluctuations in the constituent atoms as well as surrounding water solvent molecules, in addition to quantum effects such as tunneling-mediated fluctuations around the zero-point energy state. For example, local molecular fluctuations of single atoms and side groups occur over length scales <0.5 nm over a wide time scale of $\sim 10^{-15}$ to 0.1 s. Longer length scale rigid-body motions of up to ~ 1 nm for structural domains/motifs in a molecule occur over ~ 1 ns up to ~ 1 s, and larger scales motions >1 nm, such as protein unfolding events binding/unbinding of ligands to receptors, occur over time scales of ~ 100 ns up to thousands of seconds.

Measuring the evolution of molecular positions with time, as occurs in MD, is valuable in terms of generating biological insight. MD has a wide range of biophysical applications including simulating the folding and unfolding of certain biomolecules and their general stability, especially of proteins, the operation of ion channels, in the dynamics of phospholipid membranes, and the binding of molecules to *recognition sites* (e.g., ligand molecules binding to receptor complexes), in the intermediate steps involved in enzyme-catalyzed reactions, and in drug design for rationalizing the design of new pharmacological compounds (a form of *in silico drug design;* see Chapter 9). MD is still a relatively young discipline in biophysics, with the first publication of an MD-simulated biological process being only as far back as 1975. That was on the folding of a protein called "pancreatic trypsin inhibitor" known to inhibit an enzyme called *trypsin* (Levitt and Warshel, 1975).

New molecular simulation tools have also been adapted to addressing biological questions from nonbiological roots. For example, the *Ising model* of quantum statistical mechanics was developed to account for emergent properties of ferromagnetism. Here it can also be used to explain several emergent biological properties, such as modeling phase transition behaviors.

Powerful as they are, however, computer simulations of molecular behavior are only as good as the fundamental data and the models that go into them. It often pays to take a step back from the simulation results of all the different tools developed to really see if simulation predictions make intuitive sense or not. In fact, a core feature to molecular simulations is the ultimate need for "validation" by experimental tools. That is, when novel emergent behaviors are predicted from simulation, then it is often prudent to view them with a slightly cynical eye until experiments have really supported these theoretical findings.

8.2.1 GENERAL PRINCIPLES OF MD

A significant point to note concerning the structure of biomolecules determined using conventional structural biology tools (see Chapter 5), including not just proteins but also sugars and nucleic acids, is that biomolecules in a live cell in general have a highly dynamic structure, which is not adequately rendered in the pages of a typical biochemistry textbook. Ensemble-average structural determination methods of NMR, x-ray crystallography, and EM all produce experimental outputs that are biased toward the *least dynamic* structures. These investigations are also often performed using high local concentrations of the biomolecule in question far in excess to those found in the live cell that may result in tightly packed conformations (such as crystals) that do not exist naturally. However, the largest mean average signal measured is related to the most stable state that may not necessarily be the most probabilistic state in the functioning cell. Also, thermal fluctuations of the biomolecules due to the bombardment by surrounding water solvent molecules may result in considerable variability around a mean-average structure. Similarly, different dissolved ions can result in important differences in structural conformations that are often not recorded using standard structural biology tools.

MD can model the effects of attractive and repulsive forces due to ions and water molecules and of thermal fluctuations. The essence of MD is to theoretically determine the force F experienced by each molecule in the system being simulated in very small time intervals, typically around 1 fs (i.e., 10^{-15} s), starting from a predetermined set of atomic obtained from atomistic level structural biology of usually either x-ray diffraction, NMR (see Chapter 5), or sometimes *homology modeling* (discussed later in this chapter). Often, these starting structures can be further optimized initially using the energy minimization methods of MS simulations. After setting appropriate boundary conditions of system temperature and pressure, and the presence of any walls and external forces, the initial velocities of all atoms in the system are set. If the ith atom from a total of n has a velocity of magnitude V_i and mass m_i, then, at a system temperature T, the equipartition theorem (see Chapter 2) indicates, in the very simplest *ideal gas* approximation of noninteracting atoms, that

$$(8.1) \qquad \frac{3nk_{\mathrm{B}}T}{2} = \sum_{i=1}^{n} \frac{m_i v_i^2}{2}$$

The variation of individual atomic speeds in this crude ideal gas model is characterized by the *Maxwell–Boltzmann distribution*, such that the probability $p(v_{x,i})$ of atom i having a speed v_x parallel to the x-axis is given by

$$(8.2) \qquad p\left(v_{x,i}\right) = \frac{\exp\left(-m_i v_{x,i}^2/2k_{\mathrm{B}}T\right)}{\sum_{i=1}^{n}\exp\left(-m_i v_{x,i}^2/2k_{\mathrm{B}}T\right)} = \sqrt{\frac{m_i}{2\pi k_{\mathrm{B}}T}}\exp\left(-m_i v_{x,i}^2/2k_{\mathrm{B}}T\right)$$

$$= \frac{1}{\sqrt{2\pi\sigma_v^2}}\exp\left(-v_{x,i}^2/2\pi\sigma_v^2\right)$$

In other words, this probability distribution is a Gaussian curve with zero mean and variance $\sigma_y^2 = k_B T/m_i$. To determine the initial velocity, a pseudorandom probability from a uniform distribution in the range 0–1 is generated for each atom and each spatial dimensional, which is then sampled from the Maxwell–Boltzmann distribution to interpolate an equivalent speed (either positive or negative) parallel to x, y, and z for a given of set system temperature T. The net momentum P is then calculated, for example, for the x component as

$$(8.3) \qquad p_x = \sum_{i=1}^{n} m_i v_{x,i}$$

This is then used to subtract away a momentum offset from each initial atomic velocity so that the net starting momentum of the whole system is zero (to minimize computational errors due to large absolute velocity values evolving), so again for the x component

$$(8.4) \qquad v_{x,i,0} = v_{x,i,0} - \frac{p_x}{m_i}$$

and similarly, for y and z components. A similar correction is usually applied to set the net angular momentum to zero to prevent the simulated biomolecule from rotating, unless boundary conditions (e.g., in the case of explicit solvation, see in the following text) make this difficult to achieve in practice.

Then Newton's second law ($F = ma$) is used to determine the acceleration a and ultimately the velocity v of each atom, and thus where each will be, position vector of magnitude r, after a given time (Figure 8.1), and to repeat this over a total simulation time that can be anything from ~10 ps up to several hundred nanoseconds, usually imposed by a computational limit in the absence of any *coarse-graining* to the simulations, in other words, to generate *deterministic trajectories* of atomic positions as a function of time. This method involves numerical integration over time, usually utilizing a basic algorithm called the "Verlet algorithm," which results in low round-off errors. The Verlet algorithm is developed from a Taylor expansion of each atomic position:

$$(8.5) \qquad r(t + \Delta t) = r(t) + v(t)\Delta t + \frac{F(t)(\Delta t)^2}{m}\frac{1}{2} + \frac{F(t)(\Delta t)^3}{m}\frac{1}{3!} + O\left((\Delta t)^4\right)$$

Thus, the error is atomic position scales as $\sim O((\Delta t)^4)$, which is small for the femtosecond time increments normally employed, with the equivalent recursive relation for velocities having errors that scale as $\sim O((\Delta t)^2)$. The velocity form of the Verlet algorithm most commonly used in MD can be summarized as

$$
\begin{aligned}
(8.6) \qquad r(t + \Delta t) &= r(t) + v(t)\Delta t + (1/2)a(t)(\Delta t)^2 \\
v(t + \Delta t/2) &= v(t) + (1/2)a(t)\Delta t \\
a(t + \Delta t) &= -(1/m)\nabla U\left(r(t = \Delta t)\right) \\
v(t + \Delta t) &= v(t = \Delta t/2) + (1/2)a(t)\Delta t
\end{aligned}
$$

Related, though less popular, variants of this Verlet numerical integration include the *LeapFrog algorithm* (equivalent to the Verlet integration but interpolates for velocities at half time step intervals, but has a disadvantage in that velocities are not known at the same time as the atomic positions, hence the name) and *Beeman algorithm* (generates identical estimates for atomic positions as the Verlet method, but uses a slightly more accurate but computationally more costly method to estimate velocities).

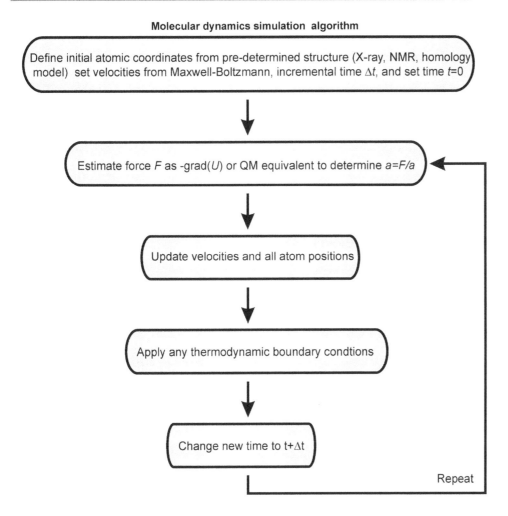

Molecular dynamics simulation algorithm

Define initial atomic coordinates from pre-determined structure (X-ray, NMR, homology model) set velocities from Maxwell-Boltzmann, incremental time Δt, and set time t=0

Estimate force F as -grad(U) or QM equivalent to determine a=F/a

Update velocities and all atom positions

Apply any thermodynamic boundary condtions

Change new time to t+Δt

Repeat

FIGURE 8.1 Molecular dynamics simulations (MDS). Simplified schematic of the general algorithm for performing MDS.

The force is obtained by calculating the grad of the potential energy function U that underlies the summed attractive and repulsive forces. No external applied forces are involved in standard MD simulations; in other words, there is no external energy input into the system, and so the system is in a state of *thermodynamic equilibrium*.

Most potential energy models used, whether quantum or classical in nature, in practice require some level of approximation during the simulation process and so are often referred to as *pseudopotentials* or *effective potentials*. However, important differences exist between MD methods depending on the length and time scales of the simulations, the nature of the force fields used, and whether surrounding solvent water molecules are included or not.

8.2.2 CLASSICAL MD SIMULATIONS

Although the forces on single molecules ultimately have a *quantum mechanical (QM)* origin—all biology in this sense one could argue is QM in nature—it is very hard to find even approximate solutions to *Schrödinger's equation* for anything but a maximum of a few atoms, let alone the ~10^{23} found in one mole of biological matter. The range of less straightforward quantum effects in biological systems, such as *entanglement, quantum coherence*, and *quantum superposition*, are discussed in Chapter 9. However, in the following section of this chapter, we discuss specific QM methods for MD simulations. But, for many applications, classical MD simulations, also known as simply *molecular mechanics (MM)*, offer a good compromise enabling biomolecule systems containing typically 10^4–10^5 atoms to be simulated for ~10–100 ns. The reduction from a quantum to a classical description entails two key assumptions. First, electron movement is significantly faster than that of atomic nuclei such that we assume they can change their relative position instantaneously. This is called the "Born–Oppenheimer approximation," which can be summarized as the total wave function being the product of the orthogonal wave functions due to nuclei and electrons separately:

(8.7) $$\psi_{total}\left(\text{nuclei,electrons}\right) = \psi(\text{nuclei})\,\psi(\text{electrons})$$

The second assumption is that atomic nuclei are treated as point particles of much greater mass than the electrons that obey classical Newtonian dynamics. These approximations lead to a unique potential energy function U_{total} due to the relative positions of electrons to nuclei. Here, the force F on a molecule is found from

(8.8) $$F = -\nabla U_{total}\left(\vec{r}\right)$$

U_{total} is the total potential energy function in the vicinity of each molecule summed from all relevant repulsive and attractive force sources experienced by each, whose position is denoted by the vector r. A parameter sometimes used to determine thermodynamic properties such as free energy is the *potential of mean force* (*PMF*) (not to be confused with the proton motive force; see Chapter 2), which is the potential energy that results in the average force calculated over all possible interactions between atoms in the system.

In practice, most classical MD simulations use relatively simple predefined potentials. To model the effects of chemical bonding between atoms, *empirical potentials* are used. These consist of the summation of independent potential energy functions associated with bonding forces between atoms, which include the covalent bond strength, bond angles, and bond dihedral potentials (a dihedral, or torsion angle, is the angle between two intersecting planes generated from the relative atomic position vectors). Nonbonding potential energy contributions come typically from van der Waals (vdW) and electrostatic forces. Empirical potentials are limited approximations to QM effects. They contain several free parameters (including equilibrium bond lengths, angles and dihedrals, vdW potential parameters, and atomic charge) that can be optimized either by fitting to QM simulations or from separate experimental biophysical measurements.

The simplest nonbonding empirical potentials consider just pairwise interactions between nearest-neighbor atoms in a biological system. The most commonly applied nonbonding empirical potential in MD simulations is the Lennard–Jones potential U_{LJ} (see Chapter 2), a version of which is given in Equation 2.10. It is often also written in the form

(8.9) $$U_{LJ} = 4\varepsilon\left(\left(\frac{\sigma}{r}\right)^{12}\left(\frac{\sigma}{r}\right)^6\right) = \varepsilon\left(\left(\frac{r_m}{r}\right)^{12}\left(\frac{r_m}{r}\right)^6\right)$$

where
    r is the distance between the two interacting atoms
    ε is the depth of the potential well
    σ is the interatomic distance that results in $U_{LJ} = 0$
    r_m is the interatomic distance at which the potential energy is a minimum (and thus the force $F = 0$) given by

(8.10) $$r_m = 2^{1/6}\left(1.22\sigma\right)$$

As discussed previously, the r^{-6} term models attractive, long-range forces due to vdW interactions (also known as dispersion forces) and has a physical origin in modeling the dipole–dipole interaction due to electron dispersion (a sixth power term). The r^{-12} term is heuristic; it models the electrostatic repulsive force between two unpaired electrons due to the Pauli exclusion principle, though as such is only used to optimize computational efficiency as the square of the r^{-6} term and has no direct physical origin.

For systems containing strong electrostatic forces (e.g., multiple ions), modified versions of *Coulomb's law* can be applied to estimate the potential energy U_{C-B} called the Coulomb–Buckingham potential consisting of the sum of a purely *Coulomb potential energy* (U_C or *electrical potential energy*) and the *Buckingham potential energy* (U_B) that models the vdW interaction:

(8.11)
$$U_C(\vec{r}) = \frac{q_1 q_2}{4\pi\varepsilon_0\varepsilon_r r}$$

$$U_B(\vec{r}) = A_B\exp\left(-B_B r\right) - \frac{U_B}{r^6}$$

$$U_{C-B}(\vec{r}) = U_B(\vec{r}) = A_B\exp\left(-B_B r\right) - \frac{U_B}{r^6} + \frac{q_1 q_2}{4\pi\varepsilon_0\varepsilon_r r}$$

where
q_1 and q_2 are the electrical charges on the two interacting ions
ε_r is the relative dielectric constant of the solvent
ε_0 is the electrical permittivity of free space

A_B, B_B, and C_B are constants, with the $-C_B/r^6$ term being the dipole–dipole interaction attraction potential energy term as was the case for the Lennard–Jones potential, but here the Pauli electrical repulsion potential energy is modeled as the exponential term $A_B\exp(-B_B r)$.

However, many biological applications have a small or even negligible electrostatic component since the *Debye length*, under physiological conditions, is relatively small compared to the length scale of atomic separations. The Debye length, denoted by the parameter κ^{-1}, is a measure of the distance over which electrostatic effects are significant:

(8.12)
$$\kappa^{-1} = \sqrt{\frac{\varepsilon_r\varepsilon_0 k_B T}{2 N_A q_e^2 I}} = \frac{1}{\sqrt{8\pi\lambda_B N_A I}}$$

where
N_A is the Avogadro's number
q_e is the elementary charge on an electron
λ is the *ionic strength* or *ionicity* (a measure of the concentration of all ions in the solution) λ_B is known as the *Bjerrum length* of the solution (the distance over which the electrical potential energy of two elementary charges is comparable to the thermal energy scale of $k_B T$)

At room temperature, if κ^{-1} is measured in nm and I in M, this approximates to

(8.13)
$$\kappa^{-1} \approx \frac{0.304}{\sqrt{I}}$$

In live cells, I is typically ~0.2–0.3 M, so κ^{-1} is ~1 nm.

In classical MD, the effects of bond strengths, angles, and dihedrals are usually empirically approximated as simple parabolic potential energy functions, meaning that the stiffness of each respective force field can be embodied in a simple single spring constant, so a typical total empirical potential energy in practice might be

(8.14)
$$U_{total} = \sum_{bound} k_r \left(r - r_{eq}\right)^2 + \sum_{angles} k_\theta \left(\theta - \theta_{eq}\right)^2 + \sum_{dthedrals} \frac{V_n}{2}\left(1 + \cos\left(n\phi - \phi_{eq}\right)\right)$$
$$+ \sum_{i<j} \left(\frac{A_{ij}}{r_{ij}^{12}} - \frac{B_{ij}}{r_{ij}^6} + \frac{q_i q_j}{4\pi\varepsilon_0\varepsilon_r r_{ij}}\right)$$

where

k_r, k_θ are consensus mean stiffness parameters relating to bond strengths and angles, respectively

r, θ, and ϕ are bond lengths, angles, and dihedrals

V_n is the dihedral energy barrier of order n, where n is a positive integer (typically 1 or 2)

r_{eq}, θ_{eq}, and ϕ_{eq} are equilibrium values, determined from either QM simulation or separate biophysical experimentation

For example, x-ray crystallography (Chapter 5) might generate estimates for r_{eq}, whereas k_r can be estimated from techniques such as Raman spectroscopy (Chapter 4) or IR absorption (Chapter 3).

In a simulation, the sum of all unique pairwise interaction potentials for n atoms indicates that the numbers of force, velocity, positions, etc., calculations required scales as $\sim O(n^2)$. That is to say, if the number of atoms simulated in system II is twice as many as those in system I, then the computation time taken for each Δt simulation step will be greater by a factor of $\sim$4. Some computational improvement can be made by *truncation*. That is, applying a cutoff for nonbonded force calculations such that only atoms within a certain distance of each other are assumed to interact. Doing this reduces the computational scaling factor to $\sim O(n)$, with the caveat of introducing an additional error to the computed force field. However, in several routine simulations, this is an acceptable compromise provided the cutoff does not imply the creation of separate ions.

In principle though, the nonbonding force field is due to a *many-body potential energy function*, that is, not simply the sum of multiple single pairwise interactions for each atom in the system, as embodied by the simple Lennard–Jones potential, but also including the effects of more than two interacting atoms, not just the single interaction between one given atom and its nearest neighbor. For example, to consider all the interactions involved between the nearest and the *next-nearest neighbor* for each atom in the system, the computation time required would scale as $\sim O(n^3)$. More complex computational techniques to account for these higher-order interactions include the particle mesh Ewald (*PME* or just *PM*) summation method (see Chapter 2), which can reduce the number of computations required by discretizing the atoms on a grid, resulting in a computation scaling factor of $\sim O(n^\alpha)$ where $1 < \alpha < 2$, or the PME variant of the *Particle–Particle–Particle–Pesh (PPPM or P³M)*.

PM is a Fourier-based Ewald summation technique, optimized for determining the potentials in many-particle simulations. Particles (atoms in the case of MD) are first interpolated onto a grid/mesh; in other words, each particle is discretized to the nearest grid point. The potential energy is then determined for each grid point, as opposed to the original positions of the particles. Obviously this interpolation introduces errors into the calculation of the force field, depending on how fine a mesh is used, but since the mesh has regular spatial periodicity, it is ideal for Fourier transformation (FT) analysis since the FT of the total potential is then a weighted sum of the FTs of the potentials at each grid point, and so the inverse FT of this weighted sum generates the total potential. Direct calculation of each discrete FT would still yield a $\sim O(m^2)$ scaling factor for computation for calculating pairwise-dependent potentials where m is the number of grid points on the mesh, but for sensible sampling m scales as $\sim O(n)$, and so the overall scaling factor is $\sim O(n^2)$. However, if the *fast Fourier transform (FFT)* is used, then the number of calculations required for an n-size dataset is $\sim n \log_e n$, so the computation scaling factor reduces to $\sim O(n \log_e n)$.

KEY POINT 8.1

In P³M, short-range forces are solved in real space; long-range forces are solved in reciprocal space.

P³M has a similar computation scaling improvement to PM but improves the accuracy of the force field estimates by introducing a cutoff radius threshold on interactions. Normal pairwise calculations using the original point positions are performed to determine the short-range forces (i.e., below the preset cutoff radius). But above the cutoff radius, the PM method is used to calculate long-range forces. To avoid aliasing effects, the cutoff radius must be less than half the mesh width, and a typical value would be ~1 nm. P³M is obviously computationally slower than PM since it scales as ~$O(wn \log n + (1 - w)n^2)$ where $0 < w < 1$ and w is the proportion of pairwise interactions that are deemed long-range. However, if w in practice is close to 1, then the computation scaling factor is not significantly different from that of PM.

Other more complex many-body empirical potential energy models exist, for example, the *Tersoff potential energy* (U_T). This is normally written as the *half potential* for a directional interaction between atom number i and j in the system. For simple pairwise potential energy models such as the Lennard–Jones potential, the pairwise interactions are assumed to be symmetrical, that is, the half potential in the $i \rightarrow j$ direction is the same as that in the $j \rightarrow i$ direction, and so the sum of the two half potentials simply equate to one single pairwise potential energy for that atomic pair. However, the Tersoff potential energy model approximates the half potential as being that due to an appropriate symmetrical half potential due U_S, for example, a U_{L-J} or U_{B-C} model as appropriate to the system, but then takes into account the *bond order* of the system with a term that models the weakening of the interaction between i and j atoms due to the interaction between the i and k atoms with a potential energy U_A, essentially through sharing of electron density between the i and k number atoms:

$$(8.15) \qquad U_T\left(\vec{r}_{ij}\right) = \frac{1}{2}\sum_{ij} U_s\left(\vec{r}_{ij}\right) + \frac{1}{2}\sum_{ij} B_{ij} U_A\left(\vec{r}_{ij}\right)$$

where the term B_{ij} is the bond order term, but is not a constant but rather a function of a coordination term G_{ij} of each bond, that is, $B_{ij} = B(G_{ij})$, such that

$$(8.16) \qquad G_{ij} = \sum_k f_c\left(r_{ik}\right) g\left(\theta_{jik}\right) f\left(r_{ij} - r_{ik}\right)$$

where
  f_c and f are functions dependent on just the relative displacements between the ith and jth and kth atoms
  g is a less critical function that depends on the relative bond angle between these three atoms centered on the ith atom

8.2.3 MONTE CARLO METHODS

Monte Carlo methods are in essence very straightforward but enormously valuable for modeling a range of biological processes, not exclusively for molecular simulations, but, for example, they can be successfully applied to simulate a range of ensemble thermodynamic properties. Monte Carlo methods can be used to simulate the effects of complex biological systems using often relatively simple underlying rules but which can enable emergent properties of that system to evolve stochastically that are too difficult to predict deterministically.

If there is one single theoretical biophysics technique that a student would be well advised to come to grips with then it is the Monte Carlo method.

Monte Carlo techniques rely on repeated random sampling to generate a numerical output for the occurrences, or not, of specific events in a given biological system. In the most common applications in biology, Monte Carlo methods are used to enable stochastic simulations. For example, a Monte Carlo algorithm will consider the likelihood that, at some time t, a given biological event will occur or not in the time interval t to $t + \Delta t$ where Δt is a small but finite discretized incremental time unit (e.g., for combined classical MD methods and Monte Carlo methods, $\Delta t \sim 10^{-15}$ s). The probability Δp for that specific biological event to occur is calculated using specific biophysical parameters that relate to that particular biological system, and this is compared against a probability p_{rand} that is generated from a pseudorandom number generator in the range 0–1. If $\Delta p > p_{rand}$, then one assumes the event has occurred, and the algorithm then changes the properties of the biological system to take account of this event. The new time in the stochastic simulation then becomes $t' = t + \Delta t$, and the polling process in the time interval t' to $t' + \Delta t$ is then repeated, and the process then iterated over the full range of time to be explored.

In molecular simulation applications, the trial probability Δp is associated with trial moves of an atom's position. Classical MD, in the absence of any stochastic fluctuations in the simulation, is intrinsically deterministic. Monte Carlo methods, on the other hand, are stochastic in nature, and so atomic positions evolve with a random element with time (Figure 8.2). Pseudorandom small displacements are made in the positions of atoms, one by one, and the resultant change ΔU in potential energy is calculated using similar empirical potential energy described in the previous section for classical MD (most commonly the Lennard–Jones potential). Whether the random atomic displacement is accepted or not is usually determined by using the *Metropolis criterion*. Here, the trial probability Δp is established using the Boltzmann factor $\exp(-\Delta U/k_B T)$, and if $\Delta p >$ the pseudorandom probability (p_{rand}), then the atomic displacement is accepted, the potential energy is adjusted and the process iterated for another atom. Since Monte Carlo methods only require potential energy calculations to compute atomic trajectories, as opposed to additional force calculations, there is a saving in computational efficiency.

This simple process can lead to very complex outputs if, as is generally the case, the biological system is also spatially discretized. That is, probabilistic polling is performed on spatially separated components of the system. This can lead to capturing several emergent properties. Depending on the specific biophysics of the system under study, the incremental time Δt can vary, but has to be small enough to ensure subsaturation levels of Δp that is, $\Delta p < 1$ (in practice, Δt is often catered so as to encourage typical values of Δp to be in the range ~0.3–0.5). However, very small values of Δt lead to excessive computational times. Similarly, to get the most biological insight ideally requires the greatest spatial discretization of the system. These two factors when combined can result in requirements of significant parallelization of computation and often necessitate *supercomputing, or high performance computing (HPC)* resources when applied to molecular simulations. However, there are still several biological processes beyond molecular simulations that can be usefully simulated using nothing more than a few lines of code in any standard engineering software language (e.g., C/C++, MATLAB˙, Fortran, Python, LabVIEW) on a standard desktop PC.

A practical issue with the Metropolis criterion is that some microstates during the finite duration of a simulation may be very undersampled or in fact not occupied at all. This is a particular danger in the case of a system that comprises two or more intermediate states of stability for molecular conformations that are separated by relatively large free energy barriers. This implies that the standard Boltzmann factor employed of $\exp(-\Delta U/k_B T)$ is in practice very small if ΔU equates to these free energy barriers. This can result in a simulation being apparently locked in, for example, one very over sampled state. Given enough time of course, a simulation would explore all microstates as predicted from the ergodic hypothesis (see Chapter 6); however, practical computational limitations may not always permit that.

To overcome this problem, a method called "umbrella sampling" can be employed. Here a weighting factor w is used in front of the standard Boltzmann term:

Monte Carlo method in molecular simulations

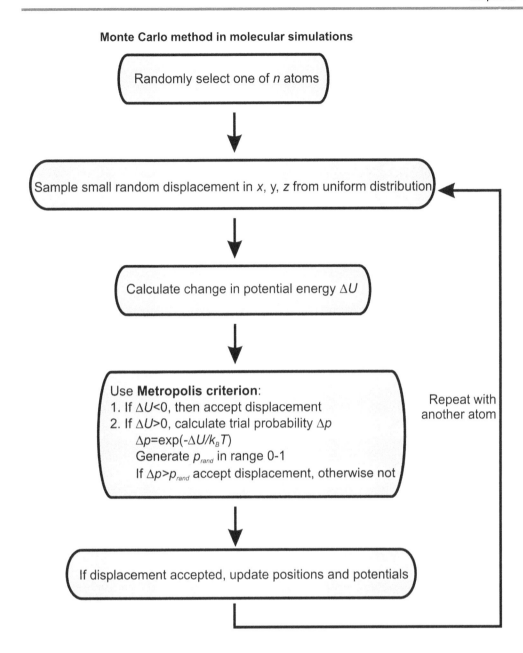

FIGURE 8.2 Monte Carlo methods in molecular simulations. Simplified schematic of a typical Monte Carlo algorithm for performing stochastic molecular simulations.

(8.17)
$$\Delta p(r_i) = w(r_i) \exp\left(\frac{-U(r_i)}{k_B T}\right)$$

Thus, for undersampled states, w would be chosen to be relatively high, and for oversampled states, w would be low, such that all microstates were roughly equally explored in the finite simulation time. Mean values of any general thermodynamic parameter A that can be measured from the simulation (e.g., the free energy change between different conformations) can still be estimated as

(8.18)
$$\langle A \rangle = \frac{\langle A/w \rangle}{\langle 1/w \rangle}$$

Similarly, the occupancy probability in each microstate can be estimated using a *weighted histogram analysis method*, which enables the differences in thermodynamic parameters for each state to be estimated, for example, to explore the variation in free energy differences between the intermediate states.

An extension to Monte Carlo MD simulations is *replica exchange MCMC sampling*, also known as *parallel tempering*. In replica exchange, multiple Monte Carlo simulations are performed that are randomly initialized as normal, but at different temperatures. However, using the Metropolis criterion, the molecular configurations at different temperatures can be dynamically exchanged. This results in efficient sampling of high and low energy conformations and can result in improved estimates to thermodynamical properties by spanning a range of simulation temperatures in a computationally efficient way.

An alternative method to umbrella sampling is the *simulated annealing* algorithm, which can be applied to Monte Carlo as well as other molecular simulations methods. Here the simulation is equilibrated initially at high temperature but then the temperature is gradually reduced. The slow rate of temperature drop gives a greater time for the simulation to explore molecular conformations with free energy levels *lower* than local energy minima found by energy minimization, which may not represent a global minimum. In other words, it reduces the likelihood for a simulation becoming locked in a local energy minimum.

8.2.4 *AB INITIO* MD SIMULATIONS

Small errors in approximations to the exact potential energy function can result in artifactual emergent properties after long simulation times. An obvious solution is therefore to use better approximations, for example, to model the actual QM atomic orbital effects, but the caveat is an increase in associated computation time. *Ab initio* MD simulations use the summed QM interatomic electronic potentials U_{QD} experienced by each atom in the system, denoted by a total wave function ψ. Here, U_{QD} is given by the action of *Hamiltonian operator* $\hat{H}$ on ψ (the kinetic energy component of the Hamiltonian is normally independent of the atomic coordinates) and the force F_{QD} by the grad of U_{QD} such that

$$U_{QD}(\vec{r}) = \hat{H}\psi$$
(8.19)
$$F_{QD} = -\nabla\left(U_{QD}(\vec{r})\right)$$

These are, in the simplest cases, again considered as pairwise interactions between the highest energy atomic orbital of each atom. However, since each electronic atomic orbital is identified by a unique value of three *quantum numbers*, n (the *principal quantum number*), ℓ (the *azimuthal quantum number*), and m_ℓ (the *magnetic quantum number*), the computation time for simulations scales as $\sim O(n^3)$. However, *Hartree–Fock* (*HF*) approximations are normally applied that reduce this scaling to more like $\sim O(n^{2.7})$. Even so, the computational expense usually restricts most simulations to ~ 10–100 atom systems with simulation times of ~ 10–100 ps.

The HF method, also referred to as the *self-consistent field* method, uses an approximation of the Schrödinger equation that assumes that each subatomic particle, *fermions* in the case of electronic atomic orbitals, is subjected to the *mean field* created by all other particles in the system. Here, the n-body fermionic wave function solution of the Schrödinger equation is approximated by the determinant of the $n \times n$ *spin–orbit matrix* called the "Slater determinant." The HF electronic wave function approximation ψ for an n-particle system thus states

(8.20)
$$\psi(\vec{r}_1, \vec{r}_2, \ldots, \vec{r}_n) \approx \frac{1}{\sqrt{n}} \begin{vmatrix} \chi_1(\vec{r}_1) & \chi_1(\vec{r}_1) & \cdots & \chi_n(\vec{r}_1) \\ \chi_1(\vec{r}_2) & \chi_2(\vec{r}_2) & \cdots & \chi_n(\vec{r}_2) \\ \vdots & \vdots & \ddots & \vdots \\ \chi_1(\vec{r}_n) & \chi_2(\vec{r}_n) & \cdots & \chi_2(\vec{r}_n) \end{vmatrix}$$

The term on the right is the determinant of the spin–orbit matrix. The χ elements of the spin–orbit matrix are the wave functions for the various orthogonal (i.e., independent) wave functions for the n individual particles.

In some cases of light atomic nuclei, such as those of hydrogen, QM MD can model the effects of the atomic nuclei as well as the electronic orbitals. A similar HF method can be used to approximate the total wave function; however, since nuclei are composed of *bosons* as opposed to *fermions*, the spin–orbit matrix *permanent* is used instead of the determinant (a matrix permanent is identical to a matrix determinant, but instead the signs in front of the matrix element product permutations being a mixture of positive and negative as is the case for the determinant, they are all positive for the permanent). This approach has been valuable for probing effects such as the quantum tunneling of hydrogen atoms, a good example being the mechanism of operation of an enzyme called "alcohol dehydrogenase," which is found in the liver and requires the transfer of an atom of hydrogen using quantum tunneling for its biological function.

KEY POINT 8.2

The number of calculations for n atoms in classical MD is $\sim O(n^2)$ for pairwise potentials, though if a cutoff is employed, this reduces to $\sim O(n)$. If next-nearest-neighbor effects (ormore) are considered, $\sim O(n^\alpha)$ calculations are required where $\alpha \geq 3$. Particle–mesh methods can reduce the number of calculations to $\sim O(n \log n)$ for classical pairwise interaction models. The number of direct calculations required for pairwise interactions in QM MD is $\sim O(n^3)$, but approximations to the total wave function can reduce this to $\sim O(n^{2.7})$.

Ab initio force fields are not derived from predetermined potential energy functions since they do not assume preset bonding arrangements between the atoms. They are thus able to model chemical bond making and breaking explicitly. Different bond coordination and hybridization states for bond making and breaking can be modeled using other potentials such as the *Brenner potential* and the *reactive force field (ReaxFF) potential*.

The use of *semiempirical potentials*, also known as *tight-binding potentials*, can reduce the computational demands in *ab initio* simulations. A semiempirical potential combines the fully QM-based interatomic potential energy derived from *ab initio* modeling using the spin–orbit matrix representation described earlier. However, the matrix elements are found by applying empirical interatomic potentials to estimate the overlap of specific atomic orbitals. The spin–orbit matrix is then diagonalized to determine the occupancy level of each atomic orbital, and the empirical formulations are used to determine the energy contributions from these occupied orbitals. Tight-binding models include the *embedded-atom method* also known as the *tight-binding second-moment approximation* and *Finnis–Sinclair model* and also include approaches that can approximate the potential energy in heterogeneous systems including metal components such as the *Streitz–Mintmire model*.

Also, there are *hybrid classical and quantum mechanical methods (hybrid QM/MM)*. Here, for example, classical MD simulations may be used for most of a structure, but the region of the structure where most fine spatial precision is essential is simulated using QM-derived force fields. These can be applied to relatively large systems containing $\sim 10^4$–10^5 atoms, but the simulation time is limited by QM and thus restricted to simulation times of ~ 10–100 ps. Examples of this include the simulation of ligand binding in a small pocket of a larger structure and the investigation of the mechanisms of catalytic activity at a specific active site of an enzyme.

KEY POINT 8.3

Classical (MM) MD methods are fast but suffer inaccuracies due to approximations of the underlying potential energy functions. Also, they cannot simulate chemical reactions involving the making/breaking of covalent bonds. QM MD methods generate very accurate spatial information, but they are computationally enormously expensive. Hybrid QM/MM offers a compromise in having a comparable speed to MM for rendering very accurate spatial detail to a restricted part of a simulated structure.

8.2.5 STEERED MD

Steered molecular dynamics (SMD) simulations, or *force probe simulations*, use the same core simulation algorithms of either MM or QM simulation methods but in addition apply external mechanical forces to a molecule (most commonly, but not exclusively, a protein) in order to manipulate its structure. A pulling force causes a change in molecular conformation, resulting in a new potential energy at each point on the pulling pathway, which can be calculated at each step of the simulation. For example, this can be used to probe the force dependence on protein folding and unfolding processes, and of the binding of a ligand to receptor, or of the strength of the molecular adhesion interactions between two touching cells. These are examples of thermodynamically *nonequilibrium* states and are maintained by the input of external mechanical energy into the system by the action of pulling on the molecule.

As discussed previously (see Chapter 2), all living matter is in a state of thermodynamic nonequilibrium, and this presents more challenges in theoretical analysis for molecular simulations. Energy-dissipating processes are essential to biology though they are frequently left out of mathematical/computational models, primarily for three reasons. First, historical approaches inevitably derive from equilibrium formulations, as they are mathematically more tractable. Second, and perhaps most importantly, in many cases, equilibrium approximations seem to account for experimentally derived data very well. Third, the theoretical framework for tackling nonequilibrium processes is far less intuitive than that for equilibrium processes. This is not to say we should not try to model these features, but perhaps should restrict this modeling only to processes that are poorly described by equilibrium models.

Applied force changes can affect the molecular conformation both by changing the relative positions of covalently bonded atoms and by breaking and making bonds. Thus, SMD can often involve elements of both classical and QM MD, in addition to Monte Carlo methods, for example, to poll for the likelihood of a bond breaking event in a given small time interval. The mathematical formulation for these types of bond state calculations relate to continuum approaches of the *Kramers theory* and are described under reaction–diffusion analysis discussed later in this chapter.

SMD simulations mirror the protocols of single-molecule pulling experiments, such as those described in Chapter 6 using optical and/or magnetic tweezers and AFM. These can be broadly divided into molecular stretches using a constant force (i.e., a *force clamp*) that, in the experiment, result in stochastic changes to the end-to-end length of the molecule being stretched, and constant velocity experiments, in which the rate of change of probe head displacement relative to the attached molecule with respect to time t is constant (e.g., the AFM tip is being ramped up and down in height from the sample surface using a triangular waveform). If the ramp speed is v and the effective stiffness of the force transducer used (e.g., an AFM tip, optical tweezers) is k, then we can model the external force F_{ext} due to potential energy U_{ext} as

(8.21)
$$U_{ext}(t) = \frac{1}{2}k\left(v(t-t_0) - \left(\vec{r}(t) - \vec{r}(t_0).\vec{u}\right)\right)^2$$
$$F_{ext}(t) = -\nabla U_{ext}(t)$$

where

t_0 is some initial time

$\vec{r}$ is the position of an atom

$\vec{u}$ is the unitary vector for the pulling direction of the force probe

U_{ext} can then be summed into the appropriate internal potential energy formulation used to give a revised total potential energy function relevant to each individual atom.

The principal issue with SMD is a mismatch of time scales and force scales between simulation outputs compared to actual experimental data. For example, single-molecule pulling experiments involve generating forces between zero and up to ~10–100 pN over a time scale of typically a second to observe a molecular unfolding event in a protein. However, the equivalent time scale in SMD is more like ~10^{-9} s, extending as high as ~10^{-6} s in exceptional cases. To stretch a molecule, a reasonable distance compared to its own length scale, for example, ~10 nm, after ~1 ns of simulation implies a probe ramp speed equivalent to ~10 m s^{-1} (or ~0.1 Å ps^{-1} in the units often used in SMD). However, in an experiment, ramp rates are limited by viscous drag effects between the probe and the sample, so speeds equivalent to ~10^{-6} m s^{-1} are more typical, seven or more orders of magnitude slower than in the SMD simulations. As discussed in the section on reaction–diffusion analysis later in this chapter, this results in a significantly lower probability of molecular unfolding for the simulations, making it nontrivial to interpret the simulated unfolding kinetics. However, they do still provide valuable insight into the key mechanistic events of importance for force dependent molecular processes and enable estimates to be made for the free energy differences between different folded intermediate states of proteins as a function of pulling force, which provides valuable biological insight.

A key feature of SMD is the importance of water-solvent molecules. Bond rupture, for example, with hydrogen bonds, in particular, is often mediated through the activities of water molecules. The ways in which the presence of water molecules are simulated in general MD are discussed as follows.

8.2.6 SIMULATING THE EFFECTS OF WATER MOLECULES AND SOLVATED IONS

The primary challenge of simulating the effects of water molecules on a biomolecule structure is computational. Molecular simulations that include an *explicit solvent* take into account the interactions of all individual water molecules with the biomolecule. In other words, the atoms of each individual water molecule are included in the MD simulation at a realistic density, which can similarly be applied to any solvated ions in the solution. This generates the most accuracy from any simulation, but the computational expense can be significant (the equivalent molarity of water is higher than you might imagine; see Worked Case Example 8.1).

The potential energy used is typically Lennard–Jones (which normally is only applied to the oxygen atom of the water molecule interacting with the biomolecule) with the addition of the Coulomb potential. Broadly, there are two explicit solvent models: the *fixed charge explicit solvent model* and the *polarizable explicit solvent model*. The latter characterizes the ability of water molecules to become electrically polarized by the nearby presence of the biomolecule and is the most accurate physical description but computationally most costly. An added complication is that there are >40 different water models used by different research groups that can account for the electrostatic properties of water (e.g., see Guillot, 2002), which include different bond angles and lengths between the oxygen and hydrogen atoms, different dielectric permittivity values, enthalpic and electric polarizability properties, and different assumptions about the number of interaction sites between the biomolecule and the water's oxygen atom through its *lone pair electrons*. However, the most common methods include *single point charge (SPC), TIP3P*, and *TIP4P* models that account for most biophysical properties of water reasonably well.

KEY POINT 8.4

A set of two paired electrons with opposite spins in an outer atomic orbital are considered a lone pair if they are not involved in a chemical bond. In the water molecule, the oxygen atom in principle has two such lone pairs, which act as electronegative sources facilitating the formation of hydrogen bonds and accounting for the fact that bond angle between the two hydrogen atoms and oxygen in a water molecule is greater than 90° (and in fact is closer to 104.5°).

To limit excessive computational demands on using an explicit solvent, *periodic boundary conditions (PBCs)* are imposed. This means that the simulation occurs within a finite, geometrically well-defined volume, and if the predicted trajectory of a given water molecule takes it beyond the boundary of the volume, it is forced to reemerge somewhere on the other side of the boundary, for example, through the point on the other side of the volume boundary that intersects with a line drawn between the position at which the water molecule moved originally beyond the boundary and the geometrical centroid of the finite volume, for example, if the simulation was inside a 3D cube, then the 2D projection of this might show a square face, with the water molecule traveling to one edge of the square but then reemerging on the opposite edge. PBCs permit the modeling of large systems, though they impose spatial periodicity where there is none in the natural system.

The minimum size of the confining volume of water surrounding the biomolecule needs must be such that it encapsulates the *solvation shell* (also known as the *solvation sphere*, also as the *hydration layer* or *hydration shell* in the specific case of a water solvent). This is the layer of water molecules that forms around the surface of biomolecules, primarily due to hydrogen bonding between water molecules. However, multiple layers of water molecules can then form through additional hydrogen bond interactions with the primary layer water molecules. The effects of a solvent shell can extend to ~1 nm away from the biomolecule surface, such that the mobility of the water molecules in this zone is distinctly lower than that exhibited in the bulk of the solution, though in some cases, this zone can extend beyond 2 nm. The time scale of mixing between this zone and the bulk solution is in the range 10^{-15} to 10^{-12} s and so simulations may need to extend to at least these time scales to allow adequate mixing than if performed in a vacuum. The *primary hydration shell method* used in classical MD with explicit solvent assumes two to three layers of water molecules and is reasonably accurate.

An *implicit solvent* uses a continuum model to account for the presence of water. This is far less costly computationally than using an explicit solvent but cannot account for any explicit interactions between the solvent and solute (i.e., between the biomolecule and any specific water molecule). In its very simplest form, the biomolecule is assumed only to interact only with itself, but the electrostatic interactions are modified to account for the solvent by assuming the value of the relative dielectric permittivity term ε_r in the Coulomb potential. For example, in a vacuum, $\varepsilon_r = 1$, whereas in water $\varepsilon_r \approx 80$.

If the straight-line joining atoms for a pairwise interaction are through the structure of the biomolecule itself, with no accessible water present, then the relative dielectric permittivity for the biomolecule itself should be used, for example, for proteins and phospholipid bilayers, ε_r can be in the range ~2–4, and nucleic acids ~8; however, there can be considerable variation deepening on specific composition (e.g., some proteins have $\varepsilon_r \approx 20$). This very simple implicit solvation model using a pure water solvent is justified in cases where the PMF results in a good approximation to the average behavior of many dynamic water molecules. However, this approximation can be poor in regions close to the biomolecule such as the solvent shell, or in discrete *hydration pockets* of biomolecules, which almost all molecules in practice have, such as in the interiors of proteins and phospholipid membranes.

The simplest formulation for an implicit solvent that contains dissolved ions is the *generalized Born* (or simply *GB*) approximation. GB is *semiheuristic* (which is a polite way of saying that it only has a physically explicable basis in certain limiting regimes), but which still provides a relatively simple and robust way to estimate long-range electrostatic interactions. The foundation of the GB approximation is the classical *Poisson–Boltzmann* (*PB*) model of continuum electrostatics. The PB model uses the Coulomb potential energy with a modified ε_r value but also considers the concentration distribution of mobile solvated ions. If U_{elec} is the electrostatic potential energy, then the *Poisson equation* for a solution of ions can be written as

$$(8.22) \qquad \nabla^2 U_{elec} = -\frac{\rho}{\varepsilon_r \varepsilon_0}$$

where ρ is the electric charge density. In the case of an SPC, ρ could be modeled as a Dirac delta function, which results in the standard Coulomb potential formulation. However, for spatially distributed ions, we can model these as having an electrical charge dq in an incremental volume of dV at a position r of

$$(8.23) \qquad dq = \rho(r)dV(r)$$

Applying the Coulomb potential to the interaction between these incremental charges in the whole volume implies

$$(8.24) \qquad U_{elec}(r_0) = \frac{1}{4\pi\varepsilon_r\varepsilon_0} \int_{allspace} \frac{\rho dV}{|r - r_0|}$$

The *Poisson–Boltzmann equation* (*PBE*) then comes from this by modeling the distribution of ion concentration C as

$$(8.25) \qquad \begin{aligned} C &= C_0\exp(-zq_e U_{elec}/k_B T) \\ \therefore \rho &= FC = FC_0\exp(-zq_e U_{elec}/k_B T) \end{aligned}$$

where
  z is the valence of the ion
  q_e is the elementary electron charge
  F is the Faraday's constant

This formulation can be generalized for other types of ions by summing their respective contributions. Substituting Equations 8.23 and 8.24 into Equation 8.25 results in a second-order *partial differential equation* (*PDE*), which is the *nonlinear Poisson–Boltzmann equation* (*NLPBE*). This can be approximated by the *linear Poisson–Boltzmann equation* (*LPBE*) if $q_e U_{elec}/k_B T$ is very small (known as the *Debye–Hückel approximation*). Adaptations to this method can account for the effects of *counter ions*. The LPBE can be solved exactly. There are also standard numerical methods for solving the NLPBE, which are obviously computationally slower than solving the LPBE directly, but the NLPBE is a more accurate electrostatic model than the approximate LPBE.

A compromise between the exceptional physical accuracy of explicit solvent modeling and the computation speed of implicit solvent modeling is to use a hybrid approach of a *multilayered solvent model*, which incorporates an additional *Onsager reaction field potential* (Figure 8.3). Here, the biomolecule is simulated in the normal way of QM or MM MD (or both for hybrid QM/MM), but around each molecule, there is a cavity within which the electrostatic interactions with individual water molecules are treated explicitly. However, outside this cavity, the solution is assumed to be characterized by a single uniform dielectric constant. The biomolecule induces electrical polarization in this outer cavity, which in turn creates a *reaction field* known as the *Onsager reaction field*. If a NLPBE is used for the implicit solvent,

Multi-layered solvent model

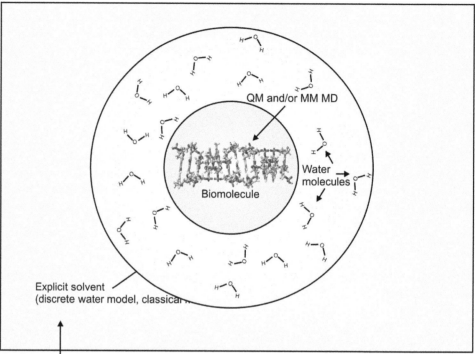

FIGURE 8.3 Multilayered solvent model. Schematic example of a hybrid molecular simulation approach, here shown with a canonical segment of a Z-DNA (see Chapter 2), which might be simulated using either a QM *ab initio* method or classical (molecular mechanics) dynamics simulations (MD) approach, or a hybrid of the two, while around it is a solvent shell in which the trajectories of water molecules are simulated explicitly using classical MD, and around this shell is an outer zone of implicit solvent in which continuum model is used to account for the effects of water.

then this can be a physically accurate model for several important properties. However, even so, no implicit solvent model can account for the effects of hydrophobicity (see Chapter 2), water viscous drag effects on the biomolecule, or hydrogen bonding within the water itself.

8.2.7 LANGEVIN AND BROWNIAN DYNAMICS

The viscous drag forces not accounted for in implicit solvent models can be included by using the methods of *Langevin dynamics* (LD). The equation of motion for an atom in an energy potential U exhibiting Brownian motion, of mass m in a solvent of viscous drag coefficient γ, involves inertial and viscous forces, but also a random stochastic element, known as the *Langevin force*, $R(t)$, embodied in the *Langevin equation*:

$$(8.26) \qquad m\ddot{r} + \gamma\dot{r} = -\nabla U(r) + \sqrt{2\pi k_B T m}R(t)$$

Over a long period of time, the mean of the Langevin force is zero but with a finite variance of $2\pi k_B T m$. Similarly, the effect on the velocity of an atom is to add a random component that has mean over long period of time of zero but a variance of $2k_B T \gamma$. Thus, for an MD simulation that involves implicit solvation, LD can be included by a random sampled component from a Gaussian distribution of mean zero and variance $2k_B T \gamma$.

The value of $\gamma/m = 1/\tau$ is a measure of the collision frequency between the biomolecule atoms and water molecules, where τ is the mean time between collisions. For example, for individual atoms in a protein, τ is in the range 25–50 ps, whereas for water molecules, τ is ~80 ps. The limit of high γ values is an *overdamped* regime in which viscous forces dominate over inertial forces. Here τ can for some biomolecule systems in water be as low as ~1 ps, which is a diffusion (as opposed to stochastic) dominated limit called *Brownian dynamics*. The equation of motion in this regime then reduces to the *Smoluchowski diffusion equation*, which we will discuss later in this chapter in the section on reaction–diffusion analysis.

One effect of applying a stochastic frictional drag term in LD is that this can be used to slow down the motions of fast-moving particles in the simulation and thus act as feedback mechanism to clamp the particle speed range within certain limits. Since the system temperature depends on particle speeds, this method thus equates to a *Langevin thermostat* (or equivalently a *barostat* to maintain the pressure of the system). Similarly, other nonstochastic thermostat algorithms can also be used, which all in effect include additional weak frictional coupling constants in the equation of motion, including the *Anderson, isokinetic/Gaussian, Nosé–Hoover*, and *Berendsen* thermostats.

8.2.8 COARSE-GRAINED SIMULATION TOOLS

There are a range of *coarse-grained* (*CG*) simulation approaches that, instead of probing the exact coordinates of every single atom in the system independently, will pool together groups of atoms as a rigid, or semirigid, structure, for example, as connected atoms of a single amino acid residue in a protein, or coarser still of groups of residues in a single structural motif in a protein. The forces experienced by the components of biological matter in these CG simulations can also be significantly simplified versions that only approximate the underlying QM potential energy. These reductions in model complexity are a compromise to achieving computational tractability in the simulations and ultimately enable larger length and time scales to be simulated at the expense of loss of fine detail in the structural makeup of the simulated biological matter.

This coarse-graining costs less computational time and enables longer simulation times to be achieved, for example, time scales up to ~10^{-5} s can be simulated. However, again there is a time scale gap since large molecular conformational changes under experimental conditions can be much slower than this, perhaps lasting hundreds to thousands of microseconds. Further coarse-graining can allow access into these longer time scales, for example, by pooling together atoms into functional structural motifs and modeling the connection between the motifs with, in effect, simple springs, resulting in a simple *harmonic potential* energy function.

Mesoscale models are at a higher length scale of coarse-graining, which can be applied to larger macromolecular structures. These, in effect, pool several atoms together to create relatively large soft-matter units characterized by relatively few material parameters, such as *mechanical stiffness, Young's modulus*, and the *Poisson ratio*. Mesoscale simulations can model the behavior of macromolecular systems potentially over a time scale of seconds, but clearly what they lack is the fine detail of information as to what happens at the level of specific single molecules or atoms. However, in the same vein of hybrid QM/MM simulations, hybrid mesoscale/CG approaches can combine elements of mesoscale simulations with smaller length scale CG simulations, and similarly hybrid CG/MM simulations can be performed, with the strategy for all these hybrid methods that the finer length scale simulation tools focus on just highly localized regions of a biomolecule, while the longer length scale simulation tool generates peripheral information about the surrounding structures.

Hybrid QM/MM approaches are particularly popular for investigating *molecular docking* processes. that is, how well or not a small ligand molecule binds to a region of another larger molecule. This process often utilizes relative simple *scoring functions* to generate rapid estimates for the goodness of fit for a docked molecule to a putative binding site to facilitate fast computational screening and is of specific interest in *in silico drug design*, that is, using computational molecular simulations to develop new pharmaceutical chemicals.

8.2.9 SOFTWARE AND HARDWARE FOR MD

Software evolves continuously and rapidly, and so this is not the right forum to explore all modern variants of MD simulation code packages; several excellent online forums exist that give up-to-date details of the most recent advances to these software tools, and the interested reader is directed to these. However, a few key software applications have emerged as having significant utility in the community of MD simulation research, whose core features have remained the same for the past few years, which are useful to discuss here. Three leading software applications have grown directly out of the academic community, including Assisted Model Building with Energy Refinement (*AMBER*, developed originally at the Scripps Research Institute, United States), Chemistry at HARvard Macromolecular Mechanics (*CHARMM* developed at the University of Harvard, United States), and GROningen MAchine for Chemical Simulations (*GROMACS*, developed at the University of Gröningen, the Netherlands).

The term "AMBER" is also used in the MD community in conjunction with "force fields" to describe the specific set of force fields used in the AMBER software application. AMBER software uses the basic force field of Equation 8.14, with presets that have been parameterized for proteins or nucleic acids (i.e., several of the parameters used in the potential energy approximation have been preset by using prior QM simulation or experimental biophysics data for these different biomolecule types). AMBER was developed for classical MD, but now has interfaces that can be used for *ab initio* modeling and hybrid QM/MM. It includes implicit solvent modeling capability and can be easily implemented on graphics processor units (GPUs, discussed later in this chapter). It does not currently support standard Monte Carlo methods but has replica exchange capability.

CHARMM has much the same functionality as AMBER. However, the force field used has more complexity, in that it includes additional correction factors:

(8.27)
$$U_{impropers} = \sum_{impropers} k_W \left(\omega - \omega_{eq} \right)^2$$

$$U_{U-B} = \sum_{Urey-Bradley} k_u \left(u - u_{eq} \right)^2$$

The addition of the *impropers potential* ($U_{impropers}$) is a dihedral correction factor to compensate for *out-of-plane bending* (e.g., to ensure that a known planar structural motif remains planar in a simulation) with k_ω being the appropriate impropers stiffness and $\boldsymbol{\omega}$ is the out-of-pane angle deviation about the equilibrium angle ω_{eq}. The *Urey–Bradley potential* (U_{U-B}) corrects for cross-term angle bending by restraining the motions of bonds by introducing a virtual bond that counters angle bending vibrations, with u a relative atomic distance from the equilibrium position u_{eq}. The CHARMM force field is physically more accurate than that of AMBER, but at the expense of greater computational cost, and in many applications, the additional benefits of the small correction factors are marginal.

GROMACS again has many similarities to AMBER and CHARMM but is optimized for simulating biomolecules with several complicated bonded interactions, such as biopolymers in the form of proteins and nucleic acids, as well as complex lipids. GROMACS can operate with a range of force fields from different simulation software including CHARMM, AMBER, and CG potential energy functions; in addition, its own force field set is called Groningen molecular simulation (GROMOS). The basic GROMOS force field is similar to that of CHARMM, but the electrostatic potential energy is modeled as a *Coulomb potential with reaction field* (U_{CRF}), which in its simplest form is the sum of the standard $\sim 1/r_{ij}$ Coulomb potential (U_C) with additional contribution from a reaction field (U_{RF}), which represents the interaction of atom i with an induced field from the surrounding dielectric medium beyond a predetermined cutoff distance R_{rf} due to the presence of atom j:

$$U_{RF} = -\sum_{ij} \frac{q_i q_j}{4\pi\varepsilon_0\varepsilon_1} \frac{C_{rf}r_{ij}^2}{R_{ij}^3}$$

(8.28)

$$C_{rf} = \frac{\left(2\varepsilon_1 - 2\varepsilon_2\right)\left(1+\kappa R_{rf}\right) - \varepsilon_2\left(\kappa R_{rf}\right)^2}{\left(\varepsilon_1 - 2\varepsilon_2\right)\left(1+\kappa R_{rf}\right) + \varepsilon_2\left(\kappa R_{rf}\right)^2}$$

where
    ε_1 and ε_2 are the relative dielectric permittivity values inside and outside the cutoff radius, respectively
    κ is the reciprocal of the Debye screening length of the medium outside the cutoff radius

The software application *NAMD* (Nanoscale Molecular Dynamics, developed at the University of Illinois at Urbana-Champaign, United States) has also recently emerged as a valuable molecular simulation tool. NAMD benefits from using a popular interactive molecular graphics rendering program called "visual molecular dynamics" (VMD) for simulation initialization and analysis, but its scripting is also compatible with other MD software applications including AMBER and CHARMM and was designed to be very efficient at scaling simulations to hundreds of processors on high-end parallel computing platforms, incorporating Ewald summation methods. The standard force field used is essentially that of Equation 8.14. Its chief popularity has been for SMD simulations, for which the setup of the steering conditions are made intuitive through VMD.

Other popular packages include those dedicated to more niche simulation applications, some commercial but many homegrown from academic simulation research communities, for example, *ab initio* simulations (e.g., *CASTEP, ONETEP, NWChem, TeraChem*, VASP), CG approaches (e.g., *LAMMPS, RedMD*), and mesoscale modeling (e.g., *Culgi*). CG tools of *oxDNA* and *oxRNA* have utility in mechanical and topological CG simulations of nucleic acids. Others for specific docking processes can be simulated (e.g., *ICM, SCIGRESS*) and molecular design (e.g., *Discovery studio, TINKER*) and folding (e.g., *Abalone, fold.it, FoldX*), with various valuable molecular visualization tools available (e.g., *Desmond, VMD*).

The physical computation time for MD simulations can take anything from a few minutes for a simple biomolecule containing ~100 atoms up to several weeks for the most complex simulations of systems that include up to ~10,000 atoms. Computational time has improved dramatically over the past decade however due to four key developments:

1 CPUs have got faster, with improvements to miniaturizing the effective minimum size of a single transistor in a CPU (~40 nm, at the time of writing) enabling significantly more processing power. Today, CPUs typically contain multiple *cores*. A core is an independent processing unit of a CPU, and even entry-level PCs and laptops use a CPU with two cores (*dual-core processors*). More advanced computers contain 4–8 core processors, and the current maximum number of cores for any processor is 16. This increase in CPU complexity is broadly consistent with *Moore's law* (Moore, 1965). Moore's law was an interpolation of reduction of transistor length scale with time, made by Gordon E. Moore, the cofounder of Intel, in which the area density of transistors in densely packed integrated circuits would continue to roughly double every two years. The maximum size of a CPU is limited by heat dissipation considerations (but this again has been improved recently by using CPU circulating cooling fluids instead air cooling). The atomic spacing of crystalline silicon is ~0.5 nm that may appear to place an ultimate limit on transistor miniaturization; however, there is increasing evidence that new transistor technology based on smaller length scale quantum tunneling may push the size down even further.

2 Improvements in simulation software have enabled far greater parallelizability to simulations, especially to route parallel aspects of simulations to different cores of the same CPU, and/or to different CPUs in a network cluster of computers. There is a limit, however, to how much of a simulation can be efficiently parallelized, since

there are inevitably bottlenecks due to one parallelized component of the simulation having to wait for the output from another before it can proceed, though of course multiple different simulations can be run simultaneously at least, for example, for simulations that are stochastic in nature and so are replicated independently several times or trying multiple different starting conditions for deterministic simulations.

3 Supercomputing resources have improved enormously, with dedicated clusters of ultrafast multiple-core CPUs coupled with locally dedicated ultrahigh bandwidth networks. The use of academic research supercomputers is in general extended to several users where supercomputing time is allocated to enable batch calculations in a block and can be distributed to different computer nodes in the supercomputing cluster.

4 Recent developments in GPU technology have revolutionized molecular simulations. Although the primary function of a GPU is to assist with the rendering of graphics and visual effects so that the CPU of a computer does not have to, a modern GPU has many features that are attractive to brute number-crunching tasks, including molecular simulations. In essence, CPUs are designed to be flexible in performing several different types of tasks, for example, involving communicating with other systems in a computer, whereas GPUs have more limited scope but can perform basic numerical calculations very quickly. A programmable GPU contains several dedicated multiple-core processors well suited to Monte Carlo methods and MD simulations with a computational power far in excess of a typical CPU. Depending on the design, a CPU core can execute up to $8\times$ 32 bit instructions per clock cycle (i.e., 256 bit per clock cycle), whereas a fast GPU used for 3D video-gaming purposes can execute $\sim 3200\times$ 32 bit instructions per clock, a bandwidth speed difference of a factor of ~ 400. A very-high-end CPU of, for example, having ~ 12 cores, has a higher clock rate of up to 2–3 GHz versus 0.7–0.8 GHz for GPUs, but even comparing coupling together four such 12-core CPUs, a single reasonable gaming GPU is faster by at least a factor of 5 and, at the time of writing, cheaper by a factor of at least an order of magnitude. GPUs can now be programmed relatively easily to perform molecular simulations, outperforming more typical multicore CPUs by a speed factor of ~ 100. GPUs have now also been incorporated into supercomputing clusters. For example, the *Blue Waters* supercomputer at the University of Urbana-Champaign is, as I write, the fastest supercomputer on any university campus and indeed one of the fastest supercomputers in the world, which can use four coupled GPUs that have performed a VMD calculation of the electrostatic potential for one frame of a MD simulation of the ribosome (an enormously complex biological machine containing over 100,000 atoms with a large length scale of a few tens of nm; see Chapter 2) in just 529 s using just one of these available GPUs, as opposed to ~ 5.2 h using on a single ultrafast CPU core.

The key advantage with GPUs is that they currently offer better performance per dollar than several of high-end CPU core applied together in a supercomputer, either over a distributed computer network or clustered together in the same machine. A GPU can be installed on an existing computer and may enable larger calculations for less money than building a cluster of computers. However, several supercomputing clusters have GPU nodes now. One caveat is that GPUs do not necessarily offer good performance on any arbitrary computational task, and writing code for a GPU can still present issues with efficient memory use.

One should also be mindful of the size of the computational problem and whether a supercomputer is needed at all. Supercomputers should really be used for very large jobs that no other machine can take on and not be used to make a small job run a bit more quickly. If you are running a job that does not require several CPU cores, you should really use a smaller computer; otherwise, you would just be hogging resources that would be better spent on something else. This idea is the same for all parallel computing, not just for problems in molecular simulation.

Another important issue to bear in mind is that for supercomputing resources you usually have a queue to wait in for your job to start on a cluster. If you care about the absolute time required to get an output from a molecular simulation, the real *door-to-door time* if you like, it may be worth waiting longer for a calculation to run on a smaller machine in order to save the queue time on a larger one.

KEY POINT 8.5

Driven by the video-gaming industry, GPUs have an improvement in speed of multicore processors that can result in reductions in computational time of two orders of magnitude compared to a multicore CPU.

8.2.10 ISING MODELS

The *Ising model* was originally developed from quantum statistical mechanics to explain the emergent properties of ferromagnetism (see Chapter 5) by using simple rules of local interactions between spin-up and spin-down states. These localized cooperative effects result in emergent behavior at the level of the whole sample. For example, the existence of either ferromagnetic (ordered) or paramagnetic (disordered) bulk emergent features in the sample, and phase transition behavior between the two bulk states, which is temperature dependent up until a critical point (in this example known as the *Curie temperature*). Many of the key features of cooperativity between smaller length scale subunits, leading to a different larger length scale emergent behavior are also present in the *Monod–Wyman–Changeux (MWC)* model, also known as the *symmetry* model, which describes the mechanism of cooperative ligand binding in some molecular complexes, for example, the binding of oxygen to the tetramer protein complex hemoglobin (see Chapter 2), as occurs inside red blood cells for oxygenating tissues.

In the MWC model, the phenomenon is framed in terms of *allostery;* that is, the binding of one subunit affects the ability of it and/or others in the complex to bind a ligand molecule, in which biochemists often characterize by the more heuristic *Hill equation* for the fraction θ of bound ligand at concentration C that has a dissociation rate constant of K_d (see Chapter 7):

$$(8.29) \qquad \theta = \frac{C^n}{K_d + C^n}$$

where n is the empirical *Hill coefficient* such that $n = 1$ indicates no cooperativity, $n < 1$ is negatively cooperative, and $n > 1$ is positively cooperative.

Hemoglobin has a highly positive cooperativity with $n = 2.8$ indicating that 2–3 of the 4 subunits in the tetramer interact cooperatively during oxygen binding, which results in a typically *sigmoidal* shaped *all-or-nothing* type of response if θ is plotted as a function of oxygen concentration. However, the Ising model puts this interaction on a more generic footing by evoking the concept of an *interaction energy* that characterizes cooperativity and thus can be generalized into several systems, biological and otherwise.

In the original Ising model, the magnetic potential energy of a sample in which the ith atom has a magnetic moment (or spin) σ_i is given by the Hamiltonian equation:

$$(8.30) \qquad \widehat{H} = K \sum_{ij} \sigma_i \sigma_j - B \sum_i \sigma_i$$

where K is a *coupling constant* with the first term summation over adjacent atoms with magnetic moments of spin +1 or −1 depending on whether they are up or down and B is an

A   **Ising model of conformational spread in magnetic domains**

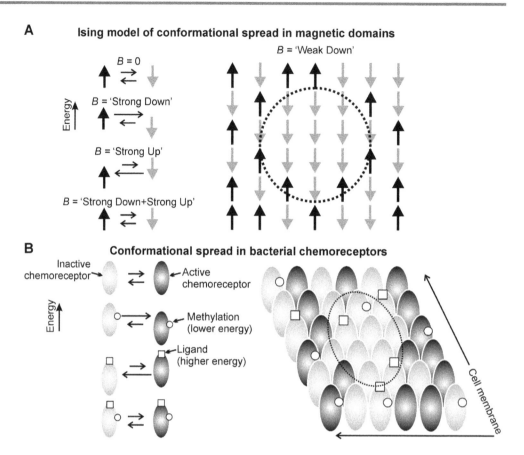

B   **Conformational spread in bacterial chemoreceptors**

FIGURE 8.4   Ising modeling in molecular simulations. (a) Spin-up (black) and spin-down (gray) of magnetic domains at low density in a population are in thermal equilibrium, such that the relative proportion of each state in the population will change as expected by the imposition of a strong external magnetic B-field, which is either parallel (lower energy) or antiparallel (higher energy) to the spin state of the magnetic domain (left panel). However, the Ising model of statistical quantum mechanics predicts an emergent behavior of conformational spreading of the state due to an interaction energy between neighboring domains, which results in spatial clustering of one state in a densely packed population as occurs in a real ferromagnetic sample, even in a weak external magnetic field (dashed circle right panel). (b) A similar effect can be seen for chemoreceptor expressed typically in the polar regions of cell membranes in certain rodlike bacteria such as *Escherichia coli*. Chemoreceptors are either in an active conformation (black), which results in transmitting the ligand detection signal into the body of the cell or an inactive conformation (gray) that will not transmit this ligand signal into the cell. These two states are in dynamic equilibrium, but the relative proportion of each can be biased by either the binding of the ligand (raises the energy state) or methylation of the chemoreceptor (lowers the energy state) (left panel). The Ising model correctly predicts an emergent behavior of conformational spreading through an interaction energy between neighboring chemoreceptors, resulting in clustering of the state on the surface of the cell membrane (dashed circle, right panel).

applied external magnetic field. The second term is an external field interaction energy, which increases the probability for spins to line up with the external field to minimize this energy. One simple prediction of this model is that at low B the nearest-neighbor interaction energy will bias nearest neighbors toward aligning with the external field, and this results in clustering of aligned spins, or *conformational spreading* in the sample of spin states (Figure 8.4a).

This same general argument can be applied to any system that involves a nearest-neighbor interaction between components that exist in two states of different energy, whose state can be biased probabilistically by an external field of some kind. At the long biological length scale, this approach can characterize flocking in birds and swarming in insects and even of social interactions in a population, but at the molecular length scale, there are several good exemplars of this behavior too. For example, bacteria swim up a concentration gradient of nutrients using a mechanism of a *biased random walk* (see later in this chapter), which involves the use of chemoreceptors that are clustered over the cell membrane. Individual chemoreceptors are two stable conformations, active (which can transmit the detected signal of a bound nutrient ligand molecule to the inside of the cell) and inactive (which cannot transmit the bound nutrient ligand signal to the inside of the cell). Active and inactive states of an isolated chemoreceptor have the same energy. However, binding of a nutrient ligand molecule lowers the energy of the inactive state, while chemical adaptation (here in the form of methylation) lowers the energy of the active state (Figure 8.4b, left panel).

However, in the cell membrane, chemoreceptors are tightly packed, and each in effect interacts with its four nearest neighbors. Because of steric differences between the active and inactive states, its energy is lowered by every neighboring chemoreceptor that is in the same conformational state but raised by every neighboring receptor in the different state. This interaction can be characterized by an equivalent nearest-neighbor interaction energy (see Shi and Duke, 1998), which results in the same sort of conformational spreading as for ferromagnetism, but now with chemoreceptors on the surface of cell membranes (Figure 8.4b, right panel), and this behavior can be captured in Monte Carlo simulations (Duke and Bray, 1999) and can be extended to molecular systems with more challenging geometries, for example, in a 1D ring of proteins as found in the bacterial flagellar motor (Duke et al., 2001).

This phenomenon of conformational spreading can often be seen in the placement of biophysics academics at conference dinners, if there is a "free" seating arrangement. I won't go into the details of the specific forces that result in increased or decreased energy states, but you can use your imagination.

KEY BIOLOGICAL APPLICATIONS: MOLECULAR SIMULATION TOOLS

Simulating multiple molecular processes including conformational changes, topology transitions, and ligand docking.

Worked Case Example 8.1: Molecular Simulations

A classical MD simulation was performed on a roughly cylindrical protein of diameter 2.4 nm and length 4.2 nm, of molecular weight 28 kDa, in a vacuum that took 5 full days of computational time on a multicore CPU workstation to simulate 5 ns.

a *Estimate with reasoning how long an equivalent simulation would take in minutes if using a particle mesh Ewald summation method. Alternatively, how long might it take if truncation was used?*

b *At best, how long a simulation in picosecond could be achieved using an ab initio simulation on the same system for a total computational time of 5 days? A conformational change involving ~20% of the structure was believed to occur over a time scale as high as ~1 ns. Is it possible to observe this event using a hybrid QM/MM simulation with the same total computational time as for part (a)?*

c *Using classical MD throughout, explicit water was then added with PBCs using a confining cuboid with a square base whose minimum distance to the surface of the protein was 2.0 nm in order to be clear of hydration shell effects. What is the molarity of water under these conditions? Using this information, suggest whether the project undergraduate student setting up the simulation will be able to witness the final result before they leave work for the day, assuming that they want to simulate 5 ns under the same conditions of system temperature and pressure and that there are no internal hydration cavities in the protein, but they decide to use a high-end GPU instead of the multicore CPU.*

(Assume that the density and molecular weight of water is 1 g/cm^{-3} and 18 Da, respectively, and that Avogadro's number is ~6.02 × 10^{23}.)

Answers

a   The mean molecular weight of an amino acid is ~137 Da, which contains a mean of ~19 atoms (see Chapter 2); thus, the number of atoms n in protein is

$$n = 19 \times (28 \times 103)/(0.137) = 3900 \text{ atoms}$$

PME method scales as $\sim O(n \log_e n)$, and classical MD scales as $\sim O(n^2)$ if no truncation used, so a difference of a factor $(n\log_e n/n^2) = (\log_e n/n)$, so computation time is

$$5\,\text{days} \times \log_e (3900/3900) = 0.011\,\text{days} = 15\,\text{min}$$

If truncation is used instead, this scales as $\sim O(n)$, so a difference of a factor $(n/n^2) = 1/n$, so computation time is

$$5\,\text{days} \times 3900 = 2\,\text{min}$$

b   The best saving in computational time for QM MD using HF approximations scales computation time as $\sim O(n^{2.7})$, so a difference of a factor $(n^{2.7}/n^2) = n^{0.7}$, so total simulation time achievable for a computational time of 5 days is

$$5\,\text{ns}/39000^{0.7} = 0.015\,\text{ns} = 15\,\text{ps}$$

For hybrid QM/MM, the simulation time is limited by QM. The total number of atoms simulated by the QM MD is $0.2 \times 3900 = 780$ atoms. The simulation time achievable for the same computational time as for part (a) is thus

$$\left(5\,\text{ns}/39000^2/780^{2.7}\right) = 1.2\,\text{ns}$$

which is >1 ns. Therefore, it should be possible to observe this conformational change.

c   The molarity is defined as the number of moles present in 1 L of the solution. The mass of 1 L of water is 1 kg, but 1 mole of water has a mass of 18 Da, or 18 g; thus, the molarity water is

$$1/(18 \times 10^{-3}) = 55.6\,\text{M}.$$

The number of water atoms (n_w) present is equal to the volume of water in liters multiplied by the molarity of water multiplied by Avogadro's number. The volume of water (V_w) present is equal to the volume of the confining cuboid minus the volume of the protein cylinder.

The square based of the confining cuboid must have a minimum edge length w of

$$w = 2.4\,\text{nm} + (2 \times 2.0\,\text{nm}) = 6.4\,\text{nm}$$

And the minimum length l of the cuboid is

$$l = 4.2\ \text{nm} + (2 \times 2.0\ \text{nm}) = 8.2\ \text{nm}$$

Thus, volume V_c of the cuboid is

$$V_c = (6.4 \times 10^{-9})^2 \times (8.2 \times 10^{-9}) = 3.2 \times 10^{-25}\ \text{m}^3 = 3.2 \times 10^{-22}\ \text{L}.$$

Similarly, the volume of the protein cylinder (V_p) is given by

$$V_p = (\pi \times (0.5 \times 2.4\ \text{nm})^2) \times 4.2\ \text{nm} = 1.9 \times 10^{-22}\ \text{L}$$

Thus,

$$V_c = (3.2 - 1.9) \times 10^{-22} = 1.3 \times 10^{-22}\ \text{L}$$

Thus,

$$n_w = 55.6 \times (6.02 \times 10^{23}) \times 1.3 \times 10^{-22} = 4350\ \text{atoms}$$

All of these atoms plus the 3900 atoms of the protein must be included explicitly in the simulation. A high-end GPU is faster than a typical multicore processor by a factor of ~100. Thus, the total computing time required is

$$(5\ \text{days}/100) \times (3900 + 4350)^2 / (3900)^2 = 0.52\ \text{days} = 12.5\ \text{h}$$

Would the project student be able to witness the results of the simulation before they go home from work? Yes, of course. Project students don't need sleep.....

8.3 MECHANICS OF BIOPOLYMERS

Many important biomolecules are polymers, for example, proteins, nucleic acids, lipids, and many sugars. Continuum approaches of polymers physics can be applied to many of these molecules to infer details concerning their mechanical properties. These include simple concepts such as the *freely jointed chain* (*FJC*) and *freely rotating chain* (*FRC*) with approximations such as the *Gaussian chain* (*GC*) and *wormlike chain* (*WLC*) approaches and how these *entropic spring* models predict responses of polymer extension to imposed force. Nonentropic sources of biopolymer elasticity to model less idealized biopolymers are also important. These methods enable theoretical estimates of a range of valuable mechanical parameters to be made and provide biological insight into the role of many biopolymers.

8.3.1 DISCRETE MODELS FOR FREELY JOINTED CHAINS AND FREELY ROTATING CHAINS

The *FJC* model, also called the "random-flight model" or "ideal chain," assumes n infinitely stiff discrete chain segments each of length b (known as the *Kuhn length*) that are freely jointed to, and transparent to, each other (i.e., parts of the chain can cross through other parts

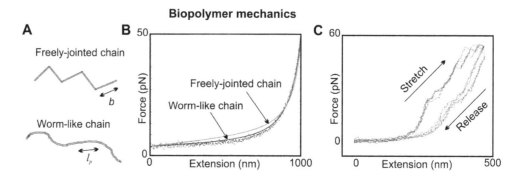

Biopolymer mechanics

FIGURE 8.5 Modeling biopolymer mechanics. (a) Schematic of freely jointed and wormlike chains (WLCs) indicating Kuhn (b) and persistence lengths (l_p), respectively. (b) Typical fits of WLC and freely jointed chain models to real experimental data obtained using optical tweezers stretching of a single molecule of the muscle protein titin (dots), which (c) can undergo hysteresis due to the unfolding of domains in each structure. (Data courtesy of Mark Leake, University of York, York, UK).

without impediment) such that the directional vector $\mathbf{r}_i$ of the ith segment is uncorrelated with all other segments (Figure 8.5a). The end-to-end distance of the chain R is therefore equivalent to a *3D random walk*. We can trivially deduce the *first moment* of the equivalent vector $\mathbf{R}$ (the time average over a long time) as

$$(8.31) \qquad \langle \mathbf{R} \rangle = \left\langle \sum_{i-1}^{n} \mathbf{r}_i \right\rangle = 0$$

Since there is no directional bias to $\mathbf{r}_i$. The *second moment* of $\mathbf{R}$ (the time average of the dot-product $\mathbf{R}^2$ over long time) is similarly given by

$$(8.32) \qquad \langle \mathbf{R}^2 \rangle = \langle \mathbf{R} \cdot \mathbf{R} \rangle = \langle R^2 \rangle = \left\langle \left(\sum_{i=1}^{n} \mathbf{r}_i \right)^2 \right\rangle = \sum_{i=1}^{n} \langle \mathbf{r}_i^2 \rangle + 2 \sum_{i=1}^{n} \sum_{j=i}^{n-i} \langle \mathbf{r}_i \cdot \mathbf{r}_{i\_j} \rangle$$

The last term equates to zero since $\mathbf{r}_i \cdot \mathbf{r}_j$ is given by $r_i r_j \cos \theta_{ij}$ where θ_{ij} is the angle between the segment vectors, and $\langle \cos \theta_{ij} \rangle = 0$ over long time for uncorrelated segments. Thus, the mean square end-to-end distance is given simply by

$$(8.33) \qquad \langle R^2 \rangle = n \langle \mathbf{r}_i^2 \rangle = nb^2$$

So the root mean square (rms) end-to-end distance scales simply with $\sqrt{n}$:

$$(8.34) \qquad \sqrt{\langle R_{FJC}^2 \rangle} = \sqrt{n} b$$

Another valuable straightforward result also emerges for the *radius of gyration* of the FJC, which is the mean square distance between all of the segments in the chain. This is identical to the mean distance of the chain from its center of mass, denoted by vector $\mathbf{R}_G$:

$$\mathbf{R}_G = \frac{1}{n}\sum_{j=1}^{n}\mathbf{R}_j$$

$$\therefore R_G^2 = \frac{1}{n}\sum_{i=1}^{n}\left\langle\left(\mathbf{R}_i - \mathbf{R}_G\right)^2\right\rangle = \frac{1}{n}\sum_{i=1}^{n}\left\langle\left(\mathbf{R}_i - \frac{1}{n}\sum_{j=1}^{n}\mathbf{R}_j\right)^2\right\rangle$$

(8.35)

$$= \frac{1}{n}\sum_{i=1}^{n}\left\langle\left(\mathbf{R}_i^2 - \frac{2\mathbf{R}_i}{n}\cdot\sum_{j=1}^{n}\mathbf{R}_j + \frac{1}{n^2}\sum_{j=1}^{n}\sum_{k=1}^{n}\mathbf{R}_j\cdot\mathbf{R}_k\right)^2\right\rangle$$

$$= \left\langle\frac{1}{n}\sum_{i=1}^{n}\mathbf{R}_i^2 - \frac{1}{n^2}\sum_{j=1}^{n}\sum_{k=1}^{n}\mathbf{R}_j\cdot\mathbf{R}_k\right\rangle$$

$$= \frac{1}{2n^2}\sum_{j=1}^{n}\sum_{k=1}^{n}\left\langle\left(\mathbf{R}_j - \mathbf{R}_k\right)^2\right\rangle$$

Using the result of Equation 8.34, we can then say

(8.36)
$$R_G^2 = \frac{1}{2n^2}\sum_{j=1}^{n}\sum_{k=1}^{n}|j-k|b^2 = \frac{b^2 n(n-2)}{6(n-1)}$$

A valuable general approximation comes from using high values of n in this discrete equation or by approximating the discrete summations as continuum integrals, which come to the same result of

(8.37)
$$R_G^2 \approx \frac{1}{2n^2}\int_0^n dj\int_0^n dk\,|j-k|b^2 = \frac{b^2}{n^2}\int_0^n dj\int_0^j dk\,(j-k) = \frac{nb^2}{6} = \frac{\left\langle R_{FJC}^2\right\rangle}{6}$$

$$\therefore \sqrt{R_G^2} \approx \frac{\sqrt{\left\langle R_{FJC}^2\right\rangle}}{\sqrt{6}} \approx 0.41\sqrt{\left\langle R_{FJC}^2\right\rangle}$$

Another useful result that emerges from similar analysis is the case of the radius of gyration of a *branched polymer*, where the branches are of equal length and joined at a central node such that if there are f such branches, the system is an f *arm star polymer*. This result is valuable for modeling biopolymers that form oligomers by binding together at one specific end, which can be reduced to

(8.38)
$$R_{G,f-\mathrm{arm}} = \frac{b^2}{2n^2}\int_0^{nif} dj\int_0^{nif} dk\,(j+k)f(f+1) + \frac{fb^2}{n^2} = \int_0^{nif} dj\int_0^j dk\,(j-k)$$

The first and second terms represent inter- and intra-arm contributions respectively, which can be evaluated as

(8.39)
$$R_{G,f-\mathrm{arm}} = \frac{nb^2}{6}\frac{(3f-2)}{f^2} = -(f)\langle R_G^2\rangle$$

Thus, at small f (= 1 or 2), α is 1, for much larger f, α is ~$3/f$ and $R_{G\,f\text{-arm}}$ decreases roughly as ~$3R_G/f$ The *FRC* has similar identical stiff segment assumptions as for the FJC; however, here the angle θ between position vectors of neighboring segments is fixed but the torsional angle ψ (i.e., the angle of twist of one segment around the axis of a neighboring segment) is free to rotate. A similar though slightly more involved analysis to that of the FJC, which considers the recursive relation between adjacent segments, leads to

(8.40)
$$\langle R^2 \rangle = nb^2 \left(1 + \frac{2\cos\theta}{1-\cos\theta} \left(1 - \frac{1-(\cos\theta)^n}{n(1-\cos\theta)} \right) \right)$$

For high values of n, this equation can be reduced to the approximation

(8.41)
$$\langle R^2 \rangle \approx nb^2 \left(\frac{1+\cos\theta}{1-\cos\theta} \right)$$

$$\therefore \sqrt{\langle R^2_{FJC} \rangle} \approx b\sqrt{n\left(\frac{2\cos\theta}{1-\cos\theta} \right)} \equiv \sqrt{\langle R^2_{FJC} \rangle} g(\theta)$$

This additional angular constraint often results in better agreement to experimental data, for example, using light scattering measurements from many proteins (see Chapter 3) due to real steric constraints due to the side groups of amino acid residues whose effective value of g is in the range ~2–3.

8.3.2 CONTINUUM MODEL FOR THE GAUSSIAN CHAIN

The GC is a continuum approximation of the FJC (which can also be adapted for the FRC) for large n (>100), applying the *central limit theorem* to the case of a 1D random walk in space. Considering n unitary steps taken on this random walk that are parallel to the x-axis, a general result for the probability $p(n, x)$ for being a distance x away from the origin, which emerges from *Stirling's approximation* to the log of the binomial probability density per unit length at large n, is

(8.42)
$$p(n,x) = \frac{1}{\sqrt{2\pi n}} \exp\left(\frac{-x^2}{2n^2} \right)$$

$$\therefore \langle x^2 \rangle = \int_{-\infty}^{+\infty} x^2 \, p \, dx = n$$

This integral is evaluated using as a standard Gaussian integral. Thus, we can say

(8.43)
$$p(n,x) = \frac{1}{\sqrt{2\pi \langle x^2 \rangle}} \exp\left(\frac{-x^2}{2\langle x^2 \rangle} \right)$$

However, a FJC random walk is in 3D, and so the probability density per unit volume $p(n, R)$ is the product of the three independent probability distributions in x, y, and z, and each step is of length b, leading to

(8.44)
$$p(n,R) = \left(\frac{3}{2\pi \langle R^2 \rangle} \right)^{3/2} \exp\left(\frac{-3R^2}{2\langle R^2 \rangle} \right) = \left(\frac{3}{2\pi nb^2} \right)^{3/2} \exp\left(\frac{-3R^2}{2nb^2} \right)$$

For example, the probability of finding the end of the polymer in a spherical shell between R and $R + dR$ is then $p(n, R) \cdot 4\pi R^2 dR$. This result can be used to predict the effective *hydrodynamic (or Stokes) radius* (R_H). Here, the effective rate of diffusion of all arbitrary sections between the ith and jth segments is additive, and since the diffusion coefficient is proportional to the reciprocal of the effective radius of a diffusing particle from the Stokes–Einstein equation (Equation 2.12), this implies

$$\frac{1}{R_H} = \left\langle \frac{1}{|\mathbf{r}_i - \mathbf{r}_j|} \right\rangle = \int_0^\infty \left(2\pi \langle |i-j| \rangle b^2/3\right)^{-3/2} \exp\left(\frac{-3r^2}{2\langle |i-j| \rangle b^2}\right).4\pi r^2 \frac{1}{r} dr$$

(8.45)
$$= \left(6/\pi\right)^{1/2} b^{-1} \langle |i-j| \rangle^{-1/2} = \left(6/\pi\right)^{1/2} b^{-1} n^{-2} \int_0^n di \int_0^n dj |i-j|^{-1/2} = \left(6/\pi\right)^{1/2} b^{-1} \left(8/3\right) n^{-1/2}$$

$$\therefore R_H = \frac{1}{8}\sqrt{\left(\frac{3\pi}{2}\right)} \sqrt{n}b = \frac{1}{8}\sqrt{\left(\frac{3\pi}{2}\right)} \sqrt{\langle R_{FJC}^2 \rangle} \approx 0.27 \sqrt{\langle R_{FJC}^2 \rangle}$$

Thus $\sqrt{\langle R_{FJC}^2 \rangle} > \sqrt{\langle R_{FJC}^2 \rangle} > R_G > R_H$.

8.3.3 WORMLIKE CHAINS

Biopolymers with relatively long segment lengths are accurately modeled as a *wormlike chain*, also known as the continuous version of the *Kratky–Porod* model. This model can be derived from the FRC by assuming small θ, with the result that

(8.46)
$$\langle R_{WLC}^2 \rangle = 2l_p R_{max} - 2l_p^2 \left(1 - \exp\left(\frac{-R_{max}}{l_p}\right)\right)$$

where
  l_p is known as the persistence length
  R_{max} is the maximum end-to-end length, which is equivalent to the contour length from the FJC of *nb*

In the limit of a very long biopolymer (i.e., $R_{max} \gg l_p$),

(8.47)
$$\langle R_{WLC}^2 \rangle \approx 2l_p R_{max} - 2l_p^2 (nb) \equiv \langle R_{FJC}^2 \rangle = nb^2$$
$$\therefore l_p \approx \frac{b}{2}$$

This is the *ideal chain limit* for a relatively compliant biopolymer. Similarly, in the limit of a short biopolymer (i.e., $R_{max} \ll l_p$),

(8.48)
$$\langle R_{WLC}^2 \rangle \approx R_{max}^2$$

This is known as the *rodlike limit* corresponding to a very stiff biopolymer.

8.3.4 FORCE DEPENDENCE OF POLYMER EXTENSION

The theory of flexible polymers described previously for biopolymer mechanics results in several different polymer conformations having the same free energy state (i.e., several different microstates). At high end-to-end extension close to a biopolymer's natural contour length, there are fewer conformations that the molecule can adopt compared to lower values. Therefore, there is an entropic deficit between high and low extensions, which is manifest as an entropic force in the direction of smaller end-to-end extension.

KEY POINT 8.6

There are fewer accessible microstates for high values of end-to-end extension compared to low values for a stretched biopolymer resulting in an entropic restoring force toward smaller extension values. The force scale required to straighten an FJC runs from zero to $\sim k_B T / b$.

The force response F of a biopolymer to a change in end-to-end extension r can be characterized in the various models of elasticity described earlier. If the internal energy of the biopolymer stays the same for all configurations (in other words that each segment in the equivalent chain is infinitely rigid), then entropy is the only contribution to the Helmholtz free energy A:

$$(8.49) \qquad A(R) = -TS(R) = k_B T \ln p(n, R)$$

where p is the radial probability density function discussed in the previous section. The restoring force experienced by a stretched molecule of end-to-end length R is then given by

$$(8.50) \qquad F = \frac{\partial A}{\partial R}$$

with a molecular stiffness k given by

$$(8.51) \qquad k = \frac{\partial F}{\partial R}$$

Using the result for p for the GC model of Equation 8.44 indicates

$$(8.52) \qquad A = \frac{3 k_B T R^2}{2 R_{max} b}$$

Therefore, the GC restoring force is given by

$$(8.53) \qquad F_{GC} = \frac{3 k_B T R}{R_{max} b}$$

Thus, the GC molecular stiffness is given by

$$(8.54) \qquad F_{GC} = \frac{3 k_B T}{n b^2} = \frac{3 k_B T}{\langle R_{FJC}^2 \rangle}$$

This implies that a GC exhibits a constant stiffness with extension, that is, it obeys *Hooke's law*, with the stiffness proportional to the thermal energy scale of $k_B T$ and to the reciprocal of b^2. This is a key weakness with the GC model; a finite force response even after the biopolymer is stretched beyond its own contour length is clearly unphysical. The FJC and WLC models are better for characterizing a biopolymer's force response with extension. Evaluating p for the FJC and WLC models is more complicated, but these ultimately indicate that, for the FJC model,

$$(8.55) \qquad R \approx R_{max} \left(\coth\left(\frac{F_{FJC} b}{k_B T}\right) - \frac{k_B T}{F_{FJC} b} \right) s$$

Similarly, for the WLC model,

$$(8.56) \qquad F_{WLC} \approx \left(\frac{k_B T}{l_p}\right)\left(\frac{1}{4\left(1 - R/R_{max}\right)^2} - \frac{1}{4} + \frac{R}{R_{max}}\right)$$

The latter formulation is most popularly used to model force-extension data generated from experimental single-molecule force spectroscopy techniques (see Chapter 6), and this approximation deviates from the exact solution by <10% for forces <100 pN. Both models predict a linear Hookean regime at low forces. Examples of both FJC and WLC fit to the same experimental data obtained from optical tweezers stretching of a single molecule of titin are shown in Figure 8.5b.

By integrating the restoring force with respect to end-to-end extension, it is trivial to estimate the work done, ΔW in stretching a biopolymer from zero up to end-to-end extension R. For example, using the FJC model, at low forces ($F < k_B T/b$ and $R < R_{max}/3$),

$$(8.57) \qquad \Delta W = \frac{3k_B T}{2nb^2}R^2 = \frac{3k_B T}{2}\left(\frac{R}{\langle R \rangle}\right)^2$$

In other words, the *free energy cost* per spatial dimension degree of freedom reaches $k_B T/2$ when the end-to-end extension reaches its mean value. This is comparable to the equivalent energy E to stretch an elastic cylindrical rod of radius r modeled by Hooke's law:

$$(8.58) \qquad \frac{\Delta R}{R} = \frac{\sigma}{Y} = \frac{F}{F_0}$$

$$\therefore E = \int_0^{\Delta R} f \mathrm{d}(\Delta R) = \frac{F_0(\Delta R)^2}{2}$$

where
  Y is the Young's modulus
  σ is the stress $F/\pi r^2$
  F_0 is $\pi r^2 Y$

However, at high forces and extensions, the work done becomes

$$(8.59) \qquad \Delta W \approx C - \frac{k_B T}{b}\ln = \left(1 - \frac{R}{R_{max}}\right)$$

where C is a constant. As the biopolymer chain extension approaches its contour length R_{max}, the *free energy cost* in theory diverges.

8.3.5 REAL BIOPOLYMERS

As discussed earlier, the GC chain model does not account well for the force response of real molecules to changes in molecular extension, which are better modeled by FJC, FRC, and WLC models. In practice, however, real biopolymers may have heterogeneity in b and l_p values, that is, there can be multiple different structural components within a molecule. We can model these effects in the force dependence as having N-such separate chains attached in parallel, both with FJC and WLC models. Thus, for N-FLC,

$$(8.60) \qquad R = \sum_{i=1}^{N} R_i = \sum_{i=1}^{N} R_{\mathrm{max},i} \left(\coth\left(\frac{Fb_i}{k_B T}\right) - \frac{k_B T}{Fb_i} \right)$$

Similarly, for N-WLC,

$$(8.61) \qquad F = \left(\frac{k_B T}{l_{p,i}}\right)\left(\frac{1}{4\left(1 - R_i/R_{\mathrm{max},i}\right)} - \frac{1}{4} + \frac{R_i}{R_{\mathrm{max},i}}\right)$$

$$R = \sum_{i=1}^{N} R_i$$

However, in practice, real biopolymers do not consist of segments that are freely jointed as in the FJC. There is real steric hindrance to certain conformations that are not accounted for in the FJC model, though to some extent these can be modeled by the additional angular constraints on the FRC model, which are also embodied in the WLC model. However, all of these models have the key assumption that different sections of the molecule can transparently pass through each other, which is obviously unphysical. More complex *excluded volume* WLC approaches can disallow such molecular transparency.

Also, the assumption that chain segments, or equivalent sections of the chain of l_p length scale, are infinitely stiff is unrealistic. In general, there may also be important enthalpic spring contributions to molecular elasticity. For example, the behavior of the chain segments can be more similar to Hookean springs, consistent with an increasing separation of constituent atoms with greater end-to-end distance, resulting in an opposing force due to either bonding interactions (i.e., increasing the interatomic separation beyond the equilibrium value, similarly for bond angle and dihedral components) and for nonbonding interactions, for example, the Lennard–Jones potential predicts a smaller repulsive force and greater vdW attractive force at increasing interatomic separations away from the equilibrium separation, and similar arguments apply to electrostatic interactions. These enthalpic effects can be added to the purely entropic components in modified *elastically jointed chain* models. For example, for a modified FJC model in which a chain is composed of freely jointed bonds that are connected by joints with some finite total enthalpic stiffness K,

$$(8.62) \qquad R = R_{\mathrm{max}} \left(\coth\left(\frac{Fb}{k_B T}\right) - \frac{k_B T}{Fb} \right)\left(1 + \frac{F}{K}\right)$$

or similarly, for an enthalpic-modified WLC model,

$$(8.63) \qquad F = \left(\frac{k_B T}{l_p}\right)\left(\frac{1}{4\left(1 - R/R_{\mathrm{max}} + F/K\right)^2} - \frac{1}{4} + \frac{R}{R_{\mathrm{max}}} - \frac{F}{K}\right)$$

Deviations exist between experimentally determined values of b and l_p parameters and those predicted theoretically on the basis of what appear to be physically sensible length scales for known structural information. For example, the length of a single amino acid in a protein, or the separation of a single nucleotide base pair in a nucleic acid, might on first inspection seem like sensible candidates for an equivalent segment of a chain. Also, there are inconsistencies between experimentally estimated values for R_G of some biopolymers compared against the size of cellular structures that contain these molecules (Table 8.1).

Table 8.1 Biopolymer Model Parameters of Real Molecules

Molecule	Number of Nucleotides or Amino Acids	Contour Length	b	l_p	R_G	Confining Structure
dsDNA; λ virus	49,000	15,000	100 nm	50 nm	0.5 μm	60 nm; capsid diameter
dsDNA; T2 virus	150,000	50,000	100 nm	50 nm	900	100 nm; capsid diameter
dsDNA; *E coli* bacteria	4.6×10^6	15 mm	100 nm	50 nm	5 μm	1–2 μm; cell width–length
dsDNA; human	4×10^9	2 m	100 nm	50 nm	0.2 mm	5–6 μm; nucleus diameter
ssDNA; STMV virus	1,063	0.7 μm	0.6 nm	2 nm	15 nm	7 nm; capsid diameter
Titin protein; Human muscle	30,000	1.7 μm	4 nm	2 nm	30 nm	3 nm; titin spacing
Titin protein; "random coil"	1,100	0.7 μm	0.4 nm	0.5 nm	5 nm	3 nm; titin spacing

Notes: Selection of polymer elasticity parameters from different types of DNA (ss/ds = single-strand/double-stranded) and muscle protein titin used in various biophysical elasticity experiments. R_G has been calculated from the measured b values. The general trend is that the estimated R_G is much larger than the length scale of their respective cellular "containers."

The persistence length values of typical biopolymers can vary from as low as ~0.2 nm (equivalent to a chain segment length of ~0.4 nm, which is the same as the length of a single amino acid residue) for flexible molecules such as denatured proteins and single-stranded nucleic acids, through tens of nanometers for stiffer molecules such as double-stranded DNA, up to several microns for very stiff protein filaments (e.g., the F-actin filament in muscle tissue has a persistence length of ~15 μm). The general trend for real biopolymers is that the estimated R_G is often much larger than the length scale of the cellular "container" of that molecule. This implies the presence of significant confining forces beyond just the purely entropic spring forces of the biomolecules themselves. These are examples of out of thermal equilibrium behavior and so require an external free energy input from some source to be maintained.

Deviations to models based on the assumption can also occur in practice due to repulsive self-avoidance forces (i.e., excluded volume effects) but also to attractive forces between segments of the biopolymer chain in the case of a poor solvent (e.g., hydrophobic-driven interactions). *Flory's mean field theory* can be used to model these effects. Here, the radius of gyration of a polymer relates to n through the *Flory exponent* υ as $R_G \sim n^\upsilon$. In the case of a so-called theta solvent, the polymer behaves as an ideal chain and υ = ½. In reality, the solvent is somewhere between extremes of being a *good solvent* (results in additional repulsive effects between segments such that the polymer conformation is an *excluded volume coil*) and υ = 3/5 and a *bad solvent* (the polymer is compacted to a sphere) and υ = 1/3.

KEY POINT 8.7

The radius of gyration of a biopolymer consisting n of segments in an equivalent chain varies as $\sim n^\upsilon$ where $1/3 \leq \upsilon \leq 3/5$.

In practice, the realistic modeling of biopolymer mechanics may involve not just the intrinsic force response of the biopolymer but also the effects of external forces derived from complicated potential energy functions, more similar to the approach taken for MD simulations (see earlier in this chapter). For example, modeling the translocation of a polymer through a nanopore, a common enough biological process, is nontrivial. The theoretical approaches that agree best with experimental data involve MD simulations combined with biopolymer mechanics, and the manner in which these simulations are constructed illustrates the general need for exceptionally fine precision not only in experimental measurement but in theoretical analysis also at this molecular level of biophysical investigation.

KEY BIOLOGICAL APPLICATIONS: BIOPOLYMER MECHANICS ANALYSIS TOOLS

Modeling molecular elasticity and force dependence of unfolding transitions.

Here, the polymer is typically modeled as either an FJC or WLC, but then a nontrivial potential energy function U is constructed corresponding to the summed effects from each segment, including contributions from a segment-bonding potential, chain bending, excluded volume effects due to the physical presence of the polymer itself not permitting certain spatial conformations, electrostatic contributions, and vdW effects between the biopolymer and nanopore wall. Then, the Langevin equation is applied that equates the force experienced by any given segment to the sum of the grad of U with an additional contribution due to random stochastic coupling to the water solvent thermal bath. By then, solving this force equation for each chain segment in $\sim 10^{-15}$ s time step predictions can be made as to position and orientation of each given segment as a function of time, to simulate the translocation process through the nanopore.

The biophysics of a matrix of biopolymers adds a further layer of complication to the theoretical analysis. This is seen, for example, in *hydrogels*, which consist of an aqueous network of concentrated biopolymers held together through a combination of solvation and electrostatic forces and long range vdW interactions. However, some simple power-law analysis makes useful predications here as to the variation of the elastic modulus G of a gel that varies as $\sim C^{2.25}$, where C is the biopolymer concentration, which illustrates a very sensitive dependence to gel stiffness for comparatively small changes in C (much of the original analytical work in this area, including this, was done by one of the pioneers of biopolymer mechanics modeling, Pierre Gilles de Gennes).

8.3.6 MODELING BIOMOLECULAR LIQUID–LIQUID PHASE SEPARATION

A recent emergence of myriad experimental studies of biomolecular liquid–liquid phase separation (LLPS, see Chapter 2) has catalyzed the development of new models to investigate how LLPS droplets form and are regulated. The traditional method to investigate liquid–liquid phase separation is *Flory–Huggins solution theory*, which models the dissimilarity in molecular sizes in polymer solutions taking into account the entropic and enthalpic changes that drive the free energy of mixing. It uses a random walk approach for polymer molecules on a lattice. To obtain the free energy, you need calculate the interaction energies for a given lattice square with its nearest neighbors, which can therefore involve "like" interactions such as polymer–polymer and solvent–solvent, or "unlike" interactions of polymer–solvent.

During phase separation, demixing results in an increase in order in reducing the number of available thermodynamic microstates in the system, thus a decrease in entropy. Therefore, the driving force in phase separation is the net gain in enthalpy that can occur in allowing like polymer and solvent molecules to interact, and much of modern theory research into LLPS lies in trying to understand specifically what causes this imbalance between entropic loss and enthalpic gain. For computational simplicity, a mean-field treatment is often used, which generates an average forcefield across the lattice to account for all neighbor effects. Flory–Huggins theory and its variants can be used to fit data from a range of experiments and construct phase diagrams, for example to predict temperature boundaries at which phase separation can occur and providing semi-quantitative explanations for the effects of ionic strength and sequence dependence of RNA and proteins have on the shape of these phase diagrams.

One weakness with traditional Flory–Huggins theory is that it fails to predict an interesting feature of real biomolecular LLPS in live cells, that droplets have a preferred length scale—Flory–Huggins theory predicts that either side of a boundary on the phase transition diagram a phase transition ultimately either completely mixes or, in the *super-saturation* side of the boundary, demixes completely after a sufficiently long time such that ultimately all of the polymer material will phase separate—in essence resulting in two physically separated states of just polymer and solvent; in other words, this would be manifest as one very large droplet inside a cell were it to occur. However, what is observed in general is a range of droplet diameters typically from a few tens of nanometers up to several hundred nanometers. One approach to account for this distribution and preferment of length scale involves modeling the effects of surface tension in droplet growth embodied in the classical nucleation

Szilard model, which allows reversible transitions to occur between droplets of different size as they exchange material through diffusion, both through droplets growing (or "ripening") and shrinking.

For a simple theoretical treatment of this effect, if N is the number of biomolecules phase separated into a droplet, then the free energy ΔG in the super-saturated regime can be modeled as $-AN + BN^{2/3}$ where A and B are positive constants that depend upon the enthalpic of interaction and energy per unit area at the interface between the droplet and the surrounding water solvent due to surface tension (see Worked Case Example 8.3). This results in a maximum value ΔG_{max} as a function of N, which thus serves as a nucleation activation barrier; from ΔG in the range $0-\Delta G_{max}$, the effect from surface tension slows down the rate of droplet growth, whereas above ΔG_{max}, droplet growth is less impaired by surface tension and more dominated by the net gain in enthalpy, at the expense of depleting the population of smaller droplets (this process is known as *Ostwald ripening* (also known as coarsening) and is a signature of liquid–liquid phase transitions).

Sviliard modeling can explain the qualitative appearance of size distributions of droplets, but it does not explain what drives the fine-tuning of the A and B parameters, which is down to molecular scale interaction forces. A valuable modeling approach which has emerged to address these questions has involved a *stickers-and-spacers* framework that has been adapted from the field of interacting polymers (Choi et al., 2020). This approach models interacting polymers as strings which contain several sticker regions separated by noninteracting spacer sequences, such that the spacers can undergo spatial fluctuations to enable interactions between stickers either from the same molecule or with neighbors. Stickers are defined at specific locations of the molecule due to the likelihood of electrostatic or hydrophobic interactions (which are dependent on the nature of the sequence of the associated polymer, typically either RNA or a peptide) to enable insight into how multivalent proteins and RNA molecules can drive phase transitions that give rise to biomolecular condensates. The reduction in complexity in modeling a polymer as a string with sticky regions avails the approach to coarse-graining computation and so has been very successful in simulating how droplets form over relatively extended durations of several microseconds even for systems containing many thousands of molecules.

8.4 REACTION, DIFFUSION, AND FLOW

Reaction–diffusion continuum mathematical models can be applied to characterize systems that involve a combination of chemical reaction kinetics and mobility through diffusional processes. This covers a wide range of phenomena in the life sciences. These mathematical descriptions can sometimes be made more tractable by first solving in the limits of being either *diffusion limited* (i.e., fast reaction kinetics, slow diffusion) or *reaction limited* (fast diffusion, slow reaction kinetics), though several processes occur in an intermediate regime in which both reaction and diffusion effects need to be considered. For example, the movement of molecular motors on tracks in general comprises both a random 1D diffusional element and a chemical reaction element that results in bias of the direction of the motor motion on the track. Other important continuum approaches include methods that characterize fluid flow in and around the biological structures.

8.4.1 MARKOV MODELS

The simplest general *reaction–diffusion equation* that considers the spatial distribution of the localization probability P of a biomolecule as a function of time t at a given point in space is as follows:

(8.64)
$$\frac{\partial P}{\partial t} = D\nabla^2 P + v(P)$$

The first term on the right-hand side of the equation is the diffusional term, with D being the diffusion coefficient. The second term v can often be a complicated function that embodies the reaction process. The form of the reaction–diffusion equation that describes Brownian diffusion of a particle in the presence of an external force (the reaction element) is also known as the *Smoluchowski equation* or *Fokker–Planck equation* (also sometimes referred to as the *Kolmogorov equation*), all of which share essentially the same core features in the mathematical modeling of Markov processes; as discussed previously in the context of localization microscopy (see Chapter 4), these processes are *memoryless* or *history independent*. Here, the function v is given by an additional drift drag component due to the action of an external force F on a particle of viscous drag coefficient γ, and the Fokker–Planck equation in 1D becomes

$$(8.65) \qquad \frac{\partial P}{\partial t} = D\frac{\partial^2 P}{\partial x^2} - \frac{1}{\gamma}\frac{\partial (FP)}{\partial x}$$

An example of such a reaction–diffusion process in biology is the movement of molecular motors on tracks, such as the muscle protein myosin moving along an F-actin filament, since they can diffuse randomly along the 1D but experience an external force due to the power stroke action of the myosin on the F-actin while a motor is in a given active state. However, in addition to this mechanical component, there is also a chemical reaction component that involves transitions into and out of a given active state of the motor. For example, if the active state of a motor is labeled as i, and a general different state is j, then the full reaction–diffusion equation becomes the Fokker–Planck equation plus the chemical transition kinetics component:

$$(8.66) \qquad \frac{\partial P}{\partial t} = D\frac{\partial^2 P_i}{\partial x^2} - \frac{1}{\gamma}\frac{\partial (FP_i)}{\partial x} + \sum_j \left(k_{ji}P_j - k_{ij}P_i\right)$$

where kj and kj_i are the equivalent off-rates from state i to j, and on-rates from state j to i, respectively. Most PDEs of reaction–diffusion equations are too complicated to be solved analytically. However, the model of the process of a single molecular motor, such as myosin, translocating on a track, such as F-actin, can be simplified using assumptions of constant translocation speed v and steady-state probabilities. These assumptions are realistic in many types of muscles, and they reduce the mathematical problem to a simple ordinary differential equation (ODE) that can be solved easily. Importantly, the result compares very well with experimentally measured dependence of muscle force with velocity of muscle contraction (see Chapter 6).

The case of motion of molecular motor translocation on a track can be reduced to the *minimal model* or *reduced model*, exemplified by myosin–actin. It assumes that there is just one binding power stroke event between the myosin head and actin with rate constant k_{on} and one release step with rate constant k_{off}. Assuming a distance d between accessible binding sites on the actin track for a myosin head, a linear spring stiffness κ of the myosin–actin link in the bound state (called the "crossbridge") and binding of myosin to actin occur over a narrow length range x_0 to $x_0 + \Delta x$, where the origin is defined as the location where the post–power stroke crossbridge is relaxed, and release is assumed to only occur for $x > 0$. Under these assumptions, the Fokker–Planck equation can be reduced to a simple ODE and solved to give the average force per crossbridge of

$$(8.67) \qquad \langle F \rangle = \frac{\kappa}{d}\left(1 - \exp\left(\frac{-k_{on}\Delta x}{v}\right)\right)\left(\frac{x_0^2\ v^2}{2\ k_s^2}\right)$$

An intuitive result from this model is that the higher the speed of translocation, the smaller the average force exerted per crossbridge, which is an obvious conclusion on the grounds of conservation of energy since there is a finite energy input due to the release of chemical potential energy from the hydrolysis of ATP that fuels this myosin motor translocation (see Chapter 2).

The key physical process is one of a *thermal ratchet*. In essence, a molecular motor is subject to Brownian motions due to thermal fluctuations and so can translocate in principle both forward and backward along its molecular track. However, the action of binding to the track coupled to an external energy input of some kind (e.g., ATP hydrolysis in the case of many molecular motors such as myosin molecules translocating on F-actin) results in molecular conformation changes, which *biases* the random motion in one particular direction, thus allowing it to ratchet along the track once averaged over long times in one favored direction as opposed to the other. But this does not exclude the possibility for having "backward" steps; simply these occur with a lower frequency than the "forward" steps due to the directional biasing of the thermal ratchet. The minimal model is of such importance that we discuss it formally in Worked Case Example 8.2.

Reaction–diffusion processes are also very important for *pattern formation* in biology (interested students should read Alan Turing's classic paper, Turing, 1952). In essence, if there is reaction–diffusion of two of more components with feedback and some external energy input, this results in *disequilibrium*, that is, stable out of thermal equilibrium behavior, embodied by the *Turing model*. With boundary conditions, this then results in oscillatory pattern formation. This is the physical basis of *morphogenesis*, namely, the diffusion and reaction of chemicals called morphogens in tissues resulting in distinctly patterned cell architectures.

Experimentally, it is difficult to study this behavior since there are often far more biomolecule components than just two involved, which make experimental observations difficult to interpret. However, a good bacterial model system exists in *Escherichia coli* bacteria. Here, the position of cell division of a growing cell is determined by the actions of just three proteins MinC, MinD, and MinE. Min is named as such because deleting a Min component results in the asymmetrical cell division in the formation of a small *mini cell* from one of the ends of a dividing cell. ATP hydrolysis is the energy input in this case.

A key feature in the Min system, which is contrasted with other different types of biological oscillatory patterns such as those purely in solution, is that one of the components is integrated into the cell membrane. This is important since the mobility in the cell membrane is up to three orders of magnitude lower than in the cytoplasmic solution phase, and so this component acts as a *time scale adapter*, which allows the period of the oscillatory behavior to be tuned to a time scale, which is much longer than the time taken to diffuse the length of a typical bacterial cell. For example, a typical protein in the cytoplasm has a diffusion coefficient of ~5 μm^2 s^{-1} so will diffuse a 1D length of few microns in ~1 s (see Equation 2.12). However, the cell division time of *E. coli* is typically a few tens of minutes, depending on the environmental conditions. This system is reduced enough in complexity to also allow synthetic Min systems to be utilized *in vitro* to demonstrate this pattern formation behavior (see Chapter 9). Genetic oscillations are another class of reaction–diffusion time-resolved molecular pattern formation. Similar time scale adapters potentially operate here, for example, in slow off-rates of transcription factors from promoters of genes extending the cycle time to hundreds of seconds.

8.4.2 REACTION-LIMITED REGIMES

In instances of comparatively rapid diffusion, the bottleneck in processes may be the reaction kinetics, and these processes are said to be reaction limited. A good example of this behavior is the turnover of subunits in a molecular complex. As discussed in Chapter 3, a popular experimental tool to probe molecular turnover is fluorescence recovery after photobleaching (FRAP).

With FRAP, an intense confocal laser excitation volume is used to photobleach a region of a fluorescently labeled biological sample, either a tissue or a single cell, and any diffusion and subsequent turnover and reincorporation of subunits into the molecular complex in that bleached zone will be manifested as a recovery in fluorescence intensity in the original bleached zone with time following the photobleach. This system can often be satisfactorily modeled as a closed reaction–diffusion environment confined to a finite volume, for example, that of a single cell, in which total content of the fluorescently labeled biomolecule of interest is in a steady state. Each specific system may have bespoke physical conditions that need to be characterized in any mathematical description; however, a typical system might involve a molecular complex that contains subunits that are either integrated into that complex or are diffusing free in the cell cytoplasm. Then, the general starting point might involve first denoting the following:

$n_F(t)$ = Number of fluorescently labeled biomolecule of a specific type under investigation, which is free in the cytoplasm at time t (≥ 0) following initial focused laser bleach

$n_B(t)$ = Number of fluorescently labeled biomolecule bound to a molecular complex identified in the bleach zone

$n_T(t)$ = Total number of fluorescently labeled biomolecules in the cell

$n_B^*(t)$ = Number of *photoactive* fluorescently labeled molecules bound to the molecular complex in the bleach zone

f = Fraction of photobleached fluorescently labeled biomolecules following initial focused laser bleach

k_1 = On-rate per fluorescently labeled biomolecule for binding to bleach-zone molecular complex

k_{-1} = Off-rate per fluorescently labeled biomolecule for unbinding from molecular complex If the fluorescently labeled molecules are in steady state, we can say

(8.68)
$$\frac{\partial n_T}{\partial t} = 0$$
$$\therefore n_T = n_F + n_B$$

Thus, n_T is a constant. The reaction–diffusion equations can be decoupled into separate diffusion and reaction components:

(8.69)
$$\frac{\partial n_T}{\partial t} = D\nabla^2 n_F$$

(8.70)
$$\frac{\partial n_B}{\partial t} = k_1 n_F - k_{-1} n_B$$

D is the effective diffusion coefficient of the fluorescently labeled biomolecule in the cell cytoplasm. Since the typical diffusion time scale τ is set by $\sim L^2/D$ where L is the typical length dimension of the cell (~ 1 μm if a model bacterial organism is used) and D for small proteins and molecular complexes in the cytoplasm is ~ 10 μm^2 s^{-1}, indicating $\tau \sim 10$ ms. However, the turnover in many molecular complexes is often over a time scale of more like ~ 1–100 s, at least two orders of magnitude slower than the diffusion time scale. Thus, this is clearly a reaction-limited regime and so the diffusion component can be ignored. At steady state just before the confocal volume photobleach, we can say that

(8.71)
$$\frac{\partial n_{B,S}}{\partial t} = 0 \therefore k_1 n_{F,S} = k_{-1} n_{B,S} \therefore k_1 = \frac{k_{-1} n_{B,S}}{n_T - n_{B,S}}$$

where $n_{B,S}$ and $n_{F,S}$ are the values of n_B and n_F, respectively, at steady state. If we assume that the binding kinetics of photoactive and photobleached labeled biomolecules are identical and that the population of bleached and nonbleached are ultimately well mixed, then

(8.72)
$$n_B^* = n_B(1-f)$$

Under general non-steady-state conditions,

$$(8.73) \qquad \frac{\partial n_B}{\partial t} = k_1 \left(n_T - n_B \right) - k_{-1} n_B$$

Solving this, substituting and rearranging then indicates

$$(8.74) \qquad n_B^* (t) = n_B (t)(1-f) = n_{B,S}^* \left(1 - (1--)\exp\left(\frac{-k_{-1} n_{T,t}}{n_T - n_{B,S}} \right) \right)$$

where α is the ratio of the bound photoactive component of the labeled biomolecule at zero time, immediately after the initial confocal volume photobleach to the bound photoactive component of Ssb-YPet at steady state. Since the fluorescence intensity $I_B(t)$ of the bound biomolecule component is proportional to the number of photoactive biomolecules, we can write

$$(8.75) \qquad I_B (t) = I_B (\infty)\left(1 - (1-\alpha)\exp\left(\frac{-t}{t_r} \right) \right)$$

where t_r is the characteristic exponential fluorescence recovery time constant, which in terms of the kinetics of molecular turnover can be seen to depend on the off-rate but not by the on-rate since the assumption of rapid diffusion means in effect that there are always available free subunits in the cytoplasm ready to bind to the molecular complex if a free binding site is available:

$$(8.76) \qquad t_r = \frac{n_T - n_{B,S}}{k_{-1} n_T}$$

In other words, by fitting an experimental FRAP trace to a recovery exponential function, the off-rate for turnover from the molecular complex can be estimated provided the total copy number of biomolecules in the cell that bounds to the complex at steady state can be quantified (in practice, this may be achieved in many cases using fluorescence intensity imaging-based quantification; see *in silico* methods section later in this chapter).

Other examples in biology of reaction-limited processes are those that simply only involve reaction kinetics, that is, there is no relevant diffusional process to consider. This occurs, for example, in molecular conformational changes and in the making and breaking processes of chemical bonds under external force. The statistical theory involves the principle of *detailed balance*.

The principle of detailed balance, or *microscopic reversibility*, states that if a system is in thermal equilibrium, then each of its degrees of freedom is separately in thermal equilibrium. For the kinetic transition between two stable states 1 and 2, the forward and reverse rate constants can be predicted using *Eyring theory* of QM oscillators. If the transition goes from a stable state 1 with free energy G_1 to a metastable intermediate state I with a higher free energy value G_I, back down to another stable state 2 with lower free energy G_2, then the transition rates are given by

$$(8.77) \qquad \begin{aligned} K_{12} &= \exp\left(\frac{-G_I - G_1}{k_B T} \right) \\[2mm] K_{21} &= v\exp\left(\frac{-G_I - G_2}{k_B T} \right) \end{aligned}$$

where v is the universal rate constant for a transition state, also given by $k_B T/h$ from Eyring theory, and has a value of $\sim 6 \times 10^{12}\,\text{s}^{-1}$ at room temperature, where h is Planck's constant. These rate equations form the basis of the *Arrhenius equation* that is well known to biochemists.

Processes exhibiting detailed balance are such that each elemental component of that process is equilibrated with its reverse process, that is, the implicit assumption is one of microscopic reversibility. In the case of single biomolecules undergoing some reaction and/ or molecular conformational change, this means the transition would go in reverse if the time coordinate of the reaction were inverted. A useful way to think about kinetic transitions from state 1 to state 2 is to consider what happens to the free energy of the system as a function of some *reaction coordinate* (e.g., the time elapsed since the start of the reaction/transition was observed). If the rate constant for $1 \rightarrow 2$ is k_{12} and for the reverse transition $2 \rightarrow 1$ is k_{21}, then the principle of detailed balance predicts that the ratio of these rate constants is given by the Boltzmann factor:

$$(8.78) \qquad \frac{k_{12}}{k_{21}} = \exp\left(\frac{-\Delta G}{k_B T}\right)$$

where ΔG is the free energy difference between states 1 and 2. For example, this process was encountered previously in Chapter 5 in probing the proportional of spin-up and spin-down states of magnetic atomic nuclei in NMR. This expression is generally true whether the system is in thermal equilibrium or not. Historically, rate constants were derived from ensemble average measurements of chemical flux from each reaction, and the ratio is k_{12}/k_{21}, which is the same as the ratio of the ensemble average concentrations of molecules in each state at equilibrium. However, using the ergodic hypothesis (see Chapter 6), a single-molecule experiment can similarly be used, such that each rate constant is given by the reciprocal of the average dwell time that the molecule spends in a given state. Therefore, by sampling the distribution of a lifetime of a single-molecule state experimentally, for example, the lifetime of a folded or unfolded state of a domain, these rate constants can in principle be determined, and thus the shape of the *free energy landscape* can be mapped out for the molecular transitions.

Note also that the implicit assumption in using the Boltzmann factor for the ratio of the forward and reverse rate constants is that the molecules involved in the rate processes are all in contact with a large thermal bath. This is usually the surrounding solvent water molecules, and their relatively high number is justification for using the Boltzmann distribution to approximate the distribution in their accessible free energy microstates. However, if there is no direct coupling to a thermal bath, then the system is below the *thermodynamic limit*, for example, if a component of a molecular machine interacts directly with other nearby components with no direct contact with water molecules or, in the case of gene regulation (see Chapter 9), where the number of repressor molecules and promoters can be relatively small, which potentially again do not involve direct coupling to a thermal bath. In these cases, the actual exact number of different microstates needs to be calculated directly, for example, using a binomial distribution, which can complicate the analysis somewhat. However, the core of the result is the same in the ratio of forward to reverse rate constants, which is given by the relative number of accessible microstates for the forward and reverse processes, respectively.

As we saw from Chapter 6, single-molecule force spectroscopy in the form of optical or magnetic tweezers or AFM can probe the mechanical properties of individual molecules. If these molecular stretch experiments are performed relatively slowly in *quasi-equilibrium* (e.g., to impose a sudden molecular stretch and then take measurements over several seconds or more before changing the molecular force again), then there is in effect microscopic reversibility at each force, meaning that the mechanical work done on stretching the molecule from one state to another is equal to the total free energy change in the system, and thus the detailed balance analysis earlier can be applied to explore the molecular kinetic process using slow-stretch molecular data.

Often however, molecular stretches using force spectroscopy are relatively rapid, resulting in hysteresis on a force-extension trace due to the events such as molecular domain unfolding (e.g., involving the breaking of chemical or other weaker molecular bonds), which drives the system away from thermal equilibrium. In the case of domains unfolding, they may not have enough time to refold again before the next molecular stretch cycle and so the transition is, in effect, irreversible, resulting in a stiffer stretch half-cycle compared to the relaxation half-cycle, (Figure 8.5c). In this instance, the *Kramers theory* can be applied. The Kramers theory is adapted to model systems in nonequilibrium states diffusing out of a potential energy well (interested readers should see the classic Kramers, 1940). The theory can be extended to making and breaking adhesive bonds (characterized originally in Bell's classic paper to model the adhesive forces between cells; see Bell, 1978). This theory can also be adapted into a two-state transition model of folded/unfolded protein domains in a single-molecule mechanical stretch-release experiment (Evans and Ritchie, 1997). This indicates that the mean expected unfolding force F_d is given by

$$(8.79) \qquad F_d = \frac{k_B T}{x_w} \log_e \left(\frac{r x_w}{k_d(0) k_B T} \right)$$

where $k_d(0)$ is the spontaneous rate of domain unfolding at zero restoring force on the molecule, assuming a uniform rate of molecular stretch r, while x_w is the *width of potential energy* for the unfolding transition. Therefore, by plotting experimentally determined F_d values, measured from either optical or magnetic tweezers or, more commonly, AFM, against the logarithm of different values of r, the value for x_w can be estimated.

A similar analysis may also be performed for refolding processes, and for typical domain motifs that undergo such unfolding/refolding transitions, a width value of ~0.2–0.3 nm is typical for the unfolding transition, while the width for the refolding transitions is an order of magnitude or more greater. This is consistent with molecular simulations suggesting that the unfolding is due to the unzipping of a few key hydrogen bonds that are of similar length scales to the estimated widths of potential energy (see Chapter 2), whereas the refolding process is far more complex requiring cooperative effects over a much longer length scales of the whole molecule. An axiom of structural biology known as *Levinthal's paradox* states that where a denatured/unfolded protein to refolded by exploring all possible refolding possibilities through the available stable 3D conformations one by one to determine which is the most stable folded conformation, this would take longer than the best estimate for the current age of the universe. Thus, refolding mechanisms typically employ short cuts that, if rendered on a plot of the free energy state of the molecule during the folding transition as a function of two orthogonal coordinates that define the shape of the molecule in some way (e.g., mean end-to-end extension of the molecule projected on the x and y axes), the free energy surface resembles a funnel, such that refolding occurs by in effect spiraling down the funnel aperture.

An important relation of statistical mechanics that links the mechanical work done W on a thermodynamic system of a nonequilibrium process to the free energy difference ΔG between the states for the equivalent equilibrium process is the *Jarzynski equality* that states

$$(8.80) \qquad \exp\left(\frac{-\Delta G}{k_B T} \right) = \left\langle \exp\left(\frac{-W}{k_B T} \right) \right\rangle$$

In other words, it relates the classical free energy difference between two states of a system to the ensemble average of *finite-time* measurements of the work performed in switching from one state to the other. Unlike many theorems in statistical mechanics, this was derived relatively recently (interested students should read Jarzynski, 1997). This is relevant to force spectroscopy of single biopolymers, since the mean work in unfolding a molecular domain is given by the work done in moving a small distance x_w against a force F_d, that is, $F_d x_w$, and so can be used as an estimator for the free energy difference, which in turn can be related to actual rate constants for unfolding.

Note that a key potential issue with routine reaction-limited regime analysis is the assumption that a given biological system can be modeled as a *well-mixed system*. In practice, in many real biological processes, this assumption is not valid due to compartmentalization of cellular components. Confinement and compartmentalization are often pervasive in biology, which needs to feature in more realistic, complex, mathematical analysis.

A commonplace example of reaction-limited processes is a chemical reaction that can be modeled using *Michaelis–Menten kinetics*. A useful application of this approach is to model the behavior of the molecular machine F1Fo ATP synthase, which is responsible for making ATP inside cells (see Chapter 2). This is an example of a cyclical model, with the process having ~100% efficiency for the molecular action of the F1Fo, requiring cooperativity between the three ATP binding sites:

1 There is an empty ATP-waiting state for one site brought about by the orientation of the rotor shaft of the F1 machine, while the other two sites are each bound to either an ATP or ADP molecule.
2 ATP binding to this site drives an 80°–90° rotation step of the rotor shaft, such that the wait time before the step is dependent on ATP concentration with a single exponential distribution indicating that a single ATP binding event triggers the event.
3 This binding triggers ATP hydrolysis and phosphate release but leaving an ADP molecule still bound in this site.
4 Unbinding ADP from this site is then coupled to another rotor shaft rotation of 30°–40°, such that the wait time does not depend on ATP concentration but has a peaked distribution, indicating that it must be due to at least two sequential steps of ~1 ms duration each (correlated with the hydrolysis and product release).
5 The 1–4 cycle repeats.

The process of F1-mediated ATP hydrolysis can be represented by a relatively simple chemical reaction in which ATP binds to F1 with rate constants k_1 and k_2 for forward and reverse processes, respectively, to form a complex ATP·F1, which then re-forms F1 with ADP and inorganic phosphate as the products in an irreversible manner with rate constant k_3. This reaction scheme follows Michaelis–Menten kinetics. Here, a general enzyme E binds reversibly to its substrate to form a metastable complex ES, which then breaks effectively irreversibly to form product(s) P while regenerating the original enzyme E. In this case, F1 is the enzyme and ATP is the substrate. If the substrate ATP is in significant excess over F1 (which in general is true since only a small F1 amount is needed as it is regenerated at the end of each reaction and can be used in subsequent reactions), this results in steady-state values of the complex F1·ATP, and it is easy to show that this results in a net rate of reaction k given by the *Michaelis–Menten equation*:

(8.81)
$$v = \frac{v_{max}[S]}{K_m + [S]}$$

where
  [S] is the substrate concentration and K_m is the Michaelis constant which is a measure of the substrate-binding affinity, so in the case of F1Fo [S] is ATP concentration
  $K_m = (k_1 + k_2)/k_1$
  k_{max} is the maximum rate of reaction given by $k_3[F_1 \cdot ATP]$ (square brackets here indicates a respective concentration)

In the case of a nonzero load on the F1 complex (e.g., it is dragging a large fluorescent F-actin filament that can be used as lever to act as a marker for rotation angle of the F1 machine), then the rate constants are weighted by the relevant Boltzmann factor $\exp(-\Delta W/k_B T)$ where

ΔW is the rotational energy that is required to work against the load to bring about rotation. The associated free energy released by hydrolysis of an ATP molecule to ADP and the inorganic phosphate P_i can also be estimated as

$$(8.82) \qquad \Delta G = \Delta G_0 + k_B T \ln\left(\frac{[ADP][P_i]}{[ATP]}\right)$$

where G_0 is the standard free energy for a hydrolysis of a single mole of ATP molecule at pH 7, equivalent to around -50 pN·nm per molecule (the minus sign indicates a spontaneously favorable reaction, and again square brackets indicate concentrations).

8.4.3 DIFFUSION-LIMITED REGIMES

In either the case of very slow diffusional processes compared to reaction kinetics or instances where there is no chemical reaction element but only diffusion, then diffusion-limited analysis can be applied. A useful parameter is the time scale of diffusion τ. For a particle of mass m experience a viscous drag γ this time scale is given by m/γ. If the time t following observation of the position of this tracked particle is significantly smaller than τ, then the particle will exhibit ballistic motion, such that its mean square displacement is proportional to $\sim t^2$. This simply implies that an average particle would not yet have had sufficient time to collide with a molecule from the surrounding solvent. In the case of t being much greater than τ, collisions with the solvent are abundant, and the motion in a heterogeneous environment is one of regular Brownian diffusion such that the mean square displacement of the particle is proportional to $\sim t$.

Regular diffusive motion can be analyzed by solving the *diffusion equation*, which is a subset of the reaction–diffusion equation but with no reaction component, and so the exact analytical solution will depend upon the geometry of the model of the biological system and the boundary conditions applied. The universal mathematical method used to solve this, however, is *separation of variables* into separate spatial and temporal components. The general diffusion equation is derived from the two Fick's equations. *Fick's first equation* simply describes the net flux of diffusing material, which in its 1D form indicates that the flux J_x of material diffusing parallel to the x-axis at rate D and concentration C is given by

$$(8.83) \qquad J_x = -D\frac{\partial C}{\partial x}$$

Fick's second equation simply indicates conservation of matter for the diffusing material:

$$(8.84) \qquad \frac{\partial C}{\partial t} = -\frac{\partial J_x}{\partial x}$$

Combining Equations 8.81 and 8.82 gives the diffusion equation, which can then also be written in terms of a probability density function P, which is proportional to C, generalized to all coordinate systems as

$$(8.85) \qquad \frac{\partial P}{\partial t} = D\nabla^2 P$$

Thus, the 1D form for purely radial diffusion at a distance r from the origin is

$$(8.86) \qquad \frac{\partial P(r,t)}{\partial t} = D\frac{1}{r^2}\frac{\partial}{\partial r}\left(r^2\frac{\partial P(r,t)}{\partial r}\right)$$

Similarly, the 1D form in Cartesian coordinates parallel to the x-axis is

(8.87)
$$\frac{\partial P(x,t)}{\partial t} = D\frac{\partial^2 P(x,t)}{\partial x^2}$$

This latter form has the most general utility in terms of modeling many real biophysical systems. Separation of variables to this equation indicates solutions of the form

(8.88)
$$P(x,t) = A(x)B(t)$$
$$\therefore \frac{1}{D}\frac{B'(t)}{B(t)} = \frac{A''(x)}{A(x)}$$

The left-hand side and right-hand side of this equation depend only on t and x, respectively, which can only be true if each is a constant. An example of this process is 1D passive diffusion of a molecule along a filament of finite length. If the filament is modeled as a line of length L and sensible boundary conditions are imposed such that P is zero at the ends of the rod (e.g., where $x = 0$ or L), this leads to a general solution of a Fourier series:

(8.89)
$$P(x,t) = \sum_{n=1}^{\infty} A_n \sin\left(\frac{\pi n x}{L}\right) \exp\left(D\left(\frac{\pi n}{L}\right)^2 t\right)$$

Another common 1D diffusion process is that of a narrow pulse of dissolved molecules at zero x and t followed by diffusive spreading of this pulse. If we model the pulse as a Dirac delta function and assume the "container" of the molecules is relatively large such that typical molecules will not encounter the container boundaries over the time scale of observation, then the solution to the diffusion equation becomes a 1D Gaussian function, such that

(8.90)
$$P(x,t) = \frac{1}{\sqrt{4\pi D t}}\exp\left(\frac{-x^2}{4Dt}\right)$$

Comparing this to a standard Gaussian function of $\sim\exp(-x^2/2\sigma^2)$, P is a Gaussian function in x of σ width equal to $\sqrt{(2Dt)}$. This is identical to the rms displacement of a particle with diffusion coefficient D after a time interval of t (see Equation 2.12). Several other geometries and boundary conditions can also be applied that are relevant to biological processes, and for these, the reader is guided to Berg (1983).

Fick's equations can also be easily modified to incorporate diffusion with an additional scaler drift component of speed v_d, yielding a modified version of the diffusion equation, the *advection–diffusion equation*, as in 1D Cartesian coordinates as

(8.91)
$$\frac{\partial P(x,t)}{\partial t} = D\frac{\partial^2 P(x,t)}{\partial x^2} - v_d\frac{\partial P(x,t)}{\partial x}$$

It can be seen that the advection–diffusion equation is a special case of the Fokker–Planck equation for which the external force F is independent of x, for example, if a biomolecule experiences a constant drift speed, then this will be manifested as a constant drag force. The resulting relation of mean square displacement with time interval for advection–diffusion is parabolic as opposed to linear as is the case for regular Brownian diffusion in the absence of drift, which is similar to the case for ballistic motion but over larger time scales.

A valuable, simple result can be obtained for the case of a particle exhibiting *diffusion-to-capture*. We can model this as, for example, a particle exhibiting regular Brownian diffusion for a certain distance before being captured by a "pure" absorber—a pure absorber is something that models the effects of many interactions between ligands and receptors such that the on-rate is significantly higher than the off-rate, and so within a certain time window of observation, a particle that touches the absorber is assumed to bind instantly and permanently. By simplifying the problem in assuming a spherically symmetrical geometry, it is easy to show that this implies (see Berg, 1983) that a particle released at a radius $r = b$ in between a pure spherical absorber at radius $r = a$ (e.g., defining a the surface of a cell organelle) and a spherical shell at $r = c$ (e.g., defining the surface of the cell membrane), such that $a < b < c$, the probability P that a particle released at $r = b$ is absorbed at $r = a$ is

(8.92)
$$P = \frac{a(c-b)}{b(c-q)}$$

So, in the limit $c \rightarrow \infty$, $P \rightarrow a/b$.

General analysis of the shape of experimentally determined mean square displacement relations for tracked diffusing particles can be used to infer the particular mode of diffusion (Figure 8.6). For example, a parabolic shape as a function of time interval over large time scales may be indicative of drift, indicating *directed diffusion*, that is, diffusion that is directionally biased by an external energy input, as occurs with molecular motors running on a track. Caution needs to be applied with such analysis however, since drift can also be due to

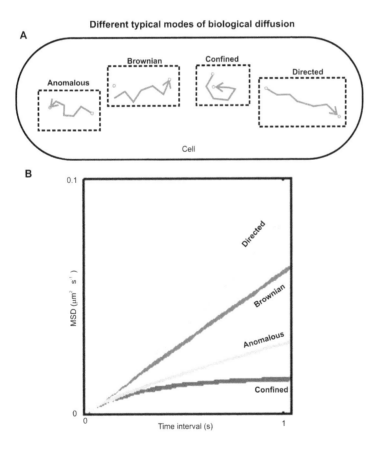

FIGURE 8.6 Different diffusion modes. (a) Schematic cell inside of which are depicted four different tracks of a particle, which exhibits four different common modes of diffusion or anomalous, Brownian confined and directed. (b) These modes have distinctly different relations on mean square displacement with time interval, such that directed is parabolic, Brownian is linear, confined is asymptotic, and anomalous varies as ~τ^{α}, where τ is the time interval and α is the anomalous diffusion coefficient such that $0 < \alpha < 1$.

slow slippage in a microscope stage, or thermal expansion effects, which are obviously then simply experimental artifacts and need to be corrected prior to diffusion analysis.

In the case of inhomogeneous fluid environments, for example, the mean square displacement of a particle after time interval τ as $\sim \tau^{\alpha}$ (see Equation 4.18) where a is the anomalous diffusion coefficient and $0 < \alpha < 1$, which depends on factors such as the crowding density of obstacles to diffusion in the fluid environment. Confined diffusion has a typically asymptotic shape with τ.

In practice, however, real experimental tracking data of diffusing particles are intrinsically stochastic in nature and also have additional source of measurement noise, which complicates the problem of inferring the mode of diffusion from the shape of the mean square displacement relation with τ. Some experimental assays involve good sampling of the position of tracked diffusing biological particles. An example of this is a protein labeled with a nanogold particle in an *in vitro* assay involving a similar viscosity environment using a tissue mimic such as collagen to that found in the living tissue. The position of the nanogold particle can be monitored as a function of time using laser dark-field microscopy (see Chapter 3). The scatter signal from the gold tag does not photobleach, and so they can be tracked for long durations allowing a good proportion of the analytical mean square displacement relation to be sampled thus facilitating inference of the type of underlying diffusion mode.

However, the issue of using such nanogold particles is that they are relatively large (tens of nm diameter, an order of magnitude larger than typical proteins investigated) as a probe and so potentially interfere sterically with the diffusion process, in addition to complications with extending this assay into living cells and tissue due to problems of specific tagging of the right protein and its efficient delivery into cells. A more common approach for live-cell assays is to use fluorescence microscopy, but the primary issue here is that the associated tracks can often be relatively truncated due to photobleaching of the dye tag and/or diffusing beyond the focal plane resulting in tracked particles going out of focus.

Early methods of mean square displacement analysis relied simply on determining a metric for linearity of the fit with respect to time interval, with deviations from this indicative of diffusion modes other than regular Brownian. To overcome issues associated with noise on truncated tracks, however, improved analytical methods now often involve aspects of *Bayesian inference*.

KEY POINT 8.8

The principle of Bayesian inference is to quantify the present state of knowledge and refine this on the basis of new data, underpinned by Bayes' theorem, emerging from the definition of conditional probabilities. It is one of the most useful statistical theorems in science.

In words, *Bayes' theorem* is simply *posterior = (likelihood × prior)/evidence*.

The definitions of these terms are as follows, which utilizes statistical nomenclature of the *conditional probability* "$P(A|B)$" meaning "the probability of A occurring given that B has occurred." The joint probability for both events A and B occurring is written as $P(A \cap B)$. This can be written as the probability of A occurring given that B has occurred:

$$(8.93) \qquad P(A \cap B) = P(B) \cdot P(A|B)$$

But similarly, this is the same as the probability of B occurring given that A has occurred:

$$(8.94) \qquad P(A \cap B) = P(A) \cdot P(B|A)$$

Thus, we can arrive at a more mathematical description of Bayes' theorem, which is

$$(8.95) \qquad P(A|B) = \frac{P(B|A) \cdot P(A)}{P(B)}$$

Thus, the $P(A)$ is the prior of A, $P(B)$ is the evidence, $P(B|A)$ is the likelihood for B given A, and $P(A|B)$ is the posterior for A given B (i.e., the probability for this occurring given prior knowledge of the probability distributions of A and B).

Bayes' theorem provides the basis for Bayesian inference as a method for the statistical testing of different hypotheses, such that $P(A|B)$ is then a specific model/hypothesis. Thus, we can say that

$$(8.96) \qquad P(\text{model}|\text{data}) = \frac{P(\text{data}|\text{model}) \cdot P(\text{model})}{P(\text{data})} = \frac{P(\text{data}|\text{model}) \cdot P(\text{model})}{\sum_{\text{all model}} P(\text{data}|\text{model})P(\text{model})}$$

$P(\text{data}|\text{model})$ is an experimentally observed dataset, $P(\text{model})$ is a prior for the given model, and $P(\text{model}|\text{data})$ is the resultant prediction for posterior probability that that model accounts for these data. Thus, the higher the values of $P(\text{model}|\text{data})$, the more likely it is that the model is a good one, and so different models can thus be compared with each other and ranked to decide which one is the most sensible to use.

For using Bayesian inference to discriminate between different models of diffusion for single-particle tracking data, we can define the following:

1 The *likelihood*, $P(d|w,M)$, where d represents the spatial tracking data from a given biomolecule and w is some general parameters of a specific diffusion model M. This is the probability distribution of the data for a given parameter from this model.
2 The *prior*, $P(w|M)$. This is the initial probability distribution function to any conditioning by the data; priors represent any initial estimate of the system, such as distribution of the parameters or the expected order of magnitude.
3 The *posterior*, $P(w|d,M)$. This is the distribution of the parameter following the conditioning by the data.
4 The evidence, $P(d|M)$. This acts as a normalization factor and is a portable unitless quantity.

One such analytical method that employs this strategy in inferring modes of diffusion from molecular tracking data is called "Bayesian ranking of diffusion" (BARD). This uses two stages in statistical inference: first, parameter inference, and second, model selection. The first stage infers the posterior distributions about each model parameter, which is defined as

$$(8.97) \qquad P(w|d,M) = \frac{P(d|w,M)P(w|M)}{P(d|M)}$$

The second stage is model selection:

$$(8.98) \qquad P(M|d) = \frac{P(d|M)P(M)}{P(d)}$$

$P(M|d)$ is a number that is the model posterior, or probability. $P(M)$ is a number that is the model prior (model priors are usually flat, indicating that the initial assumption is that all models are expected equally), $P(d|M)$ is a number that is the model likelihood, and $P(d)$ is a number that is a normalizing factor that accounts for all possible models. This now generates the posterior (i.e., probability) for a specific model. Linking the two stages is the term $P(d|M)$, the model likelihood, which is also the normalization term in the first stage. Comparing calculated normalized $P(d|M)$ values for each independent model then allows each model to be ranked from the finite set of diffusion models tried in terms of the likelihood that it accounts for the observed experimental tracking data (e.g., the mean square displacement vs τ values for a given tracked molecule).

8.4.4 FLUID TRANSPORT IN BIOLOGY

Previously in this book, we discussed several examples of fluid dynamics. The core mathematical model for all fluid dynamics processes is embodied in the *Navier–Stokes equation*. The Navier–Stokes equation results from Newton's second law of motion and conservation of mass. In a viscous fluid environment under pressure, the total force at a given position in the fluid is the sum of all external forces, F, and the divergence of the stress tensor σ. For a velocity of magnitude v in a fluid of density, ρ this leads to

$$(8.99) \qquad \rho\left(\frac{\mathrm{d}v}{\mathrm{d}t} + v \cdot \nabla v\right) = \nabla \cdot \sigma + F$$

The left-hand side of this equation is equivalent to the mass multiplied by total acceleration in Newton's second law. The stress tensor is given by the grad of the velocity multiplied by the viscous drag coefficient minus the fluid pressure, which after rearrangement leads to the traditional form of the Navier–Stokes equation of

$$(8.100) \qquad \frac{\mathrm{d}v}{\mathrm{d}t} = -(v \cdot \nabla) \cdot v - \frac{1}{\rho}\nabla p + \gamma\nabla^2 v + F$$

For many real biological processes involving fluid flow, the result is a complicated system of nonlinear PDEs. Some processes can be legitimately simplified, however, in terms of the mathematical model, for example, treating the fluid motion as an intrinsically 1D problem, in which case the Navier–Stokes equation can be reduced in dimensionality, but often some of the terms in the equation can be neglected, and the problem reduced to a few linear differential equations, and sometimes even just one, which facilitates a purely analytical solution. Many real systems, however, need to be modeled as coupled nonlinear PDEs, which makes an analytical solution difficult or impossible to achieve. These situations are often associated with the generation of fluid turbulence with additional random stochastic features, which are not incorporated into the basic Navier–Stokes equation becoming important in the emergence of large-scale pattern formation in the fluid over longer time scales. These situations are better modeled using Monte Carlo–based discrete computational simulations, in essence similar to the molecular simulation approaches described earlier in this chapter but using the Navier–Stokes equation as the relevant equation of motion instead of simply $F = ma$.

In terms of simplifying the mathematical complexity of the problem, it is useful to understand the physical origin and meaning of the three terms on the right-hand side of Equation 8.100.

The first of these is $-(v\cdot\nabla)\cdot v$. This represents the divergence on the velocity. For example, if fluid flow is directed to converge through a constriction, then the overall flow velocity will increase. Similarly, the overall flow velocity will decrease if fluid flow is divergent in the case of a widening of a flow channel.

The second term is $-\nabla p/\rho$. This represents the movement of diffusing molecules with changes to fluid pressure. For example, molecules will be forced away from areas of high-pressure changes, specifically, the tendency to move away from areas of higher pressure. If the local density of molecules is high, then a smaller proportion of molecules will be affected by changes in the local pressure gradient. Similarly, at low density, there is a greater proportion of the molecules that will be affected.

The third term is $\gamma\nabla^2 v$. This represents the effects of viscosity between neighboring diffusing molecules. For example, in a high viscosity fluid (e.g., the cell membrane), there will be a greater correlation between the motions of neighboring biomolecules than in a low viscosity fluid (e.g., the cell cytoplasm). The fourth term F as discussed is the net sum of all other external forces.

In addition to reducing mathematical descriptions to a 1D problem where appropriate, further simplifications can often involve neglecting the velocity divergence and external force terms, the governing equation of motion can often then be solved relatively easily, for example, in the case of Hagen–Poiseuille flow in a microfluidics channel (see Chapter 7) and also fluid movement involving laminar flow around biological structures that can be modeled acceptably by well-defined geometrical shapes. The advantage here is that well-defined shapes have easily derived analytical formulations for drag coefficients, for example, that of a sphere is manifested in Stokes law (see Chapter 6) or that of a cylinder that models the effects of drag on transmembrane proteins (see Chapter 3), or that of a protein or nucleic acid undergoing gel electrophoresis whose shape can be modeled as an ellipsoid (see Chapter 6). The analysis of molecular combining (see Chapter 6) may also utilize cylindrical shapes as models of biopolymers combed by the surface tension force of a retracting fluid meniscus, though better models involve characterizing the biopolymers as WLC, which is *reptating* within the confines of a virtual tube and thus treating the viscous drag forces as those on the surface of this tube.

Fluid dynamics analysis is also relevant to swimming organisms. The most valuable parameter to use in terms of determining the type of analytical method to employ to model swimming behavior is the dimensionless Reynolds number R_e (see Chapter 6). Previously, we discussed this in the context of allowing us to determine if fluid flow was turbulent or laminar, such that R_e values above a rough cutoff of ~2100 (see Chapter 7). The importance for swimming organisms is that there are two broadly distinct regimes of low Reynolds number swimmers and high Reynolds number swimmers.

High Reynolds number hydrodynamics involve in essence relatively large, fast organisms resulting in $R_e > 1$. Conversely, low Reynolds number hydrodynamics involve in essence relatively small, slow organisms resulting in $R_e < 1$. For example, a human swimming in water has a characteristic length scale of ~ 1 m, with a typical speed of ~1 m s^{-1}, indicating an R_e of ~10^6.

Low Reynolds number swimmers are typically microbial, often single-cell organisms. For example, *E. coli* bacteria have a characteristic length scale of ~1 μm but have a typical swimming speed of a few tens of microns every second. This indicates a R_e of ~10^{-5}.

High Reynolds number hydrodynamics are more complicated to analyze since the complete Navier–Stokes equation often needs to be considered. Low Reynolds number hydrodynamics is far simpler. This is because the viscous drag force on the swimming organism is far in excess of the inertial forces. Thus, when a bacterial cell stops actively swimming, then the cell motion, barring random fluid fluctuations, stops. In other words, there is no gliding.

For example, consider the periodic back-and-forth motions of an oar of a rowing boat in water, a case of high R_e hydrodynamics. Such motions will propel the boat forward. However, for low R_e hydrodynamics, this simple reciprocal motion in water will result in a net swimmer movement of zero. Thus, for bacteria and other microbes, their swimming motion needs to be more complicated. Instead of a back-and-forth, bacteria rotate a helical flagellum to propel themselves, powered by a rotary molecular machine called the flagellar motor, which is embedded in the cell membrane. This acts in effect as a helical propeller.

Bacteria utilize this effect in bacterial chemotaxis that enables them to swim up a chemical nutrient gradient to find source of food. The actual method involves a biased random walk. For example, the small length scale of *E. coli* bacteria means that they cannot operate a system of different locations of nutrient receptor on their cell surface to decide which direction to swim since there would be insufficient distance between any two receptors to discriminate subtle differences of a nutrient concentration over a ~1 μm length scale. Instead, they operate a system of alternating runs and tumbles in their swimming, such that a run involves a bacterial cell swimming in a straight line and a tumble involving the cell stopping but then undergoing a transient tumbling motion that then randomizes the direction of the cell body prior to another run that will then be in a direction largely uncorrelated with the previous run (i.e., random).

The great biological subtlety of *E. coli* chemotaxis is that the frequency of tumbling increases in a low nutrient concentration gradient but decreases in a high nutrient concentration gradient using a complicated chemical cascade and adaption system that involves feedback between chemical receptors on the cell membrane and the flagellar motor that power the cell swimming via diffusion-mediated signal transduction of a response regulator protein called CheY through the cell cytoplasm. This means that if there is an absence of food locally,

KEY BIOLOGICAL APPLICATIONS: REACTION–DIFFUSION MODELING TOOLS

Molecular mobility and turnover analysis; Molecular motor translocation modeling.

the cell will change its direction more frequently, whereas if food is abundant, the cell will tumble less frequently, ultimately resulting in a biased random walk in the average direction of a nutrient concentration gradient (i.e., in the direction of food). Remarkably, this system does not require traditional gene regulation (see Chapter 7) but relies solely on interactions between a network of several different chemotaxis proteins.

Worked Case Example 8.2: Reaction–Diffusion Analysis of Molecular Motors

In the minimal model for motor translocation on a 1D molecular track parallel to the x-axis release of a bound motor that occurs only in the region $x > 0$ and motor binding to the track that occurs along a small region x_0 to $x_0 + \Delta x$, with a mean spacing between binding sites on the track of d, the bound motor translocates along the track with a constant speed v, and the on-rate for an unbound motor to bind to the track is k_{on}, while the off-rate for a bound motor to unbind from the track is k_{off} (see Figure 8.7a).

a Calculate an ordinary differential equation of the 1D Fokker–Planck equation in steady state for the probability density function $P(x,t)$ for a bound motor.

b In reference to this equation, find expressions for the binding motor probability $P_b(x,t)$ in three different regions of $x_0 < x < x_0 + \Delta x$, $x_0 < x < 0$, and $x > 0$, and sketch a plot of P_b vs x for the case of a "fast" motor and a "slow" motor.

c If the force exerted by a crossbridge of a bound motor linked to the track is $-\kappa x$, such that it is a "forward" force for $x_0 < x < 0$ and a "backward" force for $x > 0$, calculate an expression for the mean work done W per binding site traversed and the average force $\langle F \rangle$ per binding site, and sketch a graph of the variation of $\langle F \rangle$ vs v.

Answers

a If $P(x,t)$ is the probability density function for a bound molecular motor to its track assumed to be a line along the x-axis, then the 1D Fokker–Planck equation can be rewritten as

$$(8.101) \qquad \frac{dP}{dt} = k_{on}\left(1-P\right) - k_{off}P - v\frac{dP}{dx} = 0$$

Assuming in steady state, this equation equates to zero.

Minimal model for molecular motor translocation

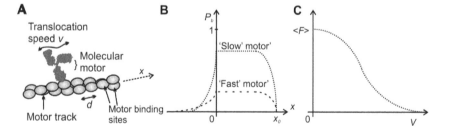

FIGURE 8.7   Reaction–diffusion analysis of molecular motor translocation. (a) Schematic of translocating motor on a 1D track. (b) Sketch of variation of motor binding probability on the track with respect to distance **x** for a "fast" and "slow" motor and (c) sketch of the predicted variation of average force per bound motor crossbridge (**F** with motor translocation speed **v**).

b The three different regions of motor behavior are equivalent to the "binding zone" ($x_0 < x < x_0 + \Delta x$), the "bound zone" ($x_0 < x < 0$), and the "unbinding zone" ($x > 0$). For the binding zone, k_{on} is constant, k_{off} is zero, and the initial boundary condition at $t = 0$ is $P = 0$. Solving Equation 8.99 under these conditions indicates a solution of

$$(8.102) \qquad P = 1 - \exp\left(\frac{-k_{on}(x - x_0)}{v}\right)$$

In the bound zone, the probability of binding will be a constant P_c and will match that in the binding zone when $x = x_0$:

$$(8.103) \qquad P_b = P_c = 1 - \exp\left(\frac{-k_{on}\Delta x}{v}\right)$$

In the unbinding zone, k_{on} is zero, k_{off} is a constant, and the initial boundary condition at $x = 0$ is $P = P_b$. Thus, solving Equation 8.99 under these conditions indicates

$$(8.104) \qquad P_b = C\exp\left(\frac{-k_{off}x}{v}\right)$$

Thus, for a motor–track interaction with the same binding/unbinding kinetics, a "slow" motor indicates a steeper magnitude of gradient of P with respect to x compared to a "fast" motor at equivalent x and has a larger value of P_c (see Figure 8.7b).

c The mean work done per binding site repeat across which the motor translocates is given by

$$(8.105) \qquad W = \int_{1\,binding\,site} F(x)P(x)dx = \int_0^{x_0}(-\kappa x P_c)dx + \int_0^{\infty}\left(-\kappa x P_c \exp\left(-\frac{k_{off}x}{v}\right)\right)dx$$

The first term is trivial; the second term can be solved by parts:

$$(8.106) \qquad W = \frac{1}{2}\kappa P_c x_0^2 - \kappa P_c \frac{v^2}{k_{off}^2}\kappa\left(1 - \exp\left(\frac{-k_{on}\Delta x}{v}\right)\right)\left(\frac{x_0^2}{2} - \frac{v^2}{k_{off}^2}\right)$$

Thus, the average force per binding site repeat is

$$(8.107) \qquad \langle F \rangle = \frac{W}{d} = \frac{\kappa}{d}\left(1 - \exp\left(-\frac{k_{on}\Delta x}{v}\right)\right)\left(\frac{x_0^2}{2} - \frac{v^2}{k_{off}^2}\right)$$

A sketch of the variation of $\langle F \rangle$ with v is shown in Figure 8.7c.

8.5 ADVANCED *IN SILICO* ANALYSIS TOOLS

Computational analysis of images is of core importance to any imaging technique, but the most basic light microscopy technique. The challenge here is to automate and objectify detection and quantitation. In particular, advances in low-light fluorescence microscopy have allowed single-molecule imaging experiments in living cells across all three domains of life to become commonplace. Many open-source computational software packages are available for image analysis, the most popular of which is *ImageJ*, which is based on a Java platform and has an extensive back catalog of software modules written by members of the user community. C/C++ is a popular computational language of choice for dedicated computationally efficient image analysis, while the commercial language MATLAB is ideal for the development of new image analysis routines as it benefits from extensive image analysis coding libraries as well as an enthusiastic user community that regularly exchanges ideas and new beta versions of code modules.

Single-molecule live-cell data are typically obtained in a very low signal-to-noise ratio regime often only marginally in excess of 1, in which a combination of detector noise, suboptimal dye photophysics, natural cell autofluorescence, and underlying stochasticity of biomolecules results in noisy datasets for which the underlying true molecular behavior is nontrivial to observe. The problem of faithful signal extraction and analysis in a noise-dominated regime is a "needle in a haystack" challenge, and experiments benefit enormously from a suite of objective, automated, high-throughput analysis tools that detect underlying molecular signatures across a large population of cells and molecules. Analytical computational tools that facilitate this detection comprise methods of robust localization and tracking of single fluorescently labeled molecules, analysis protocols to reliably estimate molecular complex stoichiometry and copy numbers of molecules in individual cells, and methods to objectively render distributions of molecular parameters. Aside from image analysis computational tools, there are also a plethora of bioinformatics computational tools that can assist with determining molecular structures in particular.

8.5.1 IMAGE PROCESSING, SEGMENTATION, AND RECOGNITION

The first stage of image processing is generally the eradication of noise. The most common way to achieve this is by applying low-pass filters in the Fourier space of the image, generated from the *discrete Fourier transform* (*DFT*), to block higher spatial frequency components characteristic of pixel noise, and then reconstructing a new image using the discrete inverse Fourier transform. The risk here is a reduction in equivalent spatial resolution. Other methods of *image denoising* operate in real space, for example, using 2D kernel convolution that convolves each pixel in the image with a well-defined 2D kernel function, with the effect of reducing the noise on each pixel by generating a weighted averaging signal output involving it and several pixels surrounding it, though as with Fourier filtering methods a caveat is often a reduction in spatial resolution manifest as increased blurring on the image, though some real space denoising algorithms incorporate components of feature detection in images and so can refine the denoising using an *adaptive kernel*, that is, to increase the kernel size in relatively featureless regions of an image but decrease it for regions suggesting greater underlying structure.

Several other standard image processing algorithms may also be used in the initial stages of image analysis. These include converting *pixel thresholding* to convert raw pixel intensity into either 0 or 1 values, such that the resulting image becomes a binary large object (BLOB). A range of different image processing techniques can then be applied to BLOBs. For example, *erosion* can trim off outer pixels in BLOB hotspots that might be associated with noise, whereas dilation can be used to expand the size of BLOBs to join proximal neighboring BLOBs into a single BLOB, the reason being again that this circumvents the artifactual effects of noise that can often have an effect to make a single foreground object feature in an image that appear to be composed of multiple separate smaller foreground objects.

Light microscopy bright-field images of cells and tissues, even using enhancement methods such as phase contrast or DIC (see Chapter 3), result in diffraction-limited blurring of the cellular boundary and of higher length scale tissue structures. Several *image segmentation* techniques are available that can determine the precise boundaries of cells and tissue structures. These enable cell bodies and structures to be separately masked out and subjected to further image analysis, in addition to generating vital statistics of cell and tissue structure dimensions. Vital statistics are valuable for single-cell imaging in model bacteria in being used to perform coordinate transformations between the Cartesian plane of the camera detector and the curved surface of a cell, for example, if tracking the diffusion of a membrane protein on the cell's surface. The simplest and most computationally efficient method involves pixel intensity thresholding that in essence draws a contour line around a cell image corresponding to a preset pixel intensity value, usually interpolated from the raw pixel data. Such methods are fast and reliable if the cell body pixel intensity distribution is homogeneous.

Often cell images are very close to each other to the extent that they are essentially touching, for example, cell-to-cell adhesions in a complex multicellular tissue or recent cell division events in the case of observation on isolated cells. In these instances simple pixel thresholding tools will often fail to segment these proximal cells. *Watershed* methods can overcome this problem by utilizing "flooding" strategies. Flooding seed points are first determined corresponding roughly to the center of putative cell objects, and the image is flooded with additional intensity added to pixel values radially from these seeds until two juxtaposed flooding wave fronts collide. This collision interface then defines the segmentation boundary between two proximal cells.

The most robust but computationally expensive cell image segmentation methods utilize prior knowledge of what the sizes and shapes of the specific cells under investigation are likely to be. These can involve Bayesian inference tools utilizing prior knowledge from previous imaging studies, and many of these approaches use *maximum entropy* approaches to generate optimized values for segmentation boundaries. Similarly, *maximum a posteriori* methods using Bayesian statistics operate by minimizing an objective energy function obtained from the raw image data to determine the most likely position of cell boundaries.

KEY POINT 8.9

Information entropy is a numerical measure that describes how uninformative a particular probability distribution is, ranging from a minimum of zero that is completely informative up to $\log(m)$, which is completely uninformative, where m is the number of mutually exclusive propositions to choose from. The principle of maximum entropy optimizes parameters in any given model on the basis that the information entropy is maximized and thus results in the most uninformative probability distribution possible since choosing a distribution that has lower entropy would assume information that is not possessed.

Many molecular structures visualized in biological specimens using high-spatial-resolution imaging techniques such as EM, AFM, and super-resolution light microscopy may exhibit a distribution on their orientation in their image data, for example, due to random orientation of the structure in the specimen itself and/or to random orientation of cells/tissues in the sample with respect to the imaging plane. To facilitate identification of such structures, there are algorithms that can rotate images and compare them with reference sources to compare in computationally efficient ways. This usually involves *maximum likelihood methods* that generate a pixel-by-pixel cross-correlation function between the candidate-rotated image and the reference source and find solutions that optimize the maximum cross-correlation, equivalent to *template matching*.

However, the most widely used method to recognize specific shapes and other topological and intensity features of images is *principal component analysis* (PCA). The general method of PCA enables identification of the principal directions in which the data vary. In essence, the *principal components*, or *normal modes*, of a general dataset are equivalent to axes that can be in multidimensional space, parallel to which there is most variation in the data. There can be several principal components depending on the number of independent/orthogonal features of the data. In terms of images, this in effect represents an efficient method of data compression, by reducing the key information parallel to typically just a few tens of principal components that encapsulate just the key features of an image.

In computational terms, the principal components of an image are found from a stack of similar images containing the same features and then calculating the eigenvectors and eigenvalues of the equivalent data covariance matrix for the image stack. An eigenvector that has the largest eigenvalue is equivalent to the direction of the greatest variation, and the eigenvector with the second largest eigenvalue is an orthogonal direction that has the next highest variation, and so forth for higher orders of variation beyond this. These eigenvectors can then be summed to generate a compressed version of any given image such that if more eigenvectors are included, then the compression is lower and the quality of the compressed image is better.

Each image of area $i \times j$ pixels may be depicted as a vector in $(i \times j)^2$ dimensional *hyperspace*, which has coordinates that are defined by the pixel intensity values. A stack of such images is thus equivalent to a cloud of points defined by the ends of these vectors in this hyperspace such that images that share similar features will correspond to points in the cloud that are close to each other. A pixel-by-pixel comparison of all images as is the case in maximum likelihood methods is very slow and computationally costly. PCA instead can reduce the number of variables describing the stack imaging data and find a minimum number of variables in the hyperspace, which appear to be uncorrelated, that is, the principal components. Methods involving *multivariate statistical analysis* (*MSA*) can be used to identify the principal components in the hyperspace; in essence, these generate an estimate for the covariance matrix from a set of images based on pairwise comparisons of images and calculate the eigenvectors of the covariance matrix.

These eigenvectors are a much smaller subset of the raw data and correspond to regions in the image set of the greatest variation. The images can then be depicted in a compressed form with relatively little loss in information since small variations are in general due to noise as the linear sum of a given set of the identified eigenvectors, referred to as an *eigenimage*.

PCA is particularly useful in objectively determining different *image classes* from a given set on the basis of the different combinations of eigenvalues of the associated eigenvectors. These different image classes can often be detected as clusters in m-dimensional space where m is the number of eigenvectors used to represent the key features of the image set. Several clustering algorithms are available to detect these, a common one being *k-means* that links points in m-dimensional space into clusters on the basis of being within a threshold nearest-neighbor separation distance. Such methods have been used, for example, to recognize different image classes in EM and cryo-EM images in particular (see Chapter 5) for molecular complexes trapped during the sample preparation process in different metastable conformational states. In thus being able to build up different image classes from a population of several images, averaging can be performed separately within each image class to generate often exquisite detail of molecular machines in different states. From simple *Poisson sampling statistics*, averaging across n such images in a given class reduces the noise on the averaged imaged by a factor of $\sim\sqrt{n}$. By stitching such averages from different image classes together, a movie can often be made to suggest actual dynamic movements involved in molecular machines. This was most famously performed for the stepping motion of the muscle protein myosin on F-actin filaments.

Algorithms involving the *wavelet transform* (*WT*) are being increasingly developed to tackle problems of denoising, image segmentation, and recognition. For the latter, WT is a complementary technique to PCA. In essence, WT provides an alternative to the DFT but decomposes an image into two separate spatial frequency components (high and low). These two components can then be combined to the given four separate resulting image outputs. The distribution of pixel intensities in each of these four WT output images represents a compressed signature of the raw image data. Filtering can be performed by blocking different regions of these outputs and then reconstructing a new image using the inverse WT in the same way as filtering of the DFT image output. Similarly various biological features can be identified such as cellular boundaries (which can be used for image segmentation) and in general image class recognition. PCA is ultimately more versatile in that the user has the choice to vary the number of eigenvectors to represent a set of images depending on what image features are to be detected; however, WT recognition methods are generally computationally more efficient and faster (e.g., "smart" missile technology for military applications often utilize WT methods). A useful compromise can often be made by generating hybrid WT/PCA methods.

8.5.2 PARTICLE TRACKING AND MOLECULAR STOICHIOMETRY TOOLS

As discussed previously, localization microscopy methods can estimate the position of the intensity centroid of a point spread function image (typically a fluorescent spot of a few hundred nanometers of width) from a single dye molecule imaged using diffraction-limited optics to pinpoint the location of the dye to a precision, which is one to two orders of magnitude smaller than the standard optical resolution limit (see Chapter 4). The way this is achieved in practice computationally is first to identify the candidate spots automatically using basic image processing as the previously to determine suitable hotspots in each image and second to define a small region of interest around each candidate spot, typically a square of 10–20 pixels edge length. The pixel intensity values in this region can then be fitted by an appropriate function, usually a 2D Gaussian as an approximation to the center of the Airy ring diffraction pattern, generating not only an estimate of the intensity centroid but also the total brightness $I(t)$ of the spot at a time t and of the intensity of the background in image in the immediate vicinity of each spot.

Spots in subsequent image frames may be linked provided that they satisfy certain criteria of being within a certain range of size and brightness of a given spot in a previous image frame and that their intensity centroids are separated by less than a preset threshold often set close to the optical resolution limit. If spots in two or more consecutive image frames satisfy these criteria, then they can be linked computationally to form a particle track.

The spatial variation of tracks with respect to time can be analyzed to generate diffusion properties as described earlier. But also the spot brightness values can be used to determine how many dye molecules are present in any given spot. This is important because many molecular complexes in biological systems have a *modular architecture*, meaning that they consist of multiple repeats of the same basic subunits. Thus, by measuring spot brightness, we can quantify the molecular stoichiometry subunits, depending upon what component a given dye molecule is labeling.

As a fluorescent-labeled sample is illuminated, it will undergo stochastic photobleaching. For simple photobleaching processes, the "on" time of a single dye molecule has an exponential probability distribution with a mean on time of t_b. This means that a molecular complex consisting of several such dye tags will photobleach as a function of time as $\sim\exp(-t/t_b)$ assuming no cooperative effects between the dye tags, such as quenching. Thus, each track $I(t)$ detected from the image data can be fitted using the following function:

$$(8.108) \qquad I(t) = I_0\exp\left(\frac{-t}{t_b}\right)$$

which can be compared with Equation 3.32 and used to determine the initial intensity I_0 at which all dye molecules are putatively photoactive. If I_D is the intensity of a single dye molecule under the same imaging conditions, then the stoichiometry S is estimated as

$$(8.109) \qquad S = \frac{I_0}{I_D}$$

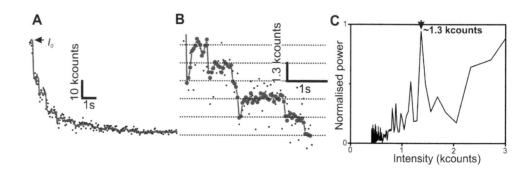

FIGURE 8.8 Measuring stoichiometry using stepwise photobleaching of fluorophores. (a) Example of a photobleach trace for a protein component of a bacterial flagellar motor called FliM labeled with the yellow fluorescent protein YPet, raw data (dots), and filtered data (line) shown, initial intensity indicated (arrow), with (b) a zoom-in of trace and (c) power spectrum of the pairwise difference distribution of these photobleaching data, indicating a brightness of a single YPet molecule of ~1.3 kcounts on the camera detector used on the microscope.

The challenge is often to estimate an accurate value for I_D. For purely *in vitro* assays, individual dye molecule can be chemically immobilized onto glass microscope coverslip surfaces to quantify their mean brightness from a population. However, inside living cells, the physical and chemical environment can often affect the brightness significantly. For example, laser excitation light can be scattered inside a cellular structure, but also the pH and the presence of ions such as chloride Cl^-, in particular, can affect the brightness of fluorescent proteins in particular (see Chapter 3).

Thus, it is important for *in vivo* fluorescence imaging to determine I_D in the same native cellular context. One way to achieve this is to extract the characteristic periodicity in intensity of each track, since steplike events occur in the intensity traces of tracks due to integer multiples of dye molecule photobleaching within a single sampling time window. Fourier spectral methods are ideal for extracting the underlying periodicity of these step events and thus estimating I_D (Figure 8.8).

Often molecular complexes will exhibit an underlying distribution of stoichiometry values. Traditional histogram methods of rendering this distribution for subsequent analysis are prone to subjectivity errors since they depend on the precise position of histogram bin edges and of the number of bins used to pool the stoichiometry data. A more objective and robust method involves *kernel density estimation* (*KDE*). This is a 1D convolution of the stoichiometry data using a Gaussian kernel whose integrated area is exactly 1 (i.e., representing a single point), and the width is a measure of the experimental error of the data measurement. This avoids the risks in particular of using too many histogram bins that suggest more multimodality in a distribution than really exists or too few bins that may suggest no multimodality when, in fact, there may well be some (Figure 8.9).

KEY POINT 8.10

KDE can be generally applied to all experimental datasets. If there are significant difficulties in estimating the experimental error in a given measurement, then you should probably not be doing that experiment. You may never need to use a histogram again!

Sometimes there will be periodicity in the observed experimental stoichiometry distributions of molecular complexes across a population, for example, due to modality of whole molecular complexes themselves in tracked spots. Such periodicity can again be determined using basic Fourier spectral methods.

Kernel density estimation

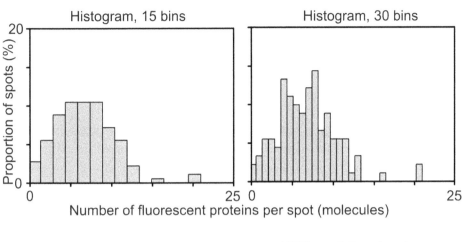

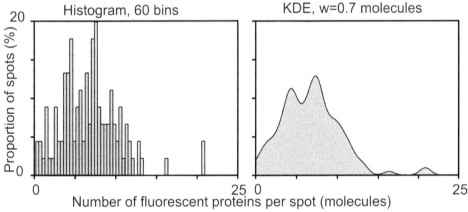

FIGURE 8.9 Using kernel density estimation to objectify stoichiometry distributions. Too few histogram bins can mask underlying multimodality in a distribution, whereas too many can give the impression of more multimodality than really exists. A kernel density estimation (here of width $w = 0.7$ molecules, equivalent to the measurement error in this case) generates an objective distribution.

8.5.3 COLOCALIZATION ANALYSIS FOR DETERMINING MOLECULAR INTERACTIONS IN IMAGES

Experiments to monitor molecular interactions have been revolutionized by multicolor fluorescence microscopy techniques (see Chapter 3). These enable one to monitor two or more different molecular components in a live biological sample labeled with different color dye that can be detected via separate color channels. The question then is to determine whether or not two such components really are interacting, in which case they will be colocalized in space at the same time. However, the situation is complicated by the optical resolution limit being several hundred nanometers, which is two orders of magnitude larger than the length scale over which molecular interactions occur.

The extent of colocalization of two given spots in separate color channels can best be determined computationally by constructing an overlap integral based on the fit outputs from the localization tracking microscopy. Each candidate spot is first fitted using a 2D Gaussian as before, and these functions are then normalized to give two Gaussian probability distribution functions g_1 and g_2, centered around (x_1, y_1) with width σ_1 and around (x_2, y_2) with width σ_2, respectively, and defined as

$$g_1(x,y) = exp\left(\frac{-\left((x-x_1)^2+(y-y_1)^2\right)}{2\sigma_1^2}\right)$$

(8.110)

$$g_2(x,y) = exp\left(\frac{-\left((x-x_2)^2+(y-y_2)^2\right)}{2\sigma_2^2}\right)$$

Their unnormalized overlap can be calculated from the integral of their product:

(8.111)
$$v_u = \iint_{allspace} g_1 g_2\, dxdy = C\exp\left(\frac{-(\Delta r)^2}{2(\sigma_1^2+\sigma_2^2)}\right)$$

where
   Δr is the distance between the centers of the Gaussians on the *xy* image plane
   C is a normalization constant equal to the maximum overlap possible for two perfectly
       overlapped spots so that

(8.112)
$$(\Delta r)^2 = (x_1-x_2)^2 + (y_1-y_2)^2$$

(8.113)
$$C = \frac{2\pi\sigma_1^2\sigma_2^2}{\sigma_1^2+\sigma_2^2}$$

The normalized overlap integral, v, for a pair of spots of known positions and widths, is then

(8.114)
$$v = \frac{v_u}{C} = \exp\left(\frac{-(\Delta r)^2}{2(\sigma_1^2+\sigma_2^2)}\right)$$

The value of this overlap integral can then be used as a robust and objective measure for the extent of colocalization. For example, for "optical resolution colocalization," which is the same as the Rayleigh criterion (see Chapter 3), values of ~0.2 or more would be indicative of putative colocalization.

One issue with this approach, however, is that if the concentration of one or the both of the molecules under investigation are relatively high (note, biologists often refer to such cells as having high *copy numbers* for a given molecule, meaning the average number of a that given molecule per cell), then there is an increasing probability that the molecules may appear to be colocalized simply by chance, but are in fact not interacting. One can argue that if two such molecules are not interacting, then after a short time their associated fluorescent spots will appear more apart from each other; however, in practice, the diffusion time scale may often be comparable to the photobleaching time scale, meaning that these chance colocalized dye molecules may photobleach before they can be detected as moving apart from each other.

It is useful to predict the probability of such chance colocalization events. Using nearest-neighbor analysis for a random distribution of particles on a 2D surface, we can determine the probability $p(r)\, dr$ that the distance from one fluorescent spot in the cell to another of the same type is between r and $r + dr$. This analysis results in effect in a value of the limiting concentration for detecting distinct spots (see Chapter 4) when averaged over a whole cell and depends upon certain geometrical constraints (e.g., whether a molecule is located in the three spatial dimension cell volume or is confined to two dimensions as in the cell membrane or one dimension as for translocation of molecular motors on a filament). As a rough guide, this is equivalent to ~10^{17} fluorescent spots per liter for the case of molecules inside the cell volume, which corresponds to a molar concentration of ~100 nM. For interested students, an elegant mathematical treatment for modeling this nearest-neighbor distance was first formulated by one of the great mathematicians of modern times (Chandrasekhar, 1943) for a similar imaging problem but involving astrophysical observations of stars and planets.

For example, consider the 2D case of fluorescent spots located in the cell membrane. The probability $p(r)dr$ must be equal to the probability that there are zero such fluorescent spots particles in the range 0–r, multiplied by the probability that a single spot exists in the annulus zone between r and $r + dr$. Thus,

$$(8.115) \qquad p(r)dr = \left(1 - \int_0^r p(r')dr'\right) \cdot 2\pi r n \, dr$$

where n is the number of fluorescent spots per unit area in the patch of cell membrane observed in the focal plane using fluorescence microscopy. Solving this equation and using the fact that this indicates that $p \to 2\pi rn$ in the limit $r \to 0$, we obtain:

$$(8.116) \qquad p(r) = 2\pi rn \exp\left(-\pi r^2 n\right)$$

Thus, the probability $p_1(w)$ that the nearest-neighbor spot separation is greater than a distance w is

$$(8.117) \qquad p_1(w) = 1 - \int_0^w p(r)dr = 1 - \int_0^w 2\pi rn \exp\left(-\pi r^2 n\right)dr = \exp(-\pi w^2 n)$$

The effective number density per unit area, n, at the focal plane is given by the number of spots N_{mem} observed in the cell membrane on average for a typical single image frame divided by the portion, ΔA, of the cell membrane imaged in focus that can be calculated from a knowledge of the depth of field and the geometry of the cell surface (see Chapter 3). (Note in the case of dual-color imaging, N_{mem} is the total summed mean number of spots for both color channels.)

The probability that a nearest-neighbor spot will be a distance less than w away is given by $1 - p_1(w) = 1 - \exp(-\pi w^2 N_{mem}/A)$. If w is then set to be the optical resolution limit (~200–300 nm), then an estimate for $1 - p_1(w)$ is obtained. Considering the different permutations of colocalization or not between two spots from *different* color channels indicates that the probability of chance colocalization p_{chance} of two different color spots is two-thirds of that value:

$$(8.118) \qquad p_{chance} = \frac{2\left(1 - \exp\left(-\pi w^2 N_{mem}/A\right)\right)}{3}$$

It is essential to carry out this type of analysis in order to determine whether the putative colocalization observed from experimental data is likely due to real molecular interactions or just to expectations from chance.

8.5.4 CONVOLUTION MODELING TO ESTIMATE PROTEIN COPY NUMBERS IN CELLS

Fluorescence images of most cells containing which specific mobile molecule has been fluorescently labeled inevitably are composed of a distinct spot component, that is, distinct fluorescent spots that have been detected and tracked and a diffusive pool component. The latter may be composed of fluorescent spots that have not been detected by automated code, for example, because the local nearest-neighbor separation distance was less than the optical resolution, but also may often comprise dimmer, faster moving spots that are individual subunits of the molecular complexes that are brighter and diffuse more slowly and thus are more likely to be detected in tracking analysis.

Convolution models can quantify this diffusive pool component on a cell-by-cell basis. This is valuable since it enables estimates to be made of the total number of fluorescently labeled molecules in a given cell if combined with information from the number and stoichiometry of tracked distinct fluorescent spots. In other words, the distribution of copy number for that molecule across a population of cell can be quantified, which is useful in its own right, but which can also be utilized to develop models of gene expression regulation.

The diffusive pool fluorescence can be modeled as a 3D convolution integral of the normalized PSF, P, of the imaging system, over the whole cell. Every pixel on the camera detector of the fluorescence microscope has a physical area ΔA with equivalent area dA in the conjugate image plane of the sample mapped in the focal plane of the microscope:

$$(8.119) \qquad dA = \frac{\Delta A}{M}$$

where M is the total magnification between the camera and the sample. The measured intensity, I', in a conjugate pixel area dA is the summation of the foreground intensity I due to dye molecules plus any native autofluorescence (I_a) plus detector noise (I_d). I is the summation of the contributions from all of the nonautofluorescence fluorophores in the whole of the cell:

$$(8.120) \qquad I\left(x_0,y_0,z_0\right)dA = \sum_{i=1}^{\text{All cell voxels}} E^{`} I_s P\left(x_i - x_0, y_i - y_0, z_i - x_0\right)$$

where

I_s is the integrated intensity of a single dye (i.e., its brightness)
ρ is the dye density in units of molecules per voxel (i.e., one pixel volume unit)
E is a function that represents the change in the laser profile excitation intensity over a cell

In a uniform excitation field, $E = 1$, for example, as approximated in wide-field illumination since cells are an order of magnitude smaller than the typical Gaussian sigma width of the excitation field at the sample. For narrow-field microscopy, the excitation intensity is uniform with height z but has a 2D Gaussian profile in the lateral xy plane parallel to the focal plane. Assuming a nonsaturating regime for fluorescent photon emission of a given dye, the brightness of that dye, assuming simple single-photon excitation, is proportional to the local excitation intensity; thus,

$$(8.121) \qquad E\left(x,y,z\right) = \exp\left[-\left(\frac{x^2 + y^2}{2\sigma_{xy}^2}\right)\right]$$

where σ_{xy} is the Gaussian width of the laser excitation field in the focal plane (typically a few microns). In Slimfield, there is an extra small z dependence also with Gaussian sigma width, which is ~2.5 that of the σ_{xy} value (see Chapter 3). Thus,

$$(8.122) \quad I\left(x_0,y_0,z_0\right)dA = \sum_{i=1}^{\text{All cell voxels}} \rho I_s \exp\left[-\left(\frac{x^2 + y^2}{2\sigma_{xy}}\right)\right] P\left(x_i - x_0, y_i - y_0, z_i - z_0\right)$$

Defining $C(x_0,y_0,z_0)$ as a numerical convolution integral and also comprising the Gaussian excitation field dependence over a cell, and assuming the time-averaged dye density is uniform in space, we can calculate the dye density, in the case of a simple single internal cellular compartment, as

$$(8.123) \qquad \rho = \frac{I(x_0,y_0,z_0)\mathrm{d}A}{C(x_0,y_0,z_0)I_s} = \frac{I'(x_0,y_0,z_0)-(I_a+I_d)}{C(x_0,y_0,z_0)I_s}$$

The mean value of the total background noise $(I_a + I_d)$ can usually be calculated from *parental* cells that do not contain any foreign fluorescent dye tags. In a more general case of several different internal cellular compartments, this model can be extended:

$$(8.124) \qquad I(x_0,y_0,z_0)\mathrm{d}A = I_s\exp\left[-\left(\frac{x^2+y^2}{2\sigma_{xy}}\right)\right] \overset{\substack{\text{Number of}\\\text{compartments}}}{\underset{j=1}{\sum}} \overset{\substack{\text{Compartment}\\\text{voxels}}}{\underset{i=1}{\sum}} \rho_j P(x_i - x_0, y_i - y_0, z_i - z_0)$$

where ρ_j is the mean concentration of the *j*th compartment, which can be estimated from this equation by least squares analysis. An important feature of this model is that each separate compartment does not necessarily have to be modeled by a simple geometrical shape (such as a sphere) but can be any enclosed 3D volume provided its boundaries are well-defined to allow numerical integration.

Distributions of copy numbers can be rendered using objective KDE analysis similar to that used for stoichiometry estimation in distinct fluorescent spots. The number of protein molecules per cell has a typically asymmetrical distribution that could be fitted well using a *random telegraph model* for gene expression that results in a gamma probability distribution $p(x)$ that has similarities to a Gaussian distribution but possesses an elongated tail region:

$$(8.125) \qquad p(x) = \frac{x^{a-1}\exp(-x/b)}{b^a\Gamma(a)}$$

where $\Gamma(a)$ is a gamma function with parameters a and b determined from the two *moments* of the gamma distribution by its mean value m and standard deviation, such that $a = m^2/\sigma^2$ and $b = \sigma^2/m$.

8.5.5 BIOINFORMATICS TOOLS

Several valuable computational *bioinformatics* tools are available, many developed by the academic community and open source, which can be used to investigate protein structures and nucleotide sequences of nucleic acid, for example, to probe for the appearance of the same sequence repeat in different sets of proteins or to predict secondary structures from the primary sequences. These algorithms operate on the basis that amino acids can be pooled in different classes according to their physical and chemical properties (see Chapter 2). Thus, to some extent, there is interchangeability between amino acids within the same class such that an ultimate secondary, tertiary, and quaternary structure may be similar. These tools are also trained using predetermined structural data from EM, NMR, and x-ray diffraction methods (see Chapter 5), enabling likely structural motifs to be identified from their raw sequence data alone. The most common algorithm used is the *basic alignment search tool* (*BLAST*).

Homology modeling (also known as *comparative modeling*) is a bioinformatics tool for determining atomic-level resolution molecular structures. It is most commonly performed on proteins operating by interpolating a sequence, usually the primary structure of amino acid residues, of a molecule with unknown structure to generate an estimate for that structure, known as the *target*, onto a similar structure of a molecule known as the *template*, which is of the same or similar structural family. The algorithms have similarity to those used in BLAST.

Protein structures are far more conserved than protein sequences among such *homologs* perhaps for reasons of *convergent evolution*; similar tertiary structures evolve from different primary structures which imparts a selective advantage on that organism (but note that primary sequences with less than 20% sequence identity often belong to different structural families). Protein structures are in general more conserved than nucleic acid structures. Alignment in the homology fits is better in regions of distinct secondary structures being forms (e.g., α-helices, β-sheets; see Chapter 2) and similarly is poorer in random coil primary structure regions.

8.5.6 STEP DETECTION

An increasingly important computational tool is the automation of the detection of steps in noisy data. This is especially valuable for data output from experimental single-molecule biophysics techniques. The effective signal-to-noise ratio is often small and so the distinction between real signal and noise is often challenging to make and needs to be objectified. Also, single-molecule events are implicitly stochastic and depend upon the underlying probability distribution for their occurrence that can often be far from simple. Thus, it is important to acquire significant volumes of signal data, and therefore, an automated method to extract real signal events is useful. The transition times between different molecular states are often short compared to sampling time intervals such that a "signal" is often manifested as a steplike change as a function of time in some physical output parameter, for example, nanometer-level steps in rapid unfolding domain events of a protein stretched under force (see Chapter 6), picoampere level steps in current in the rapid opening and closing of an ion channel in a cell membrane (see Chapter 5), and rapid steps in brightness due to photobleaching events of single dye molecules (see the previous text).

One of the simplest and robust ways to detect steps in a noisy, extended time series is to apply a running window filter that preserves the sharpness and position of a step edge. A popular method uses a simple *median filter*, which runs a window of number n consecutive data points in a time series on the data such that the running output is the median from data points included in the window. A common method to deal with a problem of potentially having $2n$ fewer data points in the output than in the raw data is to reflect the first and last n data points to the beginning and end of the time series. Another method uses the *Chung–Kennedy filter*. The Chung–Kennedy filter consists of two adjacent windows of size n run across the data such that the output switches between the two windows in being the mean value from the window that has the smallest variance (Figure 8.10). The logic here is that if one edge encapsulates a step event, then the variance in that window is likely to be higher.

Both median and Chung–Kennedy filters converge on the same expected value, though the sample variance on the expected value of a mean distribution (i.e., the square of the standard error of the mean) is actually marginally smaller than that of a median distribution; the variance on the expected value from a sampled median distribution is $\sigma^2\pi/2n$, which compares with the sample mean of σ^2/n (students seeking solace in high-level statistical theory should see Mood et al., 1974), so the error on the expected median value will be larger by a factor of $\sim\sqrt{(\pi/2)}$ or ~25%. There is therefore an advantage in using the Chung–Kennedy filter. Both filters require that the size of n is less than the typical interval between step events; otherwise, the window encapsulates multiple steps and generates nonsensible outputs, so it requires ideally some prior knowledge of likely stepping rates. These *edge-preserving filters* improve the signal-to-noise ratio for noisy time series by a factor of $\sim\sqrt{n}$. The decision of whether a putative step event is real or not can be made on the basis of the probability of the observed size of the putative step in light of the underlying noise. One way to achieve this is to perform a *Student's t-test* to examine if the mean values of the data, $<x>$, on either side over a window

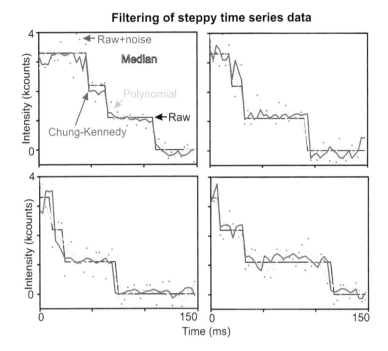

FIGURE 8.10 Filtering steppy data. Four examples of simulated photobleach steps for a tetramer complex. Raw, noise-free simulated data are shown (line), with noise added (dots), applied with either Chung–Kennedy, median, or a polynomial fit filter. The latter is not edge preserving and so blurs out the distinct step edges.

of size n of the putative step are consistent being sampled from the same underlying distribution in light of the variance values on either side of the putative step, σ^2, by estimating the equivalent t statistic:

$$(8.126) \qquad t = \frac{\langle x_{post} \rangle - \langle x_{post} \rangle}{\sqrt{\sigma_{post}^2 + \sigma_{post}^2 / n}}$$

This can then be converted into an equivalent probability, and thus the step rejected if, at some preset probability confidence limit, it is consistent with the pre- and post-means being sampled from the same distribution. Improved methods of step detection involve incorporating model-dependent features into step acceptance, for example, involving Bayesian inference. The Fourier spectral methods discussed earlier for determining the brightness of a single dye molecule on a photobleach time series improve the robustness of the step size estimate compared with direct step detection methods in real space, since they utilize all of the information included in the whole time series, but sacrifice information as to the precise time at which any individual step event occurs. These Fourier spectral methods can also be used for determining the step size of the translocation of molecular machines on tracks, for example, using optical tweezer methods (see Chapter 6), and in fact were originally utilized for such purposes.

KEY BIOLOGICAL APPLICATIONS: *IN SILICO* IMAGE ANALYSIS TOOLS

Molecular colocalization determination; Copy number estimation; Molecular stoichiometry quantitation of complexes.

8.6 RIGID-BODY AND SEMIRIGID-BODY BIOMECHANICS

Rigid-body biomechanics is concerned with applying methods of classical mechanics to biological systems. These can involve both continuum and discrete mathematical approaches. Biomechanics analysis crosses multiple length scales from the level of whole animals through to tissues, cells, subcellular structures, and molecules. Semirigid-body biomechanics encapsulates similar themes, but with the inclusion of finite compliance in the mechanical components of the system.

8.6.1 ANIMAL LOCOMOTION

At the high end of the length scale, this includes analysis of whole *animal locomotion*, be it on land, in sea, or in air. Here, the actions of muscles in moving bones are typically modeled as levers on pivots with the addition of simple linear springs (to model the action of muscles at key locations) and dampers (to model the action of friction). Several methods using these approaches for understanding human locomotion have been developed to assist in understanding of human diseases, but also much work in this area has been catalyzed by the computer video-gaming industry to develop realistic models of human motion.

Comparative biomechanics is the application of biomechanics to nonhuman animals, often used to gain insights into human biomechanics in *physical anthropology* or to simply study these animals as an end in itself. Animal locomotion includes behaviors such as running/jumping, swimming, and flying, all activities requiring an external energy to accelerate the animal's inertial mass and to oppose various combinations of opposing forces including gravity and friction. An emergent area of biophysical engineering research that utilizes the results of human models of biomechanics, in particular, is in developing artificial biological materials or *biomimetics* (see Chapter 9). This crosses over into the field of *biotribology*, which is the study of friction/wear, lubrication, and *contact mechanics* in biological systems, particularly in large joints in the human body. For example, joint implants in knees and hips rub against each other during normal human locomotion, and all lubricated by naturally produced *synovial fluid*, and biotribology analysis can be useful in modeling candidate artificial joint designs and/or engineered *cartilage* replacement material that can mimic the shock-absorbing properties in the joints of natural cartilage that has been eroded/hardened through disease/calcification effects.

8.6.2 PLANT BIOMECHANICS

Plant biomechanics is also an emerging area of research. The biological structures in plants that generate internal forces and withstand external forces are ostensibly fundamentally different from those in animals. For example, there are no *plant muscles* as such and no equivalent nervous system to enervate these nonexistent muscles anyway. However, there are similarities in the network of filament-based systems in plant cells. These are more based on the fibrous polysaccharide cellulose but have analogies to the cytoskeletal network of animal cells (see Chapter 2).

Also, although there is no established nervous system to control internal forces, there are methods of chemical- and mechanical-based signal transduction to enable complex regulation of plant forces. In addition, at a molecular level, there are several molecular motors in plants that act along similar lines to those in animal cells (see in the following text). Plant root mechanics is also a particularly emergent area of research in terms of advanced biophysical techniques, for example, using light-sheet microscopy to explore the time-resolved features of root development (see Chapter 4). In terms of analytical models, much of these have been more of the level of computational *FEA* (see in the following text).

8.6.3 TISSUE AND CELLULAR BIOMECHANICS

In terms of computational biomechanics approaches, the focus has remained on the length scale of tissues and cells. For tissue-level simulations, these involve coarse-graining the tissue into cellular units, which each obeys relatively simple mechanical rules, and treating these as elements in a FEA. Typically, the mechanical properties of each cell are modeled as single scaler parameters such as stiffness and damping factor, assumed to act homogeneously across the extent of the cell. Cell–cell boundaries are allowed to be dynamic in the model, within certain size and shape constraints.

An example of this is the *cancer, heart, and soft tissue environment* (*Chaste*) simulation toolkit developed primarily by research teams in the University of Oxford. Chaste is aimed at multiple length-scale tissue simulations, optimized to cover a range of real biomechanical tissue-level processes as well as implementing effects of electrophysiology feedback in, for example, regulating the behavior of heart muscle contraction. The simulation packages tissue- and cell-level electrophysiology, discrete tissue modeling, and soft tissue modeling. A specific emphasis of Chaste tools includes continuum modeling of cardiac electrophysiology and of cell populations, in addition to tissue homeostasis and cancer formation.

The electrical properties of the heart muscle can be modeled by a set of coupled PDEs that represents the tissue as two distinct continua either inside the cells or outside, which are interfaced via the cell membrane, which can be represented through a set of complex, nonlinear ODEs that describe transmembrane ionic current (see Chapter 6). Under certain conditions, the effect of the extracellular environment can be neglected and the model reduced to a single PDE. Simulation time scales of 10–100 ms relevant to electrical excitation of heart muscle takes typically a few tens of minutes of a computation time on a typical 32 CPU core supercomputer. Cell-population-based simulations have focused to date on the time-resolved mechanisms of bowel cancers. These include realistic features to the geometry of the surface inside the bowel, for example, specialized crypt regions.

Chaste and other biomechanical simulation models also operate at smaller length scale effects of single cells and subcellular structures and include both deterministic and stochastic models of varying degrees of complexity, for example, incorporating the transport of nutrient and signaling molecules. Each cell can be coarse-grained into a reduced number of components, which appear to have largely independent mechanical properties. For example, the nucleus can be modeled as having a distinct stiffness compared to the cytoplasm. Similarly, the cell membrane and coupled cytoskeleton. Refinements to these models include tensor approximations to stiffness values (i.e., implementing a directional dependence on the stiffness parameters).

An interesting analytical approach for modeling the dynamic localization of cells in developing tissues is to apply arguments from *foam physics*. This has been applied using numerical simulations to the data from complex experimental tissue systems, such as growth regulation of the wings of the fruit fly and fin regeneration in zebrafish. These systems are good models since they have been characterized extensively through genetics and biochemical techniques, and thus many of the key molecular components are well known. However, standard biochemical and genetics signal regulation do not explain the observed level of biological regulation that results in the ultimate patterning of cells in these tissues. What does show promise, however, is to use mechanical modeling, which in essence treats the juxtaposed cells as bubbles in a tightly packed foam.

The key point here is that the geometry of bubbles in foam has many similarities with those of cells in developing tissues, with examples including both ordered and disordered states, but also in the similarities between the topology, size, and shape of foam bubbles compared to tissue cells. A valuable model for the nearest-neighbor interactions between foam bubbles is the 2D *granocentric model* (see Miklius and Hilgenfelt, 2012); this is a packing model based purely on trigonometric arguments in which the mean free space around any particle (the problem applied to the packing of grain, hence the name) in the system is minimized. Here, the optimum packing density, even with a distribution of different cell sizes, results in each cell in a 2D array having a mean of six neighbors (see Figure 8.11a). The contact angle Φ between any two tightly packed cells can be calculated from trigonometry (in reference to Figure 8.11a) as

$$(8.127) \qquad \Phi = 2\sin^{-1}\left(1/\left(1+\sqrt{r_c/r}\right)\right)$$

Minimizing the free space around cells minimizes the distance to diffusion for chemical signals, maximizes the mechanical efficiency in mechanical signal transduction, and results in a maximally strong tissue. Hence there are logical reasons for cells in many tissues at least to pack in this way.

8.6.4 MOLECULAR BIOMECHANICS

Molecular biomechanics analysis at one level has already been discussed previously in this chapter of biopolymer modeling, which can characterize the relation between molecular force and end-to-end extension using a combination of entropic spring approximations with improvement to correct for excluded volumes and forces of enthalpic origin. In the limit of very stiff biomolecule or filaments, a rodlike approximation, in essence Hooke's law, is valid, but including also torsional effects such as in the bending cantilever approximation (see Chapter 6). In the case of a stiff, narrow rod of length L and small cross-sectional area dA, if z is a small deflection normal to the rod axis (Figure 8.11b), and we assume that strain therefore varies linearly across the rod, then

(8.128)
$$\frac{\Delta L}{L} \approx \frac{z}{R}$$

where R is the radius of curvature of the bent rod. We can then use Hooke's law and integrate the elastic energy density $(Y/2)(\Delta L/L)^2$ over the total volume of the rod, where Y is the Young's modulus of the rod material, to calculate the total bending energy E_{bend}:

(8.129)
$$\frac{E_{bend}}{L} = \frac{Y}{2} \int_{All\,rod} \left(\frac{\Delta L}{L}\right)^2 dA = \frac{Y}{2} \int_{All\,rod} \left(\frac{z}{R}\right)^2 dA = \frac{B}{2R^2}$$

where B is the bending modulus given by YI where I is the moment of inertia. For example, for a cylindrical rod of cross-sectional radius r, $I = \pi r^4/4$.

An active area of biophysical modeling also involves the mechanics of phospholipid bilayer membranes (Figure 8.11c). Fluctuations in the local curvature of a phospholipid bilayer result in changes to the free energy of the system. The free energy cost ΔG of bilayer area fluctuations ΔA can be approximated from a Taylor expansion centered around the equilibrium state of zero tension in the bilayer equivalent to area $A = A_0$. The first nonzero term in the Taylor expansion is the second-order term:

(8.130)
$$\Delta G \approx \frac{1}{2}\left(\frac{\partial^2 G}{\partial A^2}\right)_{A=A_0}$$

A mechanical parameter called the "area compressibility modulus" of the bilayer, κ_A, is defined as

(8.131)
$$\kappa_A = A_0 \left(\frac{\partial^2 G}{\partial A^2}\right)_{A=A_0}$$

Thus, the free energy per unit area, Δg, is given by

(8.132)
$$\Delta_g \approx \frac{1}{2}\kappa_A (\Delta A)^2$$

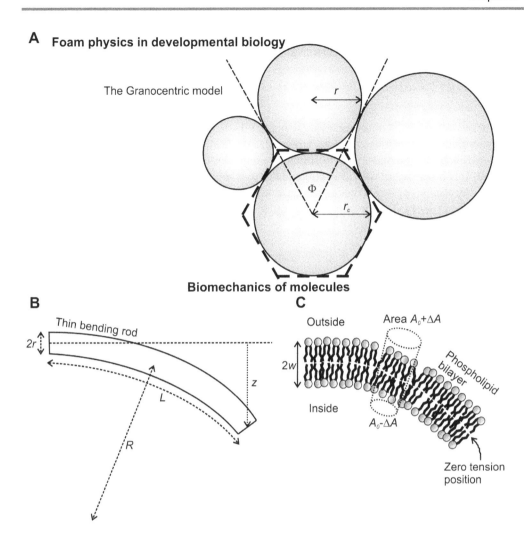

A **Foam physics in developmental biology**

The Granocentric model

r

Φ

r_c

Biomechanics of molecules

B

Thin bending rod

$2r$

L

z

R

C

Outside

Area $A_0 + \Delta A$

$2w$

Phospholipid bilayer

Inside

$A_0 - \Delta A$

Zero tension
position

FIGURE 8.11   Modeling mechanics of cells and molecules. (a) The 2D granocentric model of foam physics—cells are equivalent to the bubbles in the foam, and the contact angle Φ can be calculated using trigonometry from knowledge of the touching cell radii. The mean number of neighbors of a cell is six, as if each cell is an equivalent hexagon (dashed lines). (b) Schematic of bending of a stiff molecular rod. (c) Schematic of bending of a phospholipid bilayer.

This is useful, since we can model the free energy changes in expanding and contracting a spherical lipid vesicle (e.g., an approximation to a cell). If we say that the area on the outside of a bilayer per phospholipid polar head group (see Chapter 2) is $A_0 + \Delta A$ and that on the inside of the bilayer is $A_0 - \Delta A$ since the molecular packing density is greater on the inside than on the outside (see Figure 8.11b), then Δg is equivalent to the free energy change per phospholipid molecule. The total area S of a spherical vesicle of radius R is $4\pi R^2$; thus

$$(8.133) \qquad \frac{\Delta A}{A} = \frac{\Delta S}{S} = \frac{8\pi R \cdot \Delta R}{4\pi R^2} = \frac{2\Delta R}{R}$$

Therefore, the free energy per unit area is given by

$$(8.134) \qquad \Delta_g = \frac{4\kappa_A A_0 (\Delta R)}{R^2}$$

Equating ΔR with the membrane thickness $2w$ therefore indicates that the bending free energy cost of establishing this spherical bilayer for a membrane patch area A_0 is

$$(8.135) \qquad \Delta G = 16\pi\kappa_A A_0 w^2$$

In other words, there is no direct dependence on the vesicle radius. In practice this can be a few hundred $k_B T$ for a lipid vesicle. However, Equation 8.132 implies that relatively large changes to the packing density of phospholipid molecules have a small free energy cost ($\sim k_B T$ or less). This is consistent with phospholipid bilayers being relatively fluid structures.

FEA can also be applied at macromolecular length scales, for example, to model the effects of mesh network of cytoskeletal filaments under the cell membrane and how these respond to external mechanical perturbations. This level of *mesoscale modeling* can also be applied to the bending motions of heterogeneous semi-stiff filaments such as those found in *cilia*, which enable certain cells such as sperm to swim and which can cause fluid flow around cells in tissues.

KEY BIOLOGICAL APPLICATIONS: RIGID–SEMIRIGID BODY MODELING TOOLS

Cell pattern formation in developmental biology; Mechanical signal transduction analysis.

Worked Case Example 8.3: Nucleation of Phase-Separating Liquid Droplets

a For the formation of biomolecular liquid droplets, if the free energy ΔG is made up primarily of two components of a bulk enthalpic component due to nearest neighbor attractive interactions of phase-separating molecules and a surface tension component, which scales with the area of droplets to limit their growth, show that you can model ΔG as $-AN + BN^{2/3}$ where N is the number of phase-separated molecules inside a spherical droplet.
   What is the energy barrier and the critical number of biomolecules N_c in a droplet in terms of A and B?

c A protein of ~20 nm effective globular diameter, which was implicated in neurodegenerative disease, was observed to form liquid–liquid droplets in live mammalian cells. In separate in vitro experiments, the exothermic change in chemical potential energy upon phase separation was estimated to be 2×10^{-3} $k_B T$ per molecule, whereas the surface energy per unit area related to surface tension was equivalent to 4.5×10^{-2} $k_B T$ per molecule. Predict the radius of droplets at the nucleation activation barrier assuming tight-packing of proteins and the ratio of the number of droplets with this radius \compared to droplets with a 25% larger radius.

d When measurements were performed in a living cell, the total number of these phase-separating protein molecules was estimated to be ~7000 molecules per cell. Using super-resolution PALM on 10 different cells, from ~5000 droplets detected, a total of 352 had a diameter in the range 140–160 nm while the number of droplets whose diameter was in the range 180–190 nm was 55. Discuss these findings considering your prediction from part (c).

Answers

a The total free energy change of the bulk enthalpic interaction will be proportional to the number of molecules present N, assuming just nearest neighbor interactions, and will be negative for an attractive (i.e. exothermic) interaction, so will be $-AN$ where A is a positive constant. The free energy change associated with surface tension is proportional to droplet area. Assuming droplet density remains the same, its volume is proportional to N, so its radius is proportional to $N^{1/3}$, hence its surface area is proportional to $N^{2/3}$. Since surface tension will oppose droplet growth the associated surface tension free energy change is positive, hence $+BN^{2/3}$ where be B is a positive constant. Therefore, the net free energy change is $-AN + BN^{2/3}$.

b The most likely droplet size occurs at the maximum value of ΔG with respect to N, i.e. when the first derivative of $-AN + BN^{2/3}$ with respect to N is zero. So:
$-A+(2/3)N_c^{-1/3} = 0$, thus:

$N_c = (2B/3A)^3$

The nucleation energy barrier is given by ΔG_{max}, i.e. ΔG when $N = N_c$ or:

$\Delta G_{max} = B(2B/3A)^{2/3} - A(2B/3A) = (2B^5/3A)^{2/3} - 2B/3$

c Here $A = 2 \times 10^{-3}\, k_B T$ and $B = 4.5 \times 10^{-2}\, k_B T$ so:

$N_c = ((2 \times 4.5 \times 10^{-2})/(3 \times 2 \times 10^{-3}))^3 = 3375$ molecules
$\Delta G_{max} = (-2 \times 10^{-3} \times 3375) + (4.5 \times 10^{-2} \times 3375^{2/3}) = 3.4\, k_B T$

With tight-packing, the radius R at the critical droplet radius R_c relates to the radius of one protein R_1 by:

$N_c \times (4/3)\pi R_1^3 = (4/3)\pi R_c^3$ so:

$R_c = N_c^{(1/3)}R_1 = 3375^{1/3} \times (20/2) = 150$ nm

The probability of a given change in free energy ΔG is proportional to the Boltzmann factor $\exp(-\Delta G/k_B T)$ so the ratio of number of droplets here is:

$C = \exp(-\Delta G(R=1.25R_c))/\exp(-\Delta G_{max}(R=R_c)) = \exp(\Delta G_{max}(R=R_c) - \Delta G(R=1.25R_c))$

For droplets of radius $R=1.25R_c$, the volume greater than the critical radius droplets by a factor of 1.25^3 or ~1.95, so $N = 1.95 \times 3375$ or ~6580 molecules, giving:

$\Delta G(R=1.25R_c) = (-2 \times 10^{-3} \times 6580) + (4.5 \times 10^{-2} \times 6580^{2/3}) = 2.7\, k_B T$

So:

$C \approx \exp(3.4/2,7) \approx 3.5.$

d The range for radius of 140–160 nm is centered on the predicted critical radius of 150 nm in this case, while 180–190 nm is ~25% larger than this radius. From the analysis of part (c), we would therefore predict a greater number of droplets by a factor of ~3.5 (i.e. ~3.5 × 352 or ~1200 droplets) at the higher radius since the corresponding change in free energy is smaller for larger droplets. But here *fewer* droplets were observed at this higher radius by a factor of 352/55 or ~6. However, the total number of these protein molecules per cell is ~7000 is only a little larger than the number of molecules predicted to be present in single droplets with a radius ~25% larger than the critical droplet radius. In other words, the number of molecules present in the cell is not sufficient to sustain the growth of droplets much beyond this, which might account for the far lower proportion of larger droplets observed than predicted.

Worked Case Example 8.4: Biopolymer Compaction

The E. coli genome is a circle of, largely, B-DNA containing ~4.6 million base pairs.

a   If the DNA acts as a "floppy" molecule whose persistence length l_p is roughly 50 nm, estimate the volume that would be occupied by randomly coiled DNA assuming it is a sphere whose radius was equal to DNA radius of gyration.

b   If the DNA instead acts as a rigid rod, what is its volume?

c   SIM imaging of fluorescently labeled nucleoids of E. coli (the region inside the bacterium in which the DNA is found) suggests it is roughly ellipsoidal with long and short axes of ~1.6 μm and ~0.5 μm length, respectively. Explain if you think the DNA inside a nucleoid is a randomly coiled floppy molecule or a rigid rod.

Answers

a   The radius of gyration is given by Equation 8.37 as $\sqrt{<R^2_G>} \approx \sqrt{<R^2_{FJC}>}/\sqrt{6}$ or $(\sqrt{n},b)/\sqrt{6}$ where b is the Kuhn length or $\sim 2l_p$ for a "floppy" molecule and n is the number of equivalent segments in the freely jointed chain. Here, n is the total length L of DNA divided by b (or $2l_p$). Using the known dimensions of B-DNA (see Chapter 2) indicates:

$L = 4.6 \times 10^6 \times 0.34$ nm $= 4.6 \times 10^6 \times 0.34 \times 10^{-9} = 1.5 \times 10^{-3}$ m or 1.5 mm

$n = L/b \approx (1.5 \times 10^{-3})/(2 \times 50 \times 10^{-9}) = 15{,}000$ links in the chain. Thus:

$\sqrt{<R^2_G>} \approx \sqrt{(15{,}000)} \times (2 \times 50 \times 10^{-9})/\sqrt{6} = 5 \times 10^{-6}$ m or 5 μm. So, the random coil volume is:

$V_{rc} = (4\pi/3) \times (5 \times 10^{-6})^3 = 5.2 \times 10^{-16}$ m³ or ~520 μm³.

b   For a rigid rod, the volume is the total DNA length L multiplied by the circular cross-sectional area of B-DNA (which has a radius of ~1 nm, see Chapter 2), so the volume is:

$V_{rr} = (1.5 \times 10^{-3}) \times \pi \times (1 \times 10^{-9})^2 = 4.7 \times 10^{-21}$ m³ or ~0.005 μm³.

c   Without worrying about the exact formula for an ellipsoid, it is clear that its volume is going to be of the order of ~1 μm³. This is therefore 500-fold less than V_{rc} but 200-fold greater than V_{rr}, so the nucleoid is far more compacted than a simple random coil (indeed, it needs to be since the volume of a whole *E. coil* cell itself is only a few μm³), but it is certainly not as rigid as to be in the rodlike regime. So, the answer to this question is it is *neither*. In reality, there are so-called nucleoid associated proteins (NAPs), very similar to histones in eukaryotic cells, which serve a role to compact the bacterial nucleoid in such a way as it doesn't form a filamentous rod but rather cross-links across different segments of DNA, which can potentially be separately by hundreds and thousands of base pairs.

Worked Case Example 8.5: Bayes Theorem and COVID Testing

A lateral flow test (LFT) for COVID testing was found to have a probability also known as the sensitivity) of 85% of giving a positive result if someone is infected with one of the variants of coronavirus, but a 99% probability of correctly giving a negative result if someone is not infected (also known as the sensitivity). At the height of the pandemic in

January 2021, it was estimated that up to 1 in 40 people in England was infected with COVID-19, while by February 2021 this estimate was more like 1 in 115. If this LFT was used by someone in England, what is the probability that if they appeared to test positive on the LFT then they actually were infected with COVID-19

a *At the height of the pandemic?*
b *A month later?*
c *Comment on these results.*

Answers

This question is all about using the principle of Bayes theorem in that the probability of an event occurring, more fully known as the *posterior probability*, is determined not only by the general likelihood of the event happening but also by the probability of one or more other events also having occurred (known at the *prior probabilities*, or *priors*). In this case, the posterior is the probability that the LFT gives a positive result *given* the prior occurrence that the person is actually infected by COVID-19.

a If, for simplicity, we consider 100,000 people, then at the height of the pandemic, 1 in 40 is equivalent to 2500 people infected with COVID-19, so the remaining 97,500 are not infected. But if the LFT sensitivity and specificity are 85% and 99%, respectively, this means that 85% of the 2500 (i.e. 1955 people) and 1% of the 97,500 (i.e. 975 people) would expect to have a positive LFT results (or 2930 people in total) of whom 2500 would actually have a COVID-19 infection. So, using Bayes theorem, the posterior probability of someone with a positive LFT who also is infected with COVID-19 is 2,500/2,930 or ~85%.

b By a similar argument, 1 in 115 people is 870 people out of 100,000 infected with COVID-19, while 99,200 people are not; 85% of the 870 (i.e. 740 people) and 1% of the 99,200 (i.e. 992 people) would expect to have a positive LFT results (or 1732 people in total) of whom 740 would actually have a COVID-19 infection. So, the probably of someone with a positive LFT having COVID-19 is 2740/1732 or ~42%.

c When the community level of COVID-19 infection is relatively high, the LFT shows reasonably high probability of given a true positive result—it is accurate and reliable. However, once community infection levels of COVID-19 are significantly lower, the true positive level may reach a stage of being potentially unreliable (e.g. <50%). This is an example of the *base rate paradox* in Bayesian analysis: even though most LFTs are in themselves very sensitive and specific, if the probability of someone having COVID-19 is small then false-positives will make up most of the positive LFT results. Note, these estimates were predicated using only indicative sensitivity and specificity levels reported during the pandemic in the UK, and the answers to the questions above will vary widely depending on the actual values used. For example, sensitivity and specificity levels measured for LFTs from different studies taken from a systematic review published in August 2021 ranged from 37.7–99.2%, whereas the specificity ranged from 92.4–100% (Misty et al., 2021).

8.7 SUMMARY POINTS

■ Molecular simulations use Newton's second law of motion with varying degrees of approximations of atomic energy potentials to predict the motions of atoms as a function of time over a scale from hundreds of picoseconds for *ab initio* QM simulations to tens of nanoseconds for classical MD and tens of microseconds for CG approaches.

- Water is very important in molecular simulations and may be approximated as an explicit solvent, which is very accurate but computationally expensive, or as an implicit solvent that does not simulate details of individual models but enables longer time scales to be simulated for the same computational effort.
- The mechanics of biopolymers can be modeled using pure entropic forces, with improved modeling incorporating additional effects of enthalpic forces and excluded volumes.
- Reaction–diffusion and Navier–Stokes analysis can often be reduced in complexity for modeling important biological processes to generate tractable models, for example, for characterizing molecular motor translocation of tracks, pattern formation in cell and developmental biology, and the fluid flow around biological particles.
- Bayesian inference is an enormously powerful statistical tool that enables us to use prior knowledge to infer correct models of biological processes.
- PCA and wavelet analysis are powerful computational tools to enable automated recognition and classification of images.
- Localization microscopy data from fluorescence imaging can be objectively quantified to assess molecular stoichiometry and kinetics.

QUESTIONS

8.1 To model an interaction between a particular pair of atoms, masses m_1 and m_2, separated by a distance r an empirical potential energy function was used, $U(r) = (3/r^4 - 2/r)\alpha$, where α is a constant.

    a   Draw the variation of U with r.

    b   Show that there is a stable equilibrium separation, and calculate an expression for it.

    c   Calculate an expression for the resonant frequency about this equilibrium position.

8.2 What are the particular advantages and limitations of QM, MM, MC, and CG simulations?

8.3 Compare and contrast the techniques that can be used to reduce the computational time in molecular simulations.

8.4 Consider a typical single-molecule refolding experiment on a short filamentous protein of a molecular weight of ~14 kDa performed by holding the ends between two beads held in two low stiffness optical tweezers and allowing the tethered peptide to spontaneously refold against the imposed trapping force.

    a   Estimate the number of peptide bonds in the short protein.

    b   Each peptide bond has two independent bond angles called phi and psi, and each of these bond angles can be in one of three stable conformations based on Ramachandran diagrams (see Chapter 2). Estimate roughly how many different conformations the protein can have.

    c   If the unfolded protein refolds by exploring each conformation rapidly in ~1 ps and then subsequently exploring the next conformation if this was not the true (most stable) folded state, estimate the average length of time taken before it finds the correct stable folded conformation.

    d   In practice, unfolded proteins in cells will refold over a time scale of microseconds up to several seconds, depending on the protein. Explain why proteins in cells do not refold in the exploration manner described earlier. What do they do alternatively? (*Hint*: the best current estimate for the age of the universe is ca. 14 billion years.)

8.5 An ideal FJC consists of n rigid links in the chain each of length b that are freely hinged where they join up.

a   By treating each link as a position vector, and neglecting possible consequences of interference between different parts of the chain, derive an expression for $<x^2>$, the mean square end-to-end distance.

b   Evaluate this expression in the limits that the contour length is much greater, and much less, than b, and comment on both results.

8.6   A certain linear protein molecule was modeled as an ideal chain with 500 chain segments each equal to the average alpha-carbon spacing for a single amino acid carrying a charge of $\pm q$ at either end of each amino acid subunit where q is the unitary electron charge. What is the end-to-end length relative to its rms unperturbed length parallel to an E-field, of 12,000 V cm^{-1}?

8.7   A protein exists in just two stable conformations, 1 and 2, and undergoes reversible transitions between them with rate constants k_{12} and k_{21}, respectively, such that the free energy difference between the two states is ΔG.

a   Use detailed balance to estimate the probability that the system is in state 1 at a given time t, assuming that initially the molecule is in state 1. A diferent protein undergoes two irreversible transitions in conformation, 1 – 2 and 2 – 3, rate constants k_{12} and k_{23}, respectively.

b   What is the likelihood that a given molecule starting in state 1 initially will be in state 3 at time t?

8.8   In an experiment, the total bending free energy of a liposome was measured at ~200 $k_B T$ and the polar heads tightly packed into an area equivalent to a circle of radius of 0.2 nm each, with the length from the tip of the head group to the end of the hydrophobic tail measured at 2.5 nm. Estimate the free energy in Joules per phospholipid molecule required to double the area occupied by a single head group.

8.9   A virus was fluorescently labeled and monitored in a living cell at consecutive sampling times of interval 33 ms up to 1 s. The rms displacement was calculated for two such viral particles, giving values of [61, 75, 81, 95, 107, 112, 128, 131, 158, 167, 181, 176, 177, 182, 183, 178, 177, 180, 182, 184, 181, 179, 180, 178, 180, 182] nm and [59, 65, 66, 60, 64, 63, 58, 62, 63, 61, 64, 60, 59, 64, 62, 65, 61, 60, 63, 66, 62, 58, 60, 57, 61, 62] nm. What might this indicate about the virus diffusion?

8.10   A series of first-order biochemical reactions of molecules 1, 2, 3,…, n reacted irreversibly as $1 \leftrightarrow 2 \leftrightarrow 3 \leftrightarrow … \leftrightarrow n$ with rate constants $k_1, k_2, k_3,…,k_n$, respectively.

a   What is the overall rate constant of the process $1 \leftrightarrow n$?

b   What happens to this overall rate if the time scales for formation of one of the intermediate molecules are significantly lower than the others?

c   What implications does this have for the local concentrations of the other intermediate molecules formed? (*Hint*: such a slow-forming intermediate transition is often referred to as the "rate-limiting step.")

8.11   Write down the Brownian diffusion equation for numbers of molecules n per unit length in x that spread out by diffusion after a time t.

a   Show that a solution exists of the form $n(x,t) = \alpha t^{\frac{1}{2}}\exp(-x^2/(4Dt))$ and determine the normalization constant a if there are n_{tot} molecules in total present.

b   Calculate the mean expected values for $<x>$ and $<x^2>$, and sketch the solution of the latter for several values of $t > 0$. What does this imply for the location of the particles initially?

The longest cells in a human body are neurons running from the spinal cord to the feet, roughly shaped as a tube of circular cross-section with a tube length of ~1 m and a diameter of ~1 μm with a small localized cell body at one end that contains the nucleus. Neurotransmitter molecules are synthesized in the cell body but are required at the far end of the neuron. When neurotransmitter molecules reach the far end of the neuron, they are subsequently removed by a continuous biological process at a rate that maintains the mean concentration in the cell body at 1 mM.

c   If the diffusion coefficient is ~1000 μm^{2}s^{-1}, estimate how many neurotransmitter molecules reach the end of the cell a second due to diffusion.

In practice, however, neurotransmitters are packaged into phospholipid-bound synaptic vesicles each containing ~10,000 neurotransmitter molecules.

   d   If ~300 of such molecules are required to send a signal from the far end of the neuron to a muscle, what signal rate could be supported by diffusion alone? What problem does this present, and how does the cell overcome this? (*Hint*: consider the frequency of everyday activities involving moving your feet, such as walking.)

8.12 Phospholipid molecules tagged with the fluorescent dye rhodamine were self-assembled into a planar phospholipid bilayer *in vitro* at low surface density at a room temperature of 20°C and were excited with a 532 nm wavelength laser and imaged using high intensity narrow-field illumination (see Chapter 4) sampling at 1 ms per image to capture rhodamine fluorescence emissions peaking at a wavelength of ~560 nm wavelength, which allowed ~10 ms of fluorescence emission before each rhodamine molecule on average was irreversibly photobleached, where some molecules exhibited Brownian diffusion at a rate of roughly 7 μm²s⁻¹ while others were confined to putative lipid rafts of diameter ~100 nm.

   a   Show that for phospholipid molecules exhibiting Brownian diffusion

$$P = r/r_0 \exp\left(-\frac{r^2}{r_0^2}\right)$$ is a solution of the diffusion equation, where Pdr is the probability that the lipid lies in an annulus between r and $r + dr$ after time t.

Experiments on bulk samples suggested that raising the bilayer temperature by 20°C might disrupt the corralling effect of a lipid raft. From intensity measurements, it was estimated that in some of the rafts two fluorescently tagged lipid molecules were confined, but it was not clear whether these lipid molecules were bound to each other or not as each raft could only be resolved as a single diffraction-limited fluorescent spot. An experiment was devised such that one image frame was first recorded, the sample temperature was then increased by 20°C very rapidly taking less than 1 ms, and fluorescence imaging was continued until the sample was bleached. The experiment was performed 1000 times on separate rafts possessing two tagged phospholipid molecules. In 198 of these experiments, two distinct fluorescent spots could just be resolved near the site of the original raft at the end of the imaging, while in the rest only a single fluorescent spot could be resolved throughout.

   b   Explain with quantitative reasoning whether this supports a complete disruption or partial disruption model for the effect of raising temperature on the lipid raft.

8.13 A particle exhibiting 1D Brownian diffusion has probability P being at a distance x after a time t, which satisfies $P = t^{1/2}\exp\left(-x^2/(2\langle x^2 \rangle)\right)/(4\pi D)^{1/2}$ where $\langle x^2 \rangle$ is the mean square displacement, equal to $2Dt$. In a single-particle tracking experiment, a QD-labeled importin molecule (a protein used in trafficking molecules across the membrane of the nucleus in eukaryotes) is tracked inside a cell, being imported into the cell nucleus. The QD-importin complex was estimated as having a Stokes radius of 30 nm.

   a   What is the QD-importin complex's diffusion coefficient in the cytoplasm of the cell?

   b   After a time t, the mean square displacement through an unrestricted channel in the nuclear membrane might be expected to be $2Dt$—why not $4Dt$ or $6Dt$?

It was found that out of 850 single-particle tracks that showed evidence of translocation into the nucleus, only 180 translocated all the way through the pore complex of channel length 55 nm into the nucleus, a process which took on average 1 s.

   c   Explain with reasoning why translocation does not appear to be consistent with Brownian diffusion, and what might account for this non-Brownian behavior?

8.14 A particular mRNA molecule has a molecular weight of 100 kDa.

   a   How many nucleotide bases does it contain, assuming 6.02×10^{26} typical nucleotide molecules have a mass of 325 kg?

b   If the mRNA molecule is modeled as a GC with segment length of 4 nm, estimate the Stokes radius of the whole molecule.

The longest cell in a human body is part of a group of nerve cells with axons (tubes filled with cytoplasm) running from the spinal cord to the feet roughly shaped as a cylinder of length of ~1 m and diameter ~1 μm.

c   Estimate the diffusion coefficient of the mRNA molecule in the axon, assuming the viscosity of cytoplasm is ~30% higher than that of water. How long will it take for the mRNA molecule on average to diffuse across the width of the axon, and how long would it take to diffuse the whole length?

d   With reference to your previous answers, explain why some cells do not rely on passive (i.e., no external energy input) diffusion alone to transport mRNA, and discuss the other strategies they employ instead.

8.15   The prediction from the random telegraph model for the distribution of total number of a given type of protein molecule present in each bacterial cell is for a gamma distribution (a distribution function that has a unimodal peak with a long tail). However, if you measure the stoichiometry of certain individual molecular complexes, then the shape of that distribution can be adequately fitted using a Gaussian distribution. Explain.

8.16   Stepping rotation of the flagellar rotary motor of a marine species of bacteria was powered by a flux of Na^+ ions. The position of a bead attached to the flagellar filament was measured using back focal plane detection (see Chapter 6), which was conjugated to the rotor shaft of the motor via a hook component. The hook can be treated as a linear torsional spring exerting a torque $\Gamma_1 = \kappa\Delta\theta$ where κ is the spring constant and $\Delta\theta$ the angle of twist, and the viscous drag torque on the bead is $\Gamma_2 = \gamma d(\Delta\theta)/dt$ where γ is the viscous drag coefficient. In reducing the external Na^+ concentration to below 1 mM, the motor can be made to rotate at one revolution per second or less.

a   Estimate the maximum number of steps per revolution that might be observable if $\kappa = 1 \times 10^{-19}$ N m rad^{-1} and $\gamma = 5 \times 10^{-22}$ N m rad^{-1} s. In the experiment ~26 steps per revolution were observed, which could be modeled as thermally activated barrier hopping, that is, the local free energy state of each angular position was roughly a parabolic-shaped potential energy function but with a small incremental difference between each parabolic potential.

b   If a fraction α of all observed steps are backward, derive an expression for the sodium-motive force, which is the electrochemical voltage difference across the cell membrane due to both charge and concentration of Na^+ ions embodied by the Nernst equation (see Chapter 2), assuming each step is coupled to the transit of exactly two Na^+ ions into the cell.

c   If $\alpha = 0.2$ and the internal and external concentrations of the driving ion are 12 and 0.7 mM, respectively, what is the voltage across the cell membrane? If a constant external torque of 20 pN·nm were to be applied to the bead to resist the rotation, what would be the new observed value of α?

8.17

a   Derive the Michaelis–Menten equation for the rate of F1-mediated hydrolysis of ATP to ADP and inorganic phosphate in the limit of excess ATP, stating any assumptions you make. How does this rate relate to the rate of rotation of the F1 γ subunit rotor shaft?

b   State with mathematical reasoning how applying a load opposing the rotation of γ affects the rate of ATP hydrolysis?

c   Assuming the ATP-independent rotation step of ~30°–40° can be modeled by one irreversible rapid step due to a conformational change in the F1·ADP complex correlated with ATP hydrolysis with a rate constant k_4, followed at some time afterward by another rapid irreversible rotational step due to the release of the ADP with a rate constant k_5, derive an expression for the probability that the conversion from the pre-hydrolysis F1 · ADP state to the post-ADP-release F1 state takes a time t, and explain why this results in a peaked distribution to the experimentally measured wait time distribution for the ~30°–40° step.

8.18 The probability P that a virus at a distance d away from the center of a spherical cell of radius r will touch the cell membrane can be approximated by a simple diffusion-to-capture model that predicts $P \approx r/d$.

a By assuming a Poisson distribution to the occurrence of this touching event, show that the mean number of touches of a virus undergoes oscillating between the membrane and its release point before diffusing away completely is $(r/2d\ 2)$ $(2d-r)$.

b If a cell extracted from a tissue sample infected with the virus resembles an oblate ellipsoid of major axis of 10 μm and minor axis of 5 μm with a measured mean number of membrane touches of 4.1 ± 3.0 (±standard deviation) of a given virus measured using single-particle tracking from single-molecule fluorescence imaging, estimate how many cells there are in a tissue sample of volume 1 mL.

c If the effective Brownian diffusion coefficient of a virus is 7.5 μm^2 s^{-1}, and the incubation period (the time between initial viral infection and subsequent release of fresh viruses from an infected cell) is 1–3 days, estimate the time taken to infect all cells in the tissue sample, stating any assumptions that you make.

8.19 For 1D diffusion of a single-membrane protein complex, the drag force F is related to the speed v by $F = \gamma v$ where γ is frictional drag coefficient, and after a time t, the mean square displacement is given by $2Dt$ where D is the diffusion coefficient.

a Show that if all the kinetic energy of the diffusing complex is dissipated in moving around the surrounding lipid fluid, then D if given by the Stokes–Einstein relation $D = k_B T/\gamma$.

b For diffusion of a particular protein complex, two models were considered for the conformation of protein subunits, either a tightly packed cylinder in the membrane in which all subunits pack together to generate a roughly uniform circular cross-section perpendicular to the lipid membrane itself or a cylindrical shell model having a greater radius for the same number of subunits, in which the subunits form a ring cross-section leaving a central pore. Using stepwise photobleaching of YFP-tagged protein complexes in combination with single-particle tracking, the mean values of D were observed to vary as ~$1/N$ where N was the estimated number of protein subunits per complex. Show with reasoning whether this supports a tightly packed or a cylindrical shell model for the complex. (*Hint*: use the equipartition theorem and assume that the sole source of the complex's kinetic energy is thermal.)

8.20 In a photobleaching experiment to measure the protein stoichiometry of a molecular complex, in which all proteins in the complex were labeled by a fluorescent dye, by counting the number of photobleach steps present, a preliminary estimate using a Chung–Kennedy filter with window width of 15 data points suggested a stoichiometry of four protein molecules per complex, while subsequent analysis of the same data using a window width of eight data points suggest a stoichiometry of six protein molecules, while further analyses with window widths in the range three to seven datapoints suggested a stoichiometry of ~12 molecules while using a window width of two data points or using no Chung–Kennedy filter but just the pairwise difference of single consecutive data points suggested stoichiometries of ~15 and ~19, respectively. What's going on?

REFERENCES

KEY REFERENCE

Berg, H.C. (1983). *Random Walks in Biology*. Princeton University Press, Princeton, NJ.

MORE NICHE REFERENCES

Bell, G.I. (1978). Models for the specific adhesion of cells to cells. *Science 200*:618–627.

Chandrasekhar, S. (1943). Stochastic problems in physics and astronomy. *Rev. Mod. Phys. 15*:1–89.

Choi, J.-M. et al. (2020). Physical principles underlying the complex biology of intracellular phase transitions. *Annu. Rev. Biophys.* 49:107–133.

Duke, T.A. and Bray, D. (1999). Heightened sensitivity of a lattice of membrane receptors. *Proc. Natl. Acad. Sci. USA*. *96*:10104.

Duke, T.A., Le Novère, N., and Bray D. (2001). Conformational spread in a ring of proteins: A stochastic approach to allostery. *J. Mol. Biol. 308*:541–553.

Evans, E. and Ritchie, K. (1997). Dynamic strength of molecular adhesion bonds. *Biophys. J. 72*:1541–1555.

Guillot, B. (2002). A reappraisal of what we have learnt during three decades of computer simulations on water. *J. Mol. Liq. 101*:219–260.

Jarzynski, C. (1997). A nonequilibrium equality for free energy differences. *Phys. Rev. Lett. 78*:2690.

Kramers, H.A. (1940). Brownian motion in a field of force and the diffusion model of chemical reactions. *Physica 7*:284.

Levitt, M. and Warshel, A. (1975). Computer simulations of protein folding. *Nature 253*:694–698.

Miklius, M. and Hilgenfelt, S. (2012). Analytical results for size-topology correlations in 2D disk and cellular packings. *Phys. Rev. Lett. 108*:015502.

Misty, D. et al. (2021). A systematic review of the sensitivity and specificity of lateral flow devices in the detection of SARS-CoV-2. *BMC Infectious Diseases* 21: 828.

Mood, A.M., Graybill, F.A., and Boes, D.C. (1974). *Introduction to the Theory of Statistics*, 3rd edn. McGraw-Hill International Editions, Columbus, OH.

Moore, G.E. (1965). Cramming more components onto integrated circuits. *Proc. IEEE 86* :82–85.

Shi, Y. and Duke, T. (1998). Cooperative model of bacterial sensing. *Phys. Rev. E 58*:6399–6406.

Turing, A.M. (1952). The chemical basis of morphogenesis. *Philos. Trans. B 237*:37–72.

Emerging Biophysics Techniques

An Outlook of the Future Landscape of Biophysics Tools

Anything found to be true of *E. coli* must also be true of elephants.

—Jacques Monod, 1954 (from Friedmann, 2004)

Everything that is found to be true for *E. coli* is only sometimes true for other bacteria, let alone for elephants.

—Charl Moolman (his PhD viva, Friday, March 13, 2015, TU Delft, the Netherlands)

General Idea: Biophysics is a rapidly evolving discipline, and several emerging tools and techniques show significant promise for the future. These include systems biophysics, synthetic biology, and bionanotechnology, increasing applications of biophysics to personalizing healthcare, and biophysics approaches that extend the length scales into the smaller world quantum phenomena and the larger world of populations of organisms, which we discuss here.

9.1 INTRODUCTION

There are several core biophysical tools and techniques invented in the twentieth and early twenty-first centuries, which, although experiencing improvements and adaptations subsequent to their inception, have in their primary features at least stood the test of time. However, there are several recently emerging methods that, although less established than the core tools and technologies discussed previously in this book, still offer enormous *potential* for generating novel biological insight and/or having important applications to society. These emerging tools have largely developed from the crossover of several different scientific disciplines. For example, aspects of systems biology developed largely from computational biology themes have now been adapted to include strong physiology elements that encapsulate several experimental cellular biophysics tools.

Similarly, synthetic biology and bioengineering methods largely grew from chemical engineering concepts in the first instance but now apply multiple biophysics approaches. Developments in diagnostics for healthcare are progressing toward far greater personalization, that is, methods that can be catered toward individual patients, which in particular use biophysics methods.

9.2 SYSTEMS BIOLOGY AND BIOPHYSICS: "SYSTEMS BIOPHYSICS"

Systems biology as a discipline grew from the efforts of the nineteenth-century physiologist Claude Bernard, who developed the founding principles of *homeostasis*, namely, the phenomenon of an organism's internal environment being carefully regulated within certain limits, which optimizes its ultimate viability. Ultimately, this regulation of the physiological state involves the interaction between multiple systems in an organism, which we now know can act over multiple different length and time scales.

DOI: 10.1201/9781003336433-9

A key question that systems biology has tried to address is, "what is the correct level of abstraction at which to understand biology?" *Reductionist* approaches argue that we should be able to understand all biology from a knowledge simply of the molecules present. In a sense, this is quite obviously correct, though equally naïve, since it is not just the molecules that are important but how they interact, the order in which they interact, and the generating of higher order features from these interactions that can in turn feedback to the level of interaction with single molecules. In other words, a more *integrationist* approach is valuable, and physical scientists know this well from observing emergent behavior in many non-biological systems, complex higher length scale behavior that is difficult or impossible to predict from a simple knowledge of just the raw composition of a system that can result from the cooperativity between multiple shorter length scale elements that potentially obey relatively simple rules of interaction. As to where to draw the line in terms of what is the most appropriate level of abstraction from which to understand biology, this is still a matter of great debate.

KEY POINT 9.1

Many biologists treat the cell as the best level of abstraction from which to understand biology, though some suggest that smaller length scales, even to the level of single genes and beyond in just single sections of genes, are a more appropriated level. However, feedback clearly occurs across multiple length scales in a biological organism, and in fact in many cases between organisms and even between populations of organisms. So, the question of where exactly to draw the line presents a challenge.

Modern systems biology has now adopted a core computational biology emphasis, resulting in the development of powerful new mathematical methodologies for modeling complex biosystems by often adapting valuable algorithms from the field of systems engineering. However, it is only comparatively recently that these have been coupled to robust biophysical tools and techniques to facilitate the acquisition of far more accurate biomolecular and physiological parameters that are inputted into these models. Arguably, the biggest challenge the modern systems biology field set itself was in matching the often exquisite quality of the modeling approaches with the more challenging quality of the data input, since without having confidence in both any predictions of emergent behavior stemming from the models may be flawed. In this section, we discuss some of the key engineering modeling concepts applied to modeling interactions of components in an extended biological system, and of their coupling with modern biophysical methods, to generate a new field of *systems biophysics*.

9.2.1 CELLULAR BIOPHYSICS

The founding of systems biophysics really goes back to the early work of Hodgkin and Huxley on the squid axon (see Chapter 1), which can also be seen as the first real example of a methodical *cellular biophysics* investigation, which coupled biophysical experimental tools in the form of time-resolved electrical measurements on extracted squid nerve cells, with a mathematical modeling approach that incorporated several coupled differential equations to characterize the propagation of the electrical impulse along the nerve cell. A key result in this modeling is the establishment of feedback loops between different components in the system. Such feedback loops are a general feature of systems biology and facilitate systems regulation. For example, in the case of electrical nerve impulse propagation in the squid axon, proteins that form the ion channels in the cell membrane of the axon generate electric current through ion flux that either charges or discharges the electrical capacitance of that cell, which alters the electrical potential across the cell membrane. But similarly, the electrical potential across the cell membrane itself also controls the gating of the ion channel proteins.

Systems biophysics approaches have an enormous potential to bridge the *genotype to phenotype gap*. Namely, we now have a very good understanding of the composition, type, and number of genes from sequencing approaches in an organism (see Chapter 7) and also separately have several approaches that can act as a metric for the phenotype, or behavior, of a component of a biological process. However, traditionally it has not been easy to correlate the two in general biological processes that have a high degree of complexity. The key issue here is with the high number of interactions between different components of different biological systems.

However, a systems biophysics approach offers the potential to add significant insight here. Systems biophysics on cells incorporates experimental cellular measurements using a range of biophysics tools, with the mathematical modeling of systems biology. A useful modern example is that of chemotaxis, which can operate at the level of populations of several cells, such as in the movement of small multicellular organisms, and at the level of single cells, such as the movement of bacteria.

Chemotaxis is a form of *signal transduction*, which involves the initial detection of external chemicals by specific cell membrane receptor molecules and complexes of molecules, but which ultimately span several length and time scales. It is a remarkable process that enables whole cells to move ultimately toward a source of food or away from noxious substances. It involves the detection of external *chemoattractants* and *chemorepellents*, respectively, that act as ligands to bind either directly to receptors on the cell membrane or to adapter molecules, which then in turn bind to the receptor (which we discussed in the context of an Ising model previously in Chapter 8). Typically, binding results in a conformational change of the receptor that is transformed via several coupled chemical reaction cascades inside the cell, which, by a variety of other complex mechanisms, feed into the motility control system of that cell and result in concerted, directional cellular movement.

In prokaryotes such as bacteria, it is the means by which cells can swim toward abundant food supplies, while in eukaryotes, this process is critical in many systems that rely on concerted, coordinated responses at the level of many cells. Good examples of chemotaxis include the immune response, patterning of cells in neuron development, and the morphogenesis of complex tissues during different stages of early development in an organism.

Cells from the fungus *Dictyostelium discoideum* display a strong chemotaxis response to *cyclic adenosine monophosphate* (*cAMP*), which is mediated through a cell surface receptor complex and *G* protein–linked signaling pathway. Using fluorescently labeled Cy3-cAMP in combination with single-molecule TIRF imaging and single particle tracking on live cells, researchers were able to both monitor the dynamic localization of ligand-bound receptor clusters and measure the kinetics of ligand binding in the presence of a chemoattractant concentration gradient (Figure 9.1a).

In eukaryotic cells, such as in those of the multicellular organism *Dictyostelium*, the detection of the direction of a concentration gradient of an external chemical is made by a complex mechanism that essentially compares the rate of binding of ligand by receptors on one side to those on the other. That is, it utilizes the physical length scale of the cell to generate probes in different regions of the concentration gradient such that on the side of the higher concentration there will be a small but significant and measurable increase in the rate of ligand binding compared to the opposite side of the cell. This is in effect *spatial sampling* of the concentration gradient. Prokaryotic cells such as bacteria do not use such a mechanism because their physical length scale of $\sim 10^{-6}$ m results in far too small a difference in ligand binding rates either side of the cell above the level of stochastic noise, at least for the relatively small concentration gradients in external ligand that the cell may need to detect. Instead, the cellular strategy evolved is one of *temporal sampling* of the concentration gradient.

Bacterial sensing and reaction to their environment has been well studied using biophysical tools, in particular fluorescence microscopy in living, functional cells at a single-molecule level. Their ability to swim up a concentration gradient of a chemical attractant is well known, so these cells can clearly detect their surroundings and act appropriately. These systems are excellent models for general sensory networks in far more complex organisms—one of the great advantages of using bacteria is that their comparative low complexity allows experiments to be far more controlled and their results far more definitive in terms of

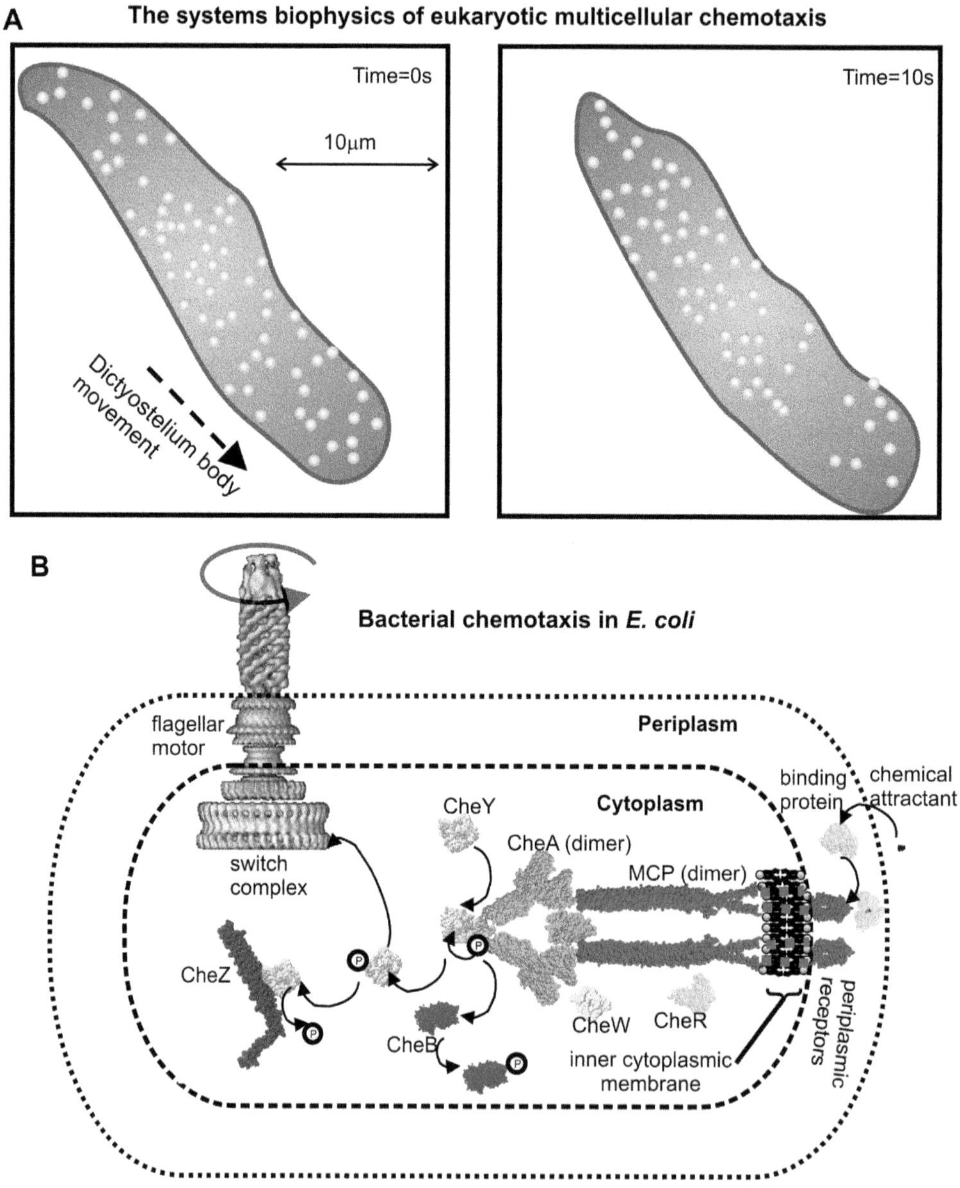

A **The systems biophysics of eukaryotic multicellular chemotaxis**

B

FIGURE 9.1 Systems biophysics of cell chemotaxis. (a) A multicellular slime mold body of species *Dictyostelium discoideum* indicating movement of the whole body (gray) in the direction of a chemoattractant gradient, which results in a redistribution of chemoattractant receptor complexes (circles) in the cell membrane after perturbing the chemoattractant concentration gradient (right panel). (b) Schematic indicating the different protein components that comprise the bacterial chemotaxis pathway in *Escherichia coli*, which results in a change in the rotational state of the flagellar motor in response to detected concentration changes in chemoattractant outside the cell.

elucidating characteristics of the key molecular components in their original biological context (see Chapter 7).

The range of signals detected by bacteria is enormous, including not only nutrients but also local oxygen concentration and the presence of toxins and fluctuations in pressure in the immediate surroundings, but the means by which signals are detected and relayed have strong generic features throughout. For similar reasons in studying single bacteria, we can increase our understanding of sensory networks in far more complicated multicellular creatures. But a key feature of bacterial chemotaxis is that different proteins in the pathway can be monitored using fluorescent protein labeling strategies (see Chapter 7) coupled with advanced single-molecule localization microscopy techniques (see Chapter 4) to monitor the spatial distribution of each component in real time as a function of perturbation in the external chemoattractant concentration, which enables systems biophysics models of the whole-cell level process of bacterial chemotaxis to be developed.

9.2.2 MOLECULAR NETWORKS

Many complex systems, both biological and nonbiological, can be represented as networks, and these share several common features. Here, the components of the system feature as *nodes*, while the interactions between the components are manifested as *edges* that link nodes. There also exist *motifs* in networks, which are commonly occurring subdomain patterns found across many different networks, for example, including feedback loops. *Modularity* is therefore also common to these motifs, in that a network can be seen to be composed of different modular units of motifs. Also, many networks in biology tend to be *scale-free networks*. This means that their degree of distribution follows a power law such that the fraction $P(k)$ of nodes in the network having k connections to other nodes satisfies

$$(9.1) \qquad\qquad P(k) = Ak^{-\gamma}$$

where
   γ is usually in the range ~2–3
   A is a normalization constant ensuring that the sum of all P values is exactly 1

Molecular networks allow regulation of processes at the cost of some redundancy in the system, but also impart *robustness to noise*. The most detailed type of molecular network involves metabolic reactions, since these involve not only reactions, substrates, and products, but also enzymes that catalyze the reactions.

The *Barabási–Albert model* is an algorithm for generating random scale-free networks. It operates by generating new edges at each node in an initial system by a method of probabilistic attachment. It is valuable here in the context of creating a synthetic, controlled network that has scale-free properties but which is a more reduced version of real, complex biological network. Thus, it can be used to develop general analytical methods for investigating scale-free network properties. Of key importance here is the robust identification of genuine nodes in a real network. There are several node clustering algorithms available, for example, the k-means algorithm alluded to briefly previously in the context of identifying different *Förster resonance energy transfer* (FRET) states in fluorescence microscopy (see Chapter 4) and to clustering of images of the same subclass in principal component analysis (PCA) (see Chapter 8).

The general k-means clustering algorithm functions to output k mean clusters from a data set of n points, such that $k < n$. It is structured as follows:

1 Initialize by randomly generating k initial clusters, each with k associated mean values, from the data set where k is usually relatively small compared to n.
2 k clusters are created by associating each data point with the nearest mean from a cluster. This can often be represented visually using partitions between the data points on a *Voronoi diagram*. This is mathematically equivalent to assigning each data point to the cluster whose mean value results in the minimum *within-cluster sum of squares* value.
3 After partitioning, the data points then calculate the new centroid value from each of the k clusters.
4 Iterate steps 2 and 3 until convergence. At this stage, rejection/acceptance criteria can also be applied on putative clusters (e.g., to insist that to be within a given cluster

all distances between data points must be less than a predetermined threshold, if not then remove these points from the putative cluster, which, after several iterations, may result in eroding that cluster entirely).

In biological applications, network theory has been applied to interactions between biomolecules, especially protein–protein interactions and metabolic interactions, including *cell signaling* and *gene regulation* in particular, as well as networks that model the organization of cellular components, disease states such as cancer, and species evolution models. Many aspects of systems biology can be explored using network theory.

Bacterial chemotaxis again is an ideal system for aligning systems biology and biophysical experimentation for the study of biomolecular interaction networks. Bacteria live in a dynamic, often harsh environment in which other cells compete for finite food resources. Sensory systems have evolved for highly efficient detection of external chemicals. The manner in which this is achieved is fundamentally different from the general operating principles of many biological systems in that no genetic regulation (see Chapter 7) as such appears to be involved. Signal detection and transduction does not involve any direct change in the amount or type of proteins that are made from the genes, but rather utilizes a network of proteins and protein complexes *in situ* to bring about this end. In essence, when one views a typical bacterium such as *E. coli* under the microscope, we see that its swimming consists of smooth runs of perhaps a few seconds mixed with cell tumbling events that last on the order of a few hundred milliseconds (see Chapter 8).

After each tumble, the cell swimming direction is randomized, so in effect each cell performs a 3D random walk. However, the key feature to bacterial chemotaxis is that if a chemical attractant is added to the solution, then the rate of tumbling drops off—the overall effect is that the cell swimming, although still essentially randomized by tumbling, is then biased in the direction of an increasing concentration of the attractant; in other words, this imparts an ability to move closer to a food source. The mechanisms behind this have been studied using optical microscopy on active, living cells, and single-molecule experiments are now starting to offer enormous insight into systems-level behavior.

Much of our experimental knowledge comes from the chemosensory system exhibited by the bacteria *E. coli* and *Salmonella enterica*, and it is worth discussing this paradigm system in reasonable depth since it illustrates some remarkable general features of signal transduction regulation that are applicable to several different systems. Figure 9.1b illustrates a cartoon of our understanding to date based on these species in terms of the approximate spatial locations and key interactions of the various molecular components of the complete system. Traveling in the direction of the signal, that is, from the outside of the cell in the first subsystem we encounter concerns the primary detection of chemicals outside the cell. Here, we find many thousands of tightly packed copies of a protein complex, which forms a chemoreceptor spanning the cell membrane (these complexes can undergo chemical modification by methyl groups and are thus described as *methyl-accepting chemotaxis proteins* [MCPs]).

The MCPs are linked via the protein CheW to the CheA protein. This component has a phosphate group bound to it, which can be shifted to another part of the same molecule. This process is known as *transautophosphorylation*, and it was found that the extent of this transautophosphorylation is increased in response to a *decrease* in local chemoattractant binding to the MCPs. Two different proteins known as CheB and CheY compete in binding specifically to this transferred phosphoryl group. Phosphorylated CheB (CheB-P) catalyzes demethylation of the MCPs and controls receptor adaptation in coordination with CheR that catalyzes MCP methylation, which thus serves as a negative feedback system to adapt the chemoreceptors to the size of the external chemical attractant signal, while phosphorylated CheY-P binds to the protein FliM on the rotary motor and causes the direction of rotation to reverse with CheZ being required for signal termination by catalyzing dephosphorylation of CheY-P back to CheY.

Biochemical reactions in molecular interaction networks can be solved computationally using the *Gillespie algorithm*. The *Gillespie algorithm* (or the *Doob–Gillespie algorithm*) generates a statistically optimized solution to a stochastic mathematical equation, for example, to simulate biochemical reactions efficiently. The algorithm is a form of Monte Carlo simulation. For example, consider two biomolecules A and B that reversibly bind to form AB, with forward and reverse rates for the process k_1 and k_{-1}. So, the total reaction rate R_{tot} is given by

$$(9.2) \qquad R_{tot} = k_1 [A][B] + k_{-1} [AB]$$

Here, square brackets indicate concentration values. This simple system here could utilize the Gillespie algorithm as follows:

1 Initialize the numbers of A and B in the system, the reaction constants, and random number generator seed.
2 Calculate the time to the next reaction by advancing the current time t of the simulation to time $t + \Delta t$, where Δt is optimized to be small enough to ensure that the forward and reverse reaction events in that time interval have a small probability of occurring (e.g., ~0.3–0.5, similar to that used in molecular MC simulations, see Chapter 8).
3 Calculate the forward and reverse reaction event deterministic probability values, p_1 and p_{-1}, respectively, as

$$(9.3) \qquad \begin{aligned} p_1 &= \frac{k_1 [A][B]}{R_{tot}} \\ p_{-1} &= \frac{k_{-1} [AB]}{R_{tot}} \end{aligned}$$

4 Compare these probabilities against pseudorandom numbers generated in the range 0–1 to decide if the reaction event has occurred or not.
5 Update the system with new values of number of A and B, etc., and iterate back to step 2.

This can clearly be generalized to far more complex reactions involving multiple different biomolecule types, provided the rate constants are well defined. There are, as we will see in the following text, several examples of rate constants that are functions of the reactant and product molecule concentrations (this implies that there is *feedback* in the system) in a nontrivial way. This adds to the computational complexity of the simulation, but these more complex schemes can still be incorporated into a modified Gillespie algorithm. What is not embodied in this approach however is any spatial information, since the assumption is clearly one of a reaction-limited regime (see Chapter 8).

KEY BIOLOGICAL APPLICATIONS: SYSTEMS BIOPHYSICS TECHNIQUES

Analysis of complex biological systems in vivo.

9.3 SYNTHETIC BIOLOGY, BIOMIMICRY, AND BIONANOTECHNOLOGY

Richard Feynman, one of the forefathers of the theory of *quantum electrodynamics* in theoretical physics, a genius, and a notorious bongo-drum enthusiast, is also viewed by many as the prophet who heralded a future era of *synthetic biology* and *bionanotechnology*. For example, he stated something that one might consider to be the entry point into the general engineering of materials, in one sentence that he thought in the English language conveyed the most information in the fewest words:

All things are made of atoms.

—Feynman (1963)

So, by implication, it's just a matter of rearranging them to make something else. Feynman also left the world in 1988 with a quote written on his university blackboard in Caltech, stating

What I cannot create, I do not understand.

It is tempting to suggest, which others have done, that Feynman had synthetic engineering approaches in mind when he wrote this, though the sentence he wrote following this, which was

Know how to solve every problem that has been solved

suggests rather that his own thoughts were toward "creating" mathematical solutions on pencil and paper, as opposed to creating novel, nonnatural materials. However, Feynman did give a lecture in 1959 titled "There's Plenty of Room at the Bottom," which suggests at least that he was very much aware of a new era of studying matter at the very small length scale in biological systems:

A biological system can be exceedingly small. Many of the cells are very tiny, but they are very active; they manufacture various substances; they walk around; they wiggle; and they do all kinds of marvelous things—all on a very small scale.

—Feynman (1959)

Feynman is perhaps not the grand inventor of synthetic biology, bioengineering, and nanotechnology, indeed the latter phrase was first coined later by the Japanese researcher Taniguchi (1974). However, he was the first physical scientist to clearly emphasize the machine-like nature of the way that biology operates, which is key to synthetic biology approaches.

Synthetic biology can be broadly defined as the study and engineering of synthetic devices for "useful purposes," which have been inspired by natural biological machines and processes. It often employs many similar biophysical tools and techniques of modern systems biology, but the key difference is that synthetic biology is really an engineering science, that is, it is about *making* something.

There is also a subtle and fascinating philosophical argument though that suggests that in many ways synthetic biology approaches can advance our knowledge of the life sciences in a more pragmatic and transparent way than more conventional hypothesis-driven investigations. In essence, the theory goes, humans are great at identifying putative *patterns* in data, and these patterns can then be packaged into a model, which is ultimately an approximation to explain these data using sound physical science principles. However, the model may be wrong, and scientists can spend much of their professional careers in trying to get it right. Whereas engineering approaches, as in synthetic biology, set a challenge for something to be made, as opposed to testing a model as such.

An example is the challenge of how to send someone to the moon, and back again. There are lots of technical problems encountered on the way, but once the ultimate challenge set has been successfully confronted, then whatever science went into tackling the challenge must contain key elements of the correct model. In other words, engineering challenges can often cut to the chase and result in a far more robust understanding of the underlying science compared to methods that explore different hypotheses within component models.

KEY POINT 9.2

Synthetic biology is an engineering science focused on generating useful, artificial things inspired by natural biology. But in setting a challenge for a "device" to be made, it can also result in a far greater level of core scientific understanding compared to purely hypothesis-led research.

Synthetic biology can operate both from a top-down or bottom-up context, for example, by stripping away components from an existing biological system to reduce it to more basic components, as exemplified by the artificial cells developed by Craig Venter (see Chapter 2). These top-down approaches can also be viewed as a *redesign* of natural biology or *reverse engineering*—adapting nature, making it more efficient for the specific set of tasks that you wish the new design to perform. Similarly, a device can be generated into larger systems by combining smaller length scale components that have been inspired by existing components in the natural world, but now in an artificial combination not seen in nature. Thus, creating a totally new artificial device, but inspired by biology, for example, in the use of oil-stabilized nanodroplets that utilize soft-matter nanopores in their outer membranes that can be joined together to create a complex array with a far greater complexity of function than a single droplet (see Chapter 6).

Although such synthetic biology devices can span multiple length scales, much current research is focused at the nanometer length scale. The motivation comes from the glacial time scales of evolution resulting in a myriad of optimized natural technological solutions to a range of challenging environmental problems. Life has existed on planet Earth for around 4 billion years in which time the process of genetic mutation combined with selective environmental pressures has resulted in highly optimized evolutionary solutions to biological problems at the molecular level. These are examples of established *bionanotechnology*. Thus, instead of attempting to design miniaturized devices completely *de novo*, it makes sense to try to learn lessons from the natural world, concerning the physical architecture of natural molecular machines and subcellular structures, and the processing structures of cellular functions, that have evolved to perform optimized biological roles.

The range of biophysics tools relevant to synthetic biology is large. These include in particular many fluorescence-based imaging detection methods and electrical measurements and control tools. Several synthetic biology methods utilize surface-based approaches, and so many surface chemistry biophysics tools are valuable. Similarly, biophysics tools that measure molecular interactions are useful in characterizing the performance of devices. Also, atomic force microscopy (AFM) has relevance here, both for imaging characterization of synthetic constructs on surfaces and also for smart positioning of specific synthetic constructs to different regions of a surface (see Chapter 6). Many of these devices have potential benefits to healthcare, and these are discussed in a separate section later in this chapter.

The general area of study of biological processes and structures with a view to generating newly inspired designs and technologies is called *biomicry. Structural coloration*, as exhibited in the well-known iridescence of the wings of many species of butterfly, is an excellent example but is also found in myriad species including the feathers of several birds, insect wing cases such as many beetles, bees, dragonflies, moths. It is also observed in fish scales, plant leaves and fruit, and the shells of several mollusks.

Many examples of this phenomenon arise from single or multilayer thin film interference and scattering effects. To understand the basic principle, some of the incident light of wavelength λ propagated in a medium of refraction index n_0 will be reflected from the surface of a thin film at angle θ_0, while a proportion will be refracted through the film of thickness d_1 and refractive index n_1 at an angle θ_1, some of which will then reflect from the opposite boundary back through the film and into the original medium. Back-reflected light will then emerge if the conditions for constructive interference are met between all the multiple reflected beams. For this single thin film system, the condition for constructive interference is given by $2n_1d_1\cos\theta_1=(m-1/2)\lambda$ where m is a positive integer.

This process can occur for an arbitrary number of layers of different thicknesses and refractive indices (Kinosita et al., 2008), and since the conditions for constructive interference are dependent on the wavelength of incident light these multilayers can be used to generate precisely tuned spectral characteristics of emergent light from a broad spectrum incident light source such as sunlight; for example, the beetle *Chrysina resplendens* contains ~120 thin layers which produce a bright iridescent gold color (see Worked Case Example 9.3). Structural coloration has an attraction compared to conventional pigment-based coloration technologies in having high brightness which does not fade, plus having iridescence and polarization effects, and a far broader spectral diversity which pigment-based systems cannot achieve including color tuneability.

There are myriad other examples of biomimicry which have emerged over the past decade, methods of developing highly advanced *structured light* (which refers to the engineering of optical fields both spatially and temporally in all of their degrees of freedom) from biological photonics crystal structures, smart syringes which cause significantly less pain by adapting structures inspired from the mouth parts of mosquitos, synthetic proteins motivated by silk produced from spiders and silk worms designed to help wound healing, natural photosynthetic machinery being adapted to develop devices to reduce global warming by capturing carbon from the atmosphere, and even faster trains conceived by using the core fluid dynamics concepts derived from the shape of the beak of the kingfisher bird (for a super and fun suite of more examples from the animal kingdom at least, check out the inspired podcasts by Patrick Aryee, 2021-22).

9.3.1 COMMON PRINCIPLES: TEMPLATES, MODULARITY, HIERARCHY, AND SELF-ASSEMBLY

There are four key principles that are largely common to synthetic biology: the use of scaffolds or templates, modularity of components of devices (and of the subunits that comprise the components), the hierarchical length scales of components used, and the process of self-assembly. Nature uses templates or scaffolds that direct the fabrication of biological structures. For example, DNA replication uses a scaffold of an existing DNA strand to make another one.

Synthetic biology components are implicitly modular in nature. Components can be transposed from one context to another, for example, to generate a modified device from different modules. A key feature here is one of *interchangeable parts*. Namely, that different modules, or parts of modules, can be interchanged to generate a different output or function.

This sort of *snap-fit* modularity implies a natural hierarchy of length scales, such that the complexity of the device scales with the number of modules used and thus with the effective length scale of the system, though this scaling is often far from linear and more likely to be exponential in nature. This hierarchical effect is not to say that these are simply materials out of which larger objects can be built, but rather that they are complete and complex systems. Modularity also features commonly at the level of molecular machines, which often comprise parts of synthetic biology devices. For example, molecular machines often contain specific protein subunits in multiple copies.

Self-assembly is perhaps the most important feature of synthetic biology. A key advantage with synthetic biology components is that many of them will assemble spontaneously from solution. For example, even a ribosome or a virus, which are examples of very complex established bionanotechnologies, can assemble correctly in solution if all key components are present in roughly the correct relative stoichiometries.

9.3.2 SYNTHESIZING BIOLOGICAL CIRCUITS

Many molecular regulation systems (e.g., see Chapter 7) can be treated as a distinct *biological circuit*. Some, as exemplified by bacterial chemotaxis, are pure *protein circuits*. A circuit level description of bacterial chemotaxis relevant to *E. coli* bacteria is given in Figure 9.2a. However, the majority of natural biological circuits involve ultimately gene regulation and are described as *gene circuits*. In other words, a genetic module has inputs (transcription factors [TFs]) and outputs (expressed proteins or peptides), and many of which can function autonomously in the sense of performing these functions when inserted at different loci in the genome. Several artificial genetic systems have now been developed, some of which have clear diagnostic potential for understanding and potentially treating human diseases. These are engineered cellular regulatory circuits in the genomes of living organisms, designed in much the same way as engineers fabricate microscale electronics technology. Examples

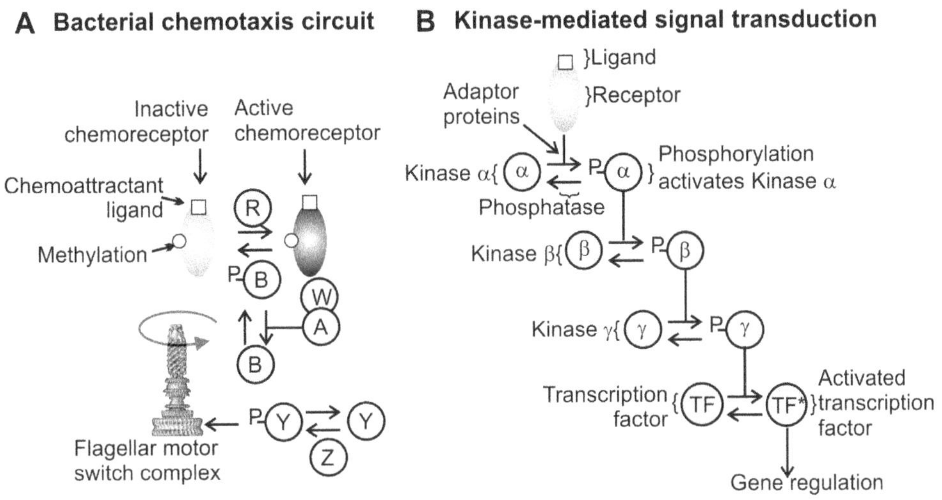

A Bacterial chemotaxis circuit

B Kinase-mediated signal transduction

FIGURE 9.2 Biological circuits. (a) Equivalent protein circuit for bacterial chemotaxis, A, B, W, Y, and Z are chemotaxis proteins CheA, CheB, CheW, CheY, and CheZ, respectively, with P indicating phosphorylation. (b) A common method of cellular signal transduction following detection of a ligand by a receptor, typically in the cell membrane, involves a cascade of serial phosphorylation steps of tyrosine kinase enzymes (depicted here by example with some genetic kinases α, β, and γ), which ultimately activates a transcription factor, which in turn can then result in regulation of a specific gene or set of genes.

include oscillators (e.g., periodic fluctuations in time of the protein output of an expressed gene), pulse generators, latches, and time-delayed responses.

One of the simplest biological circuits is that found in many gene expression systems in which the protein product from the gene expression activates further expression from the promoter of the gene in question. In this simple case, the concentration C of an expressed protein can be modeled by the following rate equation:

(9.4)
$$\dot{C} = kC - Rp$$

where
    k is the effective activation rate
    R is the concentration of repressor complexes (e.g., a relevant TF)
    p is the probability per unit time that a given promoter of a gene is occupied

The value of p can be modeled as being proportional to a Boltzmann factor $\exp(-\Delta G_p/k_B T)$ where G_p is the free energy change involved in binding a repressor molecule to the promoter of the gene. The solution to this rate equation is essentially a sigmoidal response in terms of rate of expression *versus* levels of expressed protein. In other words, the output expression rate switches from low to high over a small range of protein levels, thus acting in effect as a binary switch, which is controlled by the concentration of that particular protein in the cell (see Worked Case Example 9.1).

Ultimately, natural biological circuits have two core features in common. First, they are optimized to maximize *robustness*. By this, we mean that the output of a biological circuit is relatively insensitive to changes in biochemical parameters from cell to cell in a population. In order to ensure this, gene circuits also share the feature of involving feedback, that is, some communication between the level of output response and the level of input signal equating in effect to a gain function of the circuit, which depends on the output. Artificial biological circuits need to follow these same core design principles. Many of the components of biological circuits can be monitored and characterized using a wide range of biophysical techniques already discussed previously in this book.

KEY POINT 9.3

Biological circuits are optimized to be robust against changes in biochemical environment from cell to cell, which necessitates a gain function that depends on the output response.

For example, several natural signal transduction pathways have key components that can be adapted to be used for general synthetic biology biosensing. Many of these involve *receptor tyrosine kinases.* These are structural motifs that contain the amino acid tyrosine, and when a ligand binds to the receptors this induces a conformational change that stimulates *autophosphorylation* of the receptor (i.e., the receptor acts as an enzyme that catalyzes the binding of phosphate groups to itself). These in turn dock with a specific adapter protein and in doing so activate signaling pathways that generate specific cellular responses depending on the adapter protein. A schematic of this process is depicted in Figure 9.2b.

The point here is that the gene circuit for this response is generic, in that all that needs to be changed to turn it into a detector for another type of biomolecule are the specifics of the receptor complex and the adapter protein used, which thus offers the potential for detecting a range of different biomolecule outside cells and bringing about different, nonnative responses of the host cell. Different combinations can potentially lead to cell survival and proliferation, whereas others might lead to, for example, promoting apoptosis, or programmed cell death (see Chapter 2), which could thus have potential in destroying diseased cells in a controllable way (also, see the section in this chapter on personalizing healthcare). Similarly, the output can be tailored to express a TF that stimulates the production of a particular structural protein. For example, yeast cells have been used as a model organism to design such a system using an actin regulatory switch known as *N-WASP* to controllably manufacture F-actin filaments.

Many gene circuits also have *logic gate* features to them. For example, it is possible to engineer an AND gate using systems that require activation from two inputs to generate an output response. This is exemplified in the Y2H assay discussed previously (see Chapter 7), and there are similar gene circuit examples of OR and NOT gates. This is particularly valuable since it in principle then allows the generic design principles of electrical logic circuitry to be aligned directly with these biological circuits.

One issue with gene circuit design is the approximations used in modeling their response. For example, spatial effects are usually ignored in characterizing the behavior of gene circuits. This assumes that biochemical reactions in the system occur on time scales much slower than the typical diffusional time required for mixing of the reactants or, in other words, a reaction-limited regime (see Chapter 8). Also, stochastic effects are often ignored, that is, instead of modeling the input and output of a gene circuit as a series of discrete events, an approximation is made to represent both as continuous rather than discrete parameters. Often, in larger scale gene circuit networks, such approximations are required to generate computationally tractable simulations. However, these assumptions can often be flawed in real, extensive gene circuit networks, resulting in emergent behaviors that are sometimes difficult to predict. However, relatively simple kinetics analysis applied to gene circuits can often provide useful insight into the general output functions of a genetic module (see Worked Case Example 9.1).

In practice, there are also more fundamental biological causes for design problems of gene circuits. For example, there are sometimes cooperative effects that can occur between different gene circuits that are not embodied in a simple Boolean logic design model. One of these cooperative effects is mechanical in origin. For example, there is good evidence that mechanical perturbations in DNA can be propagated over thousands of nucleotide base pairs to change the state of a specific gene's expression, that is, to turn it "on" or "off." These effects can be measured using, for example, single-molecule force manipulation techniques such as magnetic tweezers (see Chapter 6), and there is increasing evidence that mechanical propagation is limited to the so-called topological domains in DNA, so that certain protein complexes bound to the DNA act as buffers to prevent propagation of phonons across them into adjacent domains. This is important since it implies that a simple snap-fit process may not work but requires thought as to where specifically gene circuits are engineered in a genome relative to other gene circuits.

Another biological challenge to the design of artificial gene circuits is epigenetics. Epigenetic changes to the expression of genes occur in all organisms. These are heritable changes, primarily by changes to the state of chemical methylation of the DNA, which propagate to subsequent generations without a change of the actual nucleotide sequence in the DNA (see Chapter 2). The effect of such epigenetic modifications is often different to predict from base principles, especially for an extensive network of coupled gene circuits, resulting in nontrivial design issues.

9.3.3 DNA ORIGAMI

DNA origami is a nanotechnology that utilizes the specific nucleotide base pairing of the DNA double helix to generate novel *DNA nanostructures.* This *DNA nanotechnology* was first hypothesized by Nadrian "Ned" Seeman in the 1980s (Seeman, 1982), though it was two decades later that the true potential of this speculation was first empirically confirmed using a range of biophysics tools among others. Double-stranded DNA has a persistence length of ~50 nm (see Chapter 8), implying that over a length scale of ~0–50 nm, it is in effect a stiff rod. Also, DNA has base pairing that is specific to the nucleotide types (C base pairs with G, A with T, see Chapter 2). These two factors make DNA an ideal construction material for structures at or around the nanometer length scale, in that, provided the length of each "rod" is less than ~50 nm, complex structures can be assembled on the basis of their nucleotide sequence. Since its inception toward the end of the twentieth century, these physical properties of DNA have been exploited in DNA origami to create a wide range of different *DNA nanostructures.* Biophysics tools that have been used to characterize such complex DNA nanostructure structures include fluorescence microscopy, AFM, and electron microscopy (EM) imaging, though the workhorse technique for all of this work is gel electrophoresis, which can at least confirm relatively quickly if key stages in the nanostructure assembly process have worked or not.

DNA origami offers the potential to use artificial DNA nucleotide sequences designed with the specific intention for creating novel synthetic nanoscale structures that may have useful applications (see Turberfield, 2011). It has emerged into a promising area of research both in applying single-molecule biophysics tools in characterizing DNA nanostructures, and in also utilizing the structures for further single-molecule biophysics investigations. An advantage with such structures is that, in general, they self-assemble spontaneously with high efficiency from solutions containing the correct relative stoichiometry of strand components at the correct pH and ionic strength, following controlled heating of the solution to denature or "melt" the existing double strands to single strands, and then controlled cooling allowing stable structures to self-assemble in a process called thermal annealing. Usually, sequences are designed to minimize undesirable base pairing, which can generate a range of different suboptimal structures.

In their simplest forms, DNA nanostructures include 2D array (Figure 9.3a). For example, four DNA strands, whose nucleotide sequences are designed to be *permutationally complementary* (e.g., strand 1 is complementary to strand 2 and 3, strand 2 is complementary to strand 3 and 4) can self-assemble into a stable square lattice 2D array, and similarly a three-strand combination can give rise to a hexagonal 2D array (Figure 9.3a). By using more complex complementary strand combinations, basic geometrical 3D nanostructures can be generated, such as cubes, octahedra, and tetrahedra (Figure 9.3b), as well as more complex geometries exemplified by dodecahedra and icosahedra. Typically, a range of multimers are formed in the first instance, but these can be separated into monomer nanostructure units by using repeated enzymatic treatment with specific restriction nuclease enzymes to controllably break apart the multimers (see Chapter 7).

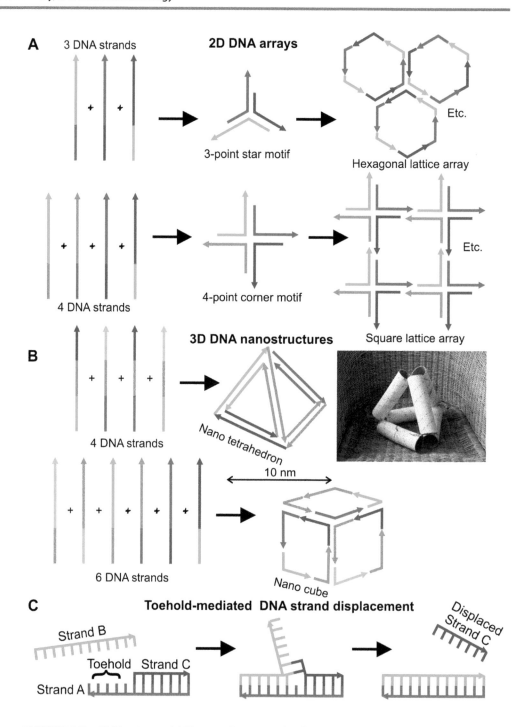

FIGURE 9.3 DNA origami. (a) Three or four strands of DNA can be designed to have complementary segments as indicated, giving rise to self-assembled 2D arrays with either hexagonal or square lattices, respectively. (b) Using more complex design involving more complementary DNA segments, 3D DNA nanostructures can be generated, such as tetrahedrons and cubes shown here (though more generally nontrivial geometrical shapes objects can also be generated). The original DNA tetrahedron design was conceived using six toilet roll tubes, some paper, and some colored bits of bendy wire (right panel: Courtesy of Richard Berry, University of Oxford, Oxford, UK). (c) Artificial DNA molecular motors all use toehold-mediated strand displacement, for example, strands A and B here are complementary, and so the addition of strand B to the solution will displace strand C from the A–C DNA nanostructure indicated.

For example (see Goodman et al., 2005), a tetrahedron can be made from four 55 nucleotide base DNA strands. Each of the six edges of the tetrahedron is composed of one of six 17-base "edge subsequences" (edge length ~7 nm), which is hybridized to its complementary segment. Each DNA strand contains three of these subsequences, or their complements, which are separated by short sequences specifically designed not to hybridize with a complementary strand, and thus act as a "hinge," to ensure that the tetrahedron vertices have flexibility to accommodate a 60° kink. Each strand runs around one of the four faces and is hybridized to the three strands running around the neighboring faces at the shared edges, and each vertex is a nicked three-arm junction, and can exist as two stereoisomers (see Chapter 2). Such a structure has the potential for acting as nanoscale brick for more extensive synthetic 3D structures.

A valuable lesson to learn for the student is the importance of basic, rudimentary thought, when it comes to the intellectual process of designing such nanostructures. When published in research articles in their final form, fancy graphics are inevitably employed. However, the initial intellectual process is often far more basic, down-to-earth, and human than this, as can be seen wonderfully exemplified from the right panel of Figure 9.3b.

It is also possible to engineer larger DNA nanostructures than the simple 3D geometrical shapes. These can include structures that are more complex than simple geometrical objects. In principle, a single strand of DNA can be used to generate such structures, referred to as the *scaffold*, though to hold it stably in place often requires several short sequences known as *staples*, which pin down certain duplex regions relative to each other, which might be liable to move relative to each other significantly otherwise. Such exotic structures have included 2D tiles, star shapes and 2D snowflake images, smiley faces, and embossed nanolettering, even a rough nanoscale map of North and South America. Many of these exotic designs, and the engineering principles used in their formulations, can be seen in the work of Caltech's Paul Rothemund in a pioneering research paper that was cited roughly 3000 times in the first 10 years since its publication, which says a great deal about the huge impact it has had to this emerging field (Rothemund, 2006).

One limit to the size of an artificial DNA structure is mismatched defects. For example, base pair interactions that do not rely on simple Watson–Crick base pairing. Although relatively uncommon, the effects over larger sections of DNA structures may be cumulative. Also, the purification methods are currently relatively low throughput, which arguably has limited extensive commercial exploitation for "useful" structures beyond the satisfaction of designing nanoscale smiley faces, though it seems tempting to imagine that these technological barriers will be reduced by future progress.

"Useful" DNA nanostructures include nanostructures that can be used as calibration tools or standards for advanced fluorescence microscopy techniques. For example, since optimized DNA nanostructures have well-defined atomic coordinates, then different color dyes can be attached as very specific locations and used as a calibration sample in FRET measurements (see Chapter 4).

Also, DNA origami can generate valuable 2D arrays that can be used as templates for the attachment of proteins. This has enormous potential for generating atomic level structural detail of membrane proteins and complexes. As discussed previously (see Chapter 7), there are technical challenges of generating stable lipid–protein interactions in a large putative crystal structure from membrane proteins, making it difficult to probe structures using x-ray crystallography. The primary alternative technique of nuclear magnetic resonance (NMR) (see Chapter 5) has associated disadvantages also. For example, it requires purified samples >95% purity in the concentration of several mg mL^{-1} typically prepared from recombinant protein to be prepared by time-consuming genetic modification of bacteria such as *E. coli* (see Chapter 7). NMR is also relatively insensitive for small proteins whose molecular weight is smaller than ~50 kDa.

The main alternative structural determination technique for membrane proteins is electron cryo-EM that allows direct imaging of biomolecules from a rapidly frozen solution supported on an electron-transparent carbon film and circumvents many of the problems associated with NMR and x-ray crystallography (see Chapter 5). However, high electron current flux in EM imaging can damage samples. Also, there are increased risks of protein sample aggregation at the high concentrations typically used, and the random orientation of particles means that analysis is limited to small groups at a time with limited potential for high-throughput analysis, though PCA has to somewhat tackled many of these issues (see Chapter 8). If instead one attaches the target protein to specifically engineered binding sites on a self-assembled 2D DNA template, this minimizes many of these issues. It also opens the possibility for 2D crystallography if proteins can be bound to the template in consistent orientations, for example, using multiple binding sites.

DNA origami can also be utilized to make dynamic as opposed to just static nanostructures can be made from DNA. These are examples of *artificial molecular motors*. A motivation to develop artificial molecular motors is for the transporting of specific biomedical cargo molecules for use in lab-on-a-chip devices (see later text). Several such devices have been constructed from DNA, inspired by the mechanisms of natural molecular motors (see Chapter 8). The key process in all DNA-based artificial molecular motors is known as *toehold-mediated strand displacement* (Figure 9.3c). Here, a single-stranded DNA toehold (also known as an *overhang* or *sticky end*) is created at the end of a double-stranded (i.e., *duplex*) segment of DNA called a "toehold." This single-stranded toehold can bind to an invading DNA strand that competes with the bound strand. Since unpaired bases have a higher effective Gibbs free energy than paired bases, then the system reaches steady state when the minimum number of unpaired bases is reached, which results in displacement of the originally bound strand. This strand displacement imparts a force on the remaining duplex structure, thus equivalent to the power stroke of natural molecular machines (see Chapter 8), with the equivalent "fuel" being the invading DNA strand. Such developments currently show promise at the point of writing this book. However, issues include artificial motors being slow and inefficient compared to native molecular motors, and DNA logic circuits are not currently as reliable as conventional electronic ones.

One interesting further application of DNA origami lies in computationally complicated optimization–minimization problems. These are exemplified by the so-called traveling salesman problem:

> Given a finite number of cities and the distances between them what is the shortest route to take such that each city is visited just once prior to returning to the starting city?

This turns out to be precisely the same problem as a single strand of DNA exploring the most optimal annealing routes for a self-assembly duplex formation process. Thus, biophysical observations of the kinetics of annealing can potentially be used as a biomolecular computational metric to complex optimization–minimization problems.

9.3.4 BIOFUELS, BIOPLASTICS, AND A GREENER ENVIRONMENT

Synthetic biology approaches have been used to engineer modified cells to generate "green" biofuels, to manufacture biodegradable plastics, and even to clean our environment (a process known as *bioremediation*). Advanced biofuels may end up being crucial to building a cleaner energy economy. With depletion of fossil fuels and decommissioning of many nuclear power stations coupled with safety and environmental concerns of those remaining, biofuel development has an appeal.

Although at an early stage of development, there are emerging signs of promising progress. For example, certain nanoparticles can increase the efficiency of biofuel production, which employ enzyme catalysis to convert cellulose from plants into smaller sugars high in their fuel value. There are also developments in *biobatteries*. These are miniaturized electrical charge storage devices that utilize biological materials. Examples are oil-stabilized *nanodroplet* arrays, discussed previously (see Chapter 6), though currently are low power and inefficient.

Other interesting designs include *nanowire* microelectrodes that can be energized by a fuel of only a few molecules of natural *redox* enzymes (Pan et al., 2008). A key challenge here is that for conventional electrical conductors, the electrical resistance varies inversely with the cross-sectional area, so that a material that obeys *Ohm's law* of electrical resistance in this way could have significantly very high electrical resistances for wires whose width is of the nanometer length scale, resulting in a highly inefficient generation of heat. Nonbiophysical methods for tackling this issue include the use of *super-conductive* materials whose electrical resistance can be made exceptionally low. Currently, these are limited in their use at relatively low temperatures (e.g., liquid nitrogen); however, it is likely that viable room-temperature superconductive material will be available in the near future.

Biophysics-inspired research into this area has included fabricating nanowires from single DNA molecule templates that are spatially periodically labeled with alternating electron-donor and electron-acceptor probes. These enable electrical conduction through a series of quantum tunneling processes, rather than by conventional electron drift. Similarly, using *molecular photonics wires* that use a DNA molecule containing multiple FRET acceptor–donor pairs along its length enables the transfer of optical information through space via FRET, thus acting as an *optical switch* at the nanometer length scale (Heilemann et al., 2004).

In 2012, the estimated average total power consumption of the human world was ~18 TW (1 TW = 1 *terawatt* = 10^{12} W), equivalent to the output of almost 10,000 *Hoover Dams*. The total power production estimated from *global photosynthesis* is more like 2000 TW, which may indicate a sensible strategy forward to develop methods to harness the energy transduction properties of natural photosynthetic systems. However, the total power available to the Earth from the sun is ~200 PW (1 PW = 1 *petawatt* = 10^{15} W)—in this sense, natural photosynthesis is arguably relatively inefficient at only extracting ~1% of the available solar energy. Solar energy that is not reflected/scattered away from plant leaves is mainly transformed to heat as opposed to being locked in high-energy chemical bonds in molecules of sugars.

One route being developed to harness this process more controllably is in the modification of natural *cyanobacteria* (photosynthetic bacteria), potentially to develop synthetic chloroplasts by introducing them into other more complex organisms. These cyanobacteria contain carboxysomes, which have a protein shell similar in size and shape to the capsid coat of viruses but contain the enzyme RuBisCo that catalyzes photosynthesis. The application of these carboxysomes to different organisms has not yet been clearly demonstrated, but chloroplasts have been imported into foreign mammalian macrophage cells and zebrafish embryos, to increase their photosynthetic yield (Agapakis et al., 2011).

Modification of bacteria has also been applied in the development of synthetic strains, which can generate alcohols as fuels, for example, *butanol*. This method manipulates the natural fermentation process by adapting the acetyl-CoA complex that forms an essential hub between many metabolic pathways, most importantly in the Krebs cycle (see Chapter 2). The key challenge is that the alcohol is toxic to the cell, so achieving commercially feasible concentrations is difficult. However, biophysics tools such as fluorescence microscopy and mass spectrometry can be used to assist in understanding how the molecular homeostasis of the acetyl-CoA is achieved and potentially can be manipulated.

9.3.5 ENGINEERING ARTIFICIAL PEPTIDES, PROTEINS, AND LARGER PROTEIN COMPLEXES

Bioengineering of peptides and proteins can take multiple forms. One is the genetic manipulation/modification of existing native proteins starting from their composition of natural amino acids. Another is the chemical conjugation of native proteins (see Chapter 7), which in part crosses over into genetic modification. And finally one can also utilize unnatural amino acids in peptide and protein synthesis to generate truly artificial products.

Engineering of proteins using genetic modifications typically involves point mutations of single amino acids to modify their function. Often, the effect is to impair functionality, for example, to impair a kinase enzyme from its normal ability to catalyze phosphorylation, but more rarely to increase functionality in some way. The techniques involved are those of standard molecular cloning (Chapter 7). Similarly, genetic modifications can include encoding tags next to genes at the level of the DNA sequence, for example, fluorescent proteins.

Modified proteins have been used in genuine bottom-up synthetic assays to build devices that utilize protein pattern formation. For example, the bacterial proteins MinC, MinD, and MinE, which are responsible for correctly determining the position of cell division in growing *E. coli* bacterial cells (see Chapter 8), have been reconstituted in an artificial noncellular system involving just microwells of different sizes and shapes. These systems require an energy input, provided by the hydrolysis of ATP, but then result in fascinating pattern formation depending on the size and shape of the microwell boundaries, even in the complete absence of cells. These systems have yet to be exploited for "useful" ends, but show enormous potential at being able to develop controllable patterns in solution from just a few protein components, which could have implications for templating of more complexing synthetic devices.

Similarly, other bacterial cell division proteins such as FtsZ have been artificially reconstituted. FtsZ is responsible for the actual constriction of the cell body during the process of cell division, through the formation of a tightening Z-ring. FtsZ can be reconstituted and fluorescently labeled in artificial liposomes with no cells present and fueled by GTP hydrolysis in this case, which can result in controlled liposome constriction that can be monitored in real time using fluorescence microscopy imaging.

The degron system is also a good example of genetic and protein-based bioengineering. As discussed previously (see Chapter 7), this uses genetic modification to insert degron tags onto specific proteins. By inducing the expression of the *sspB* adapter protein gene, these tagged proteins can then be controllably degraded in real time inside living bacterial cells. This facilitates the investigation of the biological function of these proteins using a range of biophysical techniques, but also has potential for exploitation in synthetic biology devices.

A range of unnatural amino acids have also been developed, which utilize different substituent groups to optimize their physical and chemical properties catered for specific binding environments, which require a combination of genetic and chemical methods to integrate into artificial protein structures. A subset of these is genetically encoded synthetic fluorescent amino acids, used as reporter molecules. These are artificial amino acids that have a covalently bound fluorescent tag engineered into the substituent group. The method of tagging is not via chemical conjugation but rather the amino acid is genetically coded directly into the DNA that codes for a native protein that is to be modified. This essentially involves modifying one of the *nonsense* codons that normally do not code for an amino acid (see Chapter 2). Currently, the brightness and efficiency of these fluorescent amino acids is still poor for low-light biophysical applications such as single-molecule fluorescence microscopy studies, as well as there being a limitation of the colors available, but there is certainly scope for future development.

Larger length scale synthetic biology devices also utilize macromolecular protein complexes. For example, the protein capsid of virus particles has been used. One such device has used a thin layer of M13 bacteriophage virus particle to construct a piezoelectric generator that is sufficient to operate a liquid crystal display.

9.3.6 BIOMIMETIC MATERIALS

Metamaterials are materials engineered to have properties that have not yet been found in nature. Biomimetic materials are a subset of metamaterials. They are artificial materials, but they mimic native biological materials and often require several stages of biophysical characterization of their material properties during the bioengineering design and manufacturing processes. Several existing biostructures have been used as inspiration for some of the biomimetic materials.

The development of new photonic material that is biomimetic to natural butterfly wings is a good example. Here, biomimetization of butterfly wings can be performed by a series of metal vapor deposition steps (see Chapter 7). The key step though is actually to use a native wing as a positive mold—this can be first coated in a glass material called chalcogenide to wrap closely around the butterfly structure to a precision of a few nanometers. A procedure called *plasma ashing* (exposing the sample to a high-energy ion plasma beam) can then be used to destroy the original wing but leaves the glass wrap intact. This can then be further coated using metal vapor deposition as required. The key optical feature of butterfly wings is their ability to function as very efficient photonic bandgap devices, that is, they can select very precisely which regions of the spectrum of light to transit from a broad-spectrum source. This is true not only for visible wavelength, but infrared and ultraviolet also. Biomimetic butterfly wings share these properties but have the added advantage of not being attached to a butterfly.

Another class of biomimetic materials crossover is in the field of *biocompatible materials*. These are novel, artificial materials used in *tissue engineering* and *regenerative medicine* usually to replace native damaged structures in the human body to improve health. As such, we discuss these in this chapter in the section on personalizing healthcare.

9.3.7 HYBRID BIO/BIO–BIO DEVICES

A number of artificial devices are being fabricated, which incorporate both biological and nonbiological components. For example, it is possible to use the swimming of bacteria to make a 20 μm diameter silicon-based rotor, machined with a submicron precision, and thus an example of genuine nanotechnology, rotate (Hiratsuka et al., 2006). It was the first "engine" that combined living bacteria with nonbiological, inorganic components. The bacterium used was *Mycoplasma mobile* that normally swims on the surface of soil at a few microns per second. These bacteria follow the curvature of a surface and so will on average follow the curved track around the rotor. By chemically modifying the surface of the bacteria with biotin tags, and then conjugating the rotor surface with streptavidin that has a very strong binding affinity to biotin, the cells stick to the rotor and make it spin around as they swim around its perimeter, at a rotation speed of ~2 Hz that could generate a torque of ~10^{-15} Nm.

This level of torque is roughly four orders of magnitude smaller than purely inorganic mechanical microscopic motors. However, this should also be contrasted with being five orders of magnitude larger than the torque generated by the power stroke action of the molecular motor myosin on an F-actin filament (see Chapter 8). Microorganisms, such as these bacteria, have had millions of years to evolve ingenious strategies to explore and move around. By harnessing this capability and fusing it with the high-precision technology of silicon at micro- and nanoscale fabrication, one can start to design useful hybrid devices (Figure 9.4).

Another emerging area of hybridizing bio- and nonbiocomponents is *bioelectronics*. Here, perhaps the most exciting developments involve attempts to design *biological transistors*.

The transistor was invented in 1947 by physicists John Bardeen, Walter Brattain, and William Shockley, made possible through advances in our understanding of the physics of electron mobility in semiconductor materials. *Moore's law* is a heuristic relation, relating time with the size of key integrated circuitry, most especially the transistor, suggesting that technological progress is resulting in a decrease in the size of the transistor by roughly a factor of two every two years. With the enormous developments into shrinking of the effective size of a transistor to less than ~40 nm of the present day, this has increased speed and efficiency in modern computing technology to unprecedented levels, which has affected most areas of biophysical tools.

Interestingly, the size of the modern transistor is now approaching that of assemblages of the larger biological molecular complexes, which drive many of the key processes in biology. Bioengineers have created the first biological transistor from the nucleic acids DNA and RNA, denoted as a *transcriptor* (Bonnet et al., 2013) in reference to the natural cellular process of transcribing the genetic code embodied in the DNA sequence into molecules of mRNA (see Chapter 2). The transcriptor can be viewed as a biological analog of a solid-state digital transistor in electronic circuitry. For example, transistors control electron flow, whereas transcriptors regulate the flux of RNA polymerase enzymes as they translocate

Hybrid bio/non-bio nanotechnology

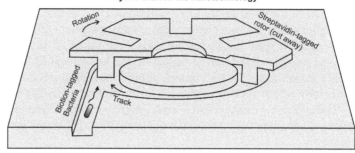

FIGURE 9.4   Synthetic biology combined with fabricated solid-state micro and nanoscale structures. Cutaway schematic of a hybrid bionanotechnology/non-bionanotechnology device. Here, a 20 μm diameter silicon dioxide rotor is pushed around by the swimming action of the bacterium *Mycoplasma mobile*, whose surface has been tagged with biotin while that of the rotor has been labeled with streptavidin. The bacterial motion is energized by glucose, and this cell motion is coupled to rotation of the rotor via chemical bonds formed between the biotin and streptavidin.

along a DNA molecule whose associated gene is being expressed. Transcriptors use combinations of enzymes called "integrases" that control RNA movement as it is fabricated from the DNA template.

Transistors amplify a relatively small current signal at the *base* input (or *gate* in the case of field effect transistors [FETs]) into a much larger current between the *emitter/collector* (or equivalent voltage between the *source/drain* in FETs). Similarly, transcriptors respond to small changes in the integrase activity, and these can result in a very large change in the flux of RNA polymerase resulting in significant increases in the level of expression of specific genes. Multiple transcriptors can be cloned on different plasmids and controlled using different chemical induction (see Chapter 7) and then combined in the same way as multiple electronic transistors to generate all the standard logic gates, but inside a living cell. Thus, this presents the opportunity for genuine *biocomputation*, that is, a biological computer inside a functional cell. An appealing potential application is toward *in situ* diagnostics and associated therapeutics of human diseases, that is, cells with a capacity to detect the presence of diseases and regulate cellular level interventions as bespoke treatment with no external intervention required (see the section on personalizing healthcare).

KEY BIOLOGICAL APPLICATIONS: SYNTHETIC BIOLOGY AND BIONANOTECHNOLOGY TOOLS

Biosensing; Smart cell–based diagnostics; Biomimetization.

Worked Case Example 9.1: Biological Circuits

A biological circuit was found to consist of a simple feedback system where protein X binds to another molecule of X to generate a dimer, concentration X_2, at forward rate k and backward rate for the reverse process k_{-2}. This dimer binds to its own promoter at forward rate k_2 and backward rate for the reverse process k_{-2} and activates the promotor at a rate k_3, whereas the monomer subunits have no effect on gene expression. Protein X is also degraded by an enzyme to form an enzyme product E·X at a forward rate k_4 with backward rate for the reverse process k_{-4}. In addition, the promotor is "leaky" meaning that it spontaneously results in expression of X at a rate k_5, which depends on the concentration of the promotor but not of X.

E·X decays irreversibly into a degraded product at a rate k_6, and X also spontaneously irreversibly degrades without the requirement of E at a rate k_7, which depends on C.

a   *What do we mean by a reaction-limited regime? Assuming a reaction-limited regime for this biological circuit, write down a rate equation for the concentration C after a time t, assuming a promotor concentration P and enzyme concentration E.*

b *Assuming that dimer formation and the binding of the dimer to a promotor is rapid, and that E is always relatively low with the enzyme saturated, indicate what shapes the separate graphs for the rate of production of X and of its degradation as a function of C take in steady state. From the shape of these graphs, what type of function might this specific biological circuit be used for in a synthetic device?*

Answers

a A reaction-limited regime is one in which the time scale that chemical reactions occur is much larger than the diffusional time required for mixing of the chemical reactants. Assuming a reaction-limited regime allows us to define simple chemical reaction equations and associated rate equations, which have no spatial dependence. For production of X we can say that

(9.5) $$\dot{C}_{production} = k_3 PX_2 + k_5 P$$

For the degradation of X, there are two components, one due to the enzyme that can be modeled using the Michaelis–Menten equation (see Chapter 8) and the other due to spontaneous degradation of X:

(9.6) $$\dot{C}_{degradation} = \left(k_6 E \frac{C}{(k_{-4} + k_6)/k_4 + C} + k_7 C \right)$$

Thus,

(9.7) $$\dot{C} = \dot{C}_{procduction} + \dot{C}_{degradation} = k_3 PX_2 + k_5 P - k_6 E \frac{C}{(k_{-4} + k_6)/k_4 + C} + k_7 C$$

However, if the enzyme is saturated, then $k_4 \gg (k_{-4} + k_6)$ and the degradation rate is approximately constant:

(9.8) $$\dot{C} = k_3 PX_2 + k_5 P - k_6 E - k_7 C$$

b The degradation rate *versus* C under these conditions is a straight line of gradient $-k_7$ and intercept $-k_6 E$. To determine the shape of the production curve, we need to first determine the steady-state concentration of the dimer in terms of C. From conservation we can say that

(9.9) $$P_{total} = P + P \cdot X_2$$

At steady state of the dimer binding to the promoter,

$$k_2 P \cdot X_2 - k_{-2} PX_2 = 0$$

$$\therefore P \cdot X_2 = \frac{k_{-2}}{k_2} (P_{total} - P \cdot X_2) X_2$$

(9.10) $$\therefore P \cdot X_2 = P_{total} \left(1 - \frac{X_2}{k_{-2}/k_2 + X_2} \right)$$

$$\therefore P \cdot P_{total} - P \cdot X = P_{total} \frac{k_{-2}/k_2}{k_{-2}/k_2 + X_2}$$

Thus,

(9.11)
$$\dot{C} = P_{total} \frac{k_5 + k_3 X_2}{k_{-2}/k_2 + X_2} - k_6 E - k_7 C$$

At steady state, the dimerization rate satisfies

(9.12)
$$k_{-4} X_2 - k_4 C^2 = 0$$

Thus,

(9.13)
$$\dot{C} = P_{total} \frac{k_3\left(k_4/k_{-4}\right)C^2}{\left(k_2/k_{-2}\right)+\left(k_4/k_{-4}\right)C^2} - k_6 E - k_7 C$$

The first term is the production rate, and the shape of this *versus* C is sigmoidal. This means that the rate switches rapidly from low to high values over a relatively short range of C. This indicates a binary switching function whose output is high or low depending on the level of C. However, there is a finite rate of degradation, which increases with C; thus, the high state of the binary switch is only transient, so in effect the biological circuit serves as a digital pulse generator controlled by the specific concentration level of the protein X.

9.4 PERSONALIZING HEALTHCARE

Personalized healthcare is a medical model that proposes to cater healthcare specifically to a unique, individual patient, as opposed to relying on generic treatments that are relevant to population-level information. For example, we know that human genomes in general vary significantly from one individual to the next (Chapter 2). Some people have a greater genetic predisposition toward certain disorders and diseases than others. Also, the responses of individual patients to different treatments can vary widely, which potentially affects the outcome of particular medical treatments.

The selection of appropriate, optimal therapies is based on specific information concerning a patient's genetic, molecular, and cellular makeup. In terms of biophysics, this has involved developments in *smart diagnostics* such as lab-on-a-chip technologies, and smart, targeted treatment, and cellular delivery methods, including *nanomedicine*. In addition, computational modeling can be combined with experimental biophysics for more intelligent *in silico drug design* catered toward specific individuals.

Developing methods to enable personalized healthcare is particularly important regarding the current global increased risks of infection, and the challenges of an increasingly aging population. For infection challenges, the past overuse of antibiotics has led to the emergence of *superbugs*, such as methicillin or vancomycin resistant *Staphylococcus aureus* (*MRSA* and *VRSA* respectively), which are resistant to many of the traditional antibiotics available. These now impose significant limitations on the successful outcomes of many surgical treatments, cancer therapies and organ transplants; personalized diagnostic biosensing to detect the suite of different infectious pathogens present in different parts of the body in specific individuals could be invaluable in developing catered drug treatments to combat these. For aging issues, the big challenges are heart disease, cancer, and dementia. Again, all these disorders are amenable to personalized biosensing—innovative, fast-response technologies which can utilize appropriate libraries of biomarkers to personalize earlier diagnosis and thereby accelerate a more tailored treatment at far earlier stages in these chronic conditions that has been available previously.

9.4.1 LAB-ON-A-CHIP AND OTHER NEW DIAGNOSTIC TOOLS

Developments in microfluidics and surface chemistry conjugation methods (see Chapter 7), photonics, micro- and bioelectronics, and synthetic biology have all facilitated the miniaturization and increased portability of smart biosensing devices. These devices are designed to detect specific features in biological samples, for example, the presence of particular types of cells and/or molecules. In doing so, this presents a diagnostic and high-throughput screening capability reduced to a very small length scale device, hence the phrase lab-on-a-chip. An ultimate aim is to develop systems in which diagnosis can be made by the detection and analysis of microliter quantities of a patient specimen, such as blood, sputum, urine, fed through a miniaturized biomolecular detection device coupled to smart microelectronics.

Typically, these devices consist of hybrid nonbiological solid-state silicon-based substrates with synthetic arrangements of biological matter, in a complex microchip arrangement that often employs controlled microfluidics to convey biological sample material in aqueous solution to one or more detection zones in the microchip. For specific detection of *biomarkers*, that is, labels that are specific to certain biomolecules or cell types, a *surface pull-down* approach is typical. Here, the surface of a detection zone is coated with a chemical rearrangement that binds specifically to one or more biomarkers in question (see Chapter 7). Once immobilized, the biological material can then be subjected to a range of biophysical measurements to detect its presence. These are all techniques that have been discussed in the previous chapters of this book.

Fluorescence detection can be applied if the biomarker can be fluorescently labeled. To achieve fluorescence excitation, devices can utilize the photonics properties of the silicon-based flow-cell substrate, for example, photonic waveguiding to enable excitation light to be guided to the detection zone, photonic bandgap filtering to separate excitation light from fluorescence emissions, and smart designs of microfabricated photonic surface geometries to generate evanescent excitation fields to increase the detection signal-to-noise ratio by minimizing signal detection from unbound biological material.

Nonfluorescence detection lab-on-a-chip biosensors are also being developed. These include detection metrics based on laser dark-field detection of nanogold particles, and label-free approaches such as evanescent field interferometry, surface plasmon resonance–type methods and Raman spectroscopy, and surface-enhanced Raman spectroscopy (see Chapter 3), also, using electrical impedance and ultrasensitive microscale quartz crystal microbalance resonators (see Chapter 6). Microcantilevers, similar to those used in AFM imaging (see Chapter 6), can similarly be used for biomolecule detection. Here, the surface of the microcantilever is chemically functionalized typically using a specific antibody. As biomolecules with specificity to the antibody bind to the cantilever surface, this equates to a small change in effective mass, resulting in a slight decrease in resonance frequency that can be detected. Typical microcantilevers have a resonance frequency of a few hundred kHz, with an associated *quality factor* (or *Q factor*) of typically 800–900. For any general resonator system, Q is defined as

$$Q = \frac{\nu_0}{\Delta \nu}$$

where
ν_0 is the resonance frequency
$\Delta \nu$ is the half-power bandwidth, which is thus ~1 kHz

For many microcantilever systems, changes of ~1 in a 1000 in v_0 can be measured across the bandwidth range, equating to ~1 Hz that corresponds to a change in mass of 10^{-15} kg (or 1 picogram, pg). This may seem a small amount, but even for a large protein of ~100 kDa molecular weight, this is equivalent to the binding of ~10^6 molecules. However, a real advantage with this method involves using multiple cantilevers with different antibody coatings inside the chip device to enable a signature for the presence of a range of different biomolecules present in a sample to be built up. Improvements in high-throughput and signature detection can be made using a similar array strategy of multiple detection zones using other biophysical detection techniques beyond microcantilevers.

Mechanical signals are also emerging as valuable metrics for cell types in biosensing, for example, using AFM to probe the stiffness of cell membranes, and also, optical force tools such as the optical stretcher to measure cell elasticity (see Chapter 6). Cell mechanics change in disease states, though for cancer the dependence is complex since different types of cancers can result in either increasing or decreasing the cell stiffness and also may have different stiffness values at different stages in tumor formation.

The faithful interpretation of relatively small biomolecule signals from portable lab-on-a-chip devices presents problems in terms of poor stability and low signal-to-noise ratios. Improvements in interpretation can be made using smart computational inference tools, for example, Bayesian inference (see Chapter 8). Although signals from individual biomarkers can be noisy, integrating the combination of multiple signals from different biomarkers through Bayesian inference can lead to greatly improved fidelity of detection. Much of this computation can be done decoupled from the hardware of the device itself, for example, to utilize smartphone technology. This enables a reduction in both the size and cost of the device and makes the prospect of such devices emerging into clinics in the near future more of a reality.

A Holy Grail in the biosensing field is the ability to efficiently and cheaply sequence single molecules of DNA. For example, the so-called $1000 genome refers to a challenge set for sequencing technologies called next-generation sequencers, which combine several biophysical techniques to deliver a whole genome sequence for a cost of $1000 or less. The U.S. biotech company Illumina has such a machine that purports to do so, essentially by first fragmenting the DNA, binding the fragments to a glass slide surface, amplifying the fragments using PCR (see Chapter 7), and detecting different fluorescently labeled nucleotide bases tagged with four different dyes in these surface clusters. Bioinformatics algorithms are used to computationally stitch the sequenced fragments together to predict the full DNA sequence. This is similar to the original *Sanger shotgun* approach for DNA sequencing, which generated DNA fragments but then quantified sequence differences by running the fragments on gel electrophoresis. However, the combination of amplification, clustering, and fluorescence-based detection results in far greater high throughput (e.g., one human genome sequenced in a few days). However, the error rate is ca. 1 in 1000 nucleotides, which may seem small but in fact results in 4 million incorrectly predicted bases in a typical genome, so there is scope for improvement.

A promising new type of sequencing technology uses ion conductance measurements through engineered nanopores, either solid state of manufactured from protein adapters such as α-hemolysin (see Chapter 6). For example, in 2012, the UK biotech company Oxford Nanopore released a disposable sequencing device using such technology that could interface with a PC via a USB port and which cost less than $1000. The device had a stated accuracy of ~4%, which makes it not sufficiently precise for some applications, but even so with a capability to sequence DNA segments up to ~1 million base pairs in length, this represents a significant step forward.

A promising area of smart *in vivo* diagnostics involves the use of synthetic biological circuits inside living cells. For example, such methods can in principle turn a cell into a biological computer that can detect changes to its environment, record such changes in memory composed of DNA, and then stimulate an appropriate cellular response, for example, send a signal to stop a cell from producing a particular hormone and/or produce more of another hormone, or to stimulate apoptosis (i.e., programmed cell death) if a cancer is detected.

One of the most promising emerging next-generation sequencing technologies is single-cell sequencing, which we discussed briefly in Chapter 7. Here, advanced microfluidics can be used to isolate a single cell on the basis of some biophysical metric. This metric is often, but not exclusively, fluorescence output from the labeling of one or more specific biomolecules in that cell, in much the same way as cells in standard fluorescence-assisted cell sorting (FACS) are isolated (see Chapter 3). The key differences here, however, are sensitivity and throughput. Single-cell sequencing demands a greater level of detection sensitivity to detect sometimes relatively small differences between individual cells, and the method is therefore intrinsically lower throughput than standard FACS since the process involves probing the biophysical metric of individual cells computationally more intensively.

For example, an individual cell may potentially be probed using single-molecule precise fluorescence imaging to infer the copy number of a specific fluorescently labeled biomolecule in that cell using a step-wise photobleaching of the fluorescent dye (see Chapter 8). That one cell can then be isolated from the rest of the population using advanced microfluidics, for example, using piezo microvalves or potentially even optical tweezers (see Chapter 6) to shunt the cell into a separate region of the smart flow cell. At this stage, the cell could, in principle, then be grown to form a clonal culture and subjected to standard bulk level sequencing technologies; however, the issue here is that such a cell population is never entirely "clonal" since there are inevitably spontaneous genetic mutations that occur at every cell division. A more definitive approach is to isolate the DNA of the one single cell and then amplify this using PCR (see Chapter 7). However, the mass of DNA from even a relatively large cell such as a human cell is just a few picograms (i.e., pg, or 10^{-12} g), which is at the very low end of copying accuracy for PCR, and so DNA replication errors during PCR amplification are much more likely with current technology available. Even so, single-cell sequencing offers genuine potential to bridge the phenotype to genotype gap. The real goal here in terms of personalized healthcare is to develop methods of very early-stage diagnosis of diseases and genetic disorders on the basis of detecting just a single cell from an individual patient sample.

Lower technology biophysics solutions to personalized diagnostics are also emerging and are especially appealing due to their low cost but high potential gain. For example, a simple card-based origami optical microscope has been developed by researchers at the UC Berkeley called the Foldscope (Cybulski et al., 2014) that can be assembled from a few simple folds of card, using just one cheap spherical microlens and an LED, as a light source produces images of sufficient quality to identify a range of different microbial pathogens up to a magnification of ~2000. But it weighs just 8 g and costs only ~$1 to make. A motivation for this cheap and low-tech device is to enable earlier diagnosis of microbial infection of patients in developing countries that may have no rapid access to microbial laboratory facilities.

9.4.2 NANOMEDICINE

The use of bionanotechnology applied to medicine is already emerging at the level of *targeted drug binding*, for example, to develop pharmaceutical treatments that destroy specific diseased cells such as those of cancers through the use of specific binding. These include radioactive nanoparticles coated with specific antibody probes to act as "killer" probes. Specific aptamers are used (see Chapter 7) to block key processes in specific diseases. Aptamers have an important advantage over molecular recognition technologies in evoking a minimal immune response, unlike the closest competing technology of antibody–antigen binding for which many antibodies evoke strong immunogenic reactions at relatively small doses. Targeted binding can also be valuable for the visualization of diseased tissue (e.g., antibody-tagged QDs can specifically bind to tumors and assist in the discrimination between healthy and nonhealthy cellular material).

Bionanotechnology is also being applied to assist in greater personalization of *targeted drug delivery*, that is, increasing the specificity and efficacy of drug actually being internalized by the cells in which they are designed to act. Established techniques in this area include the delivery of certain drug compounds into specific cell types by *piggybacking* on the normal process of endocytosis by which eukaryotic cells internalize extracellular material (see Chapter 2). Other emerging targeted delivery tools involve synthetic biological 3D nanostructures, for example, made from DNA, to act as *molecular cages* to enable the efficient delivery of a variety of drugs deep into a cell while protecting it from normal cellular degradation processes before it is released to exert its pharmacological effect.

Exciting nanomedicine developments also include *bionanoelectric devices* to interface with nerve and muscle tissue, such as in the brain and the heart. Devices used in cellular repair are sometimes referred to as *nanoscale robots*, or *nanobots*. Much recent media coverage concerning nanobots has involved speculation over the terrifying implications were they to go wrong. Such as consuming the entire Earth or at the very least turning it into a *gray goo*. However, the realistic state of the art in nanobot technology is still at a technically challenging stage; to get even natural nanoscale machines to work unmodified outside of their original biological function, let alone to generate *de novo* molecular machines that can be programmed by the user to perform cellular tasks, is hard enough, as any veteran student in this area of biophysics will attest!

The most promising emerging area of nanomedicine currently involves methods to facilitate *tissue regeneration*. This can be achieved through the use of biomimetic materials and the application of *stem cells*. Stem cells in multicellular eukaryotic organisms are cells that have not yet *differentiated*. *Differentiation* is not to be confused with the mathematical term in calculus, but here is a biological process in which cells undergo morphological and biochemical changes in their life cycle to commit to being a specific cell type, for example, a nerve cell and a muscle cell. However, prior to this stage, cells are described as stem cells and can in principle differentiate into any cell type depending on external triggers involving both the external chemical environment and mechanical signals detected by the cell from the outside world. Thus, if stem cells are transplanted from an appropriate donor source, they can in principle replace damaged tissue. In some countries, including the United States, there are still some levels of ongoing ethical debate as to the use of stem cells as a therapeutic aid to alleviate human suffering.

The development of biomimetic structures as replacements for damaged/diseased tissues has experienced many successes, such as materials either directly replace the damaged tissue and/or act as a growth template to permit stem cells to assemble in highly specific regions of space to facilitate the generation of new tissues. Biomimetic tissue replacement materials focus largely on being mimics for the structural properties of the healthy native tissue, for example, hard structural tissues such as bone and teeth and also softer structural extracellular matrix material such as collagen mimics.

Biocompatible inorganic biomimetics has focused on using materials that can be synthesized in aqueous environments under physiological conditions that exhibit chemical and structural stability. These particularly include noble metals such as gold, platinum, and palladium, as well as metal oxide semiconductors such as zinc oxide and copper(I) oxide, but also chemically inert plastics such as polyethylene, which benefit from having a low frictional drag while being relatively nonimmunogenic, and also some ceramics, and so all have applications in joint replacements. Inorganic surfaces are often precoated with short sequence peptides to encourage binding of cells from surrounding tissue.

Collagen is a key target for biomimetization. It provides a structural framework for the connective tissues in the extracellular matrix (see Chapter 2) as well as plays a key role in the formation of new bone tissue from bone producing cells (called osteoblasts) embedded in the extracellular matrix. Chemically modifying collagen, for example, by generating multiple copies of a cell binding domain, can increase the rate at which new bone growth occurs. This effect can be characterized using a range of biophysical techniques such as confocal microscopy and SEM *in vitro*. Thus, modified collagen replacement injected into the connective tissue of a patient suffering from a bone depletion disease into local regions of bone depletion, detected and quantified using x-ray and CT imaging *in vivo*, can act as a biomimetic microenvironment for osteoblasts to stimulate the regeneration of new bone tissue.

Mimicking the naturally porous and fibrous morphology of the extracellular matrix can also be utilized in biomimetization of tissues and organs. Some biomimetic tissues can be engineered in the absence of an artificial scaffold too, for example, skeletal muscle has been engineered *in vitro* using stem cells and just providing the current cellular and extracellular biochemical triggers; however, an artificial scaffold in general improves the efficiency of stem cells ultimately differentiating into regenerated tissues. For example, artificial nanofiber structures that mimic the native extracellular matrix can stimulate the adhesion of a range of different cell types and thus act as tissue-engineered scaffolds for several different tissues, which interface directly with the extracellular matrix, including bone, blood vessels, skin, muscles (including the heart muscle), the front surface of the eye (known as the cornea), and nerve tissue. Active research in this area involves the development of several new biohybrid/non-biohybrid materials to use as nanofiber scaffolds. Modification of these nanofibers involves chemically functionalizing the surface to promote binding of bioactive molecules such as key enzymes and/or various important drugs.

In wounds, infections, and burn injuries, skin tissue engineering using biomimetics approaches can be of substantial patient benefit. Nanofiber scaffolds can be both 2D and 3D. These can stimulate adhesion of injected stem cells and subsequently growth of new skin tissue. Similarly, *nanobiomimetic tissue-engineered blood vessels* can be used to promote adhesion of stem cells for stimulating growth of new blood vessels, which has been demonstrated in major blood vessels, such as the main artery of the *aorta*. This involves more complex growth of different types of cells than in skin, including both *endothelial cells* that form the structure of the wall of the artery as well as *smooth muscle cells* that perform an essential role in regulating the artery diameter and hence the blood flow rate. The efficacy of these nanofiber implants again can be demonstrated using a range of biophysics tools, here, for example, including Doppler ultrasound to probe the blood flow through the aorta and to use x-ray spectroscopy on implant samples obtained from animal model organisms *ex vivo*. Nanoscale particles and self-assembled synthetic biological nanostructures also have potential applications as scaffolds for biomimetic tissue.

KEY POINT 9.4

There has been much speculative media coverage concerning bionanotechnology and synthetic biology in general, both positive and negative in terms of potential benefits and pitfalls. However, behind the hype, significant steady advances are being made in areas of nanomedicine in particular, which have utilized the developments of modern biophysics, and may pave the way to significant future health benefits.

9.4.3 DESIGNER DRUGS THROUGH *IN SILICO* METHODS

Bioinformatics modeling and molecular simulation tools (see Chapter 8) can now be applied directly to problems of screening candidate new drugs on their desired targets to enable the so-called *in silico* drug design. These simulation methods often combine *ab initio* with classical simulation tools to probe the efficacy of molecular docking of such candidate drugs. Such virtual screening of multiple candidate drugs can provide invaluable help in homing in on the most promising of these candidates for subsequent experimental testing.

The results of these experimental assays, many of which involve biophysics techniques discussed previously in this book, can then be fed back into refined computational modeling, and the process is iterated. Similarly, however, undesirable interactions between candidate new drugs and other cellular machinery benefit from *in silico* modeling approaches, for example, to simulate toxicity effects through detrimental interactions with biomolecules that are not the primary targets of the drug.

KEY BIOLOGICAL APPLICATIONS: PERSONALIZED HEALTHCARE TOOLS

Lab-on-a-chip devices; *In silico* drug design.

9.5 EXTENDING LENGTH AND TIME SCALES TO QUANTUM AND ECOLOGICAL BIOPHYSICS

Established biophysical tools have focused mainly on the broad length scale between the single molecule and the single cell. However, emerging techniques are being developed to expand this length scale at both extremes. At the high end, this includes methods of investigating multiple cells, tissues, and even large whole organisms of animals and plants. Beyond this organismal macroscopic length scale is a larger length scale concerning populations of organisms, or *ecosystems*, which can be probed using emerging experimental and theoretical biophysical methods. In addition, there are also biophysical investigations that are being attempted at the other end of the length scale, for far smaller lengths and times, in terms of *quantum biology* effects.

9.5.1 QUANTUM BIOLOGY

Quantum mechanical effects are ubiquitous in biology. For example, all molecular orbitals have an origin in quantum mechanics. The debate about quantum biology, however, turns on this simple question:

Does it matter?

Many scientists view "everyday" quantum effects as largely trivial. For example, although the probability distribution functions for an electron's positions in time and space operate using quantum mechanical principles, one does not necessarily need to use these to predict the behavior of interacting atoms and molecules in a cellular process. Also, many quantum mechanical effects that have been clearly observed experimentally have been at low temperatures, a few Kelvin at most above absolute zero, suggesting that the relatively enormous temperatures of living matter on an absolute Kelvin scale lies in the domain of classical physics, hence the argument that quantum effects are largely not relevant to understanding the ways that living matter operates.

However, there are some valuable nontrivial counterexamples in biology that are *difficult* to explain without recourse to the physics of quantum mechanics. The clearest example of these is quantum tunneling in enzymes. The action of many enzymes is hard to explain using classical physics. Enzymes operate through a lowering of the overall free energy barrier between biochemical products and reactants by deconstructing the overall chemical reaction into a series of small intermediate reactions, the sum of whose free energy barriers is much smaller than the overall free energy barrier. It is possible in many examples to experimentally measure the reaction kinetics of these intermediate states by locking the enzyme using genetic and chemicals methods. Analysis of the reaction kinetics, however, using classical physics in general fails to explain the rapid overall reactions rates, predicting a much lower overall rate by several orders of magnitude.

The most striking example is the electron transport carrier mechanism of oxidative phosphorylation (OXPHOS) (see Chapter 2). During the operation of OXPHOS, a high-energy electron is conveyed between several different electron transport carrier enzymes, losing free energy at each step that is in effect siphoned off to be utilized in generating the biomolecule ATP that is the universal cellular energy currency. From a knowledge of the molecular structures of these electron transport carriers, we know that the shortest distance between the primary electron donor and acceptor sites is around ~0.2–0.3 nm, which condensed matter theory from classical physics is equivalent to a *Fermi energy level gap* that is equivalent to ~1 eV. However, the half reaction $2H^+ + 2e^-/H_2$ has a *reduction potential* of ca. −0.4 eV. That is, to transfer an electron from a donor to a proton acceptor, which is the case here that creates atomic hydrogen on the electron acceptor protein, results in a gain in electron energy of ~0.4 eV, but a cost in jumping across the chasm of 0.2–0.3 nm of ~1 eV. In other words, the classical prediction would be that electron transfer in this case would not occur, since it would imply a kinetic energy for the electron during the transition equivalent to ca. −0.6 eV, which is negative and thus forbidden.

However, experimentally it is known that electrons undergo these transitions (you are alive while reading this book, for example… one hopes…), and one explanation for these observations is that the electrons undergo quantum tunneling between the donor and acceptor sites, over a rapid time scale predicted to be ~10 fs (i.e., ~10^{-14} s). Quantum tunneling is also predicted to occur over longer length scales between secondary donor/acceptor sites; however, even though the time scale of tunneling transitions is within reach of the temporal resolution of electron spectroscopy measurements, no definitive experiments, at the time of writing of this book, have yet directly reported individual quantum tunneling events between electron transport carrier proteins in OXPHOS.

In a vacuum, quantum tunneling over such long distances would be highly improbable. However, in proteins, such as the electron transport carriers used in OXPHOS, the intervening soft-matter medium facilitates quantum tunneling by providing lower-energy virtual quantum states, which in effect reduce the overall height of the tunneling energy barrier, significantly increasing the physiological tunneling rates. The question is then not whether quantum tunneling is a component of the electron transport mechanism—it clearly is. Rather, it is whether these proteins evolved to utilize quantum tunneling as a selective advantage to the survival of their cell/organism host from the outset (see Chapter 2), or if the drivers of evolution were initially more aligned with classical physics with later tunneling capability providing a significant advantage. In several ways, one could argue that it is remarkable that the entire biological energy transduction machinery appears to be based on this quantum mechanical phenomenon of tunneling, given that much of biology does not require quantum mechanics at all.

A second interesting biological phenomenon, which is difficult to explain using classical physics, is the fundamental process by which plants covert the energy from sunlight into trapped chemical potential energy embedded in sugar molecules, the process of photosynthesis (see Chapter 2). Photosynthesis utilizes the absorption of photons of visible light by light-harvesting complexes, which are an array of protein and chlorophyll molecules typically embedded into organelles inside plant cells called "chloroplasts," expressed in high numbers to maximize the effective total photon absorption cross-sectional area. Since the early twenty-first century, there has been experimental evidence for coherent oscillatory electronic dynamics in these light-harvesting complexes, that is, light-harvesting complexes in an extended array, which are all in phase with regard to molecular electronic resonance.

When initially observed, it was tempting to assign this phenomenon to *quantum coherence*, a definitive and nontrivial quantum mechanical phenomenon by which the relative phases of the wave functions of particles, electrons in this case, are kept constant. However, these oscillation patterns alone are not sufficient evidence to indicate quantum mechanical effects. For example, the effect can be modeled by classical physics as one of *excitonic coupling* to generate the appearance of *coherence beating* of the electronic excitation states across an array of light-harvesting complexes.

KEY POINT 9.5

An exciton is a classical state describing a bound electron and electron hole that are coupled through the attractive electrostatic Coulomb force that has overall net zero electrical charge but can be considered as a quasiparticle, usually in insulators and semiconductors, which has a typical lifetime of around a microsecond.

It is only relatively recently that more promising evidence has emerged for room temperature quantum coherence effects in photosynthesis (O'Reilly and Olaya-Castro, 2014). When the vibrations of neighboring chlorophyll molecules match the energy difference between their electronic transitions, then a resonance effect can occur, with a by-product of a very efficient energy exchange between these electronic and vibrational modes. If the vibrational energy is significantly higher than the thermal energy scale for energy transitions equivalent to $\sim k_B T$, then this exchange, if observed experimentally, cannot be explained classically, but can be explained by modeling the exchange as due to discrete quanta of energy.

Experiments to probe the electronic–vibrational resonance in chlorophyll molecules have been performed *in vitro* using time-resolved femtosecond electron spectroscopy. Analysis of these data indicates a so-called negative joint probability for finding the chlorophyll molecules within certain specific relative positions and momenta. Classical physics predicts that these probability distributions are always positive, and therefore this represents a clear signature of quantum behavior.

That being said, the absolute thermal energy scale of $\sim k_B T$ at 300 K is still much higher than that expected for likely quantum coherence effects by several orders of magnitude. In other words, at ambient "room" temperature of biology, one might expect the surrounding molecular chaos in the cell to annihilate any quantum coherence almost instantly. But recent simulations suggest that quantum coherence may still result in a *level-broadening effect* of the free energy landscape—flattening out local minima due to thermal fluctuation noise thereby increasing the overall efficiency of energy transport—possibly by up to a factor of two.

A related recent study involves quantum coherence observed in the enzyme lysozyme (discussed previously in Chapter 5). In this study (Lundholm et al., 2015), the researchers found evidence at very low ~ 1 K temperatures for coupled electrical dipole states in crystals of lysozyme following the absorption of terahertz radiation. This observation was consistent with a previously hypothesized but hitherto undetected state of matter called a "Fröhlich condensate," a vibrational ground state of dielectric matter analogous to the Bose–Einstein condensate on bosons, in which quantum coherence results in an extended state of several coupled electric dipoles from different molecules over a long length scale equivalent in effect to the whole crystal.

Although highly speculative, theoretical physicist Roger Penrose and anesthesiologist Stuart Hameroff (Penrose et al., 2011) had previously hypothesized that microtubules inside cells might function as *quantum computing* elements, driven by quantum coherence over the long length scales of microtubule filaments. Normally, a wave function, as a superposition of all quantum states, is expected to collapse upon an interaction between the quantum system and its environment (i.e., we measure it). However, in this system the collapse was hypothesized not to occur until the quantum superpositions become physically separated in space and time, a process known as *objective reduction*, and when such quantum coherence collapses, then an instant of *consciousness* occurs. The physical cause of this hypothetical coherence in microtubules was speculated by Penrose as being due to Fröhlich condensates (Fröhlich, 1968). Previously, experimental studies failed to identify Fröhlich condensates, so this recent study, although still hotly debated, may serve to propagate this speculation of quantum-based consciousness a little further. However, at present, it is going too far to make a link between quantum mechanics and consciousness; the evidence is enormously thin. But the key thing here is that the Fröhlich condensate has been a controversy for over half a century, and now if these new experimental results support the Fröhlich hypothesis, it means that biological systems may have the potential to show quantum mechanical coherence at room temperature. This would clearly influence the organization of many biological processes and the transmission of information, in particular.

A fourth, and perhaps more mysterious, example of nontrivial quantum biology is the *geonavigation* of potentially many complex multicellular organisms. There is good experimental evidence to support the theory of *magnetoreception* in several organisms. This is the sense that allows certain organisms to detect external magnetic fields to perceive direction/altitude/location, and ultimately use this to navigate. For example, *Magnetospirilum* bacteria are *magnetotaxic* (i.e., their direction of swimming is affected by the direction of an external magnetic field), common fruit flies appear to be able to "see" magnetic fields, homing pigeons appear to respond to magnetic fields as do some mammals such as bats, and similarly salmon, newts, turtles, lobsters, honeybees even certain plants such as the model organism mouse ear cress (see Chapter 7) respond in growth direction to magnetic fields, though arguably in all of the experimental studies tried to date the size of the artificially generated magnetic field used is much larger than the native Earth's magnetic field by an order of magnitude or more.

Two competing models for magnetoreception in birds use either classical physics in alluding to a magnetic particle (the presence of small magnetic particles of *magnetite* is found in the heads of many migratory birds) or a quantum mechanical model based on the generation of radical pairs. The latter theory implicates a molecule called cryptochrome—a light-sensitive protein that absorbs a blue light photon to generate two radical pairs (molecules with a single unpaired electron).

This effect has been studied most extensively in the migratory patterns of a bird *Erithacus rubecula* (known more commonly as the *European robin*, the one you have probably seen yourself, if not in the flesh, then from its ubiquitous images on Christmas cards). Evidence has grown since the original study published in 2004 (Ritz et al., 2004) that each robin potentially utilizes a chemical compass in the form of *cryptochrome* that is expressed in certain cells in the retinas of the birds' eyes (cryptochromes have also been found in photoreceptor nerve cells in the eyes of other birds and insects and in some plant cells and bacteria).

Cryptochrome contains a *flavin* cofactor called "FADH" (the hydrogenated form of *flavin adenine dinucleotide*, the same cofactor as occurs at various stages in Krebs tricarboxylic acid cycle; see Chapter 2). One hypothesis is that molecular oxygen can enter a molecular binding pocket of the cryptochrome molecule, and a radical pair can then be formed upon the absorption of a single photon of blue light consisting of FADH$^{\cdot}$ and the molecular oxygen superoxide O_2^{-}, where "$\cdot$" indicates an unpaired electron. In this case, an expectation is, from the possible combinations of spins of the products, a proportion of 25% of pairs will be in the singlet spin state and 75% in the triplet spin state, though magnetic coupling with the radical states due to both internal and external magnetic fields, nuclear hyperfine, and Zeeman coupling, respectively, to the Earth's magnetic field (see Chapter 5), may also enable singlet/triplet interconversion.

If the radical pair is in its singlet state, the unpaired electron from the superoxide radical may transfer to the FADH radical, forming a singlet (FADH^{-A} + O_2) state, which has lower free energy and is magnetically insensitive (note that this electron transfer is not possible from the triplet state of the radical pair since this would not conserve total spin). That is, the radicals are self-quenched. However, if the superoxide radical escapes from the molecular pocket before this quenching electron transfer occurs, then this can result in an extended lifetime of the magnetically sensitive radical states for as long as the spins between the electrons of the FADH$^{\cdot}$ and the $O_2^{\cdot-}$ exhibit quantum entanglement. This presents an opportunity for an external magnetic field to affect the reaction by modulating the relative orientation of the electron spins.

If the singlet/triplet radical products have sufficiently long quantum coherence times, they may therefore be biologically detectable by as a chemical compass—in essence, the entanglement would imply that the extent of any coupling observed depends on the local magnetic field. One hypothesis is that the superoxide formed in the radical pair is particularly important in magnetoreception. This is because the oxygen radical does not exhibit hyperfine coupling, and so any observed coupling is a metric of the external magnetic field.

Quantum entanglement–based magnetoreception is speculative at several levels. First, it implies that modulation of the reaction products by a magnetic field would lead to a behavioral modulation of the bird's flight through, presumably, the bird's sense of vision combined with moving its head to scan the *B*-field landscape, for example, generating brighter or darker regions on the bird's retina. But there is no clear experimental evidence for this effect. Also, it is unclear specifically where in the bird's eye this putative radical pair reaction would actually takes place. It may well be that high-resolution biophysical structural imaging tools, such as AFM, may provide some future insight here, at least into mechanisms of free radical formation upon light absorption.

Most challenging to this theory, however, are estimates that the entangled state would need to last >100 μs to have sufficient sensitivity for use a chemical compass given the low strength of the Earth's *B*-field of ~5 mT. However, the longest entangled states observed to date experimentally in the laboratory have lifetimes of ~80 μs, seen in a special inorganic molecule from a class called *endohedral fullerenes* (or *endofullerenes* or simply *fullerenes* for short) known as "N@C60" that consists of a shell of 60 carbon atoms with a single nitrogen atom at the center and is highly optimized for extended quantum entanglement since the carbon shell provides a magnetic shield from randomized external magnetic fluctuations due to surrounding atoms, which would otherwise result in decoherence. So, currently, one would sensibly conclude that the speculation of quantum entanglement in cryptochrome molecule resulting in magnetoreception is interesting; however, the jury is definitely out!

9.5.2 FROM CELLS TO TISSUES

Although the tools and techniques of biophysics have contributed to many well-documented studies at the level of single cells, in terms of generating significant insight into a range of biological processes when studied at the level of single, isolated cells, there is a paucity of reliable experimental data, which has the equivalent quality and temporal and spatial resolutions of isolated single cells. This is true for prokaryotic single-celled organisms such as bacteria, but also single-cell eukaryotic organisms such as yeast, as well as cultured single eukaryotic cells from multicellular organisms. The latter includes cells that often appear to operate as a single cell, for example, white blood cells called macrophages that are part of the immune response and operate by engulfing foreign pathogen particles. But also, for isolated cells that would normally be expressed among a population of other neighboring cells, such as in a tissue or organ, both in the case of normal cells and for diseases such as probing cells inside a cancer tumor.

This tight packing of cells is very important. Not only do cells experience chemical cues from other surrounding cells but also often very complex mechanical signals. An equally important point to note is that this is not just true for eukaryotic cells inside tissues/organs of complex multicellular organisms but also for microbial organisms such as yeast and bacteria that are described as being "unicellular." This is because although such cells exhibit stages in their life cycles of being essentially isolated single cells, the majority of a typical cell's life is actually spent in a colony of some sort, surrounded in close proximity by other cells, often of the same type though sometimes, as is the case in many bacterial biofilms, sharing a colony with several different bacterial species.

Investigation of cellular biophysical properties but in the context of many other neighboring cells presents technical challenges. The key biophysical techniques to use involve optical microscopy, typically various forms of fluorescence imaging, and mechanical force probing and measurement methods, such as AFM. The challenge for optical microscopy is primarily that of light microscopy imaging in "deep tissues," namely, a heterogeneous spatial distribution in refractive index and increased scatter, photodamage, and background noise. However, as discussed previously (see Chapter 3), there are now technical strategies that can be applied to minimize these optical imaging issues. For example, the use of adaptive optics to counteract optical inhomogeneity, and methods such as multiphoton excitation fluorescence microscopy and light sheet imaging to minimize background noise and sample photodamage. Even, in fact, standard confocal microscopy has several merits here, though it lacks the imaging speed required to monitor many real-time biological processes. Similarly, one of AFM imaging's primary technical weaknesses was in sample damage; however, with developments in "lighter touch" torsional AFM imaging, this problem can be minimized.

Biofilms present a particularly challenging health problem when coupled to microbial infection, in being very persistent due to their resistance against many antiseptic measures and antibiotic treatments. These are particularly problematic for applications of medical prosthetics, for example, infections on the surfaces of prosthetic joints and for regions of the body where the inside and outside world collide, for example, urinary tract infections on catheters and periodontitis in teeth and gums. One reason for this is that cells in the center of a biofilm are chemically protected by the neighboring outer cells and by slimy glycocalyx structures secreted by the cells. These are known as microbial *extracellular polymeric substances* (*EPS*), but these can also combine to form a composite gel with extracellular biopolymers such as collagen, which are produced by the host organism. This composite coat allows small solutes and gases to permeate but which form an effective barrier against many detergents, antiseptics, and antibacterial toxins such as antibiotics and also to penetration by white blood cells in the host organism immune response. There is therefore enormous potential to probe the various mechanisms of biofilm chemical defense using next-generation biophysical tools.

However, there is now emerging evidence that biofilms exhibit significant mechanical robustness. This compounds the difficulty in healthcare treatment of biofilm infections since they are often very resistant to mechanical removal. This is not just at the level of needing to brush your teeth harder, but presents issues in treating infections *in vivo*, since to completely eradicate a biofilm infection potentially needs a combination of killing outer cells and weakening outer biofilms structures to the extent that, for example, viscous drag forces for *in vivo* fluid flow might then be sufficient to completely dislodge a biofilm from the surface it is infecting. Experimental measurement of the mechanical shear stress failure levels for different biofilms using a range of biophysical force probe approaches (see Chapter 6) suggests that typical physiological fluid flow levels, for example, those found in blood and urine, are in general not sufficient to dislodge biofilms.

At a coarse-grained level of approximation, the mechanical properties of a biofilm can be modeled as that of a densely packed hydrogel. As discussed previously (see Chapter 8), the variation of elastic modulus of hydrogels varies as $\sim C^{2.25}$ where C is the concentration of the effective "biopolymer." C here is more a metric for the local cell density in this case, and the implication is that under the tightly packed conditions inside a biofilm, the equivalent elastic modulus is very high. New methods to disrupt the EPS to reduce its mechanical robustness are under development currently, and a key feature of these studies is biophysical force probe investigations.

9.5.3 FROM ORGANISMS TO ECOSYSTEMS

An ecosystem is defined as the combination of a population of interacting individual organisms, whether of the same or different types, and of the environment in which these organisms reside. The connections between biophysics and ecological phenomena have only been directly investigated relatively recently, principally along the theme of *ecomechanics.* That is, how do the organisms in a population interact mechanically with each other and with their environment, and how do these mechanical interactions lead to emergent behaviors at the level of an ecosystem.

A number of model ecology systems have emerged to probe these effects. For example, the mechanics of mollusks clinging to rocks on a seabed. Modeling the mechanical forces experienced by a single mollusk is valuable of course; however, population-level modeling is essential here to faithfully characterize the system since the close packing of individual mollusks means that the effect on the local water flow due to the presence of any given mollusk will be felt by neighboring mollusks. In other words, the system has to be modeled at an ecomechanics level.

This level of organism fluid dynamics cooperativity is also exhibited by swimming organisms, for example, fish and swimming mammals such as whales create trailing vortices in the water as they move, which has the effect of sucking surrounding water into the wake of a swimmer. This in effect results in the local fluid environment being dragged into different marine environments by the swimmer. In other words, populations of large swimming organisms have an effect of increasing the mixing of sea water between different environments. This is relevant to other ecosystems since this could, for example, result in a spread in ocean acidification along migratory swimmer routes, which is greater than would be expected in the absence of this fluid drag effect.

There is also a sensible argument that this drift behavior in a school of fish results in greater hydrodynamic efficiency of the population as a whole, that is, by sticking close together, each individual fish exerts on average less energy, though a counterargument applies in that fish toward the front and edges of a school must expend more energy than those in the middle and at the back. Thus, for the theory to be credible, there would need to be something equivalent to a *social mechanism* of sharing the load, that is, fish change places, unless the fish at the front and edges benefit in other ways, for example, they will be closer to a source of food in the water. But in terms of fluid dynamics effects, this is essentially the same argument as rotating the order of cyclists in a team as they cycle together in a group.

Similar mechanical feedback is also exhibited in populations of plants. For example, the leaf canopies of trees result in difficult to predict emergent patterns in airflow experienced by other surrounding trees in a wood or forest population, sometimes extending far beyond nearest neighbor effects. This is important since it potentially results in complex patterns of rainfall over a population of trees and of the direction of dispersal of tree pollen.

Airflow considerations apply to the flocking of birds and the swarming of flies. However, here other biophysical effects are also important, such as visual cues of surrounding birds in the flock, potentially also audio cues as well. There is also evidence for cooperativity in fluid dynamics in swimmers at a much smaller length scale, for example, microbial swimmers such as bacteria. Under these low Reynolds numbers swimming conditions (see Chapter 6), the complex flow patterns created by individual microbial swimmers can potentially result in a local decrease in effective fluid viscosity ahead of the swimmer, which clearly has implications to microbial swarming at different cell densities, and how these potentially lead to different probabilities for forming microbial colonies from a population of free swimmers.

Other effects include the biophysical interactions between carbon dioxide in the atmosphere and water-based ecosystems. For example, increased levels of atmospheric carbon dioxide dissolve in the saltwater of the oceans and freshwater in rivers and lakes to result in increased acidification. This can feedback into reduced calcification of *crustaceous organisms* that possess external *calcium carbonate* shells. There is evidence that human-driven increases in atmospheric carbon dioxide (such as due to the burning of fossil fuels) can result in a local decrease in pH in the ocean by >0.1 pH unit, equivalent to an increase in proton concentration of >25%. This may have a dramatic detrimental effect on crustaceous organisms, which is particularly dangerous since marine plankton come into this category. Marine plankton are at the bottom of the ocean food chain, and so changes to their population numbers may have dramatic effects on higher organisms, such as fish numbers. Calcification also plays a significant role in the formation of coral reefs, and since coral reefs form the mechanical structure of the ecological environment of many marine species increased, ocean acidification may again have a very detrimental effect on multiple ecosystems at a very early stage in the food chain.

KEY BIOLOGICAL APPLICATIONS: EXTENDING BIOPHYSICS LENGTH SCALES

Quantum mechanics modeling of enzyme kinetics; Ecosystems biomechanics analysis.

Worked Case Example 9.2: DNA Origami

A simple single-stranded DNA origami motif was designed consisting of 21 nucleotide base pairs of sequence 3'-CCGGGCAAAAAAAAAGCCCGG-5'. The construct was subjected to thermal denaturing and annealing upon cooling at an ionic strength of 10 mM.

a   *Draw the structure of the lowest-energy time-averaged origami motif you might expect to form.*
   A blue dye EDANS [5-((2-aminoethyl)aminonaphthalene-1-sulfonic acid] was then conjugated to the 3' end and a quencher for EDANS called "Dabcyl" to the 5'.
b   *Explain with the assistance of a graph of normalized fluorescence emission intensity of the blue dye as a function of time what you might expect to observe if the thermal denaturing and annealing upon cooling is performed on this construct.*
c   *How would the graph differ if a second construct is purified that has the quencher molecule placed on the central adenine nucleotide and intensity measurements are made after the lowest free energy structure has formed? (Assume that the Förster radius for the dye–quencher pair in this case is 3.3 nm, and the persistence length of single- and double-stranded DNA at this ionic strength is ~0.7 and 50 nm, respectively.)*

Answers

a   Watson–Crick base pairing will result in a lowest energy structure, which has the maximum number of base pair interactions, which will involve 6 base pair interactions between the CCGGGC sequence at the 3′ end and the GCCCGG sequence at the 5′ end, such that the 3′ and 5′ ends are base paired directly opposite to each other. This leaves the AAAAAAAAA segment that has nothing to base pair with and so will form a single-stranded loop. The single-stranded persistence length of 0.7 nm is equivalent to roughly two base pair separation; thus, it is likely that the time-averaged structure will have the single-stranded nine adenine base pairs as a circular loop.

b   The most stable lowest free energy structure has the 3′ and 5′ ends separated by the width of the DNA double helix, which for B-DNA is 2 nm (see Chapter 2). The normalized blue dye fluorescence will have a similar distance dependence as for a FRET transition between a donor and acceptor FRET dye pair (see Chapter 4):

$$1 - \varepsilon_{max} = 1 - 1/\left(1 + \left(2/3.3\right)^6\right) = 0.047$$

However, inspection of the DNA sequence suggests that it may be possible to generate several structures that have fewer paired nucleotides bases, for example, by relative transposition at integer base pair intervals from the lowest energy structure of part (a) of up to five bases. The axial separation of individual base pairs is 0.34 nm. These imply separations of the dye–quencher pair of ~2.03, 2.11, 2.25, 2.42, and 2.62 nm, which is turn imply normalized blue dye intensity values of ~0.051, 0.064, 0.091, 0.135, and 0.200 by the same aforementioned argument, and thus a sensible plot might suggest transient transitions between these states before the stable 0.047 lowest energy state is reached.

c   If the quencher molecule is placed on the central adenine base, then the distance between it and the blue dye for the time-average stable structure assuming a circular ring of eight adenine nucleotides and six Watson–Crick base pairs is, from simple trigonometry,

$$d = \sqrt{\left(2^2 + \left(\left(6 \times 0.34\right) + \left(8 \times 0.34/2\pi\right)\right)^2\right)} = 3.18 \text{ nm}$$

The normalized blue dye intensity for this separation is thus given by

$$1 - 1/\left(1 + \left(3.18 3.3\right)^6\right) = 0.445$$

Once this lowest free energy structure is formed, the circular ring is likely to be held under reasonably high tension since the persistence length is equivalent to a two base pair separation and so large fluctuations in the circular shape, and hence the normalized blue dye intensity, are unlikely.

Worked Case Example 9.3: Structural Coloration

When a light wave is reflected at a boundary with an optical medium of a higher refractive index, it will undergo a phase change equivalent to 180°.

a   *Derive the condition $2n_1 d_1 cos\theta_1 = (m - 1/2)\lambda$ for constructive interference between the back-reflected light from the upper and lower surfaces of a thin film of thickness d, assuming the light has propagated first through air and $n_1 > 1$.*

b   *A section of a butterfly wing scale was found to contain just a single thin layer of the biopolymer protein chitin. Although the average scale size was measured at approximately 170 × 50 µm with a thickness varying from 1 to 3 µm using scanning electron microscopy, it was found that the highly colored region was restricted to a distal region of the scale which contained a series of vertical corrugations of thickness of ~175 nm, which had localized optical characteristics like a simple thin film. When a 488 nm wavelength laser was directed onto this region, back-reflected light was detected only at specific angles with the brightest being at ~63° from the normal to the surface of the corrugation. Estimate the refractive index of chitin.*

Answers

a The optical path length (OPL) (i.e. the length that light needs to travel through air to create the same phase difference it would have when traveling through another homogenous medium) for the transmitted beam from the upper to the lower surface is $dn_1/\cos\theta_1$. Similarly, the OPL back-reflected from the lower surface to the upper surface is $dn_1/\cos\theta_1$ so the total OPL for the transmitted beam to emerge back into the air is $2dn_1/\cos\theta_1$. The lateral displacement L between this emergent beam and the incident beam is from trigonometry $2d\tan\theta_1$. Using trigonometry again, the OPL for the equivalent incident beam in air of angle θ_0 reflected from the upper surface is $L\sin\theta_0 = 2d\tan\theta_1\sin\theta_0$. Thus, the optical path difference (OPD) between these two beams is:

$$OPD = 2dn_1/\cos\theta_1 - 2d\tan\theta_1\sin\theta_0$$

Using Snell's law of refraction, $n_0\sin\theta_0 = n_1\sin\theta_1$, so:

$$OPD = 2dn_1/\cos\theta_1 - 2d\tan\theta_1 n_1\sin\theta_1 = (2dn_1/\cos\theta_1)(1 - \sin^2\theta_1) = 2dn_1\cos\theta_1$$

Since $n_1 > n_0$ the phase of the incident reflected beam is shifted by 180°, equivalent to half a wavelength. Constructive interference will occur if the OPD is equivalent to some positive integer number (say m) minus a half (for this phase shift) of wavelengths. Thus:

$$2dn_1\cos\theta_1 = (m - 1/2)\lambda$$

b The brightest back-reflection will be for the first order of m (i.e. $m = 1$). Thus for 488 nm wavelength, the refraction index of the chitin is:

$$n_1 = ((1 - 1/2) \times 488 \times 10^{-9})/(175 \times 10^{-9} \times 2\cos63°) \approx 1.54$$

9.6 THE IMPACT OF AI, ML, AND DEEP LEARNING ON BIOPHYSICS AND THE PHYSICS OF LIFE

Artificial Intelligence (AI) refers to the development and application of computer systems that can perform tasks that would normally require human intelligence, such as mimicking cognitive functions involved in learning, problem-solving, perception, and decision-making. They are particularly powerful at processing large data sets, recognizing and categorizing patterns in these, and using these patterns to make predictions. Machine learning (ML) is an important subset of AI that focuses on developing algorithms and models that can learn from data without being explicitly programmed. ML algorithms learn from patterns and gold-standard examples (or *training data sets*) to classify objects, identify correlations, and make predictions. *Deep learning*, a subset of ML, utilizes *neural networks* with multiple layers to process complex data and extract hierarchical representations.

It has, of course, had a significant impact on multiple fields, including the biophysics and the physics of life:

i *Data analysis and modeling*: AI algorithms can process vast amounts of experimental data, identify patterns, and extract valuable insights. In the physics of life, AI is used to analyze complex datasets obtained from experiments such as protein folding, molecular dynamics simulations, and gene expression studies. For example, AlphaFold developed by Google DeepMind in 2018 and the improved version AlphaFold2 released in 2020 (Jumper et al 2021) show potential for genuinely disruptive impact in enabling the prediction of the 3D structure of proteins from sequence data. Both versions utilize DL methods utilizing large-scale supervised training datasets from established experimental structural biology data, but AlphaFold2 incorporates an additional attention-based neural network architecture called the "transformer," which allows for better modeling of longer-range interactions within proteins compared to the original version. The key potential is that it enables prediction of protein structures, which are very challenging to achieve using existing experimental techniques, such as membrane integrated proteins.

AI models can also be developed to simulate and predict the behavior of complex biological systems, aiding in understanding and predicting emergent biological phenomena. AI can also be used to build accurate models for biophysical systems, such as protein folding, molecular interactions, and drug design. An excellent recent example is the application of AI methods to develop new, an entirely synthetic antibiotic (Liu et al., 2023), which is effective against the Gram-negative pathogen *Acinetobacter baumannii*, a superbug prevalent in many hospitals that exhibits multidrug resistance. Here, researchers first screened 7684 small molecules, consisting of 2341 off-patent drugs and 5343 synthetic compounds, to look for evidence of growth inhibition of *A. baumannii in vitro*. These were used to train a neural network, which was then used on a chemical library of 6680 molecules selected as proof-of-concept for its structural diversity and favorable antibacterial activities, to predict potential undiscovered candidates that would inhibit *A. baumannii growth*. Of these, 240 molecules showed promisingly high scores and were then subject to laboratory testing, with nine "priority molecules" ultimately whittled down to a top candidate of abaucin, an entirely new antibiotic with narrow-spectrum activity against *A. baumannii*. A particularly telling finding from this study was that once the training dataset was developed, the *in silico* screening of 6680 molecules top reduce down to 240 likely candidates took only ~90 minutes.

ii *Bioimage analysis and pattern recognition*: AI excels in image analysis and pattern recognition tasks, which, as can be seen from the range of biophysics tools and techniques discussed in this book, are crucial in studying life and the physics of life. They can aid in *big data* tasks (i.e. those in which the data sets are too large/complex to be analysis by traditional data processing tools) such as recognition techniques to identify and cells and track them in real time, or to detect specific tissues or specific molecular structures in biological images, enabling the study of a range of biological processes at far more granular levels than was possible previously.

AI also shows clear promise in enabling general data-driven discovery by revealing previously hidden relationships from large-scale datasets and helping to identify non-intuitive correlations and patterns that may lead to new discoveries in the physics of life. This may serve to catalyze the generation of new hypotheses and new experimental designs and optimized protocols. As with all applications of AI, however, several challenges remain. Regarding biophysics and physics of life research, two of these are key. One is ethical, in that allowing non-human intervention in critical decision-making concerning new drug designs and biotech innovations could, for example, result in inventions which are potentially dangerous for unforeseen reasons, and potentially difficult to control since the AI optimization process is, at least at the time of writing, very largely a black box. The other is more nuanced but important issue comes down to aspects of performance. Ultimately, the output accuracy of any AI system is only as good as the training dataset used. An inevitable challenge will arise in the not-so-distant future whereby AI outputs themselves will be accepted as sufficiently gold-standard for training new models. But if these AI outputs have not been subjected to the rigorous experimental validation of the original training data, drift in output accuracy is inevitable. So, revolutionary as AI is for big data processing, there needs to be checks and balances in place.

9.7 SUMMARY POINTS

- Systems biology coupled with biophysics enables significant scientific insight through coupling cutting-edge computation with experimental biophysics, often in native biological systems.
- Synthetic biology enables complex self-assembled nanostructures to be fabricated using biological material such as DNA, which have applications in nanomedicine and basic biophysics investigations.
- Biological circuits offer potential to design complex biocomputers inside living cells.
- Biophysics developments have enabled miniaturization of biosensors to facilitate lab-on-a-chip technologies for improved personalized healthcare diagnostics and treatment.
- Biophysics can extend into the quantum world and into whole ecosystems.

QUESTIONS

9.1 The surface area of a light-harvesting protein complex was estimated using AFM imaging to be 350 nm^2. Sunlight of mean wavelength 530 nm and intensity equivalent to 4×10^{21} photons m^{-2} s^{-1} was directed onto the surface of cyanobacteria containing light-harvesting complexes in their cell membranes, whose energy was coupled to the pumping of protons across a cell membrane, with a steady-state protonmotive force of −170 mV. To be an efficient energy transfer, one might expect that the effective transfer time for transmembrane pumping of a single proton should be faster than the typical rotational diffusion time of a protein complex in the membrane of a few nanoseconds; otherwise, there could be energy dissipation away from the proton pump. Explain with reasoning whether this is an efficient energy transfer.

9.2 With specific, quantitative reference to length scales, rates of diffusion, and concentrations of nutrients, explain why a time-sampling strategy is more sensible for a bacterium than a space-sampling mechanism for chemotaxis.

9.3 Discuss the biological, chemical, and biophysical challenges to efficiently deliver and release a drug specifically to a given subcellular organelle that is caged inside a 3D DNA nanostructure?

9.4 What do biologists mean by "robustness" in the context of a biological circuit? Using an electrical circuit diagram approach, show how robustness can be achieved in principle using feedback. How can oscillatory behavior arise in such robust systems?

9.5 What is a transcriptor? How can multiple transcriptors be coupled to generate a bistable oscillator (i.e., an oscillating output in time between on and off states)?

9.6 If a particular single gene's expression is in steady state in a spherical bacterial cell of 1 μm diameter such that there are a total of 50 repressor molecules in the cell, which can each bind to the gene's promoter with a probability of 10% s^{-1}, and the expressed protein can directly activate its own expression at a rate of one molecule per second, calculate the cellular concentration of the expressed protein in nM units stating any assumptions you make.

9.7 How is Schrödinger's famous quantum mechanical wave equation, which governs the likelihood for finding an electron in a given region of space and time, related to the probability distribution function of a single diffusing biological molecule? Discuss the significance of this between the life and physical sciences.

9.8 A synthetic DNA molecule was designed to be used for optical data transmission by turning the DNA into a molecular photonics wire comprising five different neighboring dye molecules, with each having an increasing peak wavelength of excitation from blue, green, yellow, orange through to red. Each dye molecule was conjugated in sequence to accessible sites of the DNA, which was tethered from one end of a glass coverslip. Each dye molecule was spaced apart by a single-DNA helix pitch. The mean Förster radius between adjacent FRET pairs was known to be ~6 nm, all with similar absorption cross-sectional areas of ~10^{-16} cm^2.

  a   When a stoichiometrically equal mix of the five dyes was placed in bulk solution, the ratio of the measured FRET changes between the blue dye and the red dye was ~15%. How does that compare with what you might expect?

  b   Similar measurements on the single DNA-dye molecule suggested a blue-red FRET efficiency of ~90%. Why is there such a difference compared to the bulk measurements?

9.9 For the example data transmission synthetic biomolecule of Question 9.8, a blue excitation light of wavelength 473 nm was shone on the sample in a square wave of intensity 3.5 kW cm^{-2}, oscillating between on and off states to act as a clock pulse signature.

  a   If the thermal fluctuation noise of the last dye molecule in the sequence is roughly ~$k_B T$, estimate the maximum frequency of this clock pulse that can be successfully transmitted through the DNA-dye molecule, assuming that the emission from the donor dye at the other end of the DNA molecule was captured by an objective lens of NA 1.49 of transmission efficiency 80%, split by a dichroic mirror to remove low-wavelength components that captured 60% of the total fluorescence, filtered by an emission filter of 90% transmission efficiency, and finally imaged using a variety of mirrors and lenses of very low photon loss (<0.1%) onto an electron-multiplying charge-coupled device (CCD) detector of 95% efficiency.

  b   How would your answer be different if a light-harvesting complex could couple to the blue dye end of the photonic wire? (*Hint:* see Heilemann et al., 2004.)

9.10 The integrated intensity for a single molecule of the yellow fluorescent protein mVenus was first estimated to be 6100 ± 1200 counts (±standard deviation) on a camera detector in a single-molecule fluorescence microscope. Each protein subunit in the shell of a carboxysome was labeled with a single molecule of mVenus with biochemical experiments suggesting that the distribution of stoichiometry values of these specific carboxysomes was Gaussian with a mean of 16 and standard deviation sigma width of 10 subunit molecules. The initial integrated fluorescence intensity of a carboxysome population numbering 1000 carboxysomes was measured in the same fluorescence microscope under the same conditions as was used to estimate the single-molecule brightness of mVenus, with the stoichiometry of each carboxysome estimated by dividing the initial fluorescence intensity obtained from a single exponential fit to the photobleach trace (see Chapter 8) by the single-molecule fluorescence intensity for mVenus. Estimate with reasoning how many carboxysomes have a calculated stoichiometry, which is precise to a single molecule.

9.11 Discuss the key general technical scientific challenges to developing a lab-on-chip biosensor for use in detecting early-stage bacterial infections that are resistant to antibiotics and strategies to overcome them.

9.12 If there are no "preferred" length and time scales in biology, with complex and multi-directional feedback occurring across multiple scales, then what can the isolated study of any small part of this milieu of scales, for example, at a single-molecule level, a cellular level, a tissue level, a whole organism level, or even at a whole ecosystem level, actually tell us?

REFERENCES

KEY REFERENCE

Rothemund, P.W.K. (2006). Folding DNA to create nanoscale shapes and patterns. *Nature 44*:297–302.

MORE NICHE REFERENCES

Agapakis, C.M. et al. (2011). Towards a synthetic chloroplast. *PLOS One 6*:e18877.

Aryee, P. 2021–22. 30 animals that made us smarter. Podcast from *BBC World Service.*

Bonnet, J. et al. (2013). Amplifying genetic logic gates. *Science 340*:599–603.

Cybulski, J.S., Clements, J., and Prakash, M. (2014). Foldscope: Origami-based paper microscope. *PLOS One 9*:e98781.

Feynman, R.P. (1959). *"There's plenty of room at the bottom" lecture.* Transcript deposited at Caltech Engineering and Science Library, Vol. 23(5), February 1960, pp. 22–36, California Institute of Technology, Pasadena, CA.

Feynman, R.P. (1963). *The Feymann Lectures on Physics*, Vol. I. California Institute of Technology, Pasadena, CA, pp. 1–2.

Friedmann, H.C. (2004). From "butyribacterium" to "*E. coli*": An essay on unity in biochemistry. *Perspect. Biol. Med. 47*:47–66.

Fröhlich, H. (1968). Long-range coherence and energy storage in biological systems. *Int. J. Quant. Chem. 2*:641–649.

Goodman, R.P. et al. (2005). Rapid chiral assembly of rigid DNA building blocks for molecular nanofabrication. *Science 310*:1661–1665.

Heilemann, M. et al. (2004). Multistep energy transfer in single molecular photonic wires. *J. Am. Chem. Soc. 126*:6514–6515.

Hiratsuka, Y. et al. (2006) A microrotary motor powered by bacteria. *Proc. Natl. Acad. Sci. USA 03*:13618–13623.

Jumper, J. et al. (2021). Highly accurate protein structure prediction with AlphaFold. *Nature 596*:583–589.

Kinosita, S. et al. (2008). Physics of structural colors. *Rep. Prog. Phys.* 71:076401.

Liu, G. et al, (2023). Deep learning-guided discovery of an antibiotic targeting *Acinetobacter baumannii*. *Nature Chem. Biol.* doi: 10.1038/s41589-023-01349-8.

Lundholm, I.V. et al. (2015). Terahertz radiation induces non-thermal structural changes associated with Fröhlich condensation in a protein crystal. *Struct. Dyn. 2*:054702.

O'Reilly, E.J. and Olaya-Castro, A. (2014). Non-classicality of the molecular vibrations assisting exciton energy transfer at room temperature. *Nat. Commun. 5*:3012.

Pan, C. et al. (2008). Nanowire-based high-performance "micro fuel cells": One nanowire, one fuel cell. *Adv. Mater. 20*:1644–1648.

Penrose, R. et al. (2011). *Consciousness and the Universe: Quantum Physics, Evolution, Brain & Mind.* Cosmology Science Publishers, Center for Astrophysics, Harvard-Smithsonian, Cambridge, MA.

Ritz, T. et al. (2004). Resonance effects indicate a radical-pair mechanism for avian magnetic compass. *Nature 429*:177–180.

Seeman, N.C. (1982). Nucleic-acid junctions and lattices. *J. Theor. Biol. 99*:237–247.

Taniguchi, N. (1974). On the basic concept of "nano-technology". Proceedings of the International Conference on Production Engineering, Part II, Japan Society of Precision Engineering, Tokyo, Japan.

Turberfield, A.J. (2011). DNA nanotechnology: Geometrical self-assembly. *Nat. Chem. 3*:580–581.

Index

Milton Keynes UK
Ingram Content Group UK Ltd.
UKHW050639161024
449569UK00052B/777